Fish Nutrition and Feed Technology

Fish Nutrition and Feed Technology

Editors

Marina Paolucci
Shunsuke Koshio

MDPI • Basel • Beijing • Wuhan • Barcelona • Belgrade • Manchester • Tokyo • Cluj • Tianjin

Editors
Marina Paolucci
University of Sannio
Benevento
Italy

Shunsuke Koshio
Kagoshima University
Kagoshima
Japan

Editorial Office
MDPI
St. Alban-Anlage 66
4052 Basel, Switzerland

This is a reprint of articles from the Special Issue published online in the open access journal *Fishes* (ISSN 2410-3888) (available at: http://www.mdpi.com).

For citation purposes, cite each article independently as indicated on the article page online and as indicated below:

LastName, A.A.; LastName, B.B.; LastName, C.C. Article Title. *Journal Name* **Year**, *Volume Number*, Page Range.

ISBN 978-3-0365-7278-9 (Hbk)
ISBN 978-3-0365-7279-6 (PDF)

Contents

fishes

MDPI

Editorial

Fish Nutrition and Feed Technology

Marina Paolucci

Department of Sciences and Technologies, University of Sannio, 82100 Benevento, Italy; paolucci@unisannio.it

This Special Issue was designed to address advances in feeding and feed technology and challenges in aquaculture in order to achieve a greater understanding of its management and improve the sector. According to the FAO (Food and Agriculture Organization of the United Nations), aquaculture and fisheries provide approximately 17% of global edible animal proteins; aquaculture and fisheries are thus considered crucial sources of animal protein for human diets. Importantly, aquaculture exceeded fishery production in 2016, contributing 52% of the total aquatic foodstuff produced for human consumption in 2018 (FAO, 2020).

After years of a decreasing trend, the COVID-19 pandemic caused an increase in the number of malnourished people, with an additional 9.3 million children suffering from acute malnutrition presently. Aquaculture is a sustainable and environmentally friendly practice with significant potential to feed the rapidly increasing world population and combat malnutrition. It can provide highly nutritious food containing high-quality proteins and essential ingredients for human health. Al-Banna et al. [1], in their review, reported that dried fish could be used to address malnutrition in Bangladesh, thanks to its high protein and supply of minerals such as iron, zinc, and calcium, which are essential for children and pregnant and lactating women. These results are supported by the analysis of the nutrient composition of dried fish in the databases Web of Science, PubMed, Google Scholar, ScienceDirect, Banglajol (a Bangladesh-based database), and ResearchGate, in addition to literature searches. The authors conclude with the hope that policymakers will work together with experts in food and nutrition to promote the consumption of dried fish via encouraging the inclusion of dried fish in the diet of vulnerable populations (children and pregnant and lactating women).

Based on current fish consumption worldwide, aquaculture production is expected to increase from 82,087 Kt in 2018 to 129,000 Kt by 2050 to meet the needs of the growing population. However, the diet of aquatic species includes fishmeal (FM) and fish oil (FO) derived from wild fish stocks. Generally, sardines or anchovies are used, which are fish of little economic value but essential for maintaining the food web. Due to the constant depletion of natural resources, the awareness of the need for sustainable aquaculture has become increasingly evident and has led scientific research to look for new solutions. To increase the production of farmed fish, it is necessary to develop new feed compositions with innovative ingredients that can replace FM and FO. Zaman et al. [2] estimated that FM substitutions of 7.0% with meat meal (MM) reached the highest value of weight gain and specific growth rate in the juvenile olive flounder (*Paralichthys olivaceus*), without any negative consequence on feed consumption, protein retention, hematological parameters, or innate immune system functionality. Thus, the authors concluded that the substitution of up to 20% MM for FM is safe for juvenile olive flounder.

As pointed out by Hameed et al. [3] in their review, insects have spurred considerable interest in both fish and crustacean aquaculture, as innovative ingredients, thanks to their excellent nutritional value, although not comparable to FM. However, their nutrient composition may vary according to the species and developmental stage. In general, insects have a lower protein content with respect to FM and lack several essential amino acids. However, in their feeding experiments, Carral and Saez-Royuela [4] report that

Citation: Paolucci, M. Fish Nutrition and Feed Technology. *Fishes* **2023**, *8*, 146. https://doi.org/10.3390/fishes8030146

Received: 17 February 2023
Accepted: 27 February 2023
Published: 28 February 2023

substitutions of up to 47% FM with black soldier fly larvae (*Hermetia illucens*) did not modify the quantity of essential amino acids in the body of tench (*Tinca tinca*) juveniles.

Certainly, while the variety of ingredients that can be employed in artificial diets is virtually endless, the time and economic resources needed to test them are not. In their study, McKay and Jeff [5] compared three commercial diets for giant kokopu (*Galaxias argenteus*) larvae to identify the nutritional needs and improve their commercial aquaculture production. Indeed, they reported evident differences in groups fed different commercial diets.

In this respect, we turn to the contribution of Campbell et al. [6] reporting on the initiative carried out by the F3 Consortium Feed Innovation Network (f3fin.org), which aims to tailor diets to suit user demands and the availability of local ingredients. This process is based on the acknowledgement that a proper diet does not need to be manufactured with specific ingredients, but rather it should provide a balanced variety of ingredients to meet the nutritional requirements of the targeted species. Indeed, substitutions must take into account that the amount of protein included can influence the metabolism, immune response, and disease resistance of the target species, as reported by Huang et al. [7] in genetically improved farmed tilapia (*Oreochromis niloticus*) adapted to high temperatures, for which an overall proportion of 29.28–31.69% dietary protein was suggested.

The emerging picture is a complex one, and the mere evaluation of the protein requirement may not be sufficient to define a suitable diet. Indeed, the nutrient availability can be modified during feed manufacturing due to the fact that raw proteins are denatured during the heating process, with consequently reduced amounts of amino acids and peptides. As Cho et al. [8] report, aquafeeds available on the market are heat-processed, and therefore have a low nutritive value, making the counterpart produced without the heating process preferred as a feed source for juvenile pacific bluefin tuna (*Thunnus orientalis*)

Plant proteins represent a good FM replacement in aquaculture. Soybean is one of the most common replacements due to its cost-effectiveness and high nutritional value. However, the presence of plant ingredients has been proved to negatively impact the growth, gut and liver health and structure, intestinal microbiota, and immunological response of fish. Moreover, in order to meet the increasing demand for fish, intensive aquaculture practices are sought. It is unlikely that intensive aquaculture poses a source of stress for fish and favors the transmission of infectious and stress-related diseases. Usually, antibiotics are employed to treat diseases and combat pathogens with the consequent development of antibiotic resistance, which is a global issue in humans as well as in animal farming, including aquaculture. Research is committed to bringing to light sustainable alternatives to antibiotics by using macro- and microalgae, fungi, plant extracts, or parts of plants alone or as prebiotics and probiotics, as oral integrators to stimulate growth performances and disease resistance by boosting the immune system and antioxidant defenses. This is reported by Faheem et al. [9] in their study carried out on the inclusion of 5% *Spirulina platensis* in the juvenile grass carp (*Ctenopharyngodon idella*) diet, resulting in positive effects on their growth, digestive enzymes, antioxidant status, and innate immune system. Similarly, the integration of the crustacean decapod *Litopenaeus vannamei* with the microalgae *Haematococcus pluvialis* and the fungus *Yarrowia lipolytica* into their diet as sources of astaxanthin improved the body color as well as the growth performance, metabolism, and antioxidant status of the fish, as reported by Liu et al. [10].

The following contributions focus on this area of research. Abdel-Latif et al. [11] analyzed the impact of extract of *Astragalus membranaceus*, or mongolian milkvetch, on the growth performance, physiological response, and serum immunity of *Pangasianodon hypophthalmus* juveniles, with positive outcomes. Extract from *Macleaya cordata*, or plume poppy, positively affected the growth performance, serum parameters, and intestinal health of juvenile American eel (*Anguilla rostrata*) (Chen et al. [12]). Wang et al. [13] reported that the addition of *Phragmites australis*, or common reed, to feed can improve the liver morphology and functionality, as well as the non-specific immune response, of grass carp (*Ctenopharyngodon idellus*). Outama et al. [14] found that dietary supplementation with passionfruit (*Passiflora edulis*) peel powder positively stimulated the immune system and the antioxidant defenses in Nile tilapia (*Oreochromis niloticus*) cultured in a biofloc system.

Finally, Orso et al. [15], in their review, underline how Lamiaceae, including medical herbs such as oregano, rosemary, sage, thyme, and mint, have been increasingly used in aquaculture as feed additives due to their low cost and simple use. However, they also underline in their conclusions that various critical issues still remain and limit their use on a large scale, and that the contribution of scientific research is crucial to improve their applicability. In particular, the dosage, part of the plant used, extraction technique, and the administration method and its duration are aspects that require further investigation.

I hope that the high-quality contributions in this Special Issue stimulate insights into ongoing research and at the same time open up new lines of research.

Conflicts of Interest: The author declares no conflict of interest.

References

1. Banna, M.H.A.; Al Zaber, A.; Rahman, N.; Siddique, M.A.M.; Siddique, M.A.B.; Hagan, J.E., Jr.; Rifat, M.A.; Nsiah-Asamoah, C.N.A.; Seidu, A.-A.; Ahinkorah, B.O.; et al. Nutritional Value of Dry Fish in Bangladesh and Its Potential Contribution to Addressing Malnutrition: A Narrative Review. *Fishes* **2022**, *7*, 240. [CrossRef]
2. Zaman, M.F.U.; Li, R.; Cho, S.H. Evaluation of Meat Meal as a Replacer for Fish Meal in Diet on Growth Performance, Feed Utilization, Chemical Composition, Hematology, and Innate Immune Responses of Olive Flounder (*Paralichthys olivaceus*). *Fishes* **2022**, *7*, 343. [CrossRef]
3. Hameed, A.; Majeed, W.; Naveed, M.; Ramzan, U.; Bordiga, M.; Hameed, M.; Ur Rehman, S.; Rana, N. Success of Aquaculture Industry with New Insights of Using Insects as Feed: A Review. *Fishes* **2022**, *7*, 395. [CrossRef]
4. Carral, J.M.; Sáez-Royuela, M. Replacement of Dietary Fishmeal by Black Soldier Fly Larvae (*Hermetia illucens*) Meal in Practical Diets for Juvenile Tench (*Tinca tinca*). *Fishes* **2022**, *7*, 390. [CrossRef]
5. McKay, W.; Jeffs, A. Comparison of Three Artificial Diets for the Larviculture of Giant Kōkopu (*Galaxias argenteus*). *Fishes* **2022**, *7*, 310. [CrossRef]
6. Campbell, K.B.; McLean, E.; Barrows, F.T. In Pursuit of Fish-Free Feeds: A Multi-Species Evaluation. *Fishes* **2022**, *7*, 336. [CrossRef]
7. Huang, D.; Liang, H.; Zhu, J.; Ren, M.; Ge, X. Dietary Protein Modifies Hepatic Glycolipid Metabolism, Intestinal Immune Response, and Resistance to Streptococcus agalactiae of Genetically Improved Farmed Tilapia (GIFT: *Oreochromis niloticus*) Exposed to High Temperature. *Fishes* **2022**, *7*, 202. [CrossRef]
8. Cho, J.-H.; Kurimoto, T.; Haga, Y.; Kamimura, Y.; Itoh, A.; Satoh, S. Effects of Non-Heated and Heat Processed Krill and Squid Meal-Based Diet on Growth Performance and Biochemical Composition in Juvenile Pacific Bluefin Tuna *Thunnus orientalis*. *Fishes* **2022**, *7*, 83. [CrossRef]
9. Faheem, M.; Jamal, R.; Nazeer, N.; Khaliq, S.; Hoseinifar, S.H.; Van Doan, H.; Paolucci, M. Improving Growth, Digestive and Antioxidant Enzymes and Immune Response of Juvenile Grass Carp (*Ctenopharyngodon idella*) by Using Dietary Spirulina platensis. *Fishes* **2022**, *7*, 237. [CrossRef]
10. Liu, Y.; Zheng, L.; Xu, B.; Sagada, G.; Zhang, J.; Shao, Q. Effects of Diets with Varying Astaxanthin from *Yarrowia lipolytica* Levels on the Growth, Feed Utilization, Metabolic Enzymes Activities, Antioxidative Status and Serum Biochemical Parameters of *Litopenaeus vannamei*. *Fishes* **2022**, *7*, 352. [CrossRef]
11. Abdel-Latif, H.M.R.; Ahmed, H.A.; Shukry, M.; Chaklader, M.R.; Saleh, R.M.; Khallaf, M.A. *Astragalus membranaceus* Extract (AME) Enhances Growth, Digestive Enzymes, Antioxidant Capacity, and Immunity of *Pangasianodon hypophthalmus* Juveniles. *Fishes* **2022**, *7*, 7–319. [CrossRef]
12. Chen, R.; Huang, L.; Zhai, S. Effects of Macleaya cordata Extract on Growth Performance, Serum Biochemical Parameters, and Intestinal Health of Juvenile American Eel (*Anguilla rostrata*). *Fishes* **2022**, *7*, 229. [CrossRef]
13. Wang, R.; Lei, C.; Li, Z.; Lei, Y.; Luo, C.; Shao, L.; Huang, C.; Yang, P. Effects of a Diet of Phragmites australis instead of Triticum aestivum L. on Immune Performance and Liver Tissue Structure of *Ctenopharyngodon idellus*. *Fishes* **2022**, *7*, 378. [CrossRef]
14. Outama, P.; Linh, N.V.; Xuan, C.L.; Wannavijit, S.; Tongsiri, S.; Chitmanat, C.; Montha, N.; Van Doan, H. Passionfruit (*Passiflora edulis*) Peel Powder Stimulates the Immune and Antioxidant Defense System in Nile Tilapia, *Oreochromis niloticus*, Cultivated in a Biofloc System. *Fishes* **2022**, *7*, 233. [CrossRef]
15. Orso, G.; Imperatore, R.; Coccia, E.; Ashouri, G.; Paolucci, M. Lamiaceae as Feed Additives in Fish Aquaculture. *Fishes* **2022**, *7*, 349. [CrossRef]

fishes

Article

Replacement of Dietary Fishmeal by Black Soldier Fly Larvae (*Hermetia illucens*) Meal in Practical Diets for Juvenile Tench (*Tinca tinca*)

Jose M. Carral * and María Sáez-Royuela

Department of Animal Production, Faculty of Veterinary, University of León, 24071 León, Spain
* Correspondence: jmcarl@unileon.es; Tel.: +34-87-29110

Abstract: The development of specific diets for the juvenile stage is a main target for culture intensification of tench (*Tinca tinca*). Aquafeeds still rely heavily on the use of fishmeal (FM) but concerns about economic and ecological sustainability make the use of alternative protein sources necessary. Insect meals are considered a promising substitute to replace FM. In a 90-day experiment, 6 diets with different replacement levels of FM by partially defatted black soldier fly larvae meal (BSFLM): 0%, 15%, 30%, 45%, 60% and 75%, were tested on juvenile tench. Survival rates were high (95.8–100%) without differences between treatments. Diet with 45% FM replacement allowed for best growth performance in terms of total length (TL = 66.4 mm) and weight (W = 4.19 g), without differences with 60% and 75% of replacement. A cubic relationship was evidenced between the level of FM replacement and growth. From the regression equations, the estimated optimal level of FM replacement was 47% (356 g BSFLM kg^{-1} diet). Externally visible deformities were always under 0.05%. The whole-body lipid content of the fish had a significantly negative linear regression with BSFLM (r^2 = 0.80). The content of the essential amino acids (EAA) arginine, leucine, lysine, phenylalanine, methionine, and threonine in diets decreased with dietary BSFLM inclusion. However, it did not have a negative effect on growth performance, suggesting that EAA requirements were covered. The amount of essential amino acids in whole-body juveniles was similar independently of the diet provided. The results allow considering BSFLM as a sustainable protein source for juvenile tench feeding.

Keywords: alternative protein source; essential amino acid; growth performance; insect meal; protein replacement; *T. tinca*; whole-body composition

Citation: Carral, J.M.; Sáez-Royuela, M. Replacement of Dietary Fishmeal by Black Soldier Fly Larvae (*Hermetia illucens*) Meal in Practical Diets for Juvenile Tench (*Tinca tinca*). *Fishes* **2022**, *7*, 390. https://doi.org/10.3390/fishes7060390

Academic Editors: Marina Paolucci and Shunsuke Koshio

Received: 22 November 2022
Accepted: 13 December 2022
Published: 15 December 2022

1. Introduction

In many countries, a policy objective for aquaculture development is diversification. This term not only applies to searching for new species, but also to increasing the production of those species currently farmed in small quantities [1] such as tench, *Tinca tinca*. Tench is a freshwater fish belonging to the Cyprinidae family that originally occurred in the waters of Europe and Siberia and today is present in the inland waters of all continents [2]. Tench is commonly cultured in extensive systems where growth is often limited and production usually poor [3,4]. Considering that this species is appreciated by consumers as a tasty fish with healthy meat [5,6] and attractive for anglers, the development of intensive rearing systems is required to satisfy the demand. A major obstacle to increasing tench production is the deficit of young fishes for further growth or to restock open waters for sport fishing [7,8]. Some relevant advances have been achieved in juvenile tench rearing techniques under controlled conditions, focusing mainly on feed as an essential factor. In this sense, González-Rodríguez et al. [9] established the protein requirements for juvenile tench between 48 and 52% and later, García et al. [10] proposed a practical diet which allows good survival and growth performance, setting the basis for further nutritional studies.

Fishmeal (FM) is the most nutritious and digestible source of protein in aquatic diets, which is the biggest consumer sector with a global use in 2019 of 78% [11]. In 2031, world production of fishmeal is expected to reach 5.6 Mt of which 71% will come from forage species [12]. The environmental footprint derived from its use in feedstuffs is a main concern to reach sustainable growth [13], especially when production mainly depends on a finite wild-harvest capture [14]. In addition to environmental concerns, the upscaling demand and price for FM have awakened the search for alternative ingredients, specifically byproducts and those obtained by recycling nutrients, to support a circular economy within aquaculture [15].

Among different possibilities, there is a recent interest in insects as a food source for animal feeding. Insect culture is considered sustainable as insects do not need large areas or much water and they contribute to waste recycling [16,17]. The review by Henry et al. [18] highlighted the good potential of insect meal to replace FM in fish diets due to their high content of protein and adequate profile of essential amino acids (EAA). However, only a few species of insects have the potential to be reared at a large scale [19]. In addition, there are regulatory obstacles to use insect protein for animal feeding. The recently approved UE regulations, 2017/893/EC, 2017, and 2021/1925/UE [20,21] allow the inclusion of protein from eight species of insects in aquafeeds. Among them, the black soldier fly (*Hermetia illucens*) shows major advantages derived from its ability to turn bio-waste into larvae with valuable nutrient content [22]. Table 1 includes data on the possibilities to replace FM by black soldier fly meals (BSFLM) in freshwater and marine fish species without negative effects on growth performance. Except for the experiments performed in the turbot (*Scophthalmus maximus*) [23], in the yellowtail (*Seriola quinqueradiata*) [24], and in the gilthead seabream (*Sparus aurata*) [25], where the inclusion of BSFLM negatively affected growth performance, levels of FM replacement between 20% in the dusky kob (*Argyrosomus japonicus*) [26] and 100% were shown to be feasible. However, the dietary amounts of BSFLM usually did not reflect the percentages of FM substitution. Therefore, total replacement of FM was achieved with 600 g kg^{-1} in diets for Atlantic salmon (*Salmo salar*) [27] and 106 g kg^{-1} in diets for Jian carp (*Cyprinus carpio* var. Jian) [28]. In addition to interspecific differences, several factors must be considered to explain the variability of results, such as the nutritional value of BSFLM including insect growth substrate and further meal processing, life stage of fish, diet formulation, and nutritional composition [29].

Table 1. Fishmeal (FM) replacement level (%) and corresponding dietary amount (g kg^{-1} diet) of black soldier fly larvae meal (BSFLM) included without negative effects on growth performance in marine and freshwater fish species.

Authors	Species	Initial Weight (g)	BSFLM (g kg^{-1})	FM Replacement (%)
Marine				
Kroeckel et al. [23]	Turbot (*Scophthalmus maximus*)	54.9	-[a]	0
Ido et al. [24]	Yellowtail (*Seriola quinqueradiata*)	0.7	-	0
Karapanagiotidis et al. [25]	Gilthead seabream (*Sparus aurata*)	1.47	-[b]	0
Fabrikov et al. [30]	Gilthead seabream	6.76	109 *	30%
Madibana et al. [26]	Dusky kob (*Argyrosomus japonicus*)	25.5	200 *	20%
Magalhães et al. [31]	European seabass (*Dicentrarchus labrax*)	50.0	195	45%
Abdel-Tawwab et al. [32]	European seabass	12.1	148	50%
Wang et al. [33]	Japanese seabass (*Lateolabrax japonicus*)	14.1	192	64%
Li et al. [34]	Tongue sole (*Cynoglossus semilaevis*)	543	144	25%
Takakuwa et al. [35]	Red sea bream (*Pagrus major*)	17.9	293	41.7%
Lock et al. [36]	Atlantic salmon (*Salmo salar*)	250	100	50%
Belghit et al. [27]	Atlantic salmon	7.9	600 *	100%
Belghit et al. [37]	Atlantic salmon	1398	147 *	100%

Table 1. *Cont.*

Authors	Species	Initial Weight (g)	BSFLM (g kg^{-1})	FM Replacement (%)
Freshwater				
St-Hilaire et al. [38]	Rainbow trout (*Oncorhynchus mykiss*)	22.6	149 [c]	25%
Sealey et al. [39]	Rainbow trout	146	362 [d]	50%
Renna et al. [40]	Rainbow trout	179	400	50%
Dumas et al. [41]	Rainbow trout	45.7	131	50%
Fabrikov et al. [26]	Rainbow trout	55.0	109 *	30%
Bruni et al. [42]	Rainbow trout	54.2	450 *	100%
Caimi et al. [43]	Siberian sturgeon (*Acipenser baerii*)	24.2	185	25%
Rawski et al. [44]	Siberian sturgeon	14.4	300 *	63.1%
Muin et al. [45]	Nile tilapia (*Oreochromis niloticus*)	3.0	150	50%
Devic et al. [46]	Nile tilapia	5.7	80	70%
Wachira et al. [47]	Nile tilapia	35	336	67%
Tippayadara et al. [48]	Nile tilapia	14.8	100 *	100%
Adeoye et al. [49]	African catfish (*Clarias gariepinus*)	2.67	75	50%
Fawole et al. [50]	African catfish	14	172	75%
Vongvichith et al. [51]	Climbing perch (*Anabas testudiensis*)	4.42	290 *	100%
Kattakdad et al. [52]	Climbing perch	3.12	175	50%
Xiao et al. [53]	Yellow catfish (*Pelteobagrus fulvidraco*)	48.5	223	48%
Khieokhajonkhet et al. [54]	Goldfish (*Carassius auratus*)	9.5	210 *	50.6%
Kamalii et al., [55]	Goldfish	2.8	201	40%
Li et al. [28]	Jian carp (*Cyprinus carpio var. Jian*)	34.8	106 *	100%
Zhou et al. [56]	Jian carp	10.1	140 *	100%
Katya et al. [57]	Barramundi (*Lates calcarifer*)	6.7	308	50%
Shuda et al. [58]	Striped catfish (*Pangasianodon hypophthalmus*)	3.1	174	60%
Fabrikov et al. [30]	Tench (*Tinca tinca*)	18.3	109 *	30%

* Maximum dietary amount of BSFLM in the study. [a] FCR and feed intake were unaffected by inclusion of 332 g kg^{-1}. [b] FCR and SGR were unaffected by inclusion of 276 g kg^{-1}. [c] BSF pre-pupae not meal. [d] BSFLM enriched with offal fish substrate.

Regarding the effect of FM replacement with BSFLM on body composition, several studies concluded that the inclusion of BSFLM did not affect the final whole-body or fish fillet proximate composition. Meanwhile, others reported significant increases in dry matter and lipid [40,43] or protein content [50].

To the best of our knowledge, no studies have reported the effect of replacing FM by BSFLM in the diet of juvenile tench; therefore, the present study aimed to evaluate the effects of increasing substitution level of FM with BSFLM on survival, growth performance, incidence of externally visible deformities, and whole-body composition.

2. Materials and Methods

2.1. Ethics Statement

According to Spanish law (RD 53/2013) and an EU directive (2010/63/EU), the Ethic Committee of the University of León approved the experiment conducted in this study (Approval reference ULE_16_2015). Fish health, welfare, and the environmental conditions in the experimental tanks were checked twice daily by visual observation of animal behavior. Water quality parameters, such as oxygen saturation, temperature, and water flow were periodically measured (see Section 2.2.). The necessary number of fish to analyze whole-body composition were euthanized with an over-dose of tricaine methanesulfonate (MS222, Ortoquímica S.L., Barcelona, Spain) by prolonged immersion. At the end of the experiment, the remaining animals were transported to the fish farm where breeder fish come from.

2.2. Fish, Facilities, and Experimental Procedures

Tench larvae were obtained by hatching using artificial reproduction techniques [59] and reared in outdoor tanks. After 4 months, 540 juvenile tench from a homogenous pool were randomly distributed as groups of 30 fish in 18 fiberglass tanks (0.5 × 0.25 × 0.25 m) containing 25 L of water to obtain replicates corresponding to the different feeding treatments. Prior to distribution, 100 juveniles were anesthetized with tricaine methanesulfonate (MS-222; Ortoquímica S.L., Barcelona, Spain) to measure initial total length (TL) and weight (W). Values of 30.7 ± 0.28 mm and 0.39 ± 0.02 g (mean ± SEM) were obtained. TL was measured with a digital caliper (to the nearest 0.01 mm) and, after removing excess water with tissue paper, W was determined by a precision balance (to the nearest 0.001 g). Total biomass of each tank was weighed. Following a monofactorial design, diet was the experimental factor with three replicates per level of treatment. The juveniles were previously acclimated to experimental conditions for 4 days.

Artesian well water was supplied in an open system (flow-through system) and each tank had a water inlet (inflow 0.30 L min^{-1}) and outlet (provided with a 250 μm mesh filter) and light aeration. Measures of the incoming water quality, ammonia, nitrites, hardness, and total suspended solids were performed once a week with a spectrophotometer HACH DR2800 (Hach Lange GMBH, Vigo, Spain). Dissolved oxygen in tanks was measured with a multi-meter HACH HQ30d (Hach Lange GMBH, Vigo, Spain). Mean values of water quality were pH 7.6, hardness 5.3 German degrees (calcium 32.8 mg L^{-1}), total suspended solids 34.0 mg L^{-1}, dissolved oxygen ranged between 5.7 and 7.3 mg L^{-1}, ammonia < 0.10 mg L^{-1}, and nitrites < 0.010 mg L^{-1}.

Water temperature (measured twice a day) was 26 ± 1 °C and a 16 h light:8 h dark photoperiod was maintained throughout the experiment. Tanks were cleaned of feces and uneaten feed every two days. The experiment lasted for 90 days.

2.3. Diets and Feeding

Based on the results of our research group [9,10], different diets (50% crude protein) were formulated and prepared to test the effects of different substitution levels of FM by black soldier fly (*H. illucens*) meal. A partially defatted BSFLM from Hermetia Deutschland GmbH & Co. KG (Baruth/Mark, Germany), obtained by processing larvae reared on a vegetable byproducts substrate, was included. According to the producer, partial defatting was made by a mechanical process using high pressure and without any solvents. Proximate composition and amino acid profiles of FM and BSFLM are in Table 2.

Table 2. Proximate composition and amino acid profiles of fishmeal (FM) and black soldier fly larvae meal (BSFLM) (g kg^{-1}).

	Fishmeal (FM)	Black Soldier Fly Larvae Meal (BSFLM)
Proximate composition		
Moisture	76 ± 1.0	54.7 ± 0.9
Crude protein	668 ± 2.9	565 ± 1.5
Crude fat	96 ± 2.7	139 ± 3.4
Carbohydrates	nd	161.4 ± 3.6
Ash	162 ± 3.9	79.9 ± 1.2
Essential amino acid (EAA)		
Arginine	38.2 ± 1.0	20.7 ± 0.8
Histidine	36.0 ± 1.1	29.5 ± 1.0
Isoleucine	23.3 ± 0.9	26.7 ± 1.0
Leucine	44.3 ± 1.6	35.6 ± 1.2
Lysine	51.5 ± 1.5	26.2 ± 1.1
Methionine	19.7 ± 0.8	8.7 ± 0.3
Phenylalanine	28.0 ± 1.2	18.6 ± 0.7
Threonine	24.1 ± 0.9	20.8 ± 0.8
Tryptophan	7.1 ± 0.4	4.2 ± 0.2
Valine	28.2 ± 1.1	34.1 ± 1.3

Table 2. *Cont.*

	Fishmeal (FM)	Black Soldier Fly Larvae Meal (BSFLM)
Non-essential amino acid (NEAA)		
Alanine	43.4 ± 1.4	34.2 ± 1.2
Aspartate	67.8 ± 1.5	39.5 ± 1.3
Cysteine	6.0 ± 0.2	4.1 ± 0.2
Glutamate	85.4 ± 1.2	49.3 ± 1.2
Glycine	48.9 ± 1.6	29.0 ± 1.0
Proline	28.4 ± 1.2	34.8 ± 1.3
Serine	30.5 ± 1.1	22.5 ± 0.9
Tyrosine	23.1 ± 0.9	23.2 ± 0.8

Values are mean ± standard deviation (SD).

A total of 6 diets (nearly isonitrogenous and isoenergetic) with different replacement levels of FM by BSFLM were formulated: 0% (control), 15%, 30%, 45%, 60%, or 75%, corresponding to 0, 117, 232, 348, 464, or 579 g of BSFLM kg^{-1} diet, respectively (Table 3). Ingredients were ground in a rotary mill BRABENDER (Brabender GmbH & Co. KG, Duisburg, Germany), mixed in a mixer STEPHAN UMC5 (Stephan Food Service Equipment, Hameln, Germany) and extruded using a stand-alone extruder BRABENDER KE19/25D (Brabender GmbH & Co. KG, Duisburg, Germany) at a temperature range between 100 °C and 110 °C. Pellets (1 mm diameter) were dried during 24 h at 30 °C and after receiving a coating of cod liver oil. Fish were fed manually three times a day (at 10:00, 14:00, and 18:00 h) to apparent satiation.

Table 3. Formulation of the practical diets with different levels of replacement of fishmeal (FM) by black soldier larvae fly meal (BSFLM) (g kg^{-1} diet).

	FM Replacement (%)					
	0	15	30	45	60	75
Ingredients (g kg^{-1})						
Fishmeal [1]	645	548	451.5	355	258	162
Hermetia illucens meal [2]	-	117	232	348	464	579
Corn meal [3]	166	146	127.5	108	89	70
Dried *Artemia* cysts [4]	100	100	100	100	100	100
Carboxymethyl cellulose [5]	30	30	30	30	30	30
Cod liver oil [6]	20	20	20	20	20	20
L-ascorbyl-2-monophosphate-N [7]	5	5	5	5	5	5
Dicalcium phosphate [7]	10	10	10	10	10	10
Choline chloride [7]	3	3	3	3	3	3
Soy lecithin [8]	10	10	10	10	10	10
Sodium chloride [9]	1	1	1	1	1	1
Mineral and Vitamin premix [10]	10	10	10	10	10	10

[1] Skretting España S.A., Ctra. de la Estación s/n 09620 Cojóbar. Burgos. España. [2] Hermetia Deutschland GmbH & Co. KG, Baruth/Mark (Germany). [3] Adpan Europa S.L., ES-33186 El Berrón. Siero. Asturias. Spain. [4] INVE Aquaculture Nutrition. Hoogyeld 91. Dendermonde. Belgium. [5] Helm Iberica S.A., ES-28108 Alcobendas. Madrid. Spain. [6] Acofarma distribution S.A., ES-08223 Terrassa. Barcelona. Spain. [7] Cargill, ES-28720 Colmenar Viejo. Madrid. Spain. [8] Biover N.V., Monnikenwerve 109. B-8000 Brugge. Belgium. [9] Unión Salinera de España S.A., ES-28001 Madrid. Spain. [10] Provided mg kg^{-1} premix: inositol: 50,000; thiamin: 500; riboflavin: 800; niacin: 5000; pyridoxine: 1500; pantothenic acid: 5000; biotin: 150; folic acid: 3500; cianocobalamin: 5; retinol: 2400; α-tocopherol: 30,000; cholecalciferol: 6.25; naphthoquinone: 500; butylated hydroxytoluene: 1500; $MgSO_4$-$7H_2O$: 300,000; $ZnSO_4$-$7H_2O$: 11,000; $MnSO_4$-H_2O: 4000; $CuSO_4$-$5H_2O$: 1180; $CoSO_4$: 26; $FeSO_4$-$7H_2O$: 77,400; KI: 340; Na_2SeO_3: 68.

2.4. Chemical Analysis of Diets and Fish

Juveniles were fasted for 14 h before sampling. Samples of diets and juveniles were stored at −30 °C at the beginning and at the end of the experiment. Analyses were performed in duplicate by Analiza Calidad laboratory (Burgos, Spain) following Commission Regulation (EC) 152/2009. Moisture was determined by drying at 105 °C, crude protein was determined according to the Kjeldahl method, crude lipid was determined by extraction with light petroleum and further distillation, ash was determined by calcination at 550 °C, and gross energy was determined according to EU regulation 1169/2011. The content of nitrogen-free extract was calculated by subtracting moisture, protein, lipid, and ash content from the wet weight.

Amino acid profiles were analyzed by HPLC using AccQTag method from Waters (Milford, MA, USA). Amino acids were derivatized with 6-aminoquinolyl-N-hydrosysuccinimidyl carbamate reagent (AQC) by the method of Cohen and Michaud [60] and Cohen and De Antonis [61] and detected by Dual λ Absorbance Detector Waters 2487 from Waters (Milford, MA, USA) at 254 nm. Quantification was carried out with Empower Pro 2.0 software from Waters (Milford, MA, USA). All analyses were performed in duplicate.

2.5. Data Collection

Juvenile tench behavior was observed and registered after cleaning, feeding, and measuring the water quality parameters.

Every 30 days a sample of 15 fish per tank (45 per treatment, 50% of total) were anesthetized to be individually weighed and measured to have information about the growth performance evolution. TL and W were measured as described in Section 2.1., and afterwards juveniles were gently returned to their tanks.

At the end of the experiment, surviving fishes were anesthetized and observed one by one using a magnifying glass to detect externally visible deformities affecting spinal axis, operculum, mouth and tail fin. W and TL were measured individually and total biomass per tank was weighed. The following indices were calculated:

- Survival rate (%) = (final number of juveniles/initial number of juveniles) × 100;
- Specific growth rate, SGR (% d^{-1}) = [(ln final W − ln initial W)/days)] × 100;
- Fulton's coefficient or condition factor, K = 100 × [final W/(TL^3)];
- Biomass gain, BG (g) = (final biomass/tank − initial biomass/tank);
- Feed conversion ratio, FCR = (total feed provided per tank/BG [62].

All treatments were replicated three times and the experimental unit was a tank.

2.6. Statistical Analysis

After confirmation of normality and homogeneity of variance, statistical analysis of growth performance and whole-body composition data were conducted by one-way analysis of variance (ANOVA) and polynomial contrasts with the SPSS16.0 computer program (SPSS, Chicago, IL, USA). Significant differences between means were estimated by Tukey's multiple range test. $p < 0.05$ was used for rejection of null hypothesis.

3. Results

3.1. Diets

Proximate composition and amino acid profile of practical diets are provided in Table 4. Lipid and carbohydrate contents in diets tended to increase with the inclusion of BSFLM. On the contrary, ash content decreased. Amino acid profile was also affected since the content of the essential amino acids (EAA) such as arginine, leucine, lysine, methionine phenylalanine, and threonine decreased as the dietary content of FM did. Only isoleucine increased when BSFLM was included in diets. A reduction in total NEAA was also observed as FM replacement level increased.

Table 4. Proximate composition (g kg^{-1} diet) and amino acid profiles of practical diets with different levels of replacement of fishmeal (FM) by black soldier fly larvae meal (BSFLM).

	FM Replacement (%)					
Proximate composition (g kg^{-1})	0	15	30	45	60	70
Moisture	61.0 ± 0.4	63.1 ± 0.5	60.4 ± 0.3	60.5 ± 0.2	60.0 ± 0.3	60.7 ± 0.4
Crude protein	501 ± 3.2	503 ± 2.8	502 ± 2.0	502 ± 0.9	501 ± 1.8	500 ± 1.8
Crude lipid	107.5 ± 2.1	108.3 ± 2.0	112.5 ± 2.2	118.3 ± 2.3	118.5 ± 2.3	123 ± 2.5
Carbohydrates	200.1 ± 2.8	201.4 ± 2.7	206.8 ± 3.0	208.8 ± 3.1	215.7 ± 3.3	218.7 ± 3.2
Ash	130.4 ± 1.6	124.2 ± 1.5	118.3 ± 1.2	110.4 ± 0.9	104.8 ± 0.9	97.6 ± 0.9
Gross energy (MJ kg^{-1})	15.8 ± 0.7	15.9 ± 0.6	16.0 ± 0.6	16.2 ± 0.5	16.1 ± 0.5	16.3 ± 0.6
Essential amino acid (EAA)						
Arginine	42.3 ± 0.2	31.3 ± 0.3	28.4 ± 0.3	25.8 ± 0.4	21.3 ± 0.3	19.6 ± 0.2
Histidine	38.9 ± 0.6	38.7 ± 0.1	34.7 ± 0.2	34.5 ± 0.3	34.1 ± 0.1	32.9 ± 0.2
Isoleucine	18.6 ± 0.1	19.4 ± 0.1	20.6 ± 0.2	21.3 ± 0.2	21.7 ± 0.1	22.1 ± 0.1
Leucine	34.7 ± 0.3	33.9 ± 0.2	33.0 ± 0.2	32.2 ± 0.1	32.0 ± 0.2	31.5 ± 0.3
Lysine	38.5 ± 0.2	36.7 ± 0.5	33.5 ± 0.1	29.8 ± 0.1	27.6 ± 0.4	26.2 ± 0.3
Methionine	21.0 ± 0.2	18.4 ± 0.2	16.7 ± 0.4	14.5 ± 0.2	11.3 ± 0.3	10.5 ± 0.3
Phenylalanine	20.7 ± 0.4	19.5 ± 0.3	19.1 ± 0.1	18.4 ± 0.3	17.8 ± 0.2	17.3 ± 0.3
Threonine	19.5 ± 0.3	19.1 ± 0.1	18.8 ± 0.3	18.6 ± 0.4	17.5 ± 0.4	17.3 ± 0.3
Tryptophan	4.6 ± 0.3	4.2 ± 0.1	4.2 ± 0.2	4.0 ± 0.1	3.9 ± 0.1	3.8 ± 0.1
Valine	21.2 ± 0.2	22.3 ± 0.3	22.7 ± 0.5	23.4 ± 0.4	25.0 ± 0.1	25.3 ± 0.3
∑EAA	259	243	232	222	212	206
Non-essential amino acid (NEAA)						
Alanine	30.2 ± 0.2	29.8 ± 0.2	29.1 ± 0.4	28.6 ± 0.5	27.9 ± 0.4	27.1 ± 0.5
Aspartate	50.2 ± 0.2	49.6 ± 0.1	49.3 ± 0.4	48.2 ± 0.1	47.5 ± 0.5	46.5 ± 0.6
Cysteine	4.2 ± 0.2	4.0 ± 0.2	3.9 ± 0.3	3.5 ± 0.3	3.2 ± 0.2	3.0 ± 0.2
Glutamate	60.3 ± 0.3	57.4 ± 0.3	54.6 ± 0.4	47.5 ± 0.3	45.1 ± 0.3	43.9 ± 0.4
Glycine	33.5 ± 0.5	32.0 ± 0.3	31.3 ± 0.4	29.3 ± 0.1	28.0 ± 0.6	26.6 ± 0.5
Proline	20.9 ± 0.6	21.7 ± 0.4	23.2 ± 0.3	25.8 ± 0.5	26.6 ± 0.4	27.5 ± 0.4
Serine	21.9 ± 0.3	20.8 ± 0.5	20.4 ± 0.2	20.0 ± 0.1	19.5 ± 0.3	19.1 ± 0.2
Tyrosine	17.2 ± 0.2	17.1 ± 0.3	17.2 ± 0.2	17.4 ± 0.4	17.6 ± 0.4	17.8 ± 0.3
∑NEAA	238	232	229	220	215	211
∑EAA/∑NEAA	1.09	1.05	1.01	1.01	0.98	0.98

Values are mean ± standard deviation (SD).

3.2. *Growth Performance*

Since the beginning, juvenile tench readily accepted all practical diets independently of the amount of BSFLM. There were no statistical differences between treatments neither in length nor in weight after 30 days (mean value range: 35.0–36.9 mm and 0.74–0.85 g), whereas after 60 days juveniles fed the 45%, 60%, and 75% BSFLM diets reached higher total length than those fed the control diet. Mean TL increases of 17% and 63.6% were achieved after 30 and 60 days, respectively. From initial weight, mean W was almost 2 times higher after 30 days, while at 60 days it was almost 5 times higher.

Final values (90 days) of survival, growth, feed conversion ratio, and percentages of fish with externally visible deformities are in Table 5. Survival ranged from 95.8% to 100% without significant differences between diets. Juveniles fed on a practical diet with 45% BSFLM reached a significant ($p > 0.05$) higher total length and weight than those fed with the control diet or lower levels of FM substitution. Compared with the control diet, juveniles fed 45% BSMLF reached the higher TL, W, SGR and BG values, and lower FCR ($p < 0.05$). In comparison with the 45% BSFLM diet, a significant reduction in SGR and BG was evidenced in juveniles fed the 75% BSFLM diet. No differences in TL and W of juveniles fed 45%, 60%, and 75% replacement level diets were found. From a polynomial contrast study, a cubic relationship between TL ($r^2 = 0.69$) and W ($r^2 = 0.67$) and BSFLM content was stablished. Figures 1 and 2 show the regression curves between the level of FM substitution and TL and W, respectively. From regression equations, the optimal dietary inclusion of BSFLM was estimated in 356 g BSFLM kg^{-1} (47% of FM replacement).

Table 5. Survival, growth performance, and percentages of juvenile tench with externally visible deformities fed practical diets with different levels of substitution of fishmeal (FM) by black soldier fly larvae meal (BSFLM) over 90 days.

	FM Replacement (%)							Polynomial Contrasts			
	0	15	30	45	60	75	SEM	ANOVA	Linear	Quadratic	Cubic
Survival (%)	100	97.9	95.8	99.0	100	96.9	0.58	0.31	0.72	0.18	0.87
Total length (mm)	61.8 ^{a,b}	60.7 ^{a}	62.7 ^{a,b}	67.5 ^{c}	65.7 ^{b,c}	64.4 ^{a,b,c}	0.63	0.001	0.000	0.054	0.003
Weight (g)	3.48 ^{a,b}	3.09 ^{a}	3.53 ^{a,b}	4.19 ^{c}	3.95 ^{b,c}	3.59 ^{a,b,c}	0.10	0.001	0.006	0.003	0.001
SGR [1] (% day^{-1})	2.40 ^{a,b}	2.31 ^{a}	2.38 ^{a,b}	2.62 ^{c}	2.53 ^{b,c}	2.45 ^{a,b}	0.03	0.007	0.020	0.011	0.004
K [2]	1.38 ^{a,b}	1.30 ^{b}	1.32 ^{a,b}	1.30 ^{b}	1.28 ^{b}	1.27 ^{b}	0.01	0.005	0.000	0.228	0.183
FCR [3]	1.25 ^{a,b,c}	1.46 ^{a}	1.33 ^{a,b}	1.08 ^{c}	1.18 ^{b,c}	1.30 ^{a,b,c}	0.03	0.002	0.062	0.326	0.000
BG [4] (g)	92.5 ^{a,b}	79.2 ^{b}	90.3 ^{a,b}	115 ^{c}	106 ^{b,c}	95.7 ^{a,b}	3.12	0.001	0.000	0.078	0.000
Deformed fish	0.04	0.01	0.02	0.01	0	0.02	0.007	0.64	0.33	0.28	0.87

Values are mean. SEM correspond to pooled standard error of the mean. In the same row, values with different superscript are significantly different ($p < 0.05$). [1] Specific growth rate. [2] Condition factor. [3] Feed conversion ratio. [4] Biomass gain.

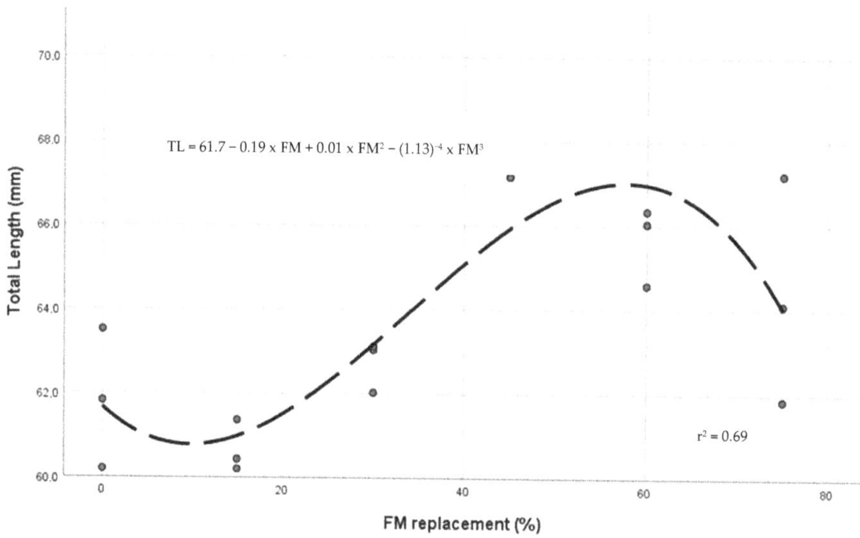

Figure 1. Regression curve between dietary BSFLM and juvenile tench total length (TL).

The condition factor (K) ranged from 1.27 to 1.38, being significantly higher in tench fed the control diet. Regarding FCR, the lower value (1.08) corresponds to fish fed with the 45% FM substitution. The percentages of fish with externally visible deformities was low, ranging from 0% to 0.04%. Body deformities affected the spinal column and caudal peduncle (break in the tail axis).

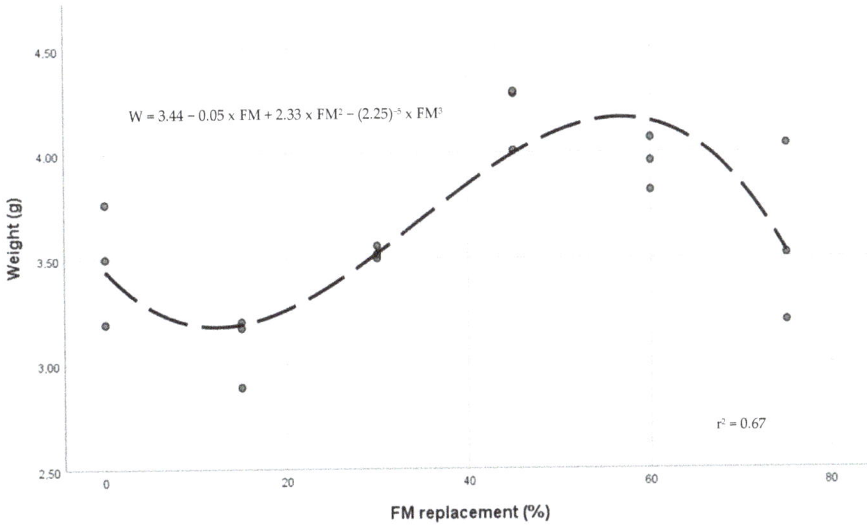

Figure 2. Regression curve between dietary BSFLM and juvenile tench weight (W).

3.3. *Juvenile Whole-Body Composition*

The proximate composition and the amino acid profiles of the whole-body of juvenile tench at the beginning and at the end of the study are provided in Table 6. The whole-body lipid content had a significantly linear negative regression ($r^2 = 0.80$), showing decreasing values as inclusion of BSFLM increased. Juvenile tench fed the 30%, 45%, 60%, and 75% BSFLM diets had lower lipid content than the control and 15% BSFLM diets. Although a content reduction in most EAA was evidenced with increasing inclusion of BSFLM in diets (see Table 5), in juvenile it was similar independently of the diet provided, except for isoleucine. With respect to non-essential amino acids (NEAA) in whole-body juvenile tench, diets affected the content of alanine, aspartate, glutamate, glycine, proline, and tyrosine.

Table 6. Proximate composition and amino acid profiles (g kg^{-1}) of the whole-body of juvenile tench fed practical diets with different levels of replacement of fishmeal (FM) by black soldier fly larvae meal (BSFLM).

		FM Replacement (%)							Polynomial Contrasts			
	Initial	0	15	30	45	60	75	SEM	ANOVA	Linear	Quadratic	Cubic
Moisture	767	732	732	753	756	757	760	3.98	0.06	0.006	0.31	0.62
Protein	133.5	160	160	159	159	162	158	1.45	0.93	0.71	0.71	0.54
Lipid	73.0	80.5 [a]	78.9 [a,b]	55.3 [b,c]	55.1 [b,c]	46.2 [c]	45.7 [c]	4.48	0.003	<0.001	0.15	0.36
Ash	24.5	26.0	25.6	23.3	24.7	26.4	27.0	1.13	0.98	0.78	0.54	0.90
Essential amino acids (EAA)												
Arginine	9.5	11.6	11.9	12.1	12.5	12.7	11.4	0.16	0.11	0.49	0.02	0.08
Histidine	2.5	4.60	4.00	4.20	4.50	4.40	4.00	0.12	0.70	0.61	0.95	0.16
Isoleucine	20.1	20.2 [a,b]	19.8 [a]	21.1 [a]	19.6 [a,b]	18.9 [b]	19.3 [a,b]	0.24	0.04	0.02	0.25	0.14
Leucine	4.9	5.80	5.90	5.40	5.20	5.70	5.40	0.12	0.66	0.35	0.54	0.97
Lysine	10.2	11.3	11.8	11.2	11.3	12.3	12.6	0.19	0.09	0.03	0.12	0.58
Methionine	3.9	5.40	4.65	4.40	4.70	4.50	4.60	0.14	0.39	0.17	0.70	0.37
Phenylalanine	11.2	12.0	13.0	12.70	12.7	13.4	12.8	0.15	0.15	0.08	0.20	0.81
Threonine	5.5	5.80	6.70	6.60	7.55	6.80	6.20	0.19	0.10	0.29	0.02	0.60
Tryptophan	0.4	0.60	0.60	0.60	0.60	0.60	0.03	0.10	1.00	1.00	1.00	1.00
Valine	6.1	6.20	7.40	6.90	7.50	6.80	6.20	0.18	0.01	0.67	0.02	0.69
∑EAA	74.3	83.5	85.7	85.2	86.1	86.1	82.5	0.43	0.06	0.35	0.08	0.12

Table 6. *Cont.*

	FM Replacement (%)								Polynomial Contrasts			
	Initial	0	15	30	45	60	75	SEM	ANOVA	Linear	Quadratic	Cubic
Non-essential amino acids (NEAA)												
Alanine	7.7	8.8 ^{a}	10.7 ^{b}	10.3 ^{a,b}	9.1 ^{a,b}	10.6 ^{a,b}	9.6 ^{a,b}	0.24	0.02	0.39	0.02	0.07
Aspartate	13.0	13.9 ^{a}	15.6 ^{b}	15.6 ^{b}	15.0 ^{a,b}	15.3 ^{a,b}	15.1 ^{a,b}	0.19	0.03	0.09	0.02	0.03
Cysteine	0.2	0.2	0.1	0.1	0.1	0.1	0.2	0.02	0.29	1.00	0.04	1.00
Glutamate	17.7	20.1 ^{a}	22.5 ^{b}	20.7 ^{a,b}	21.9 ^{b,c}	21.5 ^{a,b,c}	20.7 ^{a,b}	0.26	0.006	0.61	0.006	0.20
Glycine	2.5	8.7 ^{a}	9.8 ^{ab}	12.4 ^{d}	11.6 ^{cd}	12.2 ^{cd}	10.7 ^{b,c}	0.41	<0.001	<0.001	<0.001	0.37
Proline	4.9	5.3 ^{a}	7.2 ^{b}	7.7 ^{b}	6.5 ^{a,b}	7.9 ^{b}	6.4 ^{a,b}	0.28	0.007	0.04	0.003	0.27
Serine	6.3	6.4	6.9	6.7	7.1	7.3	6.2	0.14	0.15	0.80	0.04	0.18
Tyrosine	0.2	4.6 ^{a}	0.2 ^{b}	0.2 ^{b}	0.1 ^{b}	0.1 ^{b}	0.1 ^{b}	0.52	<0.001	<0.001	<0.001	<0.001
∑NEAA	52.5	68.0 ^{a}	73.0 ^{b}	73.7 ^{b,d}	71.4 ^{c}	75.0 ^{d}	69.0 ^{a}	0.76	0.001	0.007	<0.001	0.95
∑EAA/∑NEAA	1.41	1.23	1.17	1.16	1.20	1.15	1.19	0.01	0.35	0.37	0.18	0.39

Values are mean. SEM correspond to pooled standard error of the mean. In the same row, values with different superscript are significantly different ($p < 0.05$). Initial data were not included in the statistical analysis.

4. Discussion

Insects are part of the natural diet of many freshwater fish species and, thus, the inclusion of insect meals could be advantageous to ease feed intake [63]. Tench feed on zooplankton and other small invertebrates in natural habitats, such as some insect larvae [64–66]. Although black soldier fly is a terrestrial insect, the natural feeding habits of tench would help voluntary ingestion of diets independently of the different content of BSFLM, indicating that substitution of FM did not affect its palatability. This agrees with most experiments on partial or total FM replacement by BSFLM, where diets including insect meal were attractive to fishes. However, Kroeckel et al. [23] reported a reduction in growth in turbot which was partially attributed to a decrease in feed intake due to a lower palatability of diets including BSFLM.

In their review, Barragán-Fonseca et al. [22] reported that composition of BSFLM depends on body composition of fly larvae, which varies among the rearing substrates, but also on further processing. Black soldier fly larvae have a high fat content, reaching in non-defatted meals a lipid content average of 353.2 g kg^{-1} [67] making their inclusion in aquafeeds difficult. Thus, BSFLM is processed to obtain partially defatted meal with high protein content, which allows high inclusion levels of insect meal in fish diets without reducing the technical quality of extruded diets [41]. According to the producer, the BSFLM of this experiment was partially defatted and its high protein content (54.7%) makes possible dietary amounts of up to 579 g kg^{-1} (75% FM replacement). As the high content of saturated and monosaturated fatty acids in black soldier fly larvae is associated to a decline in growth of aquatic species [68], the defatting process could have beneficial effects in the nutritional composition of diets.

Considering that aquaculture is diverse in terms of cultured species, production systems, and culture conditions, it is difficult to establish accurate comparisons between studies of FM replacement by BSFLM. In most cases, relative data (% FM replacement) are provided but correspond to different inclusion amounts. Therefore, total FM replacement were achieved in Jin carp, with 106 g kg diet^{-1} [37] and 140 g kg diet^{-1} [38], whereas in rainbow trout [42] and Atlantic salmon [24], amounts were 450 and 600 g kg^{-1}, respectively. In this study with juvenile tench, which have similar protein requirements than rainbow trout and Atlantic salmon, high amounts of BSFLM (579 g kg diet^{-1}) did not affect survival and growth performance compared with the control diet.

The nutritional quality of black soldier fly products must be considered to accurately interpret the response of fish. Available data on the nutritional value of *H. illuscens* larvae meal showed that the EAA profile, an important indicator of protein quality [69], does not differ much between studies [22]. In agreement with Henry et al. [18] and Maurer et al. [70], the profile of EAA of BSFLM in the practical diets was similar to the FM (Table 1), with the exception of arginine, lysine, phenylalanine, and methionine (60%, 49%, 69.2%, and

33.6% less than FM, respectively). Despite the reduction in the mentioned amino acids in BSFLM diets (Table 4), juvenile growth was unaffected by inclusion of BSFLM, leading us to consider that EEA requirements were fully covered. According to Hua et al. [29], the relationship between BSFLM incorporation level and the growth response is best described by a simple negative linear equation. In this experiment, we found a cubic relationship, showing an increasing growth (TL, W, SGR, FCR, and BG) from the control diet to 45% BSFLM and, with inclusion levels above this threshold, a trend to growth reduction. The negative effects of insect meal on fish growth could be partly due to a lower nutrient digestibility [29]. Some authors speculated that the content of chitin in insect meals is closely related to a reduction in protein and fat digestibility [27,40,71]. Considering this, the reduction in growth observed in juvenile tench from inclusion BSFLM over 348 g kg $^{-1}$ could be associated to an increasing content of chitin.

Despite the defatting process, the lipid content of BSFLM was higher than in FM, determining an increase in fat in diets from 107.5 g kg^{-1} (control diet) to 123 g kg^{-1} (75% BSFLM diet). The increase in lipid content in diets is usually correlated with an increase in fish body fat [72,73], but in opposition to this statement and similarly to results reported on turbot [23] and Jian carp [28,56], whole-body tench lipid content decreased with the increase in dietary BSFLM. Kroeckel et al. [23] and Li et al. [28] hypothesized that this fact could be related to a reduction in lipid digestibility due to the content of chitin provided by the insect meal. In agreement with Ferrer-Llagostera et al. [74], the information on the effects of chitin are still discussed and further research is required to clarify its role in aquatic species.

Kamisnki et al. [3] reported one of the highest values of SGR in juvenile tench, 3.69% day^{-1}, with initial weight (W) of 0.24 g and at five-months old. In our study, SGR's ranged between 2.20 and 2.45% day^{-1} using six-month old juveniles with an initial W of 0.46 g. Considering that SGR slows down with age [74], the lower values could be partially attributed to differences in age and initial weight.

As in former experiments performed by our research team, the control diet enabled acceptable growth and high survival and no external visible deformities. External visible body deformities are mainly associated to inadequate feeding, especially during early development [75,76]. In juvenile tench, a relationship between feeding commercial diets for other species and fish with elevated condition coefficients (between 1.3 and 1.4) and presence of body deformities has been suggested [3,77–79]. Although the K factor for some diets have shown values above 1.3, the percentage of deformed fish was insignificant (<0.05) showing that diets were well balanced for tench.

From this preliminary study, BSFLM should be regarded as a good protein source for these critical early stages of tench, being feasible for inclusion in high levels of FM dietary replacement without negative effects on survival and growth. Since the high protein quality requirements during this early growth stage were covered by BSFLM, it would be expected that FM saving would increase through inclusion of this insect meal in feedstuffs for further outgrowing phases, whenever the development of efficient insect rearing and processing systems allow for market availability at reduced cost. Under this consideration, further research on the effects of dietary BSFLM in nutritional and sensory quality in tench reared to commercial size should be performed.

5. Conclusions

Compared with the control diet, 75% of fishmeal by black soldier fly larvae meal did not affect survival and growth performance of juvenile tench. Juvenile tench fed a diet where 45% of fishmeal was replaced with BSFLM reached significantly better growth performance than those fed the other diets. The optimal level of fishmeal substitution for juvenile growth performance was 47%, corresponding to 356 g of black soldier fly larvae meal kg^{-1} diet. Whole-body lipid content showed a negative linear regression with dietary black soldier fly larvae meal. Although a reduction in most essential amino acids was evidenced with increasing inclusion of black soldier fly larvae meal in diets, the content in whole-body juveniles was similar independently of the diet provided.

Author Contributions: J.M.C. contributed to funding acquisition, project administration, conceptualization, methodology, supervision, statistical analysis, and writing—original draft preparation. M.S.-R. contributed to conceptualization, methodology, supervision, statistical analysis, and writing—original draft preparation. All authors have read and agreed to the published version of the manuscript.

Funding: This study was funded by the Ministerio de Economía y Competitividad (MINECO) and the Fondo Europeo de Desarrollo Regional (FEDER) through the Research Project AGL2015-64202-R.

Institutional Review Board Statement: According to Spanish law (RD 53/2013) and an EU directive (2010/63/EU), the Ethic Committee of the University of León approved the experiment conducted in this study (Approval reference ULE_16_2015).

Informed Consent Statement: Not applicable.

Data Availability Statement: Data available on corresponding author request.

Acknowledgments: We thank the collaboration of Cargill España (Madrid) and Skretting España S.A. (Burgos) for providing us with most of the ingredients for the diets. We also thank Hermetia Deutschland GmbH & Co. KG for providing the black soldier fly meal.

Conflicts of Interest: The authors declare no conflict of interest.

References

1. FAO. Planning for Aquaculture Diversification: The Importance of Climate Change and Other drivers. In Proceedings of the FAO Technical Workshop, Bangkok, Thailand, 23–25 June 2016; Harvey, B., Soto, D., Carosfeld, J., Beveridge, M., Bartley, D.M., Eds.; FAO: Roma, Italy, 2017.
2. Freyhof, J.; Kottelat, M. *Tinca tinca*. The IUCN Red List of Threatened Species 2008, e.T21912A9339248. Available online: https://www.iucnredlist.org/species/21912/9339248 (accessed on 21 November 2022).
3. Kamiński, R.; Sikorska, J.; Wolkini, J. Diet and water temperature affect growth and body deformities in juvenile tench *Tinca tinca* (L.) reared under controlled conditions. *Aquac. Res.* **2017**, *48*, 1327–1337. [CrossRef]
4. Pula, H.J.; Trenzado, C.E.; García-Mesa, S.; Fallola, C.; Sanz, A. Effects of different culture systems on growth, immune status, and other physiological parameters of tench (*Tinca tinca* L.). *Aquaculture* **2018**, *485*, 101–110. [CrossRef]
5. Wedekind, H.; Rennert, B.; Kohlmann, K. Product quality in different strains of tench (*Tinca tinca* L.) tested under controlled environmental conditions. *J. Appl. Ichthyol.* **2003**, *19*, 174–176. [CrossRef]
6. González-Rodríguez, Á.; Celada, J.D.; Carral, J.M.; Sáez-Royuela, M.; Fuertes, J.B. Effects of varying protein level in practical diets on survival, growth, feed utilization and body composition of juvenile tench (*Tinca tinca* L.). *Aquac. Int.* **2014**, *22*, 1723–1735. [CrossRef]
7. Celada, J.D.; Aguilera, A.; García, V.; Carral, J.M.; Saez-Royuela, M.; González, R.; González, A. Rearing juvenile tench (*Tinca tinca* L.) under controlled conditions using *Artemia* nauplii as supplement to a dry diet. *Aquac. Int.* **2009**, *17*, 565–570. [CrossRef]
8. García, V.; Celada, J.D.; Carral, J.M.; Sáez-Royuela, M.; González, R.; González, Á. Decapsulated *Artemia* cysts: A suitable dietary supplement for juvenile tench (*Tinca tinca* L.). *J. Appl. Aquac.* **2010**, *22*, 57–65. [CrossRef]
9. González-Rodríguez, Á.; Celada, J.D.; Carral, J.M.; Sáez-Royuela, M.; Fuertes, J.B. Evaluation of a practical diet for juvenile tench (*Tinca tinca* L.) and substitution possibilities of fish meal by feather meal. *Anim. Feed. Sci. Technol.* **2014**, *187*, 61–67. [CrossRef]
10. García, V.; Celada, J.D.; González, R.; Carral, J.M.; Sáez-Royuela, M.; González, Á. Response of juvenile tench (*Tinca tinca* L.) fed practical diets with different protein contents and substitution levels of fish meal by soybean meal. *Aquac. Res.* **2015**, *46*, 28–38. [CrossRef]
11. EUMOFA—European Market Observatory for Fisheries and Aquaculture Products. *Fishmeal and Fish Oil: Production and Trade Flows in the UE*; Publications Office of the European Union: Luxembourg, 2021; 31p.
12. OECD/FAO. *OECD-FAO Agricultural Outlook 2022–2031*; OECD Publishing: Paris, France, 2022; 363p. [CrossRef]
13. Boyd, C.E.; D'Abramo, L.R.; Glencross, B.D.; Huyben, D.C.; Juarez, L.M.; Lockwood, G.S.; McNevin, A.A.; Tacon, A.G.J.; Teletchea, F.; Tomasso, J.R.; et al. Achieving sustainable aquaculture: Historical and current perspectives and future needs and challenges. *J. World Aquac. Soc.* **2020**, *51*, 578–633. [CrossRef]
14. Cottrell, R.S.; Blanchard, J.L.; Halpern, B.S.; Metian, M.; Froehlichet, H.E. Global adoption of novel aquaculture feeds could substantially reduce forage fish demand by 2030. *Nat. Food* **2020**, *1*, 301–308. [CrossRef]
15. Campanati, C.; Willer, D.; Schubert, J.; Aldridge, D.C. Sustainable intensification of aquaculture through nutrient recycling and circular economies: More fish, less waste, blue growth. *Rev. Fish. Sci. Aquac.* **2022**, *30*, 143–169. [CrossRef]

16. Tschirner, M.; Kloas, W. Increasing the sustainability of aquaculture systems: Insects as alternative protein source for fish diets. *GAIA-Ecol. Perspect. Sci. Soc.* **2017**, *26*, 332–340. [CrossRef]
17. Sánchez-Muros, M.-J.; Barroso, F.G.; Manzano-Agugliaro, F. Insect meal as renewable source of food for animal feeding: A review. *J. Clean. Prod.* **2014**, *65*, 16–27. [CrossRef]
18. Henry, M.; Gasco, L.; Piccolo, G.; Fountoulaki, E. Review on the use of insects in the diet of farmed fish: Past and future. *Anim. Feed. Sci. Technol.* **2015**, *203*, 1–22. [CrossRef]
19. Gasco, L.; Gai, F.; Maricchiolo, G.; Genovese, L.; Ragonese, S.; Bottari, T.; Caruso, G. Fishmeal Alternative Protein Sources for Aquaculture Feeds. In *Feeds for the Aquaculture Sector*; Current Situation and Alternative Sources; Springer Briefs in Molecular Science: Chemistry of Foods; Springer Nature: Cham, Switzerland, 2018; pp. 1–28. [CrossRef]
20. Commission Regulation (EU) 2017/893 of 24 May 2017 amending Annexes I and IV to Regulation (EC) No 999/2001 of the European Parliament and of the Council and Annexes X, XIV and XV to Commission Regulation (EU) No 142/2011 as Regards the Provisions on Processed Animal Protein. Available online: https://eur-lex.europa.eu/legal-content/EN/TXT/PDF/?uri=CELEX:32017R0893&rid=1 (accessed on 21 November 2022).
21. Commission Regulation (EU) 2021/1925 of 5 November 2021 Amending Certain Annexes to Regulation (EU) No 142/2011 as Regards the Requirements for Placing on the Market of Certain Insect Products and the Adaptation of a Containment Method. Available online: https://eur-lex.europa.eu/legal-content/EN/TXT/?uri=CELEX:32021R1925 (accessed on 21 November 2022).
22. Barragán-Fonseca, K.B.; Dicke, M.; van Loon, J.J.A. Nutritional value of the black soldier fly (*Hermetia illuscens* L.) and its suitability as animal feed—A review. *J. Insects Food Feed.* **2017**, *3*, 105–120. [CrossRef]
23. Kroeckel, S.; Harjes, A.G.E.; Roth, I.; Katz, H.; Wuertz, S.; Susenbeth, A.; Schulz, C. When a turbot catches a fly: Evaluation of a pre-pupae meal of the black soldier fly (*Hermetia illucens*) as fish meal substitute—growth performance and chitin degradation in juvenile turbot (*Psetta maxima*). *Aquaculture* **2012**, *364–365*, 345–352. [CrossRef]
24. Ido, A.; Ali, M.-F.-Z.; Takahashi, T.; Miura, C.; Miura, T. Growth of yellowtail (*Seriola quinqueradiata*) fed on a diet including partially or completely defatted black soldier fly (*Hermetia illucens*) larvae meal. *Insects* **2021**, *12*, 722. [CrossRef]
25. Karapanagiotidis, I.T.; Daskalopoulou, E.; Vogiatzis, I.; Rumbos, C.; Mente, E.; Athanassiou, C.G. Substitution of fishmeal by fly *Hermetia illuscens* prepupae meal in the diet of gilthead seabream (*Sparus aurata*). In Proceedings of the HydroMedit 2014, Volos, Greece, 13–15 November 2014; pp. 110–114.
26. Madibana, M.J.; Mwanza, M.; Lewis, B.R.; Fouché, C.H.; Toefy, R.; Mlambo, V. Black Soldier Fly Larvae Meal as a Fishmeal Substitute in Juvenile Dusky Kob Diets: Effect on feed utilization, growth performance, and blood parameters. *Sustainability* **2020**, *12*, 9460. [CrossRef]
27. Belghit, I.; Liland, N.S.; Waagbø, R.; Biancarosa, I.; Pelusio, N.; Li, Y.; Krogdahl, Å.; Lock, E.-J. Potential of insect-based diets for Atlantic salmon (*Salmo salar*). *Aquaculture* **2018**, *491*, 72–81. [CrossRef]
28. Li, S.; Ji, H.; Zhang, B.; Zhou, J.; Yu, H. Defatted black soldier fly (*Hermetia illucens*) larvae meal in diets for juvenile Jian carp (*Cyprinus carpio* var. Jian): Growth performance, antioxidant enzyme activities, digestive enzyme activities, intestine and hepatopancreas histological structure. *Aquaculture* **2017**, *477*, 62–70. [CrossRef]
29. Hua, K. A Meta-analysis of the effects of replacing fish meals with insect meals on growth performance of fish. *Aquaculture* **2020**, *530*, 735732. [CrossRef]
30. Fabrikov, D.; Vargas-García, M.d.C.; Barroso, F.G.; Sánchez-Muros, M.J.; Cacua Ortíz, S.M.; Morales, A.E.; Cardenete, G.; Tomás-Almenar, C.; Melenchón, F. Effect on intermediary metabolism and digestive parameters of the high substitution of fishmeal with insect meal in *Sparus aurata* feed. *Insects* **2021**, *12*, 965. [CrossRef]
31. Magalhães, R.; Sánchez-López, A.; Silva Leal, R.; Martínez-Llorens, S.; Oliva-Teles, A.; Peres, H. Black soldier fly (*Hermetia illucens*) pre-pupae meal as a fish meal replacement in diets for European seabass (*Dicentrarchus labrax*). *Aquaculture* **2017**, *476*, 79–85. [CrossRef]
32. Abdel-Tawwab, M.; Khalil, R.H.; Metwally, A.A.; Shakweer, M.S.; Khallaf, M.A.; Abdel-Latif, H.M.R. Effects of black soldier fly (*Hermetia illucens* L.) larvae meal on growth performance, organs-somatic indices, body composition, and hemato-biochemical variables of European sea bass, *Dicentrarchus labrax*. *Aquaculture* **2020**, *522*, 735136. [CrossRef]
33. Wang, G.; Peng, K.; Hu, J.; Yi, C.; Chen, X.; Wu, H.; Huang, Y. Evaluation of defatted black soldier fly (*Hermetia illucens* L.) larvae meal as an alternative protein ingredient for juvenile Japanese seabass (*Lateolabrax japonicus*) diets. *Aquaculture* **2019**, *507*, 144–154. [CrossRef]
34. Li, X.; Qin, C.; Fang, Z.; Sun, X.; Shi, H.; Wang, Q.; Zhao, H. Replacing dietary fish meal with defatted black soldier fly (*Hermetia illucens*) larvae meal affected growth, digestive physiology and muscle quality of Tongue Sole (*Cynoglossus semilaevis*). *Front. Physiol.* **2022**, *13*, 855957. [CrossRef] [PubMed]
35. Takakuwa, F.; Tanabe, R.; Nomura, S.; Inui, T.; Yamada, S.; Biswas, A.; Tanaka, H. Availability of black soldier fly meal as an alternative protein source to fish meal in red sea bream (*Pagrus major*, Temminck & Schlegel) fingerling diets. *Aquac. Res.* **2022**, *53*, 36–49. [CrossRef]
36. Lock, E.R.; Arsiwalla, T.; Waagbø, R. Insect larvae meal as an alternative source of nutrients in the diet of Atlantic salmon (*Salmo salar*) postsmolt. *Aquac. Nutr.* **2016**, *22*, 1202–1213. [CrossRef]
37. Belghit, I.; Liland, N.S.; Gjesdal, P.; Biancarosa, I.; Menchetti, E.; Li, Y.; Waagbø, R.; Krogdahl, Å.; Lock, E.J. Black soldier fly larvae meal can replace fish meal in diets of sea-water phase Atlantic salmon (*Salmo salar*). *Aquaculture* **2019**, *503*, 609–619. [CrossRef]

38. St-Hilaire, S.; Sheppard, C.; Tomberlin, J.K.; Irving, S.; Newton, L.; McGuire, M.A.; Mosley, E.E.; Hardy, R.W.; Sealey, W. Fly prepupae as a feedstuff for rainbow trout (*Oncorhynchus mykiss*). *J. World Aquac. Soc.* **2007**, *38*, 59–67. [CrossRef]
39. Sealey, W.M.; Gaylord, T.G.; Barrows, F.T.; Tomberlin, J.K.; McGuire, M.A.; Ross, C.; St-Hilaire, S. Sensory analysis of rainbow trout, *Oncorhynchus mykiss*, fed enriched black soldier fly prepupae, *Hermetia illucens*. *J. World Aquac. Soc.* **2011**, *42*, 34–45. [CrossRef]
40. Renna, M.; Schiavone, A.; Gai, F.; Dabbou, S.; Lussiana, C.; Malfatto, V.; Prearo, M.; Capucchio, M.T.; Biasato, I.; Biasibetti, E.; et al. Evaluation of the suitability of a partially defatted black soldier fly (*Hermetia illucens*) larvae meal as ingredient for rainbow trout (*Oncorhynchus mykiss* Walbaum) diets. *J. Anim. Sci. Biotechnol.* **2017**, *8*, 57. [CrossRef] [PubMed]
41. Dumas, A.; Raggi, T.; Barkhouse, J.; Lewis, E.; Weltzien, E. The oil fraction and partially defatted meal of black soldier fly larvae (*Hermetia illucens*) affect differently growth performance, feed efficiency, nutrient deposition, blood glucose and lipid digestibility of rainbow trout (*Oncorhynchus mykiss*). *Aquaculture* **2018**, *492*, 24–34. [CrossRef]
42. Bruni, L.; Secci, G.; Husein, Y.; Faccenda, F.; Lira de Medeiros, A.C.; Parisi, G. Is it possible to cut down fishmeal and soybean meal use in aquafeed limiting the negative effects on rainbow trout (*Oncorhynchus mykiss*) fillet quality and consumer acceptance? *Aquaculture* **2021**, *543*, 736996. [CrossRef]
43. Caimi, C.; Renna, M.; Lussiana, C.; Bonaldo, A.; Gariglio, M.; Meneguz, M.; Dabbou, S.; Schiavone, A.; Gai, F.; Elia, A.C.; et al. First insights on Black Soldier Fly (*Hermetia illucens* L.) larvae meal dietary administration in Siberian sturgeon (*Acipenser baerii* Brandt) juveniles. *Aquaculture* **2020**, *515*, 734539. [CrossRef]
44. Rawski, M.; Mazurkiewicz, J.; Kierończyk, B.; Józefiak, D. Black soldier fly full-fat larvae meal as an alternative to fish meal and fish oil in Siberian sturgeon nutrition: The effects on physical properties of the feed, animal growth performance, and feed acceptance and utilization. *Animals* **2020**, *10*, 2119. [CrossRef]
45. Muin, H.; Taufek, N.M.; Kamarudin, M.S.; Razak, S.A. Growth performance, feed utilization and body composition of Nile tilapia, *Oreochromis niloticus* (Linnaeus, 1758) fed with different levels of black soldier fly, *Hermetia illucens* (Linnaeus, 1758) maggot meal diet. *Iran. J. Fish. Sci.* **2017**, *16*, 567–577.
46. Devic, E.; Leschen, W.; Murray, F.; Little, D.C. Growth performance, feed utilization and body composition of advanced nursing Nile tilapia (*Oreochromis niloticus*) fed diets containing Black Soldier Fly (*Hermetia illucens*) larvae meal. *Aquac. Nutr.* **2018**, *24*, 416–423. [CrossRef]
47. Wachira, M.N.; Osuga, I.M.; Munguti, J.M.; Ambula, M.K.; Subramanian, S.; Tanga, C.M. Efficiency and improved profitability of insect-based aquafeeds for farming Nile tilapia fish (*Oreochromis niloticus* L.). *Animals* **2021**, *11*, 2599. [CrossRef]
48. Tippayadara, N.; Dawood, M.A.O.; Krutmuang, P.; Hoseinifar, S.H.; Doan, H.V.; Paolucci, M. Replacement of fish meal by black soldier fly (*Hermetia illucens*) larvae meal: Effects on growth, haematology, and skin mucus immunity of Nile tilapia, *Oreochromis niloticus*. *Animals* **2021**, *11*, 193. [CrossRef]
49. Adeoye, A.A.; Akegbejo-Samsons, Y.; Fawole, F.J.; Davies, S.J. Preliminary assessment of black soldier fly (*Hermetia illucens*) larval meal in the diet of African catfish (*Clarias gariepinus*): Impact on growth, body index, and hematological parameters. *J. World Aquac. Soc.* **2020**, *51*, 1024–1033. [CrossRef]
50. Fawole, F.J.; Adeoye, A.A.; Tiamiyu, L.O.; Ajala, K.I.; Obadara, S.O.; Ganiyu, I.O. Substituting fishmeal with *Hermetia illucens* in the diets of African catfish (*Clarias gariepinus*): Effects on growth, nutrient utilization, haemato-physiological response, and oxidative stress biomarker. *Aquaculture* **2020**, *518*, 734849. [CrossRef]
51. Vongvichith, B.; Morioka, S.; Sugita, T.; Phousavanh, N.; Phetsanghanh, N.; Chanthasone, P.; Pommachan, P.; Nakamura, S. Evaluation of the efficacy of aquaculture feeds for the climbing perch *Anabas testudineus*: Replacement of fishmeal by black soldier fly *Hermetia illucens* prepupae. *Fish. Sci.* **2020**, *86*, 145–151. [CrossRef]
52. Kattakdad, S.; Suratip, N.; Yuangsoi, B.; Kasamawut, K.; Udduang, S. Black soldier fly (*Hermetia illucens*) pre-pupae meal as a fish meal replacement in climbing perch (*Anabas testudineus*) diet. Aquacult. *Aquar. Conserv. Legis.* **2022**, *15*, 68–82.
53. Xiao, X.; Jin, P.; Zheng, L.; Cai, M.; Yu, Z.; Yu, J.; Zhang, J. Effects of black soldier fly (*Hermetia illucens*) larvae meal protein as a fishmeal replacement on the growth and immune index of yellow catfish (*Pelteobagrus fulvidraco*). *Aquac. Res.* **2018**, *49*, 1569–1577. [CrossRef]
54. Khieokhajonkhet, A.; Uanlam, P.; Ruttarattanamongkol, K.; Aeksiri, N.; Tatsapong, P.; Kaneko, G. Replacement of fish meal by black soldier fly larvae meal in diet for goldfish *Carassius auratus*: Growth performance, hematology, histology, total carotenoids, and coloration. *Aquaculture* **2022**, *561*, 738618. [CrossRef]
55. Kamalii, A.; Antony, C.; Ahilan, B.; Uma, A.; Prabu, E. Dietary protein replacement of fish meal with black soldier fly larvae meal: Effects on growth, whole-body composition, digestive enzyme activity, muscle-growth-related gene expression and haemato-biochemical responses of juvenile goldfish, *Carassius auratus*. *Turk. J. Fish. Aquat. Sci.* **2023**, *23*, 21837. [CrossRef]
56. Zhou, J.S.; Liu, S.S.; Ji, H.; Yu, H.B. Effect of replacing dietary fish meal with black soldier fly larvae meal on growth and fatty acid composition of Jian carp (*Cyprinus carpio* var. Jian). *Aquac. Nutr.* **2018**, *24*, 424–433. [CrossRef]
57. Katya, K.; Borsra, M.Z.S.; Ganesan, D.; Kuppusamy, G.; Herriman, M.; Salter, A.; Ali, S.A. Efficacy of insect larval meal to replace fish meal in juvenile barramundi, *Lates calcarifer* reared in freshwater. *Int. Aquat. Res.* **2017**, *9*, 303–312. [CrossRef]
58. Sudha, C.; Ahilan, B.; Felix, N.; Uma, A.; Prabu, E. Effects of dietary protein substitution of fishmeal with black soldier fly larval meal on growth and physiological responses of juvenile striped catfish, *Pangasianodon hypophthalmus*. *Aquac. Res.* **2022**, *53*, 2204–2217. [CrossRef]

59. Rodríguez, R.; Celada, J.D.; Sáez-Royuela, M.; Carral, J.M.; Aguilera, A.; Melendre, P.M. Artificial reproduction in 1-year-old tench (*Tinca tinca* L.). *J. Appl. Ichthyol.* **2004**, *20*, 542–544. [CrossRef]
60. Cohen, S.A.; Michaud, D.P. Synthesis of a fluorescent derivatizing reagent, 6-aminoquinolyl-N-hydroxysuccinimidyl carbamate, and its application for the analysis of hydrolysate amino acids via high-performance liquid chromatography. *Anal. Biochem.* **1993**, *211*, 279–287. [CrossRef] [PubMed]
61. Cohen, S.A.; De Antonis, K.M. Applications of amino acid derivatization with 6-aminoquinolyl-N-hydroxysuccinimidyl carbamate. *J. Chromatogr. A* **1994**, *661*, 25–34. [CrossRef] [PubMed]
62. Fornshell, G.; Hinshaw, J.M. Better Management Practices for Flow-Through Aquaculture Systems. In *Environmental Best Management Practices for Aquaculture*; Tucker, C.S., Hargreaves, J.A., Eds.; Blackwell Publishing: Ames, IA, USA, 2008; pp. 331–388.
63. Nogales-Mérida, S.; Gobbi, P.; Józefiak, D.; Mazurkiewicz, J.; Dudek, K.; Rawski, M.; Kierończyk, B.; Józefiak, A. Insect meals in fish nutrition. *Rev. Aquac.* **2019**, *11*, 1080–1103. [CrossRef]
64. Perrow, M.R.; Jowitt, A.J.D.; Johnsonf, S.R. Factors affecting the habitat selection of tench in a shallow eutrophic lake. *J. Fish Biol.* **1996**, *48*, 859–870. [CrossRef]
65. Pyka, J. Daily feeding cycle tench, *Tinca tinca* (L.), in larval and fry stages in the conditions of pond culture. An attempt to determine daily food ration. *Arch. Pol. Fish.* **1997**, *5*, 279–290.
66. Alaş, A.; Altindağ, A.; Yılmaz, M.; Kırpık, M.A.; Ak, A. Feeding habits of tench (*Tinca tinca* L., 1758) in Beyşehir lake (Turkey). *Turk. J. Fish. Aquat. Sci.* **2010**, *10*, 187–194. [CrossRef]
67. Lu, S.; Taethaisong, N.; Meethip, W.; Surakhunthod, J.; Sinpru, B.; Sroichak, T.; Archa, P.; Thongpea, S.; Paengkoum, S.; Purba, R.A.P.; et al. Nutritional composition of black soldier fly larvae (*Hermetia illucens* L.) and its potential uses as alternative protein sources in animal diets: A review. *Insects* **2022**, *13*, 831. [CrossRef]
68. Liland, N.S.; Araujo, P.; Xu, X.X.; Lock, E.J.; Radhakrishnan, G.; Prabhu, A.J.P.; Belghit, I. A meta-analysis on the nutritional value of insects in aquafeeds. *J. Insects Food Feed* **2021**, *7*, 743–759. [CrossRef]
69. Barroso, F.G.; de Haro, C.; Sánchez-Muros, M.-J.; Venegas, E.; Martínez-Sánchez, A.; Pérez-Bañón, C. The potential of various insect species for use as food for fish. *Aquaculture* **2014**, *422–423*, 193–201. [CrossRef]
70. Maurer, V.; Holinger, M.; Amsler, Z.; Früh, B.; Wohlfahrt, J.; Stamer, A.; Leiber, F. Replacement of soybean cake by *Hermetia illucens* meal in diets for layers. *J. Insects Food Feed* **2016**, *2*, 83–90. [CrossRef]
71. Basto, A.; Matos, I.; Valente, L.M.P. Nutritional value of different insect larvae meals as protein sources for European sea bass (*Dicentrarchus labrax*) juveniles. *Aquaculture* **2020**, *21*, 735085. [CrossRef]
72. Jobling, M. Nutrient Partitioning and the Influence of Feed Composition on Body Composition. In *Food Intake in Fish*; Houlihan, D., Boujard, T., Jobling, M., Eds.; Blackwell: New York, NY, USA, 2001; pp. 354–375.
73. Sargent, J.R.; Tocher, D.R.; Bell, J.G. The Lipids. In *Fish Nutrition*; Halver, J.E., Hardy, R.W., Eds.; Academic Press: San Diego, CA, USA, 2002; pp. 181–257.
74. Ferrer Llagostera, P.; Kallas, Z.; Reig, L.; Amores de Gea, D. The use of insect meal as a sustainable feeding alternative in aquaculture: Current situation, Spanish consumers' perceptions and willingness to pay. *J. Clean. Prod.* **2019**, *229*, 10–21. [CrossRef]
75. Lugert, V.; Thaller, G.; Tetens, J.; Schulz, C.; Krieter, J. A review on fish growth calculation: Multiple functions in fish production and their specific application. *Rev. Aquac.* **2016**, *8*, 30–42. [CrossRef]
76. Cahu, C.; Zambonino-Infante, J.; Takeuchi, T. 2003. Nutritional components affecting skeletal development in fish larvae. *Aquaculture* **2003**, *227*, 245–258. [CrossRef]
77. Fontagné, S. The Impact of Nutritional Components on Rainbow Trout. In *Control of Malformations in Fish Aquaculture: Science and Practice*; Baeverfjord, G., Helland, S., Hough, C., Eds.; RapidPress: Luxembourg, 2009; pp. 73–83.
78. Kamler, E.; Myszkowski, L.; Kaminski, R.; Korwin-Kossakowski, M.; Wolnicki, J. Does overfeeding affect tench *Tinca tinca* L. juvenile? *Aquac. Int.* **2006**, *14*, 99–111. [CrossRef]
79. Wolnicki, J.; Myszkowski, L.; Korwin-Kossakowski, M.; Kaminski, R.; Stanny, L.A. Effects of different diets on juvenile tench *Tinca tinca* (L.) reared under controlled conditions. *Aquac. Int.* **2006**, *14*, 89–98. [CrossRef]

fishes

MDPI

Article

Effects of a Diet of *Phragmites australis* instead of *Triticum aestivum* L. on Immune Performance and Liver Tissue Structure of *Ctenopharyngodon idellus*

Ronghua Wang [1,2,3], Chaobo Lei [1], Zhenyu Li [1], Yanju Lei [1,2,3], Congqiang Luo [1,2,3], Liye Shao [1,2,3,*], Chunhong Huang [1,2,3] and Pinhong Yang [1,2,3,*]

1 College of Life and Environmental Sciences, Hunan University of Arts and Science, Changde 415000, China
2 Hunan Provincial Key Laboratory for Molecular Immunity Technology of Aquatic Animal Diseases, Changde 415000, China
3 Hunan Provincial Key Laboratory for Health Aquaculture and Product Processing in Dongting Lake Area, Changde 415000, China
* Correspondence: shaoly99@163.com (L.S.); yphong@huas.edu.cn (P.Y.)

Abstract: This experiment aimed to study the effects on liver tissue structure and immune performance of grass carp *Ctenopharyngodon* idellus when the common reed *Phragmites australis* is in its diet, instead of wheat *Triticum aestivum* L. Seventy-two healthy grass carps (145.52 ± 2.56 g) were randomly divided into three groups according to their body weight. Fishes in each group were fed an essential diet with 0% (control group), 50% and 100% (test group) common reed, instead of wheat, respectively. After feeding for 41 days, the changes of serum biochemical indices, liver tissue structure and immune related indices of grass carp were detected. The results showed that, compared with the control group, the activities of serum alanine transaminase (ALT) and aspartate transaminase (AST) in the substitution groups were significantly increased ($p < 0.05$), but still at a normal level. The contents of total protein, albumin and globulin did not change significantly ($p > 0.05$). Compared with the control group, the liver cells of grass carp in the substitution groups had clear boundaries, tight arrangement and less vacuolation. The contents of serum interleukin-1 (IL-1) and complement 3 (C3) in the 100% substitution group were significantly higher than those in the control group ($p < 0.05$), and the contents of liver IL-1 and total complement (CH50) in the 100% substitution group were significantly higher than those in the control group ($p < 0.05$). The contents of IL-1, C3 in serum and IL-1, CH50 in liver in the 50% substitution group were significantly higher than those in the control group ($p < 0.05$). The mRNA relative expression levels of C3, IL-1, MHC-I and interferon (IFN) in the head-kidney, kidney, liver and spleen of grass carp were significantly affected by feeding the grass carp with different common reed substitution ratios ($p < 0.05$). In summary, common reed, instead of wheat, in feed can improve the liver tissue structure, and increase the non-specific immune response level, of grass carp.

Keywords: *Ctenopharyngodon idellus*; *Phragmites australis*; serum biochemistry; immune-related factors; immune gene

Citation: Wang, R.; Lei, C.; Li, Z.; Lei, Y.; Luo, C.; Shao, L.; Huang, C.; Yang, P. Effects of a Diet of *Phragmites australis* instead of *Triticum aestivum* L. on Immune Performance and Liver Tissue Structure of *Ctenopharyngodon idellus*. *Fishes* **2022**, *7*, 378. https://doi.org/10.3390/fishes7060378

Academic Editors: Marina Paolucci and Shunsuke Koshio

Received: 1 November 2022
Accepted: 6 December 2022
Published: 8 December 2022

1. Introduction

The common reed *Phragmites australis* is one of the most widely distributed and abundant wetland plant genera in the world [1] and found on every continent except Antarctica. The common reed plays an important role in regulating climate, inhibiting algae growth, accumulating heavy metals, preventing floods, strengthening dikes and maintaining biodiversity, with high ecological and socio-economic value [2–5]. The common reed has been used throughout history to produce non-food commodities, such as paper pulp, roofing and building materials or litter material, as well as for heating and as forage feed [6].

The Dongting Lake is one of the main producing areas of common reed in China, with a planting area of 666 km^2 and an annual output of more than 900,000 tons, accounting for about 30% of the total output in China [7]. In recent years, the main utilization mode of the common reed in the Dongting Lake area is as papermaking raw materials. However, due to environmental pollution, the paper industry has completely withdrawn from the Dongting Lake area, and the utilization rate of the common reed in the lake area has significantly reduced. A mass of common reeds rotted and accumulated, thus causing eutrophication in local areas of the lake, as well as great resource waste and ecological pollution [8]. With the continuous growth of the aquaculture industry in China, the demand for feed raw materials has increased greatly. Limited by the output of agricultural products and resource allocation, the gap of conventional feed raw materials is getting bigger and bigger.

Common reed is an excellent forage grass of Gramineae for horses, cattle, and goats, and research on its forage mainly focuses on its use as roughage or silage for breeding herbivores, in which good results have been achieved [9,10]. However, there is no research on high-value and high-efficiency utilization of common reed processed into pellet feed. The Grass carp *Ctenopharyngodon* idellus is the main freshwater breeding fish in China. Since 2013, the output of grass carp culture in China has reached more than 5 million tons. In this experiment, grass carp was taken as the research object, and the effects of different proportions of common reed instead of wheat *Triticum aestivum* L. on liver tissue structure and immune performance index of grass carp were evaluated. The results can provide some theoretical basis for the development and utilization of common reed resources in Dongting Lake.

2. Materials and Methods

2.1. Experimental Feed

Common reeds were collected in the Changde area of Dongting Lake, chopped with a 220 V straw hay cutter, dried at 56 °C and ground into powder. The moisture, crude ash, crude protein, crude fiber and crude fat contents of common reed, wheat and the diets were determined as described by AOAC (2003). The gross energy of common reed, wheat and the diets were detected by an Oxygen bomb heat meter (Calorimeter, Parr instrument Company Moline, Illinois, USA). The main effective components of common reed and wheat were shown in Table 1.

Table 1. Main active ingredients of common reed and wheat medium residue (air-dry basis%).

Items	Moisture	Crude Protein	Ash	Crude Fiber	Crude Fat	Gross Energy (MJ/kg)
Common reed	10.93	14.16	6.43	39.27	1.6	17.73
Wheat	10.61	14.18	6.61	1.88	1.7	18.24

The basic feed raw materials came from Alpha Feed, and the feed raw materials were crushed and screened, with a pore size of 0.43 mm. According to the nutrient content of common reed powder and wheat flour measured in this experiment, the grass carp compound feed in the control group was designed with VF123 feed formula software. In the experimental group, 50% and 100% common reed flour were used, instead of wheat flour, in the basic feed. See Table 2 for the composition and nutritional level of the experimental feed.

Table 2. Composition and nutrient levels of experimental diets (dry matter %).

Items	Diets		
	0	50%	100%
Ingredients			
Wheat	30	15	0
Common reed	0	15	30
Soybean meal	30	30	30
Fish meal	3	3	3
Rapeseed meal	26	26	26
$Ca(H_2PO_4)_2$	2.5	2.5	2.5
Carboxymethyl Cellulose	2	2	2
Choline chloride	0.15	0.15	0.15
Bentonite clay	1.35	1.35	1.35
Vitamin premix [1]	1	1	1
Mineral premix [2]	1	1	1
Soybean oil	3	3	3
Total	100	100	100
Nutrient levels [3]			
Crude protein	32.61	33.10	33.00
Crude fat	4.82	4.60	4.77
Crude fiber	4.84	9.24	13.87
Crude ash	9.95	11.20	12.45
Gross energy (kJ/g)	19.96	19.86	19.84

[1.] Vitamin premix (mg/kg diet): vitamin A, 120,000 IU, vitamin B_1, 200 mg, vitamin B_2, 280 mg, vitamin B_8, 240 mg, vitamin B_{12}, 0.6 mg, vitamin D_3, 40,000 IU, vitamin E 480 mg, vitamin K_3, 200 mg, biotin 1.2 mg, folic acid, 60 mg, calcium pantothenic, 720 mg, nicotinic acid, 1000 mg, vitamin C phosphate, 6850 mg, inositol, 3200 mg; [2.] Mineral premix (mg/kg diet): Cu (as copper sulfate) 160 mg, Fe (as ferrous sulfate) 4800 mg, Mn (as manganese sulfate) 800 mg, Zn (as zinc sulfate) 2000 mg, I (as potassium iodide) 40 mg, Se (as sodium selenite)4 mg, Mg (as magnesium sulfate) 400 mg, Co (as cobalt dichloride)12 mg; [3.] Nutrient levels were measured values. Nutrient contents analyzed according to AOAC (1995) protocols.

2.2. Feeding Management

Grass carp was purchased from a grass carp farm in Changde city, Hunan, China. The grass carp were acclimatized for two weeks in fiberglass tanks (diameter: 1.06 m, the volume of aquaculture water was 626L) with a flow-through system (source of water: Baima Lake in Changde; water flow velocity: 1300 mL/min) before the main culture experiment. Then, 72 healthy grass carps with an average body weight of 145.52 ± 2.56 g were randomly divided into three groups according to body weight, with no significant difference ($p > 0.05$). Each group had three replicates with eight fishes in each replicate. The feeding amount was about 3.0~5.0% of the fish body weight and adjusted according to the intake of the diet during the feeding trial, and feeding occurred at 08:00 and 15:00 every day. The feces were removed by siphoning before each feeding. The feeding status and health status of each group of experimental fish were recorded accurately every day, and the water quality was detected every week during the experimental culture. During the culture period, the water temperature was 20~26 °C, the dissolved oxygen was above 6.0 mg/L, pH was 7.5~7.8, and ammonia nitrogen under 0.5 mg/L.

2.3. Sample Collection and Processing

After culturing for 41 days and stopping feeding for one day, three fishes were randomly selected from each culturing bucket. All experimental fish were anesthetized with MS-222 (150 mg/L) (Sigma-Aldrich, Darmstadt, Germany) before sampling. Blood was taken from the tail vein, stood at 4 °C for 1 h, centrifuged for 15 min at 4000 r/min, and the serum was taken for biochemical parameters and immune factors content detection. After dissection, 0.2 g liver, spleen, kidney and head kidney were taken in each fish, frozen in liquid nitrogen and stored in a refrigerator at −80 °C for detection of immune gene expression level. One gram liver in each fish was rinsed with ice-cold PBS, and then homogenized in 9 mL PBS used JXFSTPRP-24L (Shanghai JingXin, Shanghai, China), centrifuged for

5 min at 5000 r/min, and the supernatant were stored at −80 °C for detecting the content of immune factors in liver. The liver tissues of 0.5 cm × 0.5 cm × 0.5 cm were fixed in 10% formaldehyde to observe the liver structure.

2.4. Detection of Serum Biochemical Parameters

Serum biochemical indices were tested in Liyuan Medical Testing Center of Changde City. Total protein (TP) was measured by the biuret method, albumin (ALB) was measured with bromocresol green (BCG) method, and globulin (GLB) was measured by the calculation method. Blood glucose (GLU) was measured by the hexokinase method, and Aspartate aminotransferase (AST) and Alanine aminotransferase (ALT) activities were measured by the IFCC rate method.

2.5. Observation of Liver Tissue Structure

The liver tissue was fixed in 10% formaldehyde for 24 h, then dehydrated by 30~100% ethanol gradient, made transparent with xylene and embedded in wax. Paraffin sections with a thickness of 4~5 μm were prepared, stained by means of the HE routine, sealed with neutral gum, and observed with a Leica Advanced Microscope System (DM3000). The area ration and number of hepatocyte vacuolization was visualized and quantified using ImageJ version 1.5.3 software. three areas of each liver section were quantified. By selecting ranges of pixel values in color images the pixels associated with black could be distinguished. The number of selected pixels was then quantified using a particle analysis operation and by counting the area of all bright objects (in pixels).

2.6. Detection of Immune Factor Activity and Content

Immunoglobulin M (IgM), Interleukin 1 (IL-1), Complement 3 (C3), Total Complement (50% Haemolytic Complement, CH50) and Cortisol in serum and in the liver were detected using an ELISA kit (Shanghai Enzyme Linked Biotechnology Co., Ltd., Shanghai, China).

2.7. Detection and Analysis of Immune Gene mRNA Expression Level

2.7.1. RNA Extraction and cDNA First Strand Synthesis

Total RNA was extracted from the liver, spleen, kidney and head kidney tissues, and stored at −80 °C using TRIzol reagent (Invitrogen, Carlsbad, CA, USA). The method referred to the instructions of TRIzol reagent. The integrity of the extracted RNA was detected by 1% agarose electrophoresis, and the total RNA, with good integrity, was used for the next experiment. A quantity of 2 μg total RNA was taken, and the first strand of cDNA was synthesized by using a RevertAid™ First Strand cDNA Synthesis reverse transcription kit (ThermoFisher, Carlsbad, CA, USA). The reverse transcription product was diluted 20 times and used for the analysis of immune gene mRNA expression level.

2.7.2. Fluorescence Quantitative PCR Analysis

According to the related literatures [11,12], immune genes Heat Shock Protein 70 (Hsp70), Lysozyme, Major histocompatibility complex I (MHC-I), Interferon (IFN), C3, IgM and IL-1 were screened out and detected these genes relative expression level after culture experiment. Primer synthesis was completed by Sangon Biotech Co., Ltd. (Shanghai, China), and primers sequence information is shown in Table 3. The fluorescence quantitative PCR reaction system involved the following: 2 × ChamQ Universal SYBR qPCR Master Mix (Vazyme, Nanjing, China) 10 μL, 10 μmol/L primer 0.4 μL, diluted cDNA template 9.2 μL. The reaction procedure was as follows: 95 °C for 3 min: 95 °C for 5 s, 60 °C for 10 s, 72 °C for 15 s. The melting curve ranged from 65 °C to 97 °C and increased by 0.1 °C per second. The relative expression level of genes was calculated by the $2^{-\Delta\Delta Ct}$ method. LightCycler 480 software release 1.5. 0 and Excel 2021 software were used for data processing. The β-actin gene was used as a reference gene, and the relative expression level of gene mRNA in the control group was analyzed by homogenization.

Table 3. The primers used in this study.

Primers	Primer Sequence (5′-3′)	References
IgM-F IgM-R	TGGTCATCAGGTGGCAAA GCGGCTGTCTTCCATTCT	[11]
lysozyme-F lysozyme-R	TTCGACAGCAAAACAGGACAAC GATATGATGGCAGCAATCACAGC	[11]
C3-F C3-R	AATACGCCATTCCTGAGGTTTCC CTTCCACCATTTCACTGCCACTT	[11]
IL-1-F IL-1-R	TACCGAGTCGGATGGTTCTTC TGTTATTAGCCACACCGGTCTC	[12]
IFN-I-F IFN-I-R	CGGCCGATACAGGATGATAAG TCCTCCACCTTGGCATTGTC	[12]
MHC-I-F MHC-I-R	CCTGCTAATCCTCAAGCTGTCA GCATGACACGTCACTGGAGAG	[12]
Hsp70-F Hsp70-R	GTGTCCATCCTGACCATTGA ATCTGGATTGATGCTCTTGTT	[12]
Actin-F Actin-R	GCTATGTGGCTCTTGACTTCG GGGCACCTGAACCTCTCATT	[12]

2.8. Data Analysis

SPSS 17.0 software was used for one-way ANOVA, and the Duncan method was used for significance analysis between groups. The significance level was $p < 0.05$. The results were plotted by mean ± SD, the activity and content of immune factors, and the expression level of immune-related genes used graph Pad Prism 6 software.

3. Results

3.1. Serum Biochemical Indicators

It can be seen from Table 4 that the contents of TP, ALB and GLB did not change significantly among the three groups ($p > 0.05$). The blood glucose concentration of grass carp in the substitution group was lower than that in the control group, especially in the 50% substitution group ($p < 0.05$). Except for AST activity in the 100% substitution group, serum ALT and AST activities of grass carp in the substitution group were significantly higher than those in the control group ($p < 0.05$).

Table 4. Effects of common reed on serum biochemical parameters of grass carp (N = 3).

Items	0%	50%	100%
TP (g/L)	32.97 ± 3.29	36.50 ± 1.87	34.57 ± 0.35
ALB (g/L)	15.73 ± 1.11	17.00 ± 1.04	15.70 ± 0.40
GLB/(g/L)	17.23 ± 2.18	19.50 ± 0.90	18.87 ± 0.64
GLU (mmol/L)	9.85 ± 0.81 ^{a}	6.35 ± 0.56 ^{b}	7.68 ± 1.02 ^{ab}
ALT/(U/L)	22.35 ± 2.05 ^{c}	64.60 ± 8.34 ^{b}	98.05 ± 1.48 ^{a}
AST/(U/L)	47.00 ± 3.11 ^{b}	84.10 ± 17.25 ^{a}	51.45 ± 0.92 ^{b}

Different superscript letters of peer data indicate significant difference between the two groups ($p < 0.05$), while the same or no superscript letters indicate no significant difference between the two groups ($p > 0.05$).

3.2. Histological Structure of Liver

In the control group, hepatic cell boundaries were unclear and hepatocyte vacuolation was severe, accompanied by a small amount of lymphocyte infiltration (Figure 1a,b). In the 50% and 100% substitution groups, the boundaries of hepatocytes in grass carp liver were clear, the nuclei were clearly colored and basically located in the center, and the cells were arranged neatly and were of uniform size (Figure 1c,d,e,f). The area ratio and number of hepatocyte vacuolization in the substitution group was significantly lower than that in the control group ($p < 0.05$) (Table 5).

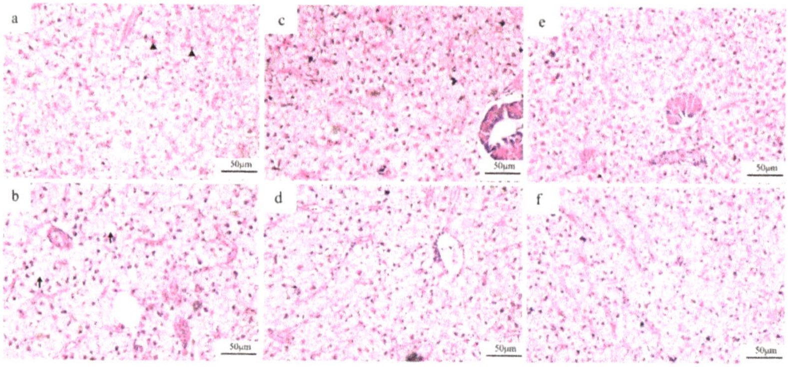

Figure 1. Effect of common reed on liver histology of grass carp (200 times). (**a**,**b**): 0% replacement group; (**c**,**d**): 50% replacement group; (**e**,**f**): 100% replacement group. The triangle in the figure indicates lymphocyte cells, and the arrow indicates vacuolar degeneration of cells.

Table 5. Histological parameters of liver from grass carp fed experimental diets.

Items	0%	50%	100%
The area ratio of hepatocyte vacuolization (%) [1]	13.26 ± 1.9^{a}	3.37 ± 0.52^{b}	2.69 ± 0.67^{b}
The number of hepatocyte vacuolization	74.33 ± 13.01^{a}	22.33 ± 2.52^{b}	32.33 ± 2.52^{b}

[1] The area ratio of hepatocyte vacuolization (%) = The area of hepatocyte vacuolization (pixels)/The area of image (pixels) × 100%. Different superscript letters of peer data indicate significant difference between the two groups ($p < 0.05$), while the same or no superscript letters indicate no significant difference between the two groups ($p > 0.05$).

3.3. Activity and Contents of Immune Factors

Compared with the control group, the contents of serum IL-1 and C3 in the 100% common reed substitution group were significantly higher ($p < 0.05$), and the activity of CH50 in the 50% and 100% common reed substitution groups were significantly lower than in the control group ($p < 0.05$). The contents of IL-1 and CH50 in the liver of the 100% common reed substitution group were significantly higher than those in the control group ($p < 0.05$), but the content of C3 was significantly lower than that in the control group ($p < 0.05$). Compared with the control group, the contents of IL-1, C3, IgM and cortisol in the serum and liver of grass carp in the 50% common reed substitution group were not significantly different ($p > 0.05$) (Figure 2).

3.4. The mRNA Expression Level of Immune Genes

Compared with the control group, the expression levels of Hsp70, IL-1, MHC-I and lysozyme mRNA in the head and kidney of grass carp in the 100% common reed substitution group were significantly up-regulated ($p < 0.05$), while the expression levels of C3 and mRNA were significantly down-regulated. The mRNA expression levels of C3, Hsp70, lysozyme, MHC-I and other genes in the 50% common reed substitution group were significantly down-regulated in the head kidney ($p < 0.05$) (Figure 3A). Compared with the control group, the mRNA expression levels of IL-1 and lysozyme in the liver of grass carp in the substitution group were significantly down-regulated ($p < 0.05$), while the mRNA expression levels of C3, MHC-I and IgM genes were not significantly different ($p > 0.05$) (Figure 3B). The mRNA expression of C3 and Hsp70 in the spleen of the 50% common reed substitution group was significantly higher than that in the control group, while the expression of MHC-I was significantly lower than that in the control group

($p < 0.05$); The mRNA expression levels of other detected genes, except the IgM gene, in grass carp spleen of the 100% common reed substitution group were significantly lower than those in the control group (Figure 3C). The mRNA expression levels of Hsp70, MHC-I and IgM genes in the kidneys of the 100% common reed substitution group were significantly higher than those of the control group ($p < 0.05$), while the mRNA expression levels of other detected genes, except C3, in the kidneys of the 50% common reed substitution group were not significantly different ($p > 0.05$) (Figure 3D).

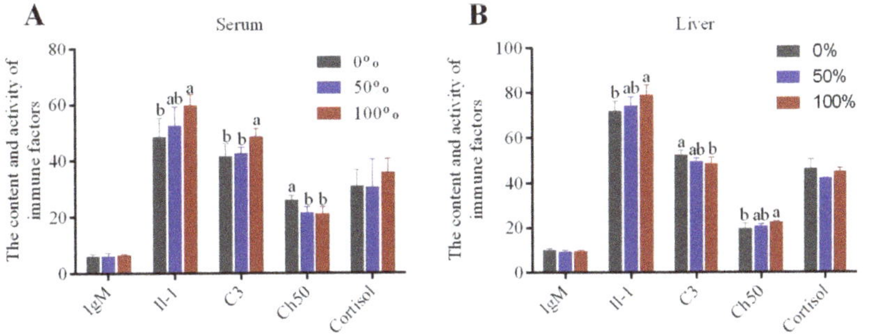

Figure 2. Effect of common reed on content and activity of immune factors in serum (**A**) and liver (**B**) of grass carp (N = 3). The units of IgM, IL-1, C3, CH50 and Cortisol were μg/mL, pg/mL, μg/mL, U/mL and ng/mL respectively. Different small letters mean significant difference among groups ($p < 0.05$), the same letters or no letters mean no significant difference ($p > 0.05$).

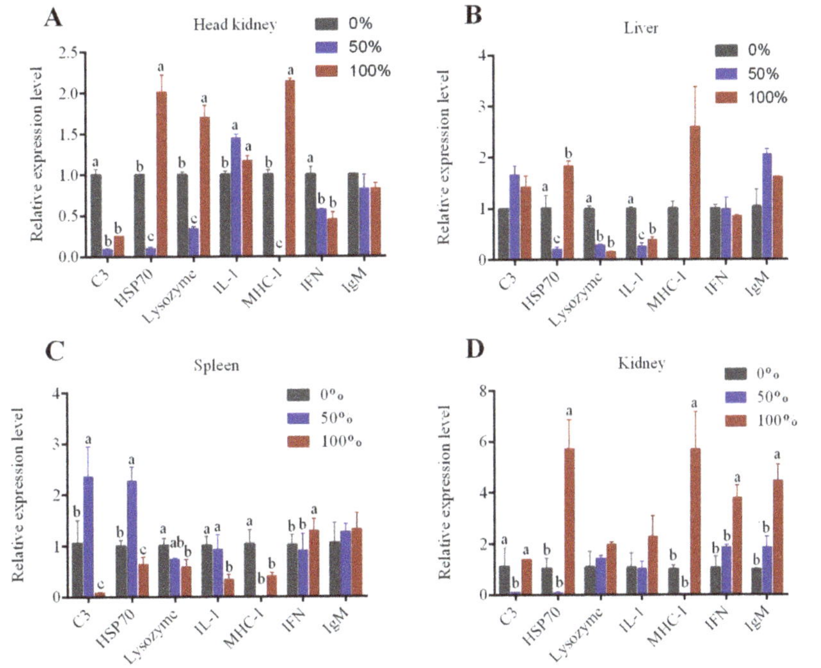

Figure 3. Effect of common reed on expression of immune genes in head kidney (**A**), liver (**B**), spleen (**C**) and kidney (**D**) of grass carp (N = 3). Different letters mean significant difference among groups ($p < 0.05$), the same letters or no letters mean no significant difference ($p > 0.05$).

4. Discussion

The liver is the most important metabolic center of fish. When fish ingest feed with unreasonable nutritional structure, it causes nutritional stress to the liver, and then damages the normal structure and functions of the liver [13,14]. In this study, we found that the cellular structure of the liver tissue of grass carp in the common reed substitution group was closely arranged, while, in the control group, lymphocytes had infiltrated, cells were vacuolated and had obvious inflammatory reaction and cell vacuolization. Therefore, from the analysis of the results of the tissue sections, adding common reed to feed was helpful to improve the health status of the liver tissue.

Fish serum biochemical indicators are widely used to evaluate the health status, nutritional status and adaptation to the environment of fish, and are good physiological, pathological and toxicological indicators. TP content in serum can reflect the metabolism of protein in the body, and increase of TP content can accelerate the deposition of protein and promote the growth of the body. ALB and GLB mainly maintain the balance of cell nutrition and blood osmotic pressure in vivo [15]. In this experiment, the TP content of the substitution group and the control group were at the normal level of total protein content for teleost fish (30–50 g/L) [16]. The TP content of the substitution group was slightly higher than that in the control group, but the difference was not significant ($p > 0.05$), indicating that the nutritional status of the substitution group was better. Blood glucose is an important index reflecting the glucose metabolism of fish, the functional state of tissues and cells and endocrine function of the whole body [17]. The normal content of blood glucose in grass carp is 2.78–12.72 mmol/L [18]. In this study, the blood glucose concentration of grass carp in the substitution group was lower than that in the control group, and the 50% substitution group was significantly lower than that in the control group, but all of them were in the normal range. Transaminase is involved in protein metabolism and synthesis [19], and the activities of ALT and AST directly reflect the protein metabolism level [20], and are affected by the quantity and quality of dietary protein [21]. In this study, the serum ALT and AST activities of grass carp in the substitution group were significantly higher than those in the control group ($p < 0.05$), but still in the normal range, though lower than those detected by Cheng et al. [22] and Chen et al. [23]. In this study, the activity of ALT and AST in the serum of the grass carp in the substitution group might be related to the amino acid composition and high cellulose content of common reed. High cellulose content affects protein absorption of grass carp.

As a lower vertebrate, the fish has a relatively perfect acquired immune response, which can resist pathogen invasion through nonspecific immune response and specific immune response [24]. Complements and lysozymes are important non-specific immune system components of the body, which play the role of dissolving bacteria and eliminating cells (bacteria) in resisting the invasion of pathogenic microorganisms [25–27]. IL-1 and IFN are important cytokines in the body, which play an important role in inducing and activating inflammatory response and regulating immune response [28]. MHC-I and IgM are important factors in the activation and effects of the specific immune system, and play an important role in antigen binding and inhibition of pathogen activity [29]. In this study, the activity and content of immune factors, such as C3, CH50, IL-1, lysozymes, IgM, and MHC-I in the serum and liver of grass carp, and the mRNA expression level of corresponding genes in different tissues, were detected to judge the effect of a common reed diet, instead of a wheat diet, on the immune performance of grass carp. The activity and content of immune factors showed that the contents of IL-1, C3 and cortisol in the serum, IL-1 and CH50 in the liver of grass carp in the substitution group increased with increase of the common reed replacement ratio. The changes of non-specific immune indices of grass carp in the 100% common reed substitution group were significantly different from those in the control group ($p < 0.05$), but there was no significant difference in IgM content ($p > 0.05$). From the change results of immune factor content, adding the proper amount of common reed to feed could enhance the nonspecific immune response of grass carp, but had no significant effect on the specific immune response. The results were consistent with the research results of Chen et al. [30], who found that rutin, an active ingredient in the common reed, could significantly improve the nonspecific immunity of grass carp. At the same time, the results of gene expression level showed that the mRNA expression level of the immune gene of grass carp in the 50% common reed substitution group was down-regulated compared with that in the control group, while the expression of Hsp70, IL-1, lysozymes, MHC-I, and IgM in head kidney and kidney of grass carp in the 100% common reed substitution group were up-regulated to varying degrees. This might be related to the abundant bioactive components in the roots, stems and leaves of the common reed, such as flavonoids, reed polysaccharides and rutin. Studies have shown that flavonoids and rutin have significant antibacterial and antioxidant properties, and increase immune response [31,32]. Therefore, combined with the changes of immune factor activity and content and gene mRNA expression level, adding a certain amount of common reed in feed can improve the nonspecific immune response level of grass carp.

5. Conclusions

In conclusion, this study indicated that the replacement of wheat with common reed could improve the liver structure and function of grass carp, and, at the same time, improve the level of non-specific immune response of grass carp to a certain extent. However, the most suitable amount of common reed in grass carp feed remains to be further studied.

Author Contributions: Conceptualization, R.W., Y.L. and C.L. (Congqiang Luo); Data curation, C.H. and L.S.; Formal analysis, R.W.; Funding acquisition, P.Y.; Methodology, R.W., C.L. (Chaobo Lei) and Z.L.; Project administration, C.L. (Chaobo Lei), Z.L., Y.L. and L.S.; Resources, L.S.; Software, Z.L.; Supervision, C.L. (Congqiang Luo); Writing—original draft, R.W.; Writing—review & editing, R.W., C.H. and P.Y. All authors have read and agreed to the published version of the manuscript.

Funding: This research was funded by The Natural Science Foundation of Hunan Province (Grant no. 2021JJ30468), The Investigation of Aquaculture Germplasm Resources in Hunan Province (Grant no. 2130122), The Research Project of Education Department of Hunan Province (Grant no. 22B0701 and 20A339), Innovation Platform project of Education Department of Hunan Province (Grant No. 19K064) and Hunan University of Arts and Science Doctoral Foundation project (Grant no. 18BSQD20).

Institutional Review Board Statement: The study was conducted in accordance with the Declaration of Helsinki, and all procedures used in this experiment were approved by the Hunan University of Arts and Science Institutional Animal Care and Use Committee and were performed in accordance with approved protocols, protocol code: JSDX-2021-025 and date of approval: 15 March 2021.

Informed Consent Statement: Not applicable.

Data Availability Statement: Not applicable.

Acknowledgments: We would like to acknowledge the contribution of the college students (Zhiwei Zhang, Fengling He and Kang Huang) at Hunan University of Arts and Science for their roles in fish husbandry and for their assistance with sampling.

Conflicts of Interest: The authors declare no conflict of interest.

References

1. Köbbing, J.F.; Thevs, N.; Zerbe, S. The Utilisation of Reed (*Phragmites australis*): A Review. *Mires Peat* **2013**, *13*, 1–14.
2. Chen, L.; Zhang, L.; Liu, Z.; Lu, F.; Feng, G.; Yan, K.; Han, G. Physiological and Ecological Responses of Hetan and Chaotan *Phragmites australis* to Salt Stress. *Acta Ecol. Sin.* **2020**, *40*, 2090–2098.
3. Meneses, N.C.; Baier, S.; Reidelstürz, P.; Geist, J.; Schneider, T. Modelling Heights of Sparse Aquatic Reed (*Phragmites australis*) Using Structure from Motion Point Clouds Derived from Rotary- And Fixed-Wing Unmanned Aerial Vehicle (Uav) Data. *Limnologica* **2018**, *72*, 10–21. [CrossRef]
4. Minkina, T.; Fedorenko, G.; Nevidomskaya, D.; Fedorenko, A.; Chaplygin, V.; Mandzhieva, S. Morphological and Anatomical Changes of *Phragmites australis* Cav. Due to the Uptake and Accumulation of Heavy Metals from Polluted Soils. *Sci. Total Environ.* **2018**, *636*, 392–401. [CrossRef]
5. Zhang, S.; Bai, J.; Wang, W.; Huang, L.; Zhang, G.; Wang, D. Heavy Metal Contents and Transfer Capacities of *Phragmites australis* and *Suaeda salsa* in the Yellow River Delta, China. *Phys. Chem. Earth Parts A/B/C* **2018**, *104*, 3–8. [CrossRef]

6. Hansson, P.; Fredriksson, H. Use of Summer Harvested Common Reed (*Phragmites australis*) as Nutrient Source for Organic Crop Production in Sweden. *Agric. Ecosyst. Environ.* **2004**, *102*, 365–375. [CrossRef]
7. Zhou, J.; Wan, R.R.; Wu, X.H.; Zhang, Y. Patterns of Long-Term Distribution of Typical Wetland Vegetation (1987–2016) and its Response to Hydrological Processes in Lake Dongting. *J. Lake Sci.* **2020**, *6*, 1723–1735.
8. Sun, Z.; Mou, X.; Sun, W. Potential Effects of Tidal Flat Variations on Decomposition and Nutrient Dynamics of *Phragmites australis*, *Suaeda salsa* and *Suaeda glauca* Litter in Newly Created Marshes of the Yellow River Estuary, China. *Ecol. Eng.* **2016**, *93*, 175–186. [CrossRef]
9. Sun, Z.; Zhou, D.; Ferreira, L.M.M.; Zhong, Q.; Lou, Y. Diet Composition, Herbage Intake and Digestibility in Inner Mongolian Cashmere Goats Grazing on Native Leymus Chinensis Plant Communities. *Livest. Sci.* **2008**, *116*, 146–155. [CrossRef]
10. Kadi, S.A.; Ouendi, M.; Bannelier, C.; Berchiche, M.; Gidenne, T. Nutritive Value of Sun-Dried Common Reed (*Phragmites australis*) Leaves and its Effect on Performance and Carcass Characteristics of the Growing Rabbit. *World Rabbit. Sci.* **2018**, *26*, 113–121. [CrossRef]
11. Xiao, T.Y.; Zhou, Z.Y.; Wang, R.H.; Li, Y.G.; Jin, S.Z.; Li, W.; Wang, H.Q. The Expression and Transmission of Immune Factors Between Generations in Maternal *Ctenopharyngodon idella* After Immunization with Gcrv Attenuated Vaccine. *J. Fish. China* **2017**, *8*, 1308–1318.
12. Chen, L.J.; Yu, E.M.; Wang, G.J.; Jun, X.; Yu, D.G.; Li, Z.F.; Zhang, K.; Gong, W.B.; Tian, J.J. Effects of Feeding with Broad Bean on the Immune Organ Ultrastructure and Expression of Immune Genes of Grass Carp. *J. China Agric. Univ.* **2019**, *6*, 92–103.
13. Zhou, W.; Rahimnejad, S.; Lu, K.; Wang, L.; Liu, W. Effects of Berberine on Growth, Liver Histology, and Expression of Lipid-Related Genes in Blunt Snout Bream (*Megalobrama amblycephala*) Fed High-Fat Diets. *Fish Physiol. Biochem.* **2019**, *45*, 83–91. [CrossRef] [PubMed]
14. Sun, S.; Wu, Y.; Yu, H.; Su, Y.; Ren, M.; Zhu, J.; Ge, X. Serum Biochemistry, Liver Histology and Transcriptome Profiling of Bighead Carp Aristichthys Nobilis Following Different Dietary Protein Levels. *Fish Shellfish Immunol.* **2019**, *86*, 832–839. [CrossRef]
15. Jahanbakhshi, A.; Imanpoor, M.R.; Taghizadeh, V.; Shabani, A. Hematological and Serum Biochemical Indices Changes Induced by Replacing Fish Meal with Plant Protein (Sesame Oil Cake and Corn Gluten) in the Great Sturgeon (*Huso huso*). *Comp. Clin. Pathol.* **2013**, *22*, 1087–1092. [CrossRef]
16. Guo, X.Z.; Liang, X.F.; Fang, L.; Yuan, X.C.; Zhou, Y.; Li, B. Effects of Non-Protein Energy Sources on Serum Biochemical Indices and Histology of Liver in Grass Carp (*Ctenopharyngodon idella*). *Acta Hydrobiol. Sin.* **2014**, *3*, 582–587.
17. Jiang, D.; Wu, Y.; Huang, D.; Ren, X.; Wang, Y. Effect of Blood Glucose Level on Acute Stress Response of Grass Carp *Ctenopharyngodon idella*. *Fish Physiol. Biochem.* **2017**, *43*, 1433–1442. [CrossRef]
18. Yang, L.; Hu, Y.; Zhu, B.; Zhong, L.; Wei, Z.H.; Tang, T.; Zhang, J.Z. Effects of *Flammulina velutipes* Stembase on Growth, Digestive Enzyme Activities, Intestinal Tissue Structure, Muscle Quality and Serum Biochemical Parameters of Grass Carp (*Ctenopharyngodon idella*). *Chin. J. Anim. Nutr.* **2021**, *8*, 4569–4579.
19. Zhu, B.; Zhong, L.; Zhao, Q.; Dai, Z.; Chen, K.; Yi, H.U. Effects of *Flammulina velutipes* Medium Residue on Growth Performance, Intestinal Tissue Structure and Serum Biochemical, Immune and Antioxidant Indexes of Grass Carp. *Chin. J. Anim. Nutr.* **2020**, *32*, 2921–2929.
20. Meton, I.; Mediavilla, D.; Caseras, A.; Canto, E.; Fernandez, F.; Baanante, I.V. Effect of Diet Composition and Ration Size on Key Enzyme Activities of Glycolysis-Gluconeogenesis, the Pentose Phosphate Pathway and Amino Acid Metabolism in Liver of Gilthead Sea Bream (*Sparus aurata*). *Br. J. Nutr.* **1999**, *82*, 223–232. [CrossRef]
21. Peres, H.; Oliva-Teles, A. Effect of the Dietary Essential Amino Acid Pattern on Growth, Feed Utilization and Nitrogen Metabolism of European Sea Bass (*Dicentrarchus labrax*). *Aquaculture* **2007**, *267*, 119–128. [CrossRef]
22. Cheng, Z.; Zhang, G.Q.; Zhou, L.; Xu, Z.; Yang, H.; Li, X.Q.; Leng, X.J. Effects of Citric Acid on Growth Performance, Nutrient Utilization, Serum Biochemical Indices and Intestinal Tissue Morphology of Grass Carp (*Ctenopharyngodon idella*). *Chin. J. Anim. Nutr.* **2021**, *8*, 4580–4591.
23. Chen, Y.; Ronghua, L.U.; Yang, G.; Zhang, Y.; Qin, C.; Hong, J.I.; Nie, G. University, HN, Effects of Replacing Soybean Oil with Black Soldier Fly (*Hermetia illucens*) Larvae Oil on the Growth Performance, Antioxidant Ability and Intestinal Microbiota of Grass Carp (*Ctenopharyngodon idella*). *J. Fish. China* **2019**, *10*, 2241–2255.
24. Smith, N.C.; Rise, M.L.; Christian, S.L. A Comparison of the Innate and Adaptive Immune Systems in Cartilaginous Fish, Ray-Finned Fish, and Lobe-Finned Fish. *Front. Immunol.* **2019**, *10*, 2292. [CrossRef]
25. Li, L.; Cardoso, J.C.R.; Félix, R.C.; Mateus, A.P.; Canário, A.V.M.; Power, D.M. Fish Lysozyme Gene Family Evolution and Divergent Function in Early Development. *Dev. Comp. Immunol.* **2021**, *114*, 103772. [CrossRef] [PubMed]
26. Song, Q.; Xiao, Y.; Xiao, Z.; Liu, T.; Li, J.; Li, P.; Han, F. Lysozymes in Fish. *J. Agric. Food Chem.* **2021**, *50*, 15039–15051. [CrossRef]
27. Demers, N.; Bayne, C. Immediate Increase of Plasma Protein Complement C3 in Response to an Acute Stressor. *Fish Shellfish Immunol.* **2020**, *107*, 411–413. [CrossRef]
28. Langevin, C.; Aleksejeva, E.; Passoni, G.; Palha, N.; Levraud, J.; Boudinot, P. The Antiviral Innate Immune Response in Fish: Evolution and Conservation of the IFN System. *J. Mol. Biol.* **2013**, *425*, 4904–4920. [CrossRef]
29. Workenhe, S.T.; Rise, M.L.; Kibenge, M.J.T.; Kibenge, F.S.B. The Fight Between the Teleost Fish Immune Response and Aquatic Viruses. *Mol. Immunol.* **2010**, *47*, 2525–2536. [CrossRef] [PubMed]
30. Chen, M.; Guoxiang, F.U.; Dong, J.; Hao, W. Effects of Rutin on Growth Performance, Body Composition, Non-Specific Immunity and Intestinal Digestive Enzyme Activities of Grass Carp (*Ctenopharyngodon idella*). *Chin. J. Anim. Nutr.* **2019**, *10*, 4868–4876.

31. Catoni, C.; Schaefer, H.M.; Peters, A. Fruit for Health: The Effect of Flavonoids on Humoral Immune Response and Food Selection in a Frugivorous Bird. *Funct. Ecol.* **2008**, *22*, 649–654. [CrossRef]
32. Guardia, T.; Rotelli, A.E.; Juarez, A.O.; Pelzer, L.E. Anti-Inflammatory Properties of Plant Flavonoids. Effects of Rutin, Quercetin and Hesperidin on Adjuvant Arthritis in Rat. *Farmaco* **2001**, *56*, 683–687. [CrossRef] [PubMed]

fishes

Article

Effects of Diets with Varying Astaxanthin from *Yarrowia lipolytica* Levels on the Growth, Feed Utilization, Metabolic Enzymes Activities, Antioxidative Status and Serum Biochemical Parameters of *Litopenaeus vannamei*

Yuechong Liu [1], Lu Zheng [1], Bingying Xu [1], Gladstone Sagada [1], Jinzhi Zhang [1] and Qingjun Shao [1,2,*]

[1] College of Animal Sciences, Zhejiang University, Hangzhou 310058, China
[2] Ocean Academy, Zhejiang University, Zhoushan 316021, China
* Correspondence: qjshao@zju.edu.cn; Tel.: +86-0571-88982200

Abstract: *Litopenaeus vannamei* was divided into seven groups (defined as diets A0–A6) and fed with diets respectively containing 0, 0.5, 1, 2, 4 and 8 g/kg *Yarrowia lipolytica* (astaxanthin content: 1.5%) and 3 g/kg *Haematococcus pluvialis* (astaxanthin content: 2%). After an eight-week feeding trial, the results reflected that different levels of *Y. lipolytica* and *H. pluvialis* could significantly increase the weight gain rate of *L. vannamei* ($p < 0.05$). The condition factor and weight gain rate of group A4 were significantly higher than those of the other groups ($p < 0.05$); the HSI significantly decreased with the increase of *Y. lipolytica* ($p < 0.05$). The addition of *Y. lipolytica* to the diet had significant effects on total protein (TP), albumin (ALB), glutathione peroxidase (GSH-Px), malonaldehyde (MDA) and total antioxidant capacity (T-AOC) ($p < 0.05$). The total protein and albumin of the A5 and A6 groups were significantly higher than those of the other groups ($p < 0.05$). The GSH-Px activity of the A5 group was the highest and the T-AOC of the A0 group was the lowest. Inducible nitric oxide synthase (I-NOS) increased with the addition of *Y. lipolytica* ($p < 0.05$). *Y. lipolytica* inclusion had no negative effect on physiological and biochemical parameters and some serum immune and antioxidant indexes ($p > 0.05$). Astaxanthin in *Y. lipolytica* had an obvious effect on body color. After cooking, the body color of the shrimp deepened with increasing *Y. lipolytica* content. The red body color of *L. vannamei* was significantly improved by adding yeasts hydrolysate 2~8 g/kg to the diet. According to the regression analysis between the level of *Y. lipolytica* added to the diets and the weight gain rates, the optimal level of *Y. lipolytica* is 4.64 g/kg.

Keywords: *Litopenaeus vannamei*; *Yarrowia lipolytica*; *Haematococcus pluvialis*; astaxanthin; growth performance; pigmentation deposit; antioxidation

Citation: Liu, Y.; Zheng, L.; Xu, B.; Sagada, G.; Zhang, J.; Shao, Q. Effects of Diets with Varying Astaxanthin from *Yarrowia lipolytica* Levels on the Growth, Feed Utilization, Metabolic Enzymes Activities, Antioxidative Status and Serum Biochemical Parameters of *Litopenaeus vannamei*. *Fishes* **2022**, *7*, 352. https://doi.org/10.3390/fishes7060352

Academic Editors: Marina Paolucci and Shunsuke Koshio

Received: 7 November 2022
Accepted: 24 November 2022
Published: 27 November 2022

1. Introduction

Litopenaeus vannamei, also known as King Prawn or Pacific White Shrimp [1], is a kind of shrimp in the East Pacific Ocean, which is usually caught or raised as food. Around the beginning of the millennium, Asia introduced this species into its aquaculture business. China, Vietnam, India and other countries have also become major producers. The marine fishing and aquaculture of *L. vannamei* are affected by weather changes and diseases [2]. The deteriorating water environment has seriously affected the shrimp industry due to stress and diseases in recent years [3,4]. Therefore, in the process of shrimp culture, it is very important to develop feed additives to improve the resistance and survival rate of shrimp for better growth and development.

Astaxanthin is a kind of ketone carotene with many uses, including as a dietary supplements and food dye [5,6]. Due to the presence of ketone functional groups and hydroxyl groups, the compound has a variety of biological properties, including antioxidation, immune regulation, anti-inflammation, disease prevention and coloring [7]. Astaxanthin is produced when algae are stressed due to lack of nutrition, increased water salinity or inordinate sunshine levels. Algae-eating animals such as red snapper, salmon, flamingo, red trout and crustaceans (i.e., shrimp, krill, crab, lobster, and crayfish) subsequently reflect red–orange pigmentation to varying degrees due to the presence of carotenoids.

Astaxanthin has been used to feed animals because of its strong antioxidant activity and safety, and because this pigment improves the sensory properties of animal products. Animals cannot synthesize carotenoids by themselves, which is why they must eventually obtain the pigment from their diets [8,9]. However, production of carotenoids through natural producers such as plants and algae faces the challenges of limited production and high planting and mining costs. Astaxanthin production through metabolically engineered artificial microbial cell factories is another promising strategy to overcome these limitations [9]. *Yarrowia lipolytica* is an aerobic, dimorphic, non-pathogenic ascomycete yeast, which has been isolated from various marine and coastal environments [10]. Due to its essential fatty acid profile, *Y. lipolytica* has been used as a dietary supplement in aquaculture [11] and is a potential commercial source of astaxanthin. However, few studies have focused on the effects of *Y. lipolytica* on *L. vannamei*. Therefore, the purpose of this work was to study the effects of astaxanthin from *Y. lipolytica* on the physiological, metabolic and hematological responses of *L. vannamei* to evaluate the effectiveness of this feed additive.

2. Materials and Methods

2.1. Experimental Diets

The *Y. lipolytica* (astaxanthin content 1.5%) was provided by Shandong Jincheng Biological Pharmaceutical Co. Ltd., China. Seven isonitrogenous (40.4% crude protein) and isoenergetic (17.3 kJ g^{-1} gross energy) diets with varying astaxanthin contents were formulated as shown in Table 1. Groups A0, A1, A2, A3, A4, and A5 had 0, 7.5, 15, 30, 60 and 120 mg/kg astaxanthin, respectively, from *Y. lipolytica*, while group A6 was supplemented with 60 mg kg^{-1} astaxanthin from *Haematococcus pluvialis* (astaxanthin content 2.0%). The content of astaxanthin in the A6 and A4 groups was the same. Fishmeal, soya protein concentrate, fermented soybean meal and wheat gluten meal were used as the major protein ingredients, while fish oil served as the main lipid source. Crystalline DL-methionine and L-arginine were added to maintain adequate and balanced levels of these essential amino acids according to the recommended level from previous studies.

Diet ingredients were crushed through an 80 mesh sieve (178-μm). After mixing evenly, the mixture was made into granular diets with a particle size of 1.0 mm and length of 2.5~4.0 mm by feed granulation mechanism. After steam curing for 10 min, they were dried indoors under air conditioning (24 °C), dehumidified by dehumidifier and fan blowing for 72 h, and then stored in a −20 °C freezer until use.

2.2. Experimental Shrimp and Feeding Trial

Juvenile *L. vannamei* (0.23 ± 0.02 g) provided by Wenzhou Qingjiang Base of Zhejiang Institute of Marine Aquaculture were used in this study. They were from the same batch, healthy and consistent in size. The feeding trial was conducted at the Marine Fisheries Research Institute of Zhejiang Province in Zhoushan, China. The shrimp were kept in an acclimatization tank (10 m × 2 m × 4.5 m) in a seawater recirculation system for two weeks. A total of 2100 healthy and disease-free juvenile shrimp of similar size were weighed and randomly divided into 42 tanks (400 L water volume) with 50 in each tank. The tanks were randomly divided into 7 groups of 6 replicates. Each diet was fed to 6 tanks of shrimp for 8 weeks. The quality parameters in the water (temperature 26 ± 2 °C; salinity 26~29 g L^{-1}; dissolved oxygen ≥5 mg L^{-1}; pH 7.6~7.8; and ammonia nitrogen <0.1 mg L^{-1}) were recorded daily and maintained throughout the acclimation and experimental periods. During the first four weeks, the shrimp were fed manually four times a day (6:00 h, 10:00 h, 14:00 h and 18:00 h), and three times a day (8:00 h, 12:00 h and 16:00 h) in the last four weeks, and observed every day to ensure that they ate the feed within two hours after feeding.

Table 1. Formulation and proximate composition of the experimental diets.

Ingredients (%)	A0	A1	A2	A3	A4	A5	A6
Fishmeal	22	22	22	22	22	22	22
Peeled soybean meal	8	8	8	8	8	8	8
Fermented soybean meal	13	13	13	13	13	13	13
Soy protein concentrate	10.5	10.5	10.5	10.5	10.5	10.5	10.5
Squid liver powder	4	4	4	4	4	4	4
Chicken meal	6	6	6	6	6	6	6
High gluten flour	22	22	22	22	22	22	22
Fish oil	2	2	2	2	2	2	2
Soy phospholipids	2	2	2	2	2	2	2
L-lysine	0.2	0.2	0.2	0.2	0.2	0.2	0.2
DL-methionine	0.2	0.2	0.2	0.2	0.2	0.2	0.2
Taurine	0.17	0.17	0.17	0.17	0.17	0.17	0.17
Sodium carboxymethyl cellulose	0.50	0.50	0.50	0.50	0.50	0.5	0.50
Carrageenan	0.20	0.20	0.20	0.20	0.20	0.2	0.20
Calcium dihydrogen phosphate	2.1	2.1	2.1	2.1	2.1	2.1	2.1
Vc phosphate	0.1	0.1	0.1	0.1	0.1	0.1	0.1
Vitamin mixture	0.3	0.3	0.3	0.3	0.3	0.3	0.3
Mineral premix	0.5	0.5	0.5	0.5	0.5	0.5	0.5
Zeolite powder	3	3	3	3	3	3	3
Antioxidant	0.1	0.1	0.1	0.1	0.1	0.1	0.1
Antifungal agent	0.1	0.1	0.1	0.1	0.1	0.1	0.1
Yarrowia lipolytica (astaxanthin 1.5%)	0	0.05	0.1	0.2	0.4	0.8	0
Haematococcus pluvialis (astaxanthin 2.0%)	0	0	0	0	0	0	0.4
Beer yeast	0.96	0.93	0.9	0.84	0.72	0.48	0.72
Alpha cellulose	2.07	2.05	2.03	1.99	1.91	1.75	1.91
Total	100	100	100	100	100	100	100
Astaxanthin (mg kg^{-1})	0	7.5	15	30	60	120	60
Proximate analysis (%)							
Crude protein	39.55	39.04	39.09	39.54	38.97	39.26	39.65
Crude lipid	7.31	7.09	7.85	7.60	7.51	7.44	7.54
Gross energy (kJ g^{-1})	16.93	16.72	17.03	17.04	16.87	16.91	17.04
Gross phosphorus	0.82	0.78	0.86	0.81	0.85	0.82	0.80

2.3. Methods of Sample Collection and Analysis

Before the start of the feeding experiment, 60 juvenile shrimp of the same size were randomly selected, weighed and stored in the −20 °C freezer for body composition analysis. At the end of the 8-week culture period, the test shrimp were starved for 24 h and then paralyzed with ice water. The shrimp were counted to obtain the survival rate (SR) and batch-weighed to obtain the data for weight gain rate (WGR) and feed conversion rate (FCR). After that, 10 shrimp were randomly weighed and measured respectively to obtain the conditioning factor (CF). Three shrimp were randomly selected for the analysis of whole-body composition, and the hemolymph of other shrimp was taken from the heart. The hemolymph was left to settle at 4 °C for 2 h, centrifuged at 8000 r min^{-1} for 15 min, and the supernatant was taken and stored in a refrigerator at −20 °C. The viscera of shrimp were taken after dissection on an ice plate, and the hepatopancreas and intestine were separated. The hepatopancreas was weighed to obtain the hepatosomatic index (HSI) and the intestine was divided into midgut and hindgut.

The proximate compositions of diets, shrimp whole-body and muscle were analyzed according to the standard protocols of the Association of Official Analytical Chemists (AOAC, 1995). The moisture content was determined by drying ground samples in an oven for 24 h. The Kjeldahl method was used to determine the crude protein, the Soxhlet extraction method was used to determine the crude fat, and the ash content was calculated after burning the sample at 550 °C for 6 h.

The supernatants of liver and intestine were obtained according to the method described by Xu et al. (2021) [12]. Diagnostic reagent kits (from Nanjing Jiancheng Bioengineering Institute, Nanjing, China) were used to determine the contents of serum nitric oxide synthase (i-NOS), triglycerides (TG), total antioxidant capacity (TAOC), total cholesterol (TCH) and total protein (TP), and the activities of gastrointestinal digestive enzyme, liver alanine aminotransferase (ALT), aspartate aminotransferase (AST), cyclooxygenase (COX), fatty acid synthetase (FAS), caspase and Nuclear Factor κB (NF-κB). The activities of catalase (CAT), superoxide dismutase (SOD), glutathione peroxidase (GSH-Px) and malondialdehyde (MDA) content in serum and liver were assayed as reported by Wang et al. (2020) [13].

Three shrimp from each tank were collected into transparent polyethylene bags and sealed. They were then heated in a 100 °C constant temperature water bath for 10 min, taken out and cooled for 5 min. Filter paper was used to wipe the water before photos taken to observe the body color. The chromatism(a*) of the shrimp body was detected by colorimeter.

The astaxanthin content in the test diet and shrimp was detected by high performance liquid chromatography (HPLC).

2.4. Statistical Analysis

SPSS 20.0 software was used for all statistical analysis, and the results were presented as the means ± SD. After one-way ANOVA was performed on the data, Duncan multiple comparison was used to test the significant difference of the mean at p level 0.05. The WGR and astaxanthin content of the whole shrimp were analyzed by regression method. Quadratic regression analysis was adopted to fit the trendline. The Shapiro Wilk test was used to verify the normality, and the Bartlett test was used to verify the homogeneity of variance.

3. Results

3.1. Growth Performance and Morphological Index

The growth performance and morphological indices of *L. vannamei* fed different *Y. lipolytica* and *H. pluvialis* are shown in Table 2. Different levels of *Y. lipolytica* and *H. pluvialis* significantly increased the weight gain rate of *L. vannamei* ($p < 0.05$). They also had a significant effect on the condition factor and HSI of *L. vannamei* ($p < 0.05$). There was no significant difference in survival rate and body length ($p > 0.05$). The quadratic regression ($y = -26.25x^2 + 243.64x + 2563.1$, $R^2 = 0.9519$) of WG against *Y. lipolytica* levels in the diets is presented in Figure 1. Xpot in the figure is the content of *Y. lipolytica* in the ration at the highest time of WG. According to the calculation of the fitting curve formula, the curve reaches the peak on the horizontal axis at 4.64 g kg^{-1}. Thus, the weight gain rate of *L. vannamei* reached a maximum at 4.64 g kg^{-1} *Y. lipolytica* inclusion level. The feeding rate, feed conversion rate, protein efficiency and protein deposition rate were not significantly influenced by various levels of *Y. lipolytica* and *H. pluvialis* ($p > 0.05$).

Table 2. Effects of *Yarrowia lipolytica* on growth performance and feed utilization of *Litopenaeus vannamei* [1].

Index [1]	A0	A1	A2	A3	A4	A5	A6
SR [2]	96.33 ± 4.80	99.60 ± 0.89	95.33 ± 5.75	95.67 ± 4.63	96.80 ± 3.03	96.67 ± 4.68	97.00 ± 2.76
WGR [3]	2502.43 ± 160.01 ^{c}	2717.21 ± 62.32 ^{b}	2839.99 ± 146.13 ^{b}	2918.19 ± 131.99 ^{ab}	3102.94 ± 51.20 ^{a}	2836.46 ± 181.21 ^{b}	2876.05 ± 143.69 ^{b}
FBL [4]	9.08 ± 0.67	9.25 ± 0.56	9.22 ± 0.54	9.34 ± 0.52	9.50 ± 0.53	9.61 ± 0.52	9.48 ± 0.55
CF [5]	0.81 ± 0.02 ^{b}	0.80 ± 0.02 ^{b}	0.82 ± 0.1 ^{b}	0.81 ± 0.01 ^{b}	0.84 ± 0.01 ^{a}	0.82 ± 0.01 ^{b}	0.78 ± 0.01 ^{c}
HIS [6]	4.99 ± 0.44 ^{ab}	5.50 ± 0.60 ^{a}	4.80 ± 0.35 ^{bc}	4.75 ± 0.39 ^{bc}	4.35 ± 0.68 ^{cd}	4.04 ± 0.41 ^{d}	4.84 ± 0.63 ^{bc}
FR [7]	4.46 ± 0.60	4.47 ± 0.83	4.49 ± 0.11	4.72 ± 0.36	4.26 ± 0.28	4.36 ± 0.29	4.32 ± 0.29
FCR [8]	1.23 ± 0.16	1.23 ± 0.23	1.23 ± 0.03	1.30 ± 0.10	1.17 ± 0.08	1.20 ± 0.08	1.19 ± 0.08
PER [9]	2.05 ± 0.29	2.08 ± 0.42	2.01 ± 0.05	1.92 ± 0.14	2.13 ± 0.14	2.08 ± 0.14	2.10 ± 0.15
PPV [10]	36.33 ± 5.02	37.75 ± 7.57	36.56 ± 0.95	35.65 ± 2.68	39.12 ± 2.60	38.97 ± 2.69	39.10 ± 2.67

[1] Mean values ± standard deviation (SD) are presented for each group ($n = 6$); values with different superscripts in the same row differ significantly ($p < 0.05$). [2] SR (survival rate, %) = 100 × (final shrimp number/initial shrimp number). [3] WGR (weight gain rate, %) = 100 × (final body weight–initial bodyweight)/initial body weight. [4] FBL (final body length, cm). [5] CF (conditioning factor) = (final body weight/final body length^3) × 100. [6] HSI (hepatosomatic index) = 100 × (liver weight/body weight). [7] FR (feeding rate, %/day) = dry feed intake/[(final weight + initial weight)/2]/days × 100. [8] FCR (feed conversion rate) = dry feed intake/weight gain. [9] PER (protein efficacy ratio) = weight gain/total protein intake. [10] PPV (protein productive value, %) = protein gain/total protein intake × 100.

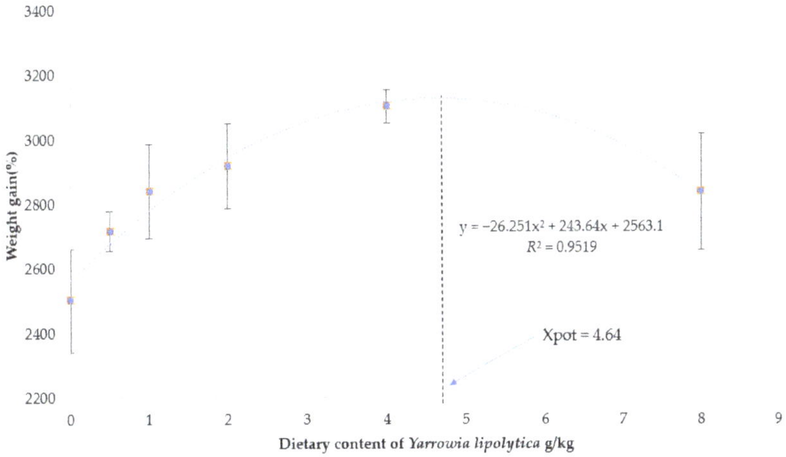

Figure 1. The relationship between the weight gain rate of shrimp and the content of *Yarrowia lipolytica* in feed based on quadratic regression analysis.

The astaxanthin content in whole shrimp (Figure 2) increased with increasing dietary astaxanthin content. Compared with the control group, astaxanthin content in the supplemented groups was significantly higher ($p < 0.05$). The whole body astaxanthin content correlated positively with the dietary levels, as the 120 mg kg^{-1} dietary inclusion group had the highest content.

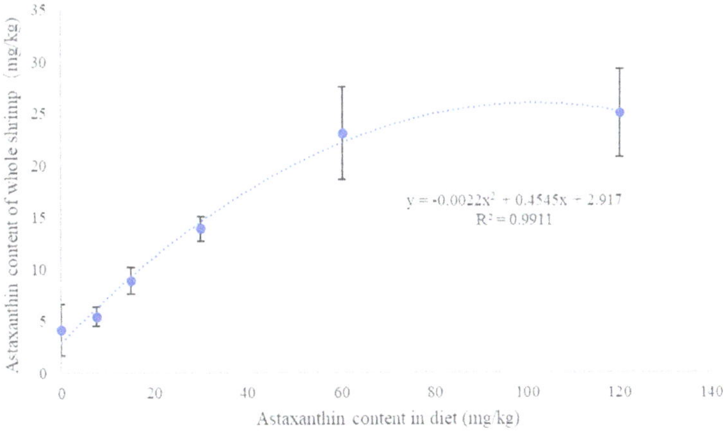

Figure 2. The relationship between the astaxanthin content of whole shrimp and astaxanthin content in diet based on quadratic regression analysis.

3.2. Effect on Body Color of Shrimp

Figure 3 shows the relationship between shrimp chromatism a* and astaxanthin content in the diet. The larger the value of a*, the redder the color. It can be seen that with the increase of astaxanthin content in the diet, a* also increases.

The pictorial results are shown in Figure 4. A high dose of *Y. lipolytica* (2~8 g/kg) in the diet significantly improved shrimp shell color (red) by colorimeter. The body color of the *H. pluvialis* group was not significantly improved.

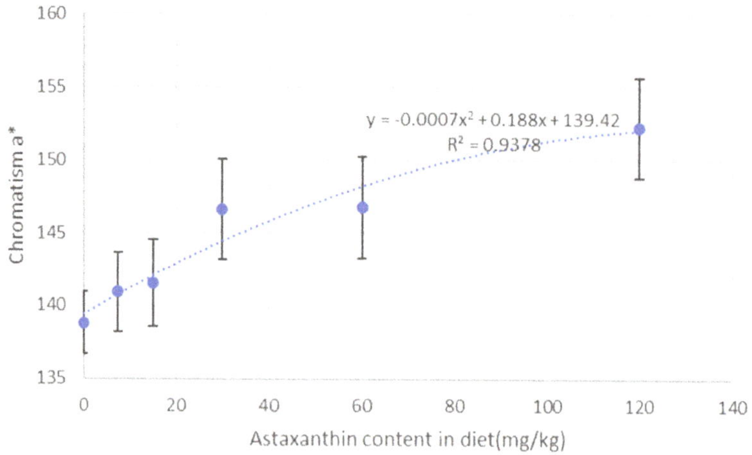

Figure 3. The relationship between the chromatism a* of shrimp and astaxanthin content in diet based on quadratic regression analysis.

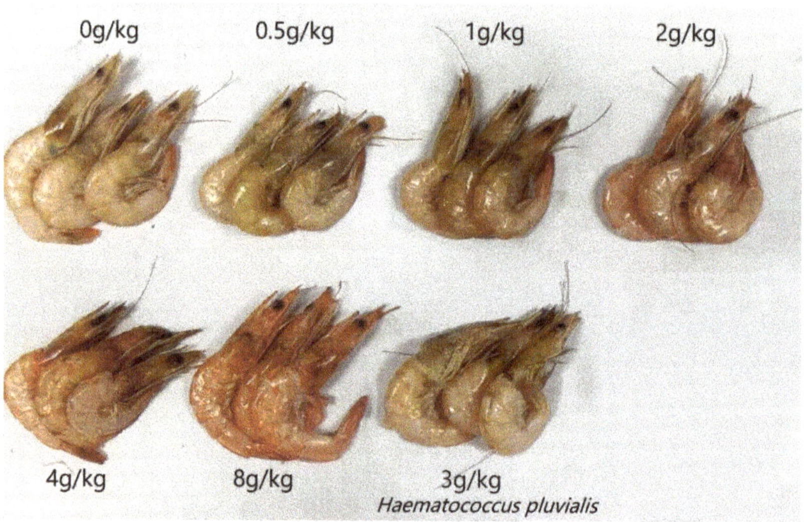

Figure 4. Comparison of body color of boiled *Litopenaeus vannamei* after 8-week growth trial.

3.3. Proximate Composition of the Whole Body and Muscles

As presented in Table 3, different amounts of *Y. lipolytica* added to the feed had significant effects on the moisture content of whole body, and the protein content of the muscle, of *L. vannamei* ($p < 0.05$). There was no significant influence on crude fat and ash ($p > 0.05$).

Table 3. Effects of *Yarrowia lipolytica* on the whole shrimp and muscle composition of *Litopenaeus vannamei* (%) [1].

Index [1]	A0	A1	A2	A3	A4	A5	A6
Whole Shrimp							
Moisture	75.78 ± 0.10 [a]	75.14 ± 0.14 [bc]	75.37 ± 0.18 [b]	74.95 ± 0.13 [c]	75.09 ± 0.18 [bc]	74.78 ± 0.11 [c]	74.69 ± 0.12 [c]
Crude protein	17.05 ± 0.21	17.53 ± 0.14	17.56 ± 0.23	17.97 ± 0.08	17.78 ± 0.16	18.10 ± 0.06	17.58 ± 0.12
Crude lipid	0.96 ± 0.02	1.01 ± 0.05	1.01 ± 0.05	0.95 ± 0.06	0.94 ± 0.04	1.01 ± 0.03	0.99 ± 0.03
Ash	3.42 ± 0.05	3.36 ± 0.05	3.40 ± 0.04	3.30 ± 0.09	3.90 ± 0.08	3.42 ± 0.06	3.32 ± 0.05
Muscle							
Moisture	76.08 ± 0.07	76.07 ± 0.13	75.54 ± 0.15	75.80 ± 0.15	76.53 ± 0.19	75.91 ± 0.15	75.81 ± 0.16
Crude protein	22.13 ± 0.14	22.19 ± 0.02	22.20 ± 0.16	22.25 ± 0.07	22.31 ± 0.20	22.50 ± 0.14	22.40 ± 0.14
Crude lipid	0.38 ± 0.01	0.40 ± 0.02	0.41 ± 0.03	0.39 ± 0.01	0.40 ± 0.02	0.41 ± 0.02	0.40 ± 0.02
Ash	1.56 ± 0.07	1.52 ± 0.04	1.51 ± 0.10	1.64 ± 0.10	1.54 ± 0.08	1.72 ± 0.03	1.65 ± 0.03

[1] Mean values ± standard deviation (SD) are presented for each group ($n = 6$); values with different superscripts in the same row differ significantly ($p < 0.05$).

3.4. Serum Biochemical and Antioxidative Indexes

The effects of *Y. lipolytica* on the main serum physiological and biochemical indexes of *L. vannamei* are shown in Table 4. The total protein and albumin of the A5 and A6 groups were significantly higher than those in other groups. The GSH-Px activity of A5, and the MDA content of A1 groups were the highest, whilst the A1 group had the lowest T-AOC. The total protein and albumin contents, and GSH-Px activity, increased with the increasing inclusion of *Y. lipolytica* and reached the highest values in the group A5. MDA content decreased with the increase of *Y. lipolytica*, and the T-AOC of group A4 was higher than that of the other groups. No significant differences were observed in the other physiological and biochemical indexes among the groups ($p > 0.05$).

Table 4. Effects of *Yarrowia lipolytica* on main biochemical, immune and antioxidant indexes in serum of *Litopenaeus vannamei*.

Index [1]	A0	A1	A2	A3	A4	A5	A6
TP [2]	20.80 ± 1.49 [c]	27.22 ± 4.33 [b]	27.99 ± 2.82 [b]	25.38 ± 4.44 [b]	28.46 ± 6.85 [b]	34.74 ± 2.13 [a]	35.05 ± 1.81 [a]
ALB [3]	10.12 ± 0.84 [c]	12.39 ± 1.36 [b]	12.34 ± 0.71 [b]	11.94 ± 1.66 [b]	12.80 ± 2.40 [b]	14.53 ± 0.49 [a]	15.44 ± 0.99 [a]
SOD [4]	19.80 ± 4.87	21.69 ± 3.94	19.84 ± 1.72	21.10 ± 4.78	19.92 ± 6.44	19.54 ± 3.22	19.62 ± 5.87
MDA [5]	12.48 ± 2.25 [a]	10.69 ± 1.04 [b]	9.58 ± 2.36 [b]	9.18 ± 3.53 [c]	9.11 ± 2.57 [c]	7.97 ± 1.14 [d]	9.10 ± 2.01 [c]
GSH-Px [6]	127.51 ± 17.72 [c]	140.07 ± 38.40 [b]	153.80 ± 41.35 [b]	153.09 ± 41.48 [b]	153.85 ± 56.20 [b]	181.29 ± 42.21 [a]	156.93 ± 19.64 [b]
CAT 7	0.68 ± 0.19	0.72 ± 0.13	0.72 ± 0.17	0.78 ± 0.19	0.73 ± 0.29	0.65 ± 0.13	0.63 ± 0.19
T-AOC [8]	0.19 ± 0.04 [c]	0.24 ± 0.04 [b]	0.25 ± 0.03 [b]	0.29 ± 0.13 [b]	0.34 ± 0.15 [a]	0.31 ± 0.05 [b]	0.30 ± 0.06 [b]
LZM [9]	254.39 ± 32.22	238.60 ± 81.99	263.16 ± 23.06	201.75 ± 28.57	208.77 ± 21.49	233.33 ± 47.27	228.07 ± 45.96
AKP [10]	0.34 ± 0.16	0.28 ± 0.08	0.35 ± 0.16	0.25 ± 0.05	0.27 ± 0.08	0.29 ± 0.04	0.29 ± 0.06
PPO [11]	9.06 ± 1.08	9.78 ± 2.66	9.17 ± 3.72	9.56 ± 1.61	10.50 ± 5.18	12.50 ± 3.90	11.83 ± 3.05
AST [12]	1.81 ± 0.70	1.96 ± 0.42	1.86 ± 0.35	1.83 ± 0.91	2.12 ± 1.49	1.93 ± 0.69	1.84 ± 0.59
TG [13]	0.86 ± 0.19	0.76 ± 0.21	0.79 ± 0.15	0.61 ± 0.17	0.62 ± 0.16	0.70 ± 0.16	0.75 ± 0.13
TC [14]	1.38 ± 0.22	1.47 ± 0.38	1.44 ± 0.32	1.22 ± 0.46	1.32 ± 0.37	1.42 ± 0.31	1.26 ± 0.28

[1] Mean values ± standard deviation (SD) are presented for each group ($n = 6$); values with different superscripts in the same row differ significantly ($p < 0.05$). [2] TP (total protein, g L^{-1}). [3] ALB (albumin, g L^{-1}). [4] SOD (superoxide dismutase, U mL^{-1}). [5] MDA (malonaldehyde, nmol mL^{-1}). [6] GSH-Px (glutathione peroxidase, U mL^{-1}). [7] CAT (catalase, U mL^{-1}). [8] T-AOC (Total antioxidant capacity, mmol L^{-1}). [9] LZM (lysozyme, U mL^{-1}). [10] AKP (alkaline phosphatase, king unit 100 mL^{-1}). [11] PPO (polyphenol oxidase, U mL^{-1}). [12] AST (aspartate aminotransferase, U L^{-1}). [13] TG (triglyceride, mmol L^{-1}). [14] TC (total cholesterol, mmol L^{-1}).

3.5. Immune and Antioxidant Indexes of Hepatopancreas

The effects of *Y. lipolytica* on hepatopancreatic immunity and antioxidant indexes of *L. vannamei* are shown in Table 5. The content of nitric oxide synthase increased significantly with the increase of *Y. lipolytica* ($p < 0.05$). The activity of caspase9 in A3 and A4 groups were significantly lower than that in the other groups, and the activity of GSH-Px in A3 group was significantly lower than that in the other groups ($p < 0.05$).

Table 5. Effects of *Yarrowia lipolytica* on antioxidation and immune indexes in hepatopancreas of *Litopenaeus vannamei*.

Index [1]	A0	A1	A2	A3	A4	A5	A6
MDA [2]	3.82 ± 1.21	3.55 ± 0.77	3.69 ± 1.53	3.42 ± 0.29	3.89 ± 0.90	4.44 ± 1.04	4.56 ± 0.92
CAT [3]	0.15 ± 0.02	0.24 ± 0.08	0.23 ± 0.10	0.18 ± 0.06	0.23 ± 0.09	0.27 ± 0.07	0.18 ± 0.05
SOD [4]	9.04 ± 1.21	9.51 ± 1.70	10.96 ± 1.05	7.86 ± 1.18	8.84 ± 1.76	8.58 ± 2.07	10.25 ± 2.41
GSH-Px [5]	63.22 ± 9.54 [b]	62.98 ± 6.69 [b]	65.31 ± 12.71 [b]	56.20 ± 8.35 [c]	71.43 ± 16.60 [b]	59.03 ± 4.47 [b]	85.60 ± 15.66 [a]
i-NOS [6]	1.75 ± 0.34 [c]	2.70 ± 0.40 [b]	2.37 ± 0.57 [bc]	2.76 ± 0.67 [b]	2.94 ± 0.28 [b]	3.84 ± 0.38 [a]	3.05 ± 0.84 [b]
NK-κB [7]	0.42 ± 0.05	0.40 ± 0.02	0.41 ± 0.02	0.38 ± 0.10	0.41 ± 0.09	0.44 ± 0.03	0.49 ± 0.06
COX-2 [8]	2.67 ± 0.23	2.37 ± 0.15	2.31 ± 0.27	2.28 ± 0.45	2.44 ± 0.70	2.33 ± 0.08	2.75 ± 0.39
Caspase-3 activation [9]	0.61 ± 0.11	0.64 ± 0.14	0.73 ± 0.12	0.54 ± 0.21	0.55 ± 0.25	0.51 ± 0.30	1.02 ± 0.66
Caspase-9 activation [10]	0.70 ± 0.16 [a]	0.68 ± 0.18 [a]	0.61 ± 0.16 [a]	0.40 ± 0.17 [c]	0.31 ± 0.17 [c]	0.44 ± 0.09 [b]	0.50 ± 0.06 [b]

[1] Mean values ± standard deviation (SD) are presented for each group ($n = 6$); values with different superscripts in the same row differ significantly ($p < 0.05$). [2] MDA (malonaldehyde, nmol mL^{-1}). [3] CAT (catalase, U mL^{-1}). [4] SOD (superoxide dismutase, U mL^{-1}). [5] GSH-Px (glutathione peroxidase, U mL^{-1}). [6] i-NOS (Inducible nitric oxide synthase, U mL^{-1}). [7] NK-κB (nuclear transcription factor-κB, ng mgprot^{-1}). [8] COX-2 (cyclooxygenase-2, ng mgprot^{-1}). [9] Caspase-3 activation: when the substrate is saturated, 1 nmol AC ietd PNA can be sheared at 37 °C to produce 1 nmol PNA of Caspase-3. [10] Caspase-9 activation: when the substrate is saturated, 1 nmol AC ietd PNA can be sheared at 37 °C to produce 1 nmol PNA of caspase-9.

3.6. Gastrointestinal Digestive Enzyme Activities

The effect of *Y. lipolytica* on the intestinal digestive enzyme activity of *L. vannamei* is shown in Table 6. There was no significant difference in intestinal digestive enzymes among the groups ($p > 0.05$).

Table 6. Effects of *Yarrowia lipolytica* on digestive enzyme activities in intestinal tissues of *Litopenaeus vannamei* ($n = 6$).

Index [1]	A0	A1	A2	A3	A4	A5	A6
Trypsin (U mgprot^{-1})	12.77 ± 1.82	11.19 ± 1.11	11.76 ± 1.10	12.09 ± 1.07	11.10 ± 1.99	12.64 ± 1.65	12.38 ± 1.40
Lipase (U gprot^{-1})	5.88 ± 0.94	5.84 ± 0.91	5.05 ± 1.46	6.30 ± 0.84	5.80 ± 1.50	6.59 ± 1.05	5.85 ± 2.07
Amylase (U mgprot^{-1})	7.01 ± 1.91	7.54 ± 1.38	7.60 ± 1.06	6.20 ± 1.05	6.92 ± 1.99	8.60 ± 1.19	8.28 ± 1.13

[1] Mean values ± standard deviation (SD) are presented for each group ($n = 6$); values with different superscripts in the same row differ significantly ($p < 0.05$).

4. Discussion

In this study, the survival rate of all treatments of *L. vannamei* was higher than 95%, and there was no significant difference among all treatments, which showed that *L. vannamei* could grow normally and achieve good survival at the level of 0–120 mg/kg astaxanthin. A significant correlation between tissue carotenoid concentration and survival has been observed [14]. Wade et al. (2017) [15] believe that when the content of carotenoids in the body is higher than a certain level, the survival rate will not be affected. Otherwise, it will be damaged below this level [16]. Therefore, higher level of tissue carotenoid beyond the minimum requirement can promote growth.

The WG improved with increasing *Y. lipolytica* contents in the diets up to A4 (4 g/kg) and then reduced, with shrimp fed the A4 diet showing the best growth performance. Previous studies have shown that dietary synthetic astaxanthin and algal astaxanthin have beneficial effects on various species of shrimp. There is evidence that dietary astaxanthin can increase body weight and improve survival rate in *L. vannamei* [16,17], *Macrobrachium rosenbergii* [18], *Paralithodes camtschaticus* [19], *Procambarus clarkii* [20], and *Marsupenaeus japonicus* [21]. A study showed that in sea water culture, when fed a diet supplemented with 50 mg/kg astaxanthin, *L. vannamei* had higher growth (10%) and survival (17%) [16]. Current research supports the positive effects of astaxanthin supplements on the growth and survival performance of *L. vannamei* [22]. Carotenoids have been shown to improve the efficient use of nutrients, and to play a significant role in the intermediate metabolism of aquatic animals, thus promoting growth [2,23]. In addition, astaxanthin can shorten the molting cycle interval of crustaceans and inhibit nicotinamide adenine dinucleotide phosphate (NADPH) in carotenoids, thus reducing energy consumption and accelerating the optimal growth of these aquatic animals [21,25].

The results showed the ability of *Y. lipolytica* to reduce the hepatosmatic index of *L. vannamei* beyond a 7.5 mg/kg dietary inclusion level. At the same astaxanthin level, the hepatosmatic index of A6 was significantly higher than that of A4. This indicates that the addition of *Y. lipolytica* can inhibit the accumulation of lipids to a certain extent. Astaxanthin has a similar effect on fish, and can improve lipid metabolism and reduce oxidative stress and apoptosis induced by fat rich diet as shown in *Micropterus salmoides* [26].

After high temperature treatment, the body color of *L. vannamei* became darker red with the increase of *Y. lipolytica* content. The body color and flesh color of aquatic animals depends on the intake of carotenoids. Astaxanthin can be used for pigmentation of fish in aquaculture [27,28]. People's interest in natural pigments stems from the increasing consciousness and the trend of consumption concepts towards natural products [29]. Astaxanthin is very important in the culture of crustaceans such as the giant tiger prawn (*Penaeus monodon*) [30]. This carotenoid can reverse the blue syndrome found in cultured shrimp lacking pigment. A diet containing 50–100 g/kg astaxanthin recovered the pigmentation of shrimp within 4 weeks [7]. There was no dramatic difference in the whole-body components of *L. vannamei* among the treatments, indicating that astaxanthin did not affect the shrimp body composition. This is in accordance with the results of Niu et al. (2009) [17]. However, the astaxanthin content of shrimp increased significantly with increasing dietary astaxanthin supplementation, similar to the trend in growth, antioxidative ability and immune response [15,31].

The content of serum total protein (TP) is a sensitive indicator reflecting the protein absorption and metabolism of animals [32,33]. The TP and ALB (Albumin) of the treatments with *Y. lipolytica* and *H. pluvialis* was significantly higher than that of A0. This is consistent with other studies, which supplemented astaxanthin in diets of *Procambarus clarkii* [20], *L. vannamei* [16,17,34], and *Marsupenaeus japonicus* [21]. The TP of group A6 was higher than that of other treatments, indicating that astaxanthin from *H. pluvialis* promoted better protein absorption.

Increasing the dietary astaxanthin level remarkably increased T-AOC compared with the control group, and T-AOC in the A4 group was higher than that in other groups. This is consistent with previous studies on *Penaeus monodon* [21,30] and *L. vannamei* [24]. T-AOC refers to the total antioxidant level, and is composed of various antioxidant enzymes and substances such as vitamin C, vitamin E and carotene, which protect cells and the body from oxidative stress caused by reactive oxygen free radicals. MDA comes from lipid oxidation and is an important indicator of oxidative harm in the body. The content of MDA in the serum presented a downward trend from A0 to A5, suggesting that the oxidized lipids were decreasing with the increase in *Y. lipolytica* level in the diets. The decrease of MDA content in serum is closely related to the increase of GSH-PX activity [33]. GSH-Px can effectively remove all organic lipid peroxides [35]. In this experiment, dietary astaxanthin markedly increased T-AOC and decreased MDA in serum, which supported the results of GSH-Px, indicating that the shrimp fed astaxanthin supplement had lower oxidative stress. There were no significant differences in the activities of SOD and CAT in the serum and liver, which can be attributed to different antioxidant enzymes having competitive responses to different degrees of antioxidant stress [36]. I-NOS is a key variable of oxidative stress, and the results showed the ability of dietary astaxanthin to increase the activity. This is consistent with previous studies on mice [37].

Caspase is a group of proteinases with analogous structure in the cytoplasm. It is implicated in apoptosis of eukaryotic cells and plays an important role in the regulation of cell growth, differentiation and apoptosis. Caspase-9 can be activated by signal stimulation through self-splicing, which then causes a Caspase cascade reaction. With the addition of *Y. lipolytica*, Caspase-9 activity decreased significantly, indicating that astaxanthin can reduce apoptosis to some extent, consistent with studies reported on mice [38]. These findings suggest that, while the shrimp pigmentation increased with increasing dietary astaxanthin concentration, some antioxidant indices may not be affected in a dose dependent manner.

5. Conclusions

Under the experimental conditions, the growth of *L. vannamei* was significantly promoted by adding different levels of *Y. lipolytica* yeast to the diet, and the body color (red) of *L. vannamei* was significantly improved by adding high doses (2~8 g/kg) of *Y. lipolytica* yeast. The addition of *Y. lipolytica* in the feed improved some antioxidant indexes, and no adverse effects on the feed utilization, serum biochemical indexes and immune parameters of *L. vannamei* were observed. It is concluded that *Y. lipolytica* can be used as an effective feed additive for *L. vannamei*, with an optimal dose of 0.5~8.0 g/kg. By regression analysis of weight gain rate, the best recommended dose of *Y. lipolytica* in the diet of *L. vannamei* is 4.64 g/kg.

Author Contributions: Y.L. wrote the paper under the supervision of Q.S.; G.S. revised the manuscript; L.Z., B.X. and J.Z. collected the data and helped analyze the results. All authors have read and agreed to the published version of the manuscript.

Funding: This work was supported by the Ministry of Science and Technology, China (Project No.: 2020YFD0900801).

Institutional Review Board Statement: All experimental procedures complied with the guidelines of the Care and Use of Laboratory Animals of the National Institutes of Health (NIH), and ethics approval was granted by the laboratory animal welfare and ethics committee of Zhejiang University (Ethical approval code: ZJU 20710).

Data Availability Statement: The raw data supporting the conclusions of this article will be made available by the authors, without undue reservation.

Acknowledgments: We are very grateful for the allocation of a rearing system and logistical support from the China-Norwegian Joint Laboratory of Nutrition and Feed for Marine Fish (Xixuan Island, Zhoushan City, Zhejiang Province) and the Key Lab of Mariculture and Breeding of Zhejiang Province. Also, thank you to Kai Chen, Fabrice Arnaud Tegomo in our lab, and Tan Peng of Zhejiang Ocean University for their assistance.

Conflicts of Interest: The authors declare no competing interest.

References

1. Ekezie, F.-G.C.; Sun, D.-W.; Cheng, J.-H. Altering the IgE binding capacity of king prawn (*Litopenaeus vannamei*) tropomyosin through conformational changes induced by cold argon-plasma jet. *Food Chem.* **2019**, *300*, 125143. [CrossRef]
2. Zhang, J.; Liu, Y.-J.; Tian, L.-X.; Yang, H.-J.; Liang, G.-Y.; Yue, Y.-R.; Xu, D.-H. Effects of dietary astaxanthin on growth, antioxidant capacity and gene expression in Pacific white shrimp *Litopenaeus vannamei*. *Aquac. Nutr.* **2013**, *19*, 917–927. [CrossRef]
3. Flegel, K.M. Physicians, finder's fees and free, informed consent. *Can. Med. Assoc. J.* **1997**, *157*, 1373–1374.
4. Sirirustananun, N.; Chen, J.-C.; Lin, Y.-C.; Yeh, S.-T.; Liou, C.-H.; Chen, L.-L.; Sim, S.S.; Chiew, S.L. Dietary administration of a *Gracilaria tenuistipitata* extract enhances the immune response and resistance against *Vibrio alginolyticus* and white spot syndrome virus in the white shrimp *Litopenaeus vannamei*. *Fish Shellfish. Immunol.* **2011**, *31*, 848–855. [CrossRef]
5. Choi, S.; Koo, S. Efficient Syntheses of the Keto-carotenoids Canthaxanthin, Astaxanthin, and Astacene. *J. Org. Chem.* **2005**, *70*, 3328–3331. [CrossRef] [PubMed]
6. Margalith, P.Z. Production of ketocarotenoids by microalgae. *Appl. Microbiol. Biotechnol.* **1999**, *51*, 431–438. [CrossRef] [PubMed]
7. Mularczyk, M.; Michalak, I.; Marycz, K. Astaxanthin and other Nutrients from *Haematococcus pluvialis*—Multifunctional Applications. *Mar. Drugs* **2020**, *18*, 459. [CrossRef]
8. Kildegaard, K.R.; Adiego-Pérez, B.; Belda, D.D.; Khangura, J.K.; Holkenbrink, C.; Borodina, I. Engineering of *Yarrowia lipolytica* for production of astaxanthin. *Synth. Syst. Biotechnol.* **2017**, *2*, 287–294. [CrossRef]
9. Zhu, H.-Z.; Jiang, S.; Wu, J.-J.; Zhou, X.-R.; Liu, P.-Y.; Huang, F.-H.; Wan, X. Production of High Levels of 3*S*, 3′*S*-Astaxanthin in *Yarrowia lipolytica* via Iterative Metabolic Engineering. *J. Agric. Food Chem.* **2022**, *70*, 2673–2683. [CrossRef]
10. Barth, G.; Gaillardin, C. Physiology and genetics of the dimorphic fungus *Yarrowia lipolytica*. *FEMS Microbiol. Rev.* **1997**, *19*, 219–237. [CrossRef]
11. Berge, G.; Hatlen, B.; Odom, J.M.; Ruyter, B. Physical treatment of high EPA *Yarrowia lipolytica* biomass increases the availability of n-3 highly unsaturated fatty acids when fed to Atlantic salmon. *Aquac. Nutr.* **2013**, *19*, 110–121. [CrossRef]

12. Xu, B.; Liu, Y.; Chen, K.; Wang, L.; Sagada, G.; Tegomo, A.F.; Yang, Y.; Sun, Y.; Zheng, L.; Ullah, S.; et al. Evaluation of Methanotroph (*Methylococcus capsulatus*, Bath) Bacteria Meal (FeedKind®) as an Alternative Protein Source for Juvenile Black Sea Bream, *Acanthopagrus schlegelii*. *Front. Mar. Sci.* **2021**, *8*, 778301. [CrossRef]
13. Wang, L.; Sun, Y.; Xu, B.; Sagada, G.; Chen, K.; Xiao, J.; Zhang, J.; Shao, Q. Effects of berberine supplementation in high starch diet on growth performance, antioxidative status, immune parameters and ammonia stress response of fingerling black sea bream (*Acanthopagrus schlegelii*). *Aquaculture* **2020**, *527*, 735473. [CrossRef]
14. Chien, Y.-H.; Jeng, S.-C. Pigmentation of kuruma prawn, *Penaeus japonicus* Bate, by various pigment sources and levels and feeding regimes. *Aquaculture* **1992**, *102*, 333–346. [CrossRef]
15. Wade, N.M.; Gabaudan, J.; Glencross, B.D. A review of carotenoid utilisation and function in crustacean aquaculture. *Rev. Aquac.* **2017**, *9*, 141–156. [CrossRef]
16. Flores, M.; Díaz, F.; Medina, R.; Re, A.D.; Licea, A. Physiological, metabolic and haematological responses in white shrimp *Litopenaeus vannamei* (Boone) juveniles fed diets supplemented with astaxanthin acclimated to low-salinity water. *Aquac. Res.* **2007**, *38*, 740–747. [CrossRef]
17. Niu, J.; Tian, L.-X.; Liu, Y.-J.; Yang, H.-J.; Ye, C.-X.; Gao, W.; Mai, K.-S. Effect of Dietary Astaxanthin on Growth, Survival, and Stress Tolerance of Postlarval Shrimp, *Litopenaeus vannamei*. *J. World Aquac. Soc.* **2009**, *40*, 795–802. [CrossRef]
18. Kumar, V. Effect of dietary astaxanthin on growth and immune response of Giant freshwater prawn *Macrobrachium rosenbergii* (de man). *Asian Fish. Sci.* **2009**, *22*, 61–69. [CrossRef]
19. Daly, B.; Swingle, J.S.; Eckert, G.L. Dietary astaxanthin supplementation for hatchery-cultured red king crab, *Paralithodes camtschaticus*, juveniles. *Aquac. Nutr.* **2012**, *19*, 312–320. [CrossRef]
20. Cheng, Y.; Wu, S. Effect of dietary astaxanthin on the growth performance and nonspecific immunity of red swamp crayfish *Procambarus clarkii*. *Aquaculture* **2019**, *512*, 734341. [CrossRef]
21. Wang, W.; Ishikawa, M.; Koshio, S.; Yokoyama, S.; Dawood, M.A.; Zhang, Y. Effects of dietary astaxanthin supplementation on survival, growth and stress resistance in larval and post-larval kuruma shrimp, *Marsupenaeus japonicus*. *Aquac. Res.* **2018**, *49*, 2225–2232. [CrossRef]
22. Fang, H.; He, X.; Zeng, H.; Liu, Y.; Tian, L.; Niu, J. Replacement of Astaxanthin With Lutein in Diets of Juvenile *Litopenaeus vannamei*: Effects on Growth Performance, Antioxidant Capacity, and Immune Response. *Front. Mar. Sci.* **2021**, *8*, 1834. [CrossRef]
23. Baron, M.; Davies, S.; Alexander, L.; Snellgrove, D.; Sloman, K.A. The effect of dietary pigments on the coloration and behaviour of flame-red dwarf gourami, *Colisa lalia*. *Anim. Behav.* **2008**, *75*, 1041–1051. [CrossRef]
24. Ettefaghdoost, M.; Haghighi, H. Impact of different dietary lutein levels on growth performance, biochemical and immuno-physiological parameters of oriental river prawn (*Macrobrachium nipponense*). *Fish Shellfish. Immunol.* **2021**, *115*, 86–94. [CrossRef] [PubMed]
25. Mao, X.; Guo, N.; Sun, J.; Xue, C. Comprehensive utilization of shrimp waste based on biotechnological methods: A review. *J. Clean. Prod.* **2017**, *143*, 814–823. [CrossRef]
26. Xie, S.; Yin, P.; Tian, L.; Yu, Y.; Liu, Y.; Niu, J. Dietary Supplementation of Astaxanthin Improved the Growth Performance, Antioxidant Ability and Immune Response of Juvenile Largemouth Bass (*Micropterus salmoides*) Fed High-Fat Diet. *Mar. Drugs* **2020**, *18*, 642. [CrossRef]
27. Kang, C.D.; Han, S.J.; Choi, S.P.; Sim, S.J. Fed-batch culture of astaxanthin-rich *Haematococcus pluvialis* by exponential nutrient feeding and stepwise light supplementation. *Bioprocess Biosyst. Eng.* **2010**, *33*, 133–139. [CrossRef]
28. Shah, M.R.; Liang, Y.; Cheng, J.J.; Daroch, M. Astaxanthin-Producing Green Microalga *Haematococcus pluvialis*: From Single Cell to High Value Commercial Products. *Front. Plant Sci.* **2016**, *7*, 531. [CrossRef]
29. Koller, M.; Muhr, A.; Braunegg, G. Microalgae as versatile cellular factories for valued products. *Algal Res.* **2014**, *6*, 52–63. [CrossRef]
30. Wang, W.; Liu, M.; Fawzy, S.; Xue, Y.; Wu, M.; Huang, X.; Yi, G.; Lin, Q. Effects of Dietary *Phaffia rhodozyma* Astaxanthin on Growth Performance, Carotenoid Analysis, Biochemical and Immune-Physiological Parameters, Intestinal Microbiota, and Disease Resistance in *Penaeus monodon*. *Front. Microbiol.* **2021**, *12*, 762689. [CrossRef]
31. Babin, A.; Moreau, J.; Moret, Y. Storage of Carotenoids in Crustaceans as an Adaptation to Modulate Immunopathology and Optimize Immunological and Life-History Strategies. *BioEssays* **2019**, *41*, e1800254. [CrossRef] [PubMed]
32. Ding, L.; Zhang, L.; Wang, J.; Ma, J.; Meng, X.; Duan, P.; Sun, L.; Sun, Y. Effect of dietary lipid level on the growth performance, feed utilization, body composition and blood chemistry of juvenile starry flounder (*Platichthys stellatus*). *Aquac. Res.* **2009**, *41*, 1470–1478. [CrossRef]
33. Zhang, J.; Zhou, F.; Wang, L.-L.; Shao, Q.; Xu, Z.; Xu, J. Dietary Protein Requirement of Juvenile Black Sea Bream, *Sparus macrocephalus*. *J. World Aquac. Soc.* **2010**, *41*, 151–164. [CrossRef]
34. Chuchird, N.; Rorkwiree, P.; Rairat, T. Effect of dietary formic acid and astaxanthin on the survival and growth of Pacific white shrimp (*Litopenaeus vannamei*) and their resistance to *Vibrio parahaemolyticus*. *SpringerPlus* **2015**, *4*, 440. [CrossRef] [PubMed]
35. Jin, M.; Yuan, Y.; Lu, Y.; Ma, H.; Sun, P.; Li, Y.; Qiu, H.; Ding, L.; Zhou, Q. Regulation of growth, tissue fatty acid composition, biochemical parameters and lipid related genes expression by different dietary lipid sources in juvenile black seabream, *Acanthopagrus schlegelii*. *Aquaculture* **2017**, *479*, 25–37. [CrossRef]
36. Wang, L.; Zhang, W.; Gladstone, S.; Ng, W.-K.; Zhang, J.; Shao, Q. Effects of isoenergetic diets with varying protein and lipid levels on the growth, feed utilization, metabolic enzymes activities, antioxidative status and serum biochemical parameters of black sea bream (*Acanthopagrus schlegelii*). *Aquaculture* **2019**, *513*, 734397. [CrossRef]

37. Bi, J.; Cui, R.; Li, Z.; Liu, C.; Zhang, J. Astaxanthin alleviated acute lung injury by inhibiting oxidative/nitrative stress and the inflammatory response in mice. *Biomed. Pharmacother.* **2017**, *95*, 974–982. [CrossRef]
38. Bhuvaneswari, S.; Yogalakshmi, B.; Sreeja, S.; Anuradha, C.V. Astaxanthin reduces hepatic endoplasmic reticulum stress and nuclear factor-kappa B-mediated inflammation in high fructose and high fat diet-fed mice. *Cell Stress Chaperones* **2014**, *19*, 183–191. [CrossRef]

fishes

Article

Evaluation of Meat Meal as a Replacer for Fish Meal in Diet on Growth Performance, Feed Utilization, Chemical Composition, Hematology, and Innate Immune Responses of Olive Flounder (*Paralichthys olivaceus*)

Md. Farid Uz Zaman [1], Ran Li [1] and Sung Hwoan Cho [2,*]

1 Department of Convergence Study on the Ocean Science and Technology, Korea Maritime and Ocean University, Busan 49112, Republic of Korea
2 Division of Marine Bioscience, Korea Maritime and Ocean University, Busan 49112, Republic of Korea
* Correspondence: chosunh@kmou.ac.kr

Abstract: This study aims to evaluate the dietary replacement effect of various levels of fish meal (FM) with meat meal (MM) on the growth, feed utilization, chemical composition, hematological parameters, and innate immune responses of olive flounder. A total of 360 juvenile fish (initial weight of 14.7 g) were randomly assigned to 18 flow-through containers. The control (MM0) diet included 65% FM. Then, 10%, 20%, 30%, 40%, and 50% FM in the MM0 diet were replaced with MM, referred to as the MM10, MM20, MM30, MM40, and MM50 diets, respectively. The fish were hand-fed to satiation daily for 56 days. Weight gain, the specific growth rate, the feed efficiency ratio, and the protein efficiency ratio of fish fed the MM0 diet were statistically greater than those of fish fed the MM30, MM40, and MM50 diets, but not statistically different from those of fish fed the MM10 and MM20 diets. To incite the maximum values of weight gain and the specific growth rate (SGR) of the fish, an estimated 7.0% of FM substitution with MM in diets was required according to regression analysis. However, the feed consumption, protein retention, hematological parameters, and innate immune (superoxide dismutase and lysozyme activities) responses of the fish were not statistically impacted by the dietary replacement levels of MM for FM. In conclusion, the feed ingredient grade of MM can substitute FM by up to 20% in the diet without causing any negative impact on the growth, feed consumption, feed utilization, or innate immune responses of olive flounder.

Keywords: olive flounder; fish meal replacement; meat meal; regression analysis; innate immune response

Citation: Zaman, M.F.U.; Li, R.; Cho, S.H. Evaluation of Meat Meal as a Replacer for Fish Meal in Diet on Growth Performance, Feed Utilization, Chemical Composition, Hematology, and Innate Immune Responses of Olive Flounder (*Paralichthys olivaceus*). *Fishes* **2022**, *7*, 343. https://doi.org/10.3390/fishes7060343

Academic Editors: Marina Paolucci and Shunsuke Koshio

Received: 21 October 2022
Accepted: 22 November 2022
Published: 23 November 2022

1. Introduction

Aquaculture production is now commonly used as a significant source of animal protein for human consumption. With an annual growth rate of about 6%, aquaculture has emerged as the most rapidly growing agricultural sector in recent decades [1]. According to statistics, world aquaculture production reached 87.5 million metric tons in 2020 and is anticipated to exceed 106 million metric tons by 2030 [2]. Olive flounder (*Paralichthys olivaceus*, Temminck and Schlegel, 1846), popularly known as bastard halibut, is one of the main commercial aquaculture fish species in the Eastern Asian countries, such as the Republic of Korea (*hereafter*, Korea) and Japan [3]. In 2021, olive flounder aquaculture production in Korea was 41,776 metric tons, accounting for 46.7% of the total aquaculture production in that year [4].

With the expansion of global fish aquaculture, the demand for fish feeds is increasing dramatically. The most expensive nutrient in fish feeds is protein, with inclusion levels typically ranging from 30 to 50% [5]. Fish meal (FM) has been employed as the primary protein source in aquafeeds due to its high protein and well-balanced amino acid (AA) content [6]. Nevertheless, the decreased supply has resulted in a decline in FM production, which has led to prices of ca. USD 1500 per metric ton [7]. The high price of FM as well as the adverse environmental effects of fish farming sites may limit the anticipated expansion of aquaculture.

Some studies [5,8] have suggested minimizing the addition of FM to fish feeds based on concerns regarding economic and environmental sustainability. Therefore, it is necessary to identify and look for a replacement that is inexpensive, readily available, and has well-balanced AA profiles for FM in fish feeds [9,10]. Recently, there has been an increasing number of studies on a novel substitute for FM in fish feeds and research will continue to expand worldwide. There has been growing interest in using various animal and plant sources in fish feeds to alleviate the pressure from FM usage [9,11–13]. Nevertheless, when utilized as sources of protein for fish diets, most plant protein sources have several issues, including imbalanced AA, declined palatability, as well as a variety of antinutritional factors [5,6,14]. These problems could also adversely impact feed intake, digestion, nutrient absorption, and the health conditions of fish [6,14]. Therefore, a candidate must have certain characteristics to be a feasible alternative source of FM in fish diets, such as being rich in protein, having a low price, worldwide availability, and no adverse impact on fish health.

Meat meal (MM) is a by-product meal generated by slaughterhouses or meat processing plants from land animals. It is not only rich in protein but also reasonable in price, and has been widely used in the production of various fish diets during the last few decades [9,15–20]. Therefore, it is predicted that feed costs could be lowered by replacing FM with MM in the diets of olive flounder and other carnivorous fish species. Reference [18] indicated that 60% of white FM could be successfully substituted with pet-grade MM (80% crude protein and 12% crude lipid) with supplementation of the essential AAs of lysine, methionine, and tryptophan, which are likely to lack or be deficient in MM in an olive flounder diet at a mean rearing temperature of 20 °C. Reference [15] reported that a 40% replacement of FM with feed ingredient-grade MM in a juvenile olive flounder diet could be implemented at a mean temperature of 12.3 °C during the winter season. Our previous study [16] also found that FM up to 40% in a diet without the supplementation of AAs could be successfully replaced with pet-grade MM (80.3% crude protein and 13.9% crude lipid) without deteriorating the growth of olive flounder. Therefore, the suitability of MM as a replacement for FM in fish feeds appears to vary depending on the fish species, MM quality, supplementation of AAs, and other experimental conditions, including the water temperature. In addition, the substitutability of FM with MM in olive flounder feeds is still controversial.

The current study, therefore, aims to evaluate the effect of replacing various levels of FM with MM in olive flounder feed on the growth, feed consumption, feed utilization, hematology, chemical composition, and innate immune responses of olive flounder.

2. Materials and Methods

2.1. Diet Formulation

Feed ingredient-grade MM (65.0% crude protein and 10.9% crude lipid) was purchased from Daekyung Oil & Transportation Co. Ltd. (Busan Metropolitan City, Korea). Six experimental diets were formulated based on the nutritional requirements of olive flounder (Table 1) [21,22]. Sixty-five percent FM (anchovy meal/sardine meal = 1:1) and 12% dehulled soybean meal were included as the primary sources of protein in the control (MM0) diet. The MM0 diet also contained 16.5% wheat flour as the carbohydrate source, and 2% fish oil and soybean oil each as the lipid sources. The graded levels (10%, 20%, 30%, 40%, and 50%) of FM were replaced by MM, referred to as MM10, MM20, MM30, MM40, and MM50, respectively. All experimental diets were formulated to be isonitrogenous at 55.0% and isolipidic at 10.5%. All ingredients of the diets were thoroughly blended with water at a ratio of 3:1. After that, the mixed ingredients were pressure-pelleted with a laboratory pellet extruder and dried at 40 °C in an electronic drying machine (SI-2400,

SIN IL Drying Machine Co. Ltd., Daegu Metropolitan City, Korea) for 24 h. Finally, all experimental pellets were kept in a refrigerator at −20 °C until use.

Table 1. Ingredients of the experimental diets (%, DM basis).

	Experimental Diets					
	MM0	MM10	MM20	MM30	MM40	MM50
Ingredient (%, DM)						
Fish meal (FM) [1]	65.0	58.5	52.0	45.5	39.0	32.5
Meat meal (MM) [2]		7.4	14.7	22.0	29.4	36.7
Dehulled soybean meal	12.0	12.0	12.0	12.0	12.0	12.0
Wheat flour	16.5	15.8	15.2	14.6	13.9	13.3
Fish oil [3]	2.0	2.0	2.0	2.0	2.0	2.0
Soybean oil	2.0	1.8	1.6	1.4	1.2	1.0
Vitamin mix [4]	1.0	1.0	1.0	1.0	1.0	1.0
Mineral mix [5]	1.0	1.0	1.0	1.0	1.0	1.0
Choline	0.5	0.5	0.5	0.5	0.5	0.5
Nutrients (%, DM)						
Dry matter	94.3	94.1	94.0	94.3	94.2	94.3
Crude protein	55.3	55.3	55.6	55.1	55.5	55.3
Crude lipid	10.3	10.5	10.9	10.4	10.8	10.7
Ash	12.4	12.5	13.0	12.9	13.4	12.5

[1] Fish meal (FM) (crude protein: 72.1%, crude lipid: 9.1%, ash: 16.0%) was a blend of anchovy meal and sardine meal at a ratio of 1:1 (USD 1.35/kg FM, USD 1 = KRW 1336 (Korean currency)). [2] Meat meal (MM) (crude protein: 65.0%, crude lipid: 10.9%, ash: 17.9%) was purchased from Daekyung Oil & Transportation Co., Ltd. (Busan Metropolitan City, Korea) (USD 0.89/kg MM). [3] Fish oil was purchased from Ewha Oil and Fat Industry Co. Ltd. (Busan Metropolitan City, Korea). [4] Vitamin mix contained the following amounts, which were diluted in cellulose (g/kg mix): L-ascorbic acid, 200; α-tocopheryl acetate, 20; thiamine hydrochloride, 5; riboflavin, 8; pyridoxine, 2; niacin, 40; Ca-D-pantothenate, 12; myo-inositol, 200; D-biotin, 0.4; folic acid, 1.5; p-amino benzoic acid, 20; K_3, 4; A, 1.5; D_3, 0.003; cyanocobalamin, 0.003. [5] Mineral mix contained the following ingredients (g/kg mix): NaCl, 7; $MgSO_4 \cdot 7H_2O$, 105; $NaH_2PO_4 \cdot 2H_2O$, 175; KH_2PO_4, 224; $CaH_4(PO_4)_2 \cdot H_2O$, 140; ferric citrate, 17.5; $ZnSO_4 \cdot 7H_2O$, 2.8; Ca-lactate, 21.8; CuCl, 0.2; $AlCl_3 \cdot 6H_2O$, 0.11; KIO_3, 0.05; $Na_2Se_2O_3$, 0.007; $MnSO_4 \cdot H_2O$, 1.4; $CoCl_2 \cdot 6H_2O$, 0.07.

2.2. Experimental Fish and Culturing Conditions

Juvenile olive flounder were acquired from the Seoul hatchery (Taean-gun, Chungcheongnam-do, Korea). Prior to the feeding experiment, fish were acclimated to the culturing conditions for 14 days. The fish were fed with a commercial extruded pellet containing 55.0% crude protein and 8.0% crude lipid (National Federation of Fisheries Cooperative Feed, Gyeongsangnam-do, Korea) during the acclimatization period. After the 14-day acclimatization period, 360 juvenile fish (initial weight of 14.7 g) were randomly allocated into 18 50-L flow-through containers (water volume: 40 L). Each container was stocked with 20 juvenile fish, and all experimental diets were fed to triplicate groups of fish. Sand-filtered seawater was supplied to each container throughout the whole feeding trial, and the flow rate of seawater into each container was 4.4 L/min. The aeration provided to each container by the sand aerator and the photoperiod was in accordance with natural conditions. The water quality was monitored daily using a multiple water quality meter (AZ-8603, AZ Instrument, Taiwan). During the feeding trial, the water temperature varied from 17.8 to 27.1 °C (mean ± SD: 22.0 ± 1.61 °C), the dissolved oxygen varied from 9.2 to 10.2 mg/L, the salinity ranged from 32.0 to 32.5 g/L, and the pH varied from 7.9 to 8.1. The fish were hand-fed to satiation twice a day (08:00 and 17:00), seven days a week for 56 days. All containers were cleaned, and dead fish were removed immediately when observed.

2.3. Sample Collection and Biological Measurements of the Experimental Fish

Following the completion of the feeding trial, all surviving fish were fasted for 24 h, and then anesthetized with tricaine methanesulfonate (MS222) at a concentration of 100 mg/L. All fish from each container were counted and weighed collectively. Eight anesthetized fish were randomly chosen from each container to measure their condition factor (CF), viscerosomatic index (VSI), and hepatosomatic index (HSI). The growth metrics of the fish were calculated using the following equations: specific growth rate (SGR, %/day) = [(Ln final weight of fish − Ln initial weight of fish) × 100]/days of feeding (56 days); feed efficiency ratio (FER) = (total final weight of fish + total weight of dead fish − total initial weight of fish)/total feed consumption; protein efficiency ratio (PER) = weight gain of fish/protein fed; protein retention (PR, %) = protein gain of fish × 100/protein fed; CF (g/cm^3) = body weight (g) × 100/total length $(cm)^3$; VSI (%) = visceral weight × 100/body weight; HSI (%) = liver weight × 100/body weight.

2.4. Hematological Analysis of Olive Flounder

Three anesthetized fish were chosen at random from each container, and blood samples were collected from the caudal veins with a heparinized syringe. The blood samples were centrifuged for 10 min at 2700× *g*, and the obtained plasma was stored at −70 °C for analysis of the aspartate aminotransferase (AST), alanine aminotransferase (ALT), alkaline phosphatase (ALP), total bilirubin (T-BIL), total cholesterol (T-CHO), triglyceride (TG), total protein (TP), and albumin (ALB) using an automatic chemistry system (Fuji Dri-Chem NX500i, Fujifilm, Tokyo, Japan).

2.5. Innate Immune Responses of Olive Flounder

Three anesthetized fish were chosen at random from each container, and blood samples were collected from the caudal veins. Serum samples were separated by centrifugation at 2700× *g* for 10 min and stored at −70 °C. Following the manufacturer's recommendations, the percentage reaction inhibition rate of the enzyme with water-soluble tetrazolium dye (WST-1) substrate and xanthine oxidase was used to measure the activity of superoxide dismutase (SOD). Following a 20-min reaction period at 37 °C, absorbance at 450 nm (the absorbance wavelength for the colored product of the WST-1 reaction with superoxide) was used to monitor each endpoint. The percentage inhibition was normalized by mg of protein and presented as SOD activity units.

A lysozyme turbidimetric assay was carried out in accordance with the study of [23]. In brief, 100 µL of the test serum was mixed with 1.9 mL of a suspension of *Micrococcus lysodeikticus* (0.2 mg/mL; Sigma, St. Louis, MO, USA) in a 0.05 M sodium phosphate buffer at pH 6.2. The reaction was conducted at 25 °C, and the absorbance at 530 nm was measured using a spectrophotometer after 0 and 60 min. A unit of lysozyme activity was defined as the amount of enzyme required to produce a 0.001/min reduction in absorbance.

2.6. Determination of the Chemical Composition of Olive Flounder and Experimental Diets

The rest of the surviving fish (≥3) from each container and the experimental diets were homogenized and used for chemical analysis. The Association of Official Analytical Chemists (AOAC)'s standard procedures [24] were used to measure the chemical composition of the experimental diets and whole-body fish. Acid digestion with the Kjeldahl method (Kjeltec 2100 distillation unit, Foss Tecator, Hoganas, Sweden) was used to analyze the crude protein content (N × 6.25). The crude lipid content was measured by the ether extraction method using the Soxtec TM 2043 Fat Extraction System (Foss Tecator, Hoganas, Sweden). The moisture was estimated using an oven-drying process at 105 °C for 24 h. The ash content was measured after 4 h of combustion in a muffle furnace at 550 °C.

2.7. Statistical Analysis

The differences among the means of the treatments were tested by one-way ANOVA and Duncan's multiple range test [25] using the SPSS program (version 26.0, Chicago, IL, USA). Before statistical analysis, all percentage data were transformed into arcsine. Furthermore, regression analysis was performed between the weight gain, SGR, FER, and PER of the fish as the dependent variables and the replaced levels of MM for FM in the diets as the independent variable.

3. Results

3.1. Performance of Fish in the Feeding Trial

The survival (≥95%) of the fish was not noticeably impacted ($p > 0.3$) by the dietary replacement levels of FM with MM (Table 2). However, the weight gain and SGR of the fish fed the MM0 and MM10 diets were statistically greater ($p < 0.001$ for both) than those of the fish fed the MM30, MM40, and MM50 diets, but not statistically different ($p > 0.05$) from those of the fish fed the MM20 diet. Regarding the orthogonal polynomial contrast, the weight gain and SGR of the fish showed significant linear relationships ($p < 0.0001$ for both) with the dietary replacement of MM for FM. The best fitting models between the dietary substitution levels of MM for FM and weight gain ($Y = 0.00009395X^3 - 0.007644X^2 + 0.09283X + 40.6283$, $R^2 = 0.7865$, $p < 0.001$) and SGR ($Y = 0.000003159X^3 - 0.0002562X^2 + 0.003123X + 2.3647$, $R^2 = 0.7755$, $p < 0.001$) were observed (Table 3). To incite the maximum values of weight gain and the SGR of the fish, an estimated 7.0% of FM substitution with MM was required in the diets.

Table 2. Survival (%), weight gain (g/fish), and specific growth rate (SGR) of olive flounder fed experimental diets for 56 days.

Experimental Diets	Initial Weight (g/fish)	Final Weight (g/fish)	Survival (%)	Weight Gain (g/fish)	SGR [1] (%/day)
MM0	14.7 ± 0.00	55.3 ± 0.35	97.5 ± 1.14	40.6 ± 0.35 [a]	2.36 ± 0.011 [a]
MM10	14.7 ± 0.00	55.6 ± 0.49	96.7 ± 1.67	40.9 ± 0.50 [a]	2.37 ± 0.016 [a]
MM20	14.7 ± 0.00	54.7 ± 0.40	100.0 ± 0.00	40.0 ± 0.40 [ab]	2.34 ± 0.013 [ab]
MM30	14.7 ± 0.01	54.0 ± 0.49	95.0 ± 2.89	39.3 ± 0.49 [b]	2.32 ± 0.016 [bc]
MM40	14.7 ± 0.01	52.7 ± 0.37	98.3 ± 1.67	38.0 ± 0.37 [c]	2.28 ± 0.013 [c]
MM50	14.7 ± 0.01	52.7 ± 0.31	100.0 ± 0.00	38.0 ± 0.31 [c]	2.28 ± 0.010 [c]
p-value			>0.3	<0.001	<0.001

Values (means of triplicate ± SE) in the same column with different letters are significantly different ($p < 0.05$).
[1] SGR (%/day) = [(ln final weight of fish − ln initial weight of fish) × 100]/days of feeding trial (56 days).

Table 3. Regression analysis of dietary substitution level of fish meal with meat meal as independent variable vs. parameters (weight gain, specific growth rate (SGR), feed efficiency ratio (FER), and protein efficiency ratio (PER) of olive flounder) as dependent variables.

	Orthogonal Polynomial Contrast [a]			Regression Analysis			
Dependent Variables	Linear	Quadratic	Cubic	Equation	*p*-Value	R^2	Y_{max} (%)
Weight gain	0.0001	0.3824	0.0894	$Y = 0.00009395X^3 - 0.007644X^2 + 0.09283X + 40.6283$	<0.001	0.7865	X = 7.0
SGR	0.0001	0.4056	0.0929	$Y = 0.000003159X^3 - 0.0002562X^2 + 0.003123X + 2.3647$	<0.001	0.7755	X = 7.0
FER	0.0001	0.8686	0.1050	$Y = 0.000002158X^3 - 0.0001594X^2 + 0.001406X + 1.0158$	<0.001	0.7540	X = 4.9
PER	0.0034	0.2491	0.0457	$Y = 0.000006433X^3 - 0.0004363X^2 + 0.004538X + 1.7929$	<0.009	0.5290	X = 6.0

[a] If statistical significance ($p < 0.05$) was detected, the model that fit best with the data was selected Y_{max} (%) indicates the meat meal substitution level for fish meal in diets to achieve the maximum for each dependent variable.

The feed consumption of the fish was not statistically influenced ($p > 0.1$) by the dietary replacement of FM with MM (Table 4). The FERs of the fish fed the MM0 and MM10 diets were statistically higher ($p < 0.002$) than those of the fish fed the MM30, MM40, and MM50 diets, but not noticeably different ($p > 0.05$) from that of the fish fed the MM20 diet. The PERs of the fish fed the MM0, MM10, and MM20 diets were statistically higher ($p < 0.02$) than those of the fish fed the MM30 and MM40 diets, but not noticeably different ($p > 0.05$) from that of the fish fed the MM50 diet. In terms of the orthogonal polynomial contrast, the FERs and PERs of the fish demonstrated statistical linear relationships ($p < 0.0001$ and $p < 0.0034$, respectively) with the dietary replacement of MM for FM. The best-fitting models between the dietary substitution levels of MM for FM and the FER ($Y = 0.000002158X^3 -$

$0.0001594X^2 + 0.001406X + 1.0158$, $R^2 = 0.7540$, $p < 0.001$) and the PER ($Y = 0.000006433X^3 - 0.0004363X^2 + 0.004538X + 1.7929$, $R^2 = 0.5290$, $p < 0.009$) were observed. To achieve the maximum values of the FER and PER of the fish, the estimated rates of 4.9% and 6.0%, respectively, of FM substitution with MM were required in the diets. The PR of the fish was not noticeably influenced ($p > 0.05$) by the dietary replacement of FM with MM. The CF of the fish fed the MM0 diet was statistically higher ($p < 0.04$) than that of the fish fed the MM20 and MM40 diets, but not noticeably different ($p > 0.05$) from that of the fish fed the MM10, MM30, and MM50 diets. However, neither the VSI nor the HSI of the fish was statistically altered ($p > 0.9$ and $p > 0.4$, respectively) by the dietary replacement of FM with MM.

Table 4. Feed consumption (g/fish), feed efficiency ratio (FER), protein efficiency ratio (PER), protein retention (PR), condition factor (CF), viscerosomatic index (VSI), and hepatosomatic index (HSI) of olive flounder fed the experimental diets for 56 days.

Experimental Diets	Feed Consumption (g/fish)	FER [1]	PER [2]	PR [3]	CF [4]	VSI [5]	HIS [6]
MM0	41.0 ± 0.62	1.02 ± 0.008 [a]	1.79 ± 0.018 [a]	33.90 ± 0.938	1.29 ± 0.002 [a]	3.16 ± 0.002	1.22 ± 0.001
MM10	41.4 ± 0.70	1.02 ± 0.011 [a]	1.79 ± 0.013 [a]	31.49 ± 1.397	1.29 ± 0.001 [ab]	3.16 ± 0.000	1.22 ± 0.001
MM20	40.2 ± 0.08	1.00 ± 0.010 [ab]	1.79 ± 0.018 [a]	31.72 ± 1.915	1.28 ± 0.000 [b]	3.16 ± 0.000	1.22 ± 0.000
MM30	42.3 ± 1.22	0.97 ± 0.011 [bc]	1.68 ± 0.035 [b]	28.81 ± 2.669	1.28 ± 0.000 [ab]	3.16 ± 0.001	1.22 ± 0.000
MM40	40.3 ± 0.68	0.96 ± 0.009 [c]	1.70 ± 0.031 [b]	29.38 ± 2.291	1.28 ± 0.000 [b]	3.16 ± 0.000	1.22 ± 0.000
MM50	39.6 ± 0.04	0.96 ± 0.009 [c]	1.73 ± 0.016 [ab]	29.55 ± 2.187	1.28 ± 0.001 [ab]	3.16 ± 0.000	1.22 ± 0.000
p-value	>0.1	< 0.002	< 0.02	>0.3	<0.04	>0.9	>0.4

Values (means of triplicate ± SE) in the same column with different letters are significantly different ($p < 0.05$). [1] Feed efficiency ratio (FER) = (total final weight of fish + total weight of dead fish − total initial weight of fish)/total feed consumption. [2] Protein efficiency ratio (PER) = weight gain of fish/protein fed. [3] Protein retention (PR) = protein gain × 100/protein fed. [4] Condition factor (CF, g/cm^3) = body weight (g) × 100/total length (cm)^3. [5] Viscerosomatic index (VSI, %) = visceral weight × 100/body weight. [6] Hepatosomatic index (HSI, %) = liver weight × 100/body weight.

3.2. Proximate Composition of the Whole-Body Olive Flounder

The moisture contents of the whole bodies of the fish fed the MM50 diet were statistically higher ($p < 0.04$) than those of the fish fed the MM0 and MM30 diets, but not noticeably different ($p > 0.05$) from those of the fish fed the MM10, MM20, and MM40 diets (Table 5). The crude protein content of the whole-body fish varied from 15.8 to 16.5% and the ash content varied from 3.4 to 3.7%, but these parameters were not noticeably altered ($p > 0.9$ and $p > 0.3$, respectively) by the dietary replacement of FM with MM. The fish fed the MM40 and MM50 diets showed statistically higher ($p < 0.002$) crude lipid contents than the fish fed all other diets.

Table 5. Proximate composition (%) of the whole body of olive flounder fed experimental diets for 56 days.

Experimental Diets	Moisture	Crude Protein	Crude Lipid	Ash
MM0	74.7 ± 0.11 [b]	16.5 ± 0.09	3.0 ± 0.01 [b]	3.4 ± 0.01
MM10	75.6 ± 0.84 [ab]	16.3 ± 0.66	3.0 ± 0.05 [b]	3.4 ± 0.15
MM20	76.6 ± 0.62 [ab]	16.3 ± 0.68	3.1 ± 0.05 [b]	3.6 ± 0.08
MM30	75.0 ± 0.16 [b]	16.4 ± 0.18	3.0 ± 0.02 [b]	3.5 ± 0.09
MM40	76.5 ± 0.49 [ab]	15.9 ± 0.76	3.2 ± 0.03 [a]	3.4 ± 0.13
MM50	77.2 ± 0.88 [a]	15.8 ± 0.81	3.2 ± 0.02 [a]	3.7 ± 0.14
p-value	<0.04	>0.9	<0.002	>0.3

Values (means of triplicate ± SE) in the same column with different letters are significantly different ($p < 0.05$).

3.3. Plasma Chemistry of Olive Flounder

The plasma AST varied from 18.1 to 18.7 U/L, the ALT varied from 5.8 to 5.9 U/L, the ALP varied from 156.8 to 157.9 U/L, the T-CHO varied from 239.1 to 241.0 mg/dL, the

TG varied from 386.6 to 389.4 mg/dL, and the ALB varied from 0.8 to 1.0 g/dL (Table 6). In addition, the T-BIL and TP rates were 0.3 mg/dL and 3.9 g/dL, respectively. None of the plasma parameters of the fish were noticeably impacted ($p > 0.05$) by the experimental diets.

Table 6. Hematological parameters of olive flounder fed the experimental diets for 56 days.

Experimental Diets	AST (U/L)	ALT (U/L)	ALP (U/L)	T-BIL (mg/dL)	T-CHO (mg/dL)	TG (mg/dL)	TP (g/dL)	ALB (g/dL)
MM0	18.1 ± 0.61	5.9 ± 0.34	156.8 ± 8.38	0.3 ± 0.03	239.1 ± 3.93	389.1 ± 4.24	3.9 ± 0.05	1.0 ± 0.03
MM10	18.2 ± 0.70	5.8 ± 0.22	157.9 ± 4.01	0.3 ± 0.03	241.0 ± 5.18	388.3 ± 5.77	3.9 ± 0.06	0.9 ± 0.04
MM20	18.3 ± 0.50	5.9 ± 0.20	156.8 ± 4.57	0.3 ± 0.02	240.7 ± 5.42	389.4 ± 7.37	3.9 ± 0.06	1.0 ± 0.04
MM30	18.7 ± 0.73	5.9 ± 0.31	156.9 ± 6.17	0.3 ± 0.03	240.3 ± 4.73	387.7 ± 6.29	3.9 ± 0.05	0.9 ± 0.03
MM40	18.3 ± 0.93	5.8 ± 0.22	156.8 ± 4.68	0.3 ± 0.03	240.7 ± 5.25	387.7 ± 5.52	3.9 ± 0.04	0.9 ± 0.05
MM50	18.3 ± 0.65	5.8 ± 0.22	157.4 ± 4.84	0.3 ± 0.02	240.0 ± 5.14	386.6 ± 6.58	3.9 ± 0.05	0.8 ± 0.06
p-value	>0.9	>0.9	>0.9	>0.1	>0.9	>0.9	>0.9	>0.3

Values (means of triplicate ± SE) in the same column with different letters are significantly different ($p < 0.05$). AST, analysis of aspartate aminotransferase; ALT, alanine aminotransferase; ALP, alkaline phosphatase; T-BIL, total bilirubin; T-CHO, total cholesterol; TG, triglyceride; TP, total protein; and ALB, albumin.

3.4. Innate Immune Responses of Olive Flounder

None of the SOD or lysozyme activities of olive flounder were noticeably altered ($p > 0.6$ and $p > 0.07$, respectively) by the dietary replacement of FM with MM (Table 7).

Table 7. Lysozyme activity and SOD of olive flounder fed the experimental diets for 56 days.

Experimental Diets	SOD Activity (%)	Lysozyme Activity (U/mL)
MM0	70.01 ± 1.648	0.144 ± 0.016
MM10	68.06 ± 0.549	0.121 ± 0.024
MM20	68.85 ± 1.256	0.092 ± 0.005
MM30	67.96 ± 0.632	0.089 ± 0.012
MM40	68.60 ± 0.541	0.094 ± 0.004
MM50	67.96 ± 0.353	0.085 ± 0.015
p-value	>0.6	>0.07

Values (means of triplicate ± SE) in the same column with different letters are significantly different ($p < 0.05$). SOD, superoxide dismutase.

4. Discussion

MM is an economically cheaper source of protein than FM and widely accessible in meat processing plants and animal slaughterhouses. In the last few decades, huge amounts of by-products, such as skin, viscera, heads, feet, and blood have become available as a result of the modern methods of animal meat processing [26]. Recycling wastes from meat processing plants and animal slaughterhouses has economical, biological, and environmental benefits. The nutritional contents of MM, being rich in protein and lipids, makes it suitable as an ingredient in feeds for carnivorous fish [27]. This may imply that MM has the nutritional potential as a substitute for FM in olive flounder diet because it requires relatively high protein and lipid content in diets [28,29]. In fact, the suitability of terrestrial animal by-product meal as a replacement for FM in the diets of various fish species has been evaluated over the last few decades [15,17,30–32]. In addition, MM as a potential replacement for FM has been known in a few fish species, such as olive flounder, barramundi (*Lates calcarifer*), rockfish (*Sebastes schlegeli*), and Australian silver perch (*Bidyanus bidyanus*) [18,19,33,34].

The lack of differences in the growth performance (weight gain and SGR), feed consumption, and feed utilization (FER, PER, and PR) of olive flounder fed the MM0, MM10, and MM20 diets in the current study proved that MM could be successfully replaced for FM up to 20% in a 65% FM-based diet of olive flounder without deterioration in the growth, feed consumption, or feed utilization. According to the third-order regression analysis, however, the ideal replacement levels of MM for FM in diets for the utmost growth (weight gain and SGR) of olive flounder were estimated to be 7.0% for both. Therefore, considering the economic views of the present study, replacing 20% FM with MM in the diet seems to be the most recommendable for olive flounder farmers in terms of feed cost.

The FER of olive flounder deteriorated when more than a 30% FM replacement with MM in the diet was made. In addition, the PERs of the fish fed the MM30 and MM40 diets were also inferior to the fish fed the MM0 diet in the current study. The consistent results show that the FERs and PERs of olive flounder were impacted by the dietary replacement of FM with MM [15,18]. Similarly, the feed utilization of gibel carp and Japanese sea bass (*Lateolabrax japonicus*) was influenced by the dietary replacement of FM with various animal protein sources [35,36]. The dietary replacement of FM with MM influenced both the feed consumption and feed utilization (FER, PER, and PR) of olive flounder, and the dietary replacement of 20% FM with MM achieved the best growth performance, highest feed consumption, and highest FER [16].

The maximum replacement level (20%) of FM with MM in the diet of olive flounder in the current study was relatively low compared to substitution levels reported in the same fish species in other studies. Reference [18] demonstrated that substitution of FM up to 60% with pet-grade MM with the supplementation of AAs had no negative impact on the growth, hematological parameters, or chemical composition of juvenile olive flounder (initial weight of about 3 g) when 80% FM was included. Recently, [16] unveiled that up to 40% of FM could be successfully replaced with pet-grade MM without the supplementation of AAs, producing no adverse effects on the growth performance, feed utilization, or non-specific immune responses of fish when juvenile olive flounder (initial weight of 9.2 g) were fed with a 65% FM-based diet or one of the diets replacing 10%, 20%, 40%, 60%, 80%, or 100% FM with MM for 56 days. During the winter season, up to 40% of FM could be successfully replaced with feed ingredient-grade MM without causing a reduction in the growth performance of olive flounder in a 10-week feeding trial [15]. Considering these results and the results obtained from the present study, the substitutability of FM with MM in olive flounder feed was highly influenced by the quality (or grade) of MM, the supplementation of EAAs, and other experimental conditions, such as the fish size, water temperature, and duration of the feeding trial. Furthermore, [37] indicated that porcine MM can substitute up to 35% of FM in a 42.4% FM-based diet without causing any adverse impact on the growth, chemical composition, FER, or PER of Pacific white shrimp (*Litopenaeus vannamei*). However, further study on the substitutability of FM with MM in olive flounder extruded pellets should be tested in commercial scale farms before practical implementation. The VSI and HSI of fish were not altered by the dietary substitution of MM for FM at the end of the 56-day feeding trial. Likewise, references [17,38] have also reported that the HSI and VSI were not noticeably impacted by the substitution of FM with chicken by-product meal and meat and bone meal in fish feeds.

The moisture and crude lipid contents of the whole bodies of olive flounder were influenced by the dietary replacement levels of FM with MM in the current study. The fish fed the MM50 diet showed the highest moisture and lipid contents in whole-body fish. Similarly, references [18,39] found that the chemical composition of olive flounder was altered by the dietary replacement of FM with MM and dehulled soybean meal, respectively. On the contrary, there were also some conflicting studies showing that the proximate composition of olive flounder was not impacted by the replacements of FM with various alternative protein sources in diets [10,16,38,40].

Hematological parameters are used to indicate the health conditions, physiological and nutritional status, and pathological changes in fish [41]. In the current study, however, there were no considerable differences in any of the hematological parameters measured, demonstrating that the dietary replacement of MM for FM might not alter the hematological status of olive flounder. Likewise, the dietary replacement of MM for FM did not change the hematological parameters of olive flounder [15,16]. In addition, reference [33] also demonstrated that the dietary replacement of FM with MM did not cause any difference in the total protein or glucose of rockfish.

The innate immune system is essential for disease resistance in fish. Compared to mammals, fish rely more heavily on innate defense mechanisms [42]. SOD and lysozyme activities are well-known markers of stress response and disease resistance [43]. SOD is a free radical-neutralizing enzyme that can protect animal tissue from damage [44]. Lysozyme activity is a crucial humoral indicator of the immune system that acts as a main defensive component against invasive microorganisms in fish [45]. It can catalyze the hydrolysis of β (1–4) glycosidic bonds of bacterial cell walls and plays an essential role in the defense against bacterial infections [46]. However, neither the SOD nor lysozyme activity of olive flounder was altered by the partial dietary replacement of FM with MM in the current study, demonstrating that the replacement of FM with MM in olive flounder feed had no adverse influence on their innate immune responses. Similarly, the dietary replacement of FM with various animal and plant protein sources did not bring about any adverse impact on the immune responses of olive flounder or red sea bream [9,11,16]. Unlike this study, however, the serum SOD and lysozyme activities of tilapia (*Oreochromis niloticus* × *O. aureus*) tended to decrease with increased dietary replacement levels of FM with soybean meal because the immune system of tilapia might be influenced by the certain constituents of soybean meal [47].

5. Conclusions

Up to 20% of FM could be successfully replaced with feed ingredient-grade MM in feed without causing adverse impacts on the growth, feed consumption, feed utilization, or innate immune responses of olive flounder. The optimum replacement levels of FM with MM for the maximum weight gain and SGR of olive flounder were estimated to be 7.0% for both.

Author Contributions: M.F.U.Z.; Data Curation, Formal Analysis, Investigation, Visualization. R.L.; Formal Analysis, Investigation, Visualization, Writing—Original Draft. S.H.C.; Conceptualization, Methodology, Writing—Review and Editing, Funding Acquisition. All authors have read and agreed to the published version of the manuscript.

Funding: This work was supported by the National Research Foundation of Korea (NRF) grant funded by the Korean Government (MSIT) (No. 2020R1A2C1009903).

Institutional Review Board Statement: Not applicable.

Data Availability Statement: The data are available upon request from the authors.

Conflicts of Interest: The authors declare no conflict of interest.

References

1. FAO. Meeting the Sustainable Development Goals. In *The State of World Fisheries and Aquaculture 2018*; FAO: Rome, Italy, 2018. Available online: https://www.fao.org/documents/card/es/c/I9540EN (accessed on 26 September 2022).
2. FAO. Towards Blue Transformation. In *The State of World Fisheries and Aquaculture 2022*; FAO: Rome, Italy, 2022. [CrossRef]
3. Kim, K.W.; Kim, K.D.; Han, H.S.; Won, S.; Moniruzzaman, M.; Lee, J.H.; Bai, S.C. Evaluation of the dietary protein requirement of a selectively bred (F-5 generation) strain of olive flounder, *Paralichthys olivaceus*. *Turk. J. Fish. Aquat. Sci.* **2017**, *17*, 1009–1016. [CrossRef] [PubMed]
4. KOSIS. Korean Statistical Information Service, Korea. 2022. Available online: https://kosis.kr/eng/ (accessed on 26 September 2022).
5. Aragão, C.; Gonçalves, A.T.; Costas, B.; Azeredo, R.; Xavier, M.J.; Engrola, S. Alternative Proteins for Fish Diets: Implications beyond Growth. *Animals* **2022**, *12*, 1211. [CrossRef] [PubMed]
6. Glencross, B.D.; Baily, J.; Berntssen, M.H.; Hardy, R.; MacKenzie, S.; Tocher, D.R. Risk assessment of the use of alternative animal and plant raw material resources in aquaculture feeds. *Rev. Aquac.* **2020**, *12*, 703–758. [CrossRef]
7. FAO. A Quarterly Update on World Seafood Markets. In *Globefish Highlights January 2020 ISSUE, with Jan.–Sep. 2019 Statistics*; FAO: Rome, Italy, 2020. [CrossRef]
8. Hamidoghli, A.; Won, S.; Farris, N.W.; Bae, J.; Choi, W.; Yun, H.; Bai, S.C. Solid state fermented plant protein sources as fish meal replacers in whiteleg shrimp *Litopaeneus vannamei*. *Anim. Feed Sci. Technol.* **2020**, *264*, 114474. [CrossRef]
9. Choi, W.; Hamidoghli, A.; Rivero, C.J.; Bae, J.; Lee, S.; Lee, B.J.; Hur, S.; Han, H.; Choi, Y.H.; Bai, S.C. Animal and plant proteins as alternative ingredients in diets for sub-adult olive flounder *Paralichthys olivaceus* at farm conditions. *Aquac. Res.* **2022**, *53*, 2739–2749. [CrossRef]

10. Kim, J.; Cho, S.H.; Kim, T.; Hur, S.W. Substitution effect of fish meal with various sources of animal by-product meals in feed on growth, feed utilization, body composition, haematology and non-specific immune response of olive flounder (*Paralichthys olivaceus*, Temminck & Schlegel, 1846). *Aquac. Res.* **2021**, *52*, 2802–2817. [CrossRef]
11. Dossou, S.; Koshio, S.; Ishikawa, M.; Yokoyama, S.; Dawood, M.A.O.; Basuini, M.F.; El-Hais, A.M.; Olivier, A. Effect of partial replacement of fish meal by fermented rapeseed meal on growth, immune response and oxidative condition of red sea bream juvenile, *Pagrus major*. *Aquaculture* **2018**, *490*, 228–235. [CrossRef]
12. Lu, F.; Haga, Y.; Satoh, S. Effects of replacing fish meal with rendered animal protein and plant protein sources on growth response, biological indices, and amino acid availability for rainbow trout *Oncorhynchus mykiss*. *Fish. Sci.* **2015**, *81*, 95–105. [CrossRef]
13. Millamena, O.M.; Golez, N.V. Evaluation of processed meat solubles as replacement for fish meal in diet for juvenile grouper *Epinephelus coioides* (Hamilton). *Aquac. Res.* **2001**, *32*, 281–287. [CrossRef]
14. Gatlin, D.M., III; Barrows, F.T.; Brown, P.; Dabrowski, K.; Gaylord, T.G.; Hardy, R.W.; Herman, E.; Hu, G.; Krogdahl, A.; Nelson, R.; et al. Expanding the utilization of sustainable plant products in aquafeeds: A review. *Aquac. Res.* **2007**, *38*, 551–579. [CrossRef]
15. Cho, S.H.; Lee, S.-M.; Park, B.-H.; Park, I.-S.; Choi, C.Y.; Lee, S.-M.; Min, B.H.; Hur, S.-B.; Lim, Y.S. Effect of Partial Dietary Substitution of Meat Meal for Fish Meal on the Growth and Body Composition of the Juvenile Olive Flounder *Paralichthys olivaceus*. *J. Fish. Sci. Technol.* **2005**, *8*, 138–141. [CrossRef]
16. Ha, M.S.; Cho, S.H.; Kim, T. Dietary substitution of fish meal by meat meal: Effects on juvenile olive flounder (*Paralichthys olivaceus*) growth performance, feed utilization, haematology, biochemical profile and disease resistance against *Streptococcus iniae*. *Aquac. Nutr.* **2021**, *27*, 1888–1902. [CrossRef]
17. Moutinho, S.; Peres, H.; Serra, C.; Martínez-Llorens, S.; Tomás-Vidal, A.; Jover-Cerdá, M.; Oliva-Teles, A. Meat and bone meal as partial replacement for fish meal in diets for gilthead seabream (*Sparus aurata*) juveniles: Growth, feed efficiency, amino acid utilization, and economic efficiency. *Aquaculture* **2017**, *468*, 271–277. [CrossRef]
18. Sato, T.; Kikuchi, K. Meat Meal as a Protein Source in the Diet of Juvenile Japanese Flounder. *Fish. Sci.* **1997**, *63*, 877–880. [CrossRef]
19. Williams, K.C.; Barlow, C.G.; Rodgers, L.J.; Ruscoe, I. Potential of meat meal to replace fish meal in extruded dry diets for barramundi, *Lates calcarifer* (Bloch). I. Growth performance. *Aquac. Res.* **2003**, *34*, 23–32. [CrossRef]
20. Yang, Y.; Xie, S.; Cui, Y.; Lei, W.; Zhu, X.; Yang, Y.; Yu, Y. Effect of replacement of dietary fish meal by meat and bone meal and poultry by-product meal on growth and feed utilization of gibel carp, Carassius auratus gibelio. *Aquac. Nutr.* **2004**, *10*, 289–294. [CrossRef]
21. Lee, S.-M.; Cho, S.H.; Kim, K.-D. Effects of Dietary Protein and Energy Levels on Growth and Body Composition of Juvenile Flounder (*Paralichthys olivaceus*). *J. World Aquac. Soc.* **2000**, *31*, 306–315. [CrossRef]
22. Kim, K.W.; Wang, X.J.; Bai, S.C. Optimum dietary protein level for maximum growth of juvenile olive flounder *Paralichthys olivaceus* (Temminck et Schlegel). *Aquac. Res.* **2002**, *33*, 673–679. [CrossRef]
23. Lange, S.; Guđmundsdottir, B.K.; Magnadottir, B. Humoral immune parameters of cultured Atlantic halibut (*Hippoglossus hippoglossus* L.). *Fish Shellfish. Immunol.* **2001**, *11*, 523–535. [CrossRef]
24. AOAC. *Official Methods of Analysis*, 18th ed.; Association of Official Analytical Chemists: Arlington, VA, USA, 2005.
25. Duncan, D.B. Multiple Range and Multiple *F* Tests. *Biometrics* **1955**, *11*, 1–42. [CrossRef]
26. Jayathilakan, K.; Sultana, K.; Radhakrishna, K.; Bawa, A.S. Utilization of byproducts and waste materials from meat, poultry and fish processing industries: A review. *J. Food Sci. Technol.* **2012**, *49*, 278–293. [CrossRef] [PubMed]
27. Chi, S.; Wang, W.; Tan, B.; Dong, X.; Yang, Q.; Liu, H.; Zhang, S. The Apparent Digestibility Coefficients of 13 Selected Animal Feedstuff for Cobia, *Rachycentron canadum*. *J. World Aquac. Soc.* **2017**, *48*, 280–289. [CrossRef]
28. Cho, S.H. Effect of Dietary Protein and Lipid Levels on Compensatory Growth of Juvenile Olive Flounder (*Paralichthys olivaceus*) Reared in Suboptimal Temperature. *Asian Australas. J. Anim. Sci.* **2011**, *24*, 407–413. [CrossRef]
29. Kim, K.-D.; Kang, Y.J.; Lee, H.M.; Kim, K.-W.; Jang, M.-S.; Choi, S.-M.; Lee, S.-M.; Cho, S.H. Effects of Dietary Protein and Lipid Levels on Growth and Body Composition of Subadult Olive Flounder, *Paralichthys olivaceus*, at a Suboptimal Water Temperature. *J. World Aquac. Soc.* **2010**, *41*, 263–269. [CrossRef]
30. Kureshy, N.; Davis, D.A.; Arnold, C.R. Partial Replacement of Fish Meal with Meat-and-Body Meal, Flash-Dried Poultry By-Product Meal, and Enzyme-Digested Poultry By-Product Meal in Practical Diets for Juvenile Red Drum. *N. Am. J. Aquac.* **2000**, *62*, 266–272. [CrossRef]
31. Millamena, O.M. Replacement of fish meal by animal by-product meals in a practical diet for grow-out culture of grouper *Epinephelus coioides*. *Aquaculture* **2002**, *204*, 75–84. [CrossRef]
32. Yu, H.; Zhang, Q.; Cao, H.; Tong, T.; Huang, G.; Li, W. Replacement of fish meal by meat and bone meal in diets for juvenile snakehead *Ophiocephalus argus*. *Fish. Sci.* **2015**, *81*, 723–729. [CrossRef]
33. Lee, Y.H.; Lee, S.M. The use of meat meal as a dietary protein source replacing fish meal in juvenile rockfish *Sebastes schlegeli*. *J. Aquac.* **2005**, *18*, 92–97.
34. Stone, D.A.; Allan, G.L.; Parkinson, S.; Rowland, S.J. Replacement of fish meal in diets for Australian silver perch, *Bidyanus bidyanus*: III. Digestibility and growth using meat meal products. *Aquaculture* **2000**, *186*, 311–326. [CrossRef]

35. Hu, L.; Yun, B.; Xue, M.; Wang, J.; Wu, X.; Zheng, Y.; Han, F. Effects of fish meal quality and fish meal substitution by animal protein blend on growth performance, flesh quality and liver histology of Japanese seabass (*Lateolabrax japonicus*). *Aquaculture* **2013**, *372–375*, 52–61. [CrossRef]
36. Yang, Y.; Xie, S.; Cui, Y.; Zhu, X.; Lei, W.; Yang, Y. Partial and total replacement of fishmeal with poultry by-product meal in diets for gibel carp, *Carassius auratus gibelio* Bloch. *Aquac. Res.* **2006**, *37*, 40–48. [CrossRef]
37. Hernández, C.; Olvera-Novoa, M.A.; Aguilar-Vejar, K.; González-Rodríguez, B.; de la Parra, I.A. Partial replacement of fish meal by porcine meat meal in practical diets for Pacific white shrimp (*Litopenaeus vannamei*). *Aquaculture* **2008**, *277*, 244–250. [CrossRef]
38. Ha, M.S.; Lee, K.W.; Kim, J.; Yun, A.; Jeong, H.S.; Lee, M.J.; Baek, S.I.; Cho, S.H.; Kim, K.W.; Lim, S.G.; et al. Dietary substitution effect of fish meal with chicken by-product meal on growth, feed utilization, body composition, haematology and non-specific immune responses of olive flounder (*Paralichthys olivaceus*). *Aquac. Nutr.* **2021**, *27*, 315–326. [CrossRef]
39. Kim, Y.C.; Yoo, G.Y.; Wang, X.; Lee, S.; Shin, I.S.; Bai, S.C. Long Term Feeding Effects of Dietary Dehulled Soybean Meal as a Fish Meal Replacer in Growing Olive Flounder *Paralichthys olivaceus*. *Asian Australas. J. Anim. Sci.* **2008**, *21*, 868–872. [CrossRef]
40. Seong, M.; Lee, S.; Lee, S.; Song, Y.; Bae, J.; Chang, K.; Bai, S.C. The effects of different levels of dietary fermented plant-based protein concentrate on growth, hematology and innate immune responses in juvenile olive flounder, *Paralichthys olivaceus*. *Aquaculture* **2018**, *483*, 196–202. [CrossRef]
41. Ahmed, I.; Reshi, Q.M.; Fazio, F. The influence of the endogenous and exogenous factors on hematological parameters in different fish species: A review. *Aquac. Int.* **2020**, *28*, 869–899. [CrossRef]
42. Anderson, D.P. Immunostimulants, adjuvants, and vaccine carriers in fish: Applications to aquaculture. *Annu. Rev. Fish Dis.* **1992**, *2*, 281–307. [CrossRef]
43. Wu, Y.-R.; Gong, Q.-F.; Fang, H.; Liang, W.-W.; Chen, M.; He, R.-J. Effect of Sophora flavescens on non-specific immune response of tilapia (GIFT *Oreochromis niloticus*) and disease resistance against *Streptococcus agalactiae*. *Fish Shellfish Immunol.* **2013**, *34*, 220–227. [CrossRef]
44. McCord, J.M.; Fridovich, I. Superoxide dismutase. An enzymatic function for erythrocuprein (hemocuprein). *J. Biol. Chem.* **1969**, *244*, 6049–6055. [CrossRef]
45. Dawood, M.A.; Koshio, S.; Ishikawa, M.; Yokoyama, S. Immune responses and stress resistance in red sea bream, Pagrus major, after oral administration of heat-killed *Lactobacillus plantarum* and vitamin C. *Fish Shellfish Immunol.* **2016**, *54*, 266–275. [CrossRef]
46. Adel, M.; Yeganeh, S.; Dadar, M.; Sakai, M.; Dawood, M.A. Effects of dietary Spirulina platensis on growth performance, humoral and mucosal immune responses and disease resistance in juvenile great sturgeon (*Huso huso* Linnaeus, 1754). *Fish Shellfish. Immunol.* **2016**, *56*, 436–444. [CrossRef] [PubMed]
47. Lin, S.; Luo, L. Effects of different levels of soybean meal inclusion in replacement for fish meal on growth, digestive enzymes and transaminase activities in practical diets for juvenile tilapia, *Oreochromis niloticus* × *O. aureus*. *Anim. Feed. Sci. Technol.* **2011**, *168*, 80–87. [CrossRef]

fishes

Article

Astragalus membranaceus Extract (AME) Enhances Growth, Digestive Enzymes, Antioxidant Capacity, and Immunity of *Pangasianodon hypophthalmus* Juveniles

Hany M. R. Abdel-Latif [1,*], Hamada A. Ahmed [2], Mustafa Shukry [3], Md Reaz Chaklader [4], Rasha M. Saleh [5] and Mohamed A. Khallaf [6]

1 Department of Poultry and Fish Diseases, Faculty of Veterinary Medicine, Alexandria University, Alexandria 22758, Egypt
2 Department of Nutrition and Veterinary Clinical Nutrition, Faculty of Veterinary Medicine, Damanhour University, Damanhour 22511, Egypt
3 Department of Physiology, Faculty of Veterinary Medicine, Kafrelsheikh University, Kafrelsheikh 33516, Egypt
4 Department of Primary Industries and Regional Development, Fleet Street, Fremantle, WA 6160, Australia
5 Department of Physiology, Faculty of Veterinary Medicine, Mansoura University, Mansoura 35516, Egypt
6 Department of Aquatic Animals Medicine and Management, Faculty of Veterinary Medicine, University of Sadat City, Sadat City 32897, Egypt
* Correspondence: hmhany@alexu.edu.eg

Citation: Abdel-Latif, H.M.R.; Ahmed, H.A.; Shukry, M.; Chaklader, M.R.; Saleh, R.M.; Khallaf, M.A. *Astragalus membranaceus* Extract (AME) Enhances Growth, Digestive Enzymes, Antioxidant Capacity, and Immunity of *Pangasianodon hypophthalmus* Juveniles. *Fishes* **2022**, 7, 319. https://doi.org/10.3390/fishes7060319

Academic Editors: Marina Paolucci and Shunsuke Koshio

Received: 21 September 2022
Accepted: 30 October 2022
Published: 3 November 2022

Abstract: The present study evaluated the impacts of powdered *Astragalus membranaceus* extract (AME) on the growth, physiological responses, and serum immunity of *Pangasianodon hypophthalmus* juveniles. Four test diets were formulated to include varying AME levels as 0.0 (control), 1.5 (AME1.5), 3.0 (AME3.0), and 4.5 (AME4.5) g/kg. Fish weighing approximately 11.50 g were stocked into four triplicate groups and hand-fed on the test diets three times daily for two months. At 60 days post-feeding, the growth performance, including weight gain and the specific growth rate, was increased quadratically ($R^2 > 0.90$) with increasing AME inclusion levels. An improvement in the feed intake and feed conversion ratio were also noticed in groups fed at different AME levels. The whole-body and amino acid composition were unaffected by the test diets. A significant quadratic trend in the digestive enzymes (lipase, α-amylase, and protease) was found along with increasing AME inclusion levels. Liver enzymes associated with liver functions were improved by AME dietary inclusion levels. Meanwhile, the blood urea nitrogen, uric acid, and creatinine values were unaffected by AME dietary inclusion. On the other hand, serum immunity (lysozyme and total Igs) was elevated with a significant quadratic trend along with increasing AME dietary inclusion levels. Liver MDA levels decreased with increasing AME levels. Liver CAT, GPx, and SOD enzyme activities demonstrated a significant increasing trend along with dietary AME inclusion. The aforementioned effects of dietary AME on *P. hypophthalmus* health underpinned the potentiality of AME to be used as a phyto-additive to improve the functionality of aquafeed.

Keywords: herbal medicines; growth; digestive enzymes; striped catfish; immunity

1. Introduction

Pangasianodon hypophthalmus is generally known as striped catfish, pangas catfish, or pangasiid catfish. It is a freshwater fish popularly farmed in several Asian countries [1,2]. Its high tolerance to environmental conditions, along with its higher growth rate, lower operational expenditures, and relatively higher economic returns, make it a favorable candidate for finfish aquaculture in several countries [3,4]. This fish is usually cultivated in high stocking density [2], which may increase its susceptibility to various infectious agents, especially bacterial pathogens [5]. However, to achieve the intensification of production, researchers have engaged, in recent years, to find novel immunostimulants to enhance the immunity of cultured species and support antibiotic- and chemotherapeutic-free and disease-resistant sustainable aquaculture. Using plant herbal extracts as immunostimulants has been considered a common practice to improve the aquafeed's functionality [6–8]. Several researchers have assessed a variety of plant- or herbal extract-based diets to promote the growth and immunity of *P. hypophthalmus*. For example, *Zingiber officinale*, *Euphorbia hirta*, *Phyllanthus amarus*, *Azadirachta indica*, and *Allium sativum* were proven to be potential immunostimulants for *P. hypophthalmus* [9,10]. Moreover, *Andrographis paniculata* extract also improved the immune responses and disease resistance of juvenile pangas catfish [11]. Our recent study also reported the positive impacts of *Silybum marianum* extract on the growth, immune performance, antioxidant condition, and intestine morphometry of this valuable fish species [8].

Astragalus membranaceus (AM) is a frequently used medicinal plant in China. It has well-known immunostimulatory, antioxidant, general tonic, and hepato-protective properties due to its phyto-components of *Astragalus* polysaccharides, flavonoids, phenolics, alkaloids, and saponins [12–14]. There is a plethora of studies on the beneficial roles and applications of AM as a functional feed supplement in several finfish and shellfish aquaculture. For example, feeding *Oreochromis niloticus* on a diet supplemented with 0.1% powdered AME extract increased its resistance to *Aeromonas hydrophila* [15]. Dietary AM plants also promoted the growth performance, antioxidant responses, immunity, and increased stress tolerance of yellow perch (*Perca flavescens*) [16]. Furthermore, dietary application of AM (1.5, 3, and 4.5%) promoted growth and antioxidative status, as well as modulated cold water stress tolerance in bluegill sunfish (*Lepomis macrochirus*) [17]. Dietary supplementation with a dried AM plant (0.25% and 0.5%) also enhanced the growth, immunity, and antioxidant status of *Litopenaeus vannamei* [18]. However, a recently published study demonstrated that dietary supplementation with 0.1–0.8% AM root extract did not improve the growth, hepatic morphology, or immunity of hybrid groupers (*Epinephelus lanceolatus* ♂ × *E. fuscoguttatus* ♀) [19].

Astragalus polysaccharides (ASP) are dietary supplements that are the subject of extensive research and recent publications in finfish and shrimp species. Diet supplementation with 0.05% and 0.10% ASP promoted growth, digestive enzyme activities, and intestinal morphology of larval large yellow croaker (*Larimichthys crocea*) [20]. Dietary ASP (50 and 150 mg/kg) ameliorated the growth, antioxidant activity, and immunity of turbot (*Scophthalmus maximus*) [21]. In addition, dietary ASP obtained from dried roots of *A. membranaceus* (100 mg/kg) enhanced the growth and immunity of *Carassius auratus* juveniles [22]. Moreover, dietary 0.01% ASP improved the growth, gut health condition, and anti-viral immunity of *Danio rerio* [23]. Furthermore, dietary ASP (1 g/kg) also enhanced the immunity, intestinal microbiota, and disease resistance of *Ctenopharyngodon idella* [24]. It was recently reported that dietary ASP (100, 200, and 300 mg/kg) significantly improved the growth, antioxidant responses, and immunity of *Catla catla* [25]. In addition, dietary supplementation with ASP obtained from dried sliced roots of *A. membranaceus* (30 g/kg) also promoted the growth and improved disease resistance of *L. vannamei* [26]. Lately, it has been demonstrated that dietary 0.15% ASP and 0.30% ASP significantly improved the immunity, disease resistance and attenuated toxicity signs in Nile tilapia [27,28].

Various reports regarding emerging issues have jeopardized the aquaculture industry, such as the elaboration of antibiotic- and therapeutics-resistant bacteria and the deposition of antibiotic and chemotherapeutic compounds in the aquaculture products [29]. These concerns have led to efforts to find suitable, effective, economic, and eco-friendly alternatives such as dietary herbal extracts [6–8]. Herein, though several researchers have conducted several experimental trials on *P. hypophthalmus* after its introduction in Egypt [4,8], there are no known reports on the effects of AME as a feed additive for striped catfish juveniles. Consequently, this study was intended to assess the supplementing impacts of powdered AME on the growth performances, serum immunity, hepato-renal functions, and antioxidant responses of *P. hypophthalmus* juveniles to test its suitability for improving the functionality of striped catfish diets and maintaining sustainable aquaculture.

2. Materials and Methods

2.1. Herbal Extract and Analysis of Its Bioactive Constituents

Powdered *Astragalus membranaceus* extract (AME) was procured from Free Trade Egypt Co., Alexandria, EGY. The GC-MS method analyzed the bio-active constituents in the ethanolic extract of AME, as illustrated by Gomathi et al. [30]. Bioactive components were characterized and identified by mass spectral and relative retention time (RT) comparisons with the WILEY 09 and NIST14 Mass Spectral databases and libraries. The flavonoid compounds in AME were determined by HPLC analysis following Mattila et al. [31], whilst the phenolic content of AME ethanolic extract was measured in accordance with the methods outlined by Öztürk et al. [32].

2.2. Fish and Adaptation Conditions

At a fish hatchery in Borg El Arab, Alexandria, EGY, three hundred juvenile healthy *P. hypophthalmus* were raised in spherical black fiberglass tanks with 1000 L of water capacity for fifteen days before starting the experiments (for adaptation). During this period, fish were fed daily on a well-balanced reference diet (30% crude protein (CP), Aller Aqua Co., October City, Egypt). The feed ingredients of this diet were formulated to contain all the nutritional necessities for rearing the fish in accordance with NRC guidelines [33] (Table 1).

Table 1. Ration elements, nutrients, and chemical composition (% on dry matter (DM) basis) of the reference diet.

Ration Elements	% DM Basis
Fish meal (FM; 72% CP) [1]	10.0
Soybean meal (SBM; 46% CP) [2]	34.0
Corn gluten meal (CGM; 60% CP)	3.5
Rice bran	14.0
Yellow corn meal	15.0
Wheat bran	9.0
Wheat flour	13.0
Sunflower oil	0.70
Vitamin and Mineral premix [3]	0.30
Di-calcium phosphate	0.50
Total	100
Chemical composition (% on DM basis)	
Dry matter (DM)	90.19
Crude protein (CP)	30.82
Ether extract (EE)	6.93
Ash	6.59
Crude fiber (CF)	8.81
Nitrogen-free extract (NFE) [4]	46.85
Gross energy (GE; KJ/g diet DM) [5]	18.06
Protein to energy ratio (P/E ratio) [6]	17.06

Formerly published in our study [8].[1] Danish FM (contains 72.0% CP), procured from TripleNine Fish Protein, DK-6700 (Esbjerg, Denmark).[2] Egyptian soybean flour (contains 46.0% CP), procured from Cargill Trading Egypt Co. (Katameya, Cairo, Egypt).[3] AGRI-VET CO. for manufacturing Vitamins and Feed additives (10th of Ramadan City A2, Egypt). The vitamin premix mixture (contains per 1 kg): Vitamin A (17,000 IU); Vitamin D3 (2400 IU); Vitamin E (240 mg); Vitamin K3 (11 mg); Vitamin B1 (24 mg); Vitamin B2 (52 mg); Vitamin B3 (275 mg); Vitamin B6 (25 mg); Vitamin B12 (0.05 mg); Vitamin C (220 mg); Folic acid (15 mg); Calcium d–pantothenate (55 mg); Biotin (1.5 mg); Inositol (125 mg), and Choline chloride (2500 mg). The mineral mixture (contains per 1 kg) is composed of the following: Iron (74.50 mg/kg); Copper (12.5 mg/kg); Manganese (200 mg/kg); Zinc (80 mg/kg); Iodine (2 mg/kg); Selenium (0.330 mg/kg), and Cobalt (1.5 mg/kg).[4] NFE = 100 − (CP + EE + CF + Ash). [5] GE was estimated on the basis of 23.60, 39.40, and 17.20 kJ/g of CP, EE, and NFE, respectively following NRC guidelines [33].[6] P/E ratio was estimated as mg crude protein/KJ.

2.3. Formulation of AME-Based Diets

The powdered AME was fully blended with the reference diet at varying inclusion levels of 0.0, 1.5, 3.0 and 4.5 g/kg diet to make 4 experimental diets termed AME0.0 (control), AME1.5, AME3.0, and AME4.5, respectively. For 15 min, we thoroughly combined and pulverized all of the feed ingredients. A suitable amount of sunflower oil and warm distilled water (20.0 °C) were added to the feed ingredients to make a dough. The formulated feed pellets of 2.0 mm in diameter and 2.0 mm in length were produced by passing the dough through a meat mincer. The prepared experimental diets were subjected to the sun to dry, then preserved in Ziplock bags, labeled with the group names, and stored in the refrigerator (−18 °C) until used for feeding the fish groups during the feeding trial.

2.4. Fish Rearing and Experimental Design

Juvenile striped catfish with an initial body weight of (11.50 ± 0.5 g) were allocated into four triplicate groups and were cultivated into twelve rectangular glass aquaria. Each glass aquarium was sized 1.0 m × 0.90 m × 0.75 m with a 100 L water capacity. There were 30 individuals per group (10 fish per aquarium). Throughout the duration of the experiment, fish were raised in these aquariums for two months. The lighting schedule was configured to a 12 h cycle of light and dark, whereas fish reared in the day sunlight for 12 h and reared in the night for 12 h using Medium Bi-Pin Fluorescent Lamps with a Power of 36.0 Watt. Two air stones attached to the air pumps supported each aquarium to maintain enough aeration. On the previously formulated test diets, fish were hand-fed three times a day (6:00 a.m., 1:00 p.m., and 8:00 p.m.) with 3% of their wet body weight. Equal meals were given to each group during each feeding time. Every two days, the water column in each replica was exchanged, and well-aerated water was used (for around 33% of the water column). Fecal matter and remaining food particles were siphoned off during water exchange to avoid deterioration of the water quality parameters.

2.5. Water Quality Measurements

Using the HI9829 multiparameter equipment (HANNA instruments, EGY), the daily evaluation of dissolved oxygen (DO) and pH values was performed. A water thermometer measured the water's temperature (°C). A pH meter was used to check the pH levels (HANNA 8424, Hungary). A DREL portable spectrophotometer 2000 was used to measure the amounts of nitrite (NO_2; mg/L) and unionized ammonia (NH_3; mg/L) (HACH Co., Loveland, CO, USA). DO, water temperature, pH, NO_2, and NH_3 values were kept at 6.6 ± 0.4 mg/L, 29.0 ± 1 °C, 8.0 ± 0.05, 0.03 ± 0.01, and 0.02 ± 0.01 mg/L, respectively.

2.6. Determination of Growth, Feed Utilization, and Survival Rates

The final fish weight (FW) was determined by dividing the total fish weight in each aquarium by their number. The following equations were used to estimate the growth, feed utilization and fish survival:

Weight gain (WG; g) = Final weight (FW) − Initial weight (IW);

Weight gain percentage (WG%) = 100 × (FW − IW)/IW;

Specific growth rate (SGR; %/day) = [Ln FW − Ln IW] × 100/60 (experiment period in days);

Feed intake (FI; g feed/fish/day) = The total amount of diets used by fish all over the whole feeding period;

Feed conversion ratio (FCR) = FI (g)/WG (g);

Fish survival rates (SR; %) = [Fish number per group after the feeding trial/their initial number] × 100.

2.7. Proximate Composition of the Whole-Body and Amino Acid Retention

After the end of the feeding trial, three fish per replicate (9 fish per group) were frozen at −20 °C then transferred to the laboratory to analyze the whole-body proximate chemical analysis and amino acid retention. According to AOAC procedures, the proximate chemical composition of the entire fish body, including crude protein (CP), moisture (%), ether extract (EE), and ash (%), was assessed [34]. The composition analysis of the essential amino acids (EAAs) and non-essential amino acids (% of total amino acid) retention in the whole-body of the treated fish was conducted using Amino Acid Analyzer (SupNIR-2700 series) following the guidelines provided by the manufacturer.

2.8. Sampling Procedures

All fish groups fasted for one day after the feeding trial so that blood and liver samples could be taken at a uniform time. Clove oil (50 µL/L; Algomhuria company, Alexandria, Egypt) was used to induce anesthesia. The serum was separated from blood samples. Liver samples and homogenates were taken from fish in a sterile environment.

Serum Collection and Preparation of Tissue Homogenates

Nine fish were sampled for each group. Blood was sampled from their caudal vessels into sterile Eppendorf tubes without using anticoagulant. Blood was centrifuged at 3000 rpm for 10 min at 4 °C, and then the serum samples were kept at −20 °C until used. The liver samples (9 per group) were sampled aseptically and kept on ice. Liver samples were homogenized, centrifuged for 15 min at 5000 rpm at 4 °C, and then supernatants were collected in sterile test tubes and stored at −20 °C until they were used to measure liver antioxidants, while sediments were discarded.

2.9. Serum Biochemical Assays

Specific diagnostic kits (Biodiagnostic Co., Cairo, Egypt) were used to measure enzymatic tests at a wavelength of 540 nm for enzymes such as aspartate transaminase (AST), alanine transaminase (ALT), and alkaline phosphatase (ALP) according to the protocols outlined in [35,36]. The digestive enzymes were assessed in serum samples by diagnostic kits (Cusabio Biotech Co. Ltd., Wuhan, China) following the procedures provided by the supplier. Amylase, lipase, and protease enzyme activities were determined by the methods represented in [37–39]. Following the protocols outlined, blood urea nitrogen, creatinine, and uric acid levels were measured using diagnostic kits (Biodiagnostic Co., Cairo, Egypt) [40–42].

2.10. Serum Immunity Parameters

Serum lysozyme (LZ) activity was measured using a turbidimetric technique with a *Micrococcus lysodeikticus* suspension (Sigma-Aldrich, St. Louis, MI, USA) [43]. LZ activities in serum were determined using a standard curve established from LZ extracted from chicken egg white (Sigma-Aldrich, USA). The serum samples' total immunoglobulin content was measured following the manufacturer's instructions of the diagnostic kits (Cusabio Biotech Co. Ltd., China) [44,45].

2.11. Hepatic Antioxidant Biomarkers

Diagnostic kits were used to check the levels of the hepatic antioxidant enzymes as catalase (CAT), superoxide dismutase (SOD), and glutathione peroxidase (GPx) (MyBioSource Inc., San Diego, CA, USA) following the manufacturer's guidelines [46–48]. Malondialdehyde (MDA; the lipid peroxidation marker) was measured in liver homogenate using the thiobarbituric acid (TBA) technique at OD 532 nm [49,50].

2.12. Statistical Analysis

Results were analysed as a function of dietary supplementation of AME using complex regression models. Best-fitted models were applied with checking inbuilt options for the normality od residuals using the D'Agostino–Pearson Omnibus Test. Adjusted R squares < 0.2 were rejected in favour of simpler models.

3. Results

3.1. The Phyto-Components, Flavonoids, and Phenolics Present in AME Supplement

GC-MS spectra of AME showed the peaks that exhibited the main constituents of AME as identified by GC-MS analysis (Supplementary Materials). GC-MS chromatogram signifies the separated bioactive constituents of AME. The compounds have been found and cross-referenced with their counterparts in the WILEY 09 and NIST14 mass spectrum databases. The compound names, retention time (RT), Area %, molecular formula, and molecular weight are described in detail (Supplementary Materials). Table 2 shows the HPLC analysis of the concentration (μg/mL), and RT of phenolics and flavonoids detected in the AME feed supplement. An ample number of flavonoids (such as rutin, catechin, quercetin, kaempferol, luteolin, chrysoeriol, naringin, and apigenin) and phenolic compounds (such as syringic acid, caffeic acid, ferulic acid, protocatechuic acid, gallic acid, ellagic acid, p-coumaric acid, resveratrol, vanillic acid, and gentisic acid) have been found in the AME that has been used in the present experiment (Table 2).

Table 2. HPLC results of the flavonoids and phenolics present in *Astragalus membranaceus* extract (AME) used in the present study.

Phytochemicals	RT (min)	Concentration (μg/mL)
Flavonoids		
Rutin	4.6	5.23
Catechin	12.01	4.14
Quercetin	6.9	9.21
Kaempferol	8.1	4.05
Luteolin	9.0	6.09
Chrysoeriol	15.0	19.08
Naringin	5.2	4.12
Apigenin	10.0	12.45
Phenolics		
Syringic acid	3.0	8.69
Caffeic acid	4.7	3.22
Ferulic acid	6.8	5.66
Protocatechuic acid	7.8	3.31
Gallic acid	9.0	3.77
Ellagic acid	11.0	6.33
p-Coumaric acid	4.0	6.14
Resveratrol	13.8	7.01
Vanillic acid	14.6	6.79
Gentisic acid	2.0	6.55

3.2. Growth Performance, Feed Utilization, and Survival Rates

The growth performance, including WG (Figure 1A) and SGR (Figure 1B) of juvenile *P. hypophthalmus*, were increased linearly ($R^2 > 0.90$) with increasing AME inclusion levels, as determined by second-order polynomial regression analysis. A significant linear increase in FI (Figure 1C) and decrease in FCR (Figure 1D) following quadratic regression were found in fish fed with all AME inclusion levels compared to the AME0.0 diet. Interestingly, after the end of the feeding experiment, the fish survival rates (SR%) were not considerably altered between the experimental and the CONT groups. It was found that the SR% was 100% in all groups including the CONT group.

3.3. Whole-Body Proximate Analysis and Amino acid Composition

At the end of the trial, AME dietary supplementation did not influence the body composition, including moisture, CP, EE, and ash of juvenile *P. hypophthalmus*, manifested by no statistically significant variation and linear trend (Table 3). The effects of AME-based diets on the essential and non-essential amino acid retention in the whole-body of juvenile *P. hypophthalmus* are exemplified in Table 4. Results showed that none of the tested diets influenced the essential and non-essential amino acid composition, manifested by no statistically significant variations (Table 4).

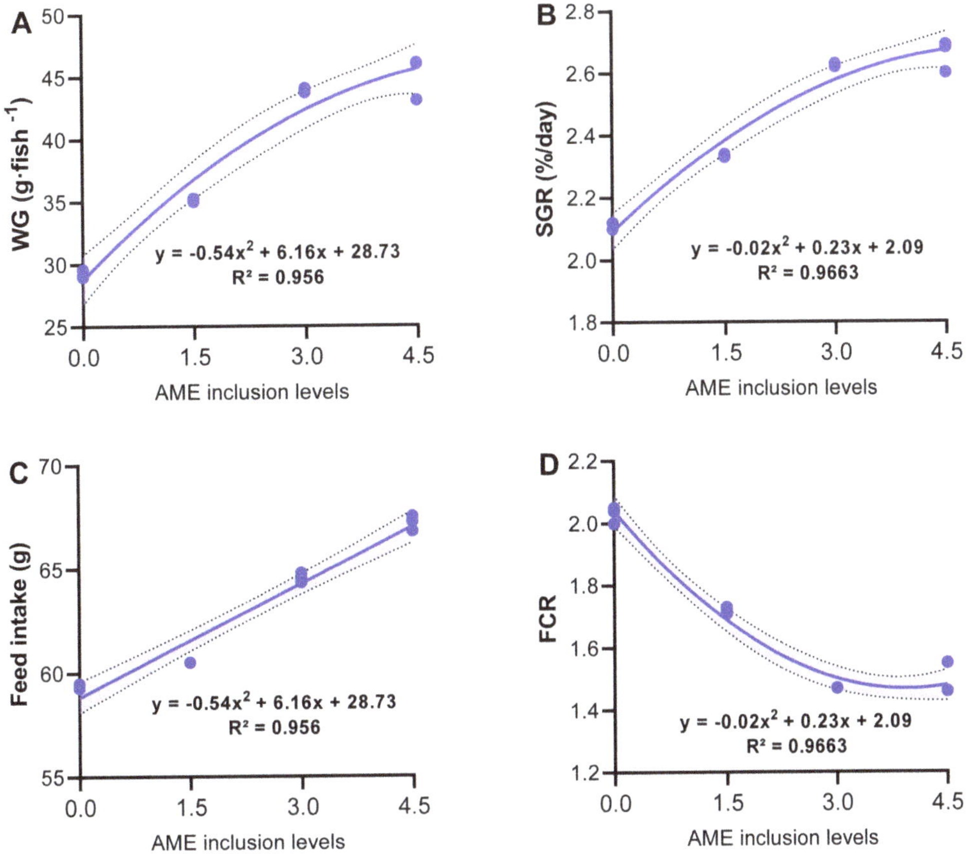

Figure 1. Growth performance including weight gain (WG) (**A**) and specific growth rate (SGR) (**B**) and feed utilization including feed intake (FI) (**C**) and feed conversion ratio (FCR) (**D**) of *P. hypophthalmus* fed diets with different AME inclusion levels. Data lines denote best fit models for the data. Equations and R square in the figures demonstrates the relationship between the measured parameters and AME inclusion levels. Individual marker indicates the number of technical replicates (n = 3).

Table 3. Whole body composition analysis of striped catfish juveniles fed diets supplemented with AME inclusion of levels after 2 months feeding trial.

Parameters	AST0.0	AST1.5	AST3.0	AST4.5
Moisture (%)	73.20 ± 0.20	73.55 ± 0.71	74.47 ± 0.75	73.69 ± 0.64
Crude protein (%)	14.24 ± 0.21	14.68 ± 0.21	14.58 ± 0.29	14.63 ± 0.52
Ether extract (%)	4.60 ± 0.17	4.62 ± 0.24	4.66 ± 0.20	4.54 ± 0.21
Ash (%)	3.10 ± 0.03	3.17 ± 0.03	3.15 ± 0.02	3.17 ± 0.03

Table 4. Effects of dietary AME levels on the amino acid composition (% of total amino acids) retention in the whole-body of striped catfish juveniles.

Amino Acid Content	Experimental Groups			
	AME0.0	AME1.5	AME3.0	AME4.5
Essential amino acids (% of total amino acids)				
Threonine	2.86 ± 0.82	2.07 ± 0.01	2.11 ± 0.06	2.16 ± 0.03
Valine	3.11 ± 0.69	2.56 ± 0.02	2.54 ± 0.07	2.65 ± 0.08
Methionine	1.00 ± 0.06	1.10 ± 0.01	1.12 ± 0.05	1.16 ± 0.08
Phenylalanine	1.67 ± 0.14	1.84 ± 0.03	1.93 ± 0.10	2.00 ± 0.19
Lysine	3.25 ± 0.11	3.33 ± 0.03	3.42 ± 0.12	3.51 ± 0.13
Leucine	3.18 ± 0.07	3.27 ± 0.02	3.35 ± 0.11	3.46 ± 0.11
Histidine	0.99 ± 0.04	1.14 ± 0.04	1.11 ± 0.04	1.12 ± 0.01
Arginine	3.41 ± 0.02	3.44 ± 0.01	3.53 ± 0.07	3.52 ± 0.01
Tryptophan	0.31 ± 0.01	0.34 ± 0.003	0.36 ± 0.04	0.41 ± 0.04
Non-essential amino acids (% of total amino acids)				
Aspartic Acid	4.55 ± 0.04	4.68 ± 0.03	4.70 ± 0.10	4.77 ± 0.04
Alanine	3.63 ± 0.61	4.25 ± 0.02	4.23 ± 0.02	4.22 ± 0.11
Isoleucine	1.63 ± 0.09	1.72 ± 0.02	1.81 ± 0.08	1.87 ± 0.15
Serine	2.20 ± 0.10	2.08 ± 0.04	2.15 ± 0.03	2.16 ± 0.07
Glutamate	6.67 ± 0.10	6.50 ± 0.10	6.62 ± 0.10	6.76 ± 0.17
Glycine	5.76 ± 0.15	5.58 ± 0.02	5.71 ± 0.04	5.52 ± 0.14
Tyrosine	1.32 ± 0.06	1.23 ± 0.03	1.31 ± 0.04	1.29 ± 0.03
Proline	2.81 ± 0.11	2.70 ± 0.01	2.79 ± 0.03	2.75 ± 0.07
Cysteine	0.40 ± 0.01	0.40 ± 0.01	0.42 ± 0.01	0.45 ± 0.02

Values are depicted as means ± standard error of means of three technical replicates.

3.4. Digestive Enzymes

The effect of different supplemental AME levels on the different serum digestive enzymes of juvenile *P. hypophthalmus*, including lipase E, α-amylase E, and protease E, is presented in Figure 2. Lipase E (Figure 2A), α-amylase E (Figure 2B), and protease E (Figure 2C) activities have been increased with a significant quadratic trend in fish fed diets supplied with different AME inclusion levels when compared with the AME0.0 group.

3.5. Serum Biochemical Variables

A panel of serum biochemistry in juvenile *P. hypophthalmus* fed diets with different AME inclusion levels is presented in Figure 3. Liver enzymes (AST, ALT, and ALP) (Figure 3A–C) were improved by the dietary AME inclusion levels, manifested by a significant quadratic decreasing trend in the levels of those enzymes in fish fed with different AME levels. Meanwhile, there was no relationship between the blood urea nitrogen (Figure 3D), uric acid (Figure 3E), and creatinine (Figure 3F) and the AME inclusion levels.

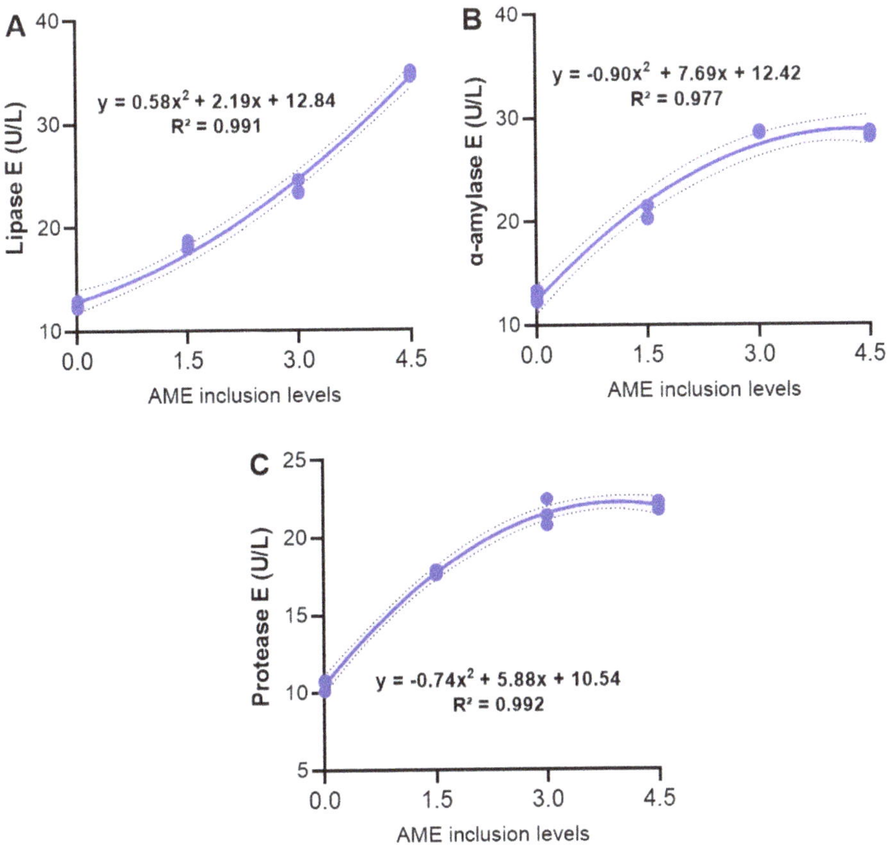

Figure 2. Serum lipase E (**A**), α-amylase E (**B**), and protease E (**C**) of juvenile *P. hypophthalmus* fed diets with different AME inclusion levels over the feeding trial period of 56 days. Data lines denote best fit models for the data. Equations and R square in the figures demonstrate the relationship between the measured parameters and AME inclusion levels. Individual marker indicates the number of technical replicates (n = 3). Individual marker is the mean of two biological replicates.

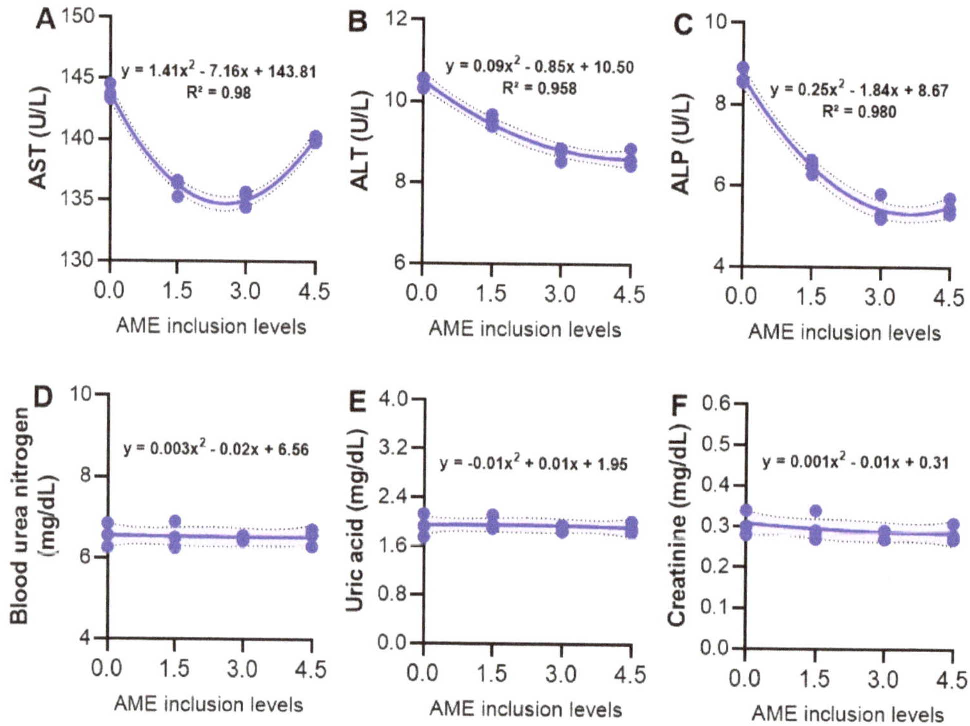

Figure 3. Serum biochemistry including AST (**A**), ALT (**B**), ALP (**C**), blood urea nitrogen (**D**), uric acid (**E**), and creatinine (**F**) of juvenile *P. hypophthalmus* fed diets with different AME inclusion levels over the feeding trial period of 56 days. Data lines denote best fit models for the data. Equations and R square in the figures demonstrate the relationship between the measured parameters and AME inclusion levels. Individual marker indicates the number of technical replicates (n = 3). Individual marker is the mean of two biological replicates.

3.6. Serum Immunity and Hepatic Antioxidant Activity

The impacts of dietary AME on serum immune response and antioxidant activity of juvenile *P. hypophthalmus* is demonstrated in Figure 4. Serum immunity, including lysozyme (Figure 4A) and total Igs (Figure 4B), has been elevated considerably with a strong quadratic trend in fish fed with different AME levels when compared with the control. Hepatic MDA levels (Figure 4C) were decreased with increasing AME levels. Hepatic CAT (Figure 4D) and SOD (Figure 4E) have been increased with a quadratic trend with the inclusion of AME in fish diets. Hepatic GPx (Figure 4F) activity showed a similar result.

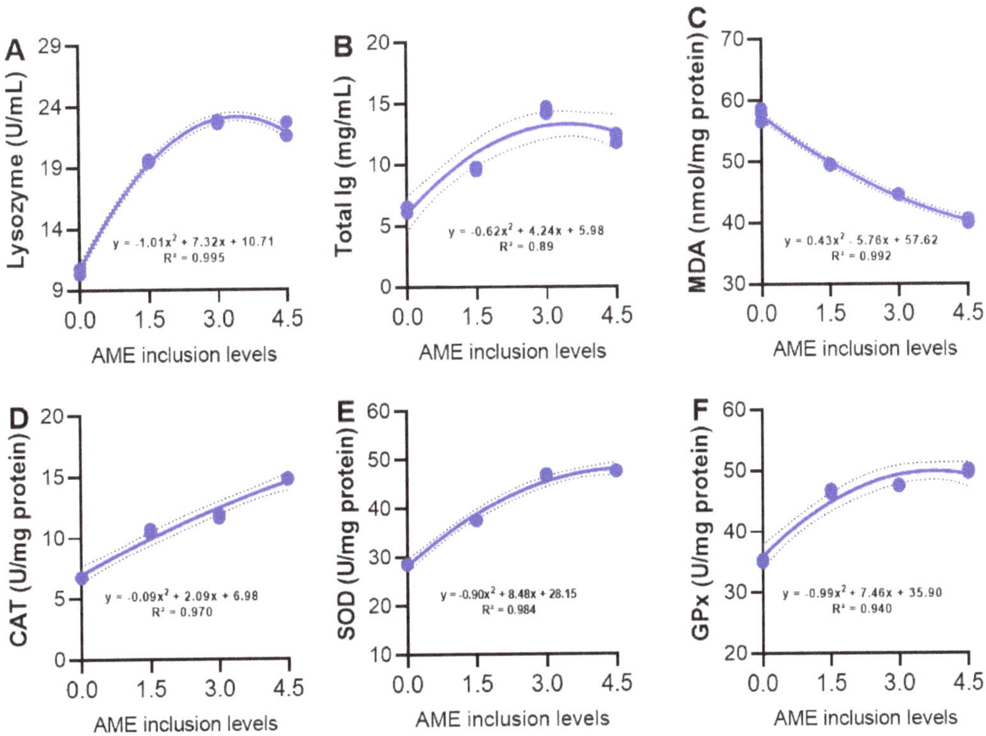

Figure 4. Serum immune responses (lysozyme and total Ig) (**A**,**B**) and hepatic antioxidant activity (**C**–**F**) of juvenile *P. hypophthalmus* fed diets with different AME inclusion levels over the feeding trial period of 56 days. Data lines denote best fit models for the data. Equations and R square in the figures demonstrate the relationship between the measured parameters and AME inclusion levels. Individual marker indicates the number of technical replicates (n = 3). Individual marker is the mean of two biological replicates.

4. Discussion

The observed enhancement in the growth performance, in terms of WG and SGR, in juvenile *P. hypophthalmus* in response to AME-based diets was the result of the improvements recorded in the FI and FCR values. The growth and feed utilization results were concomitant with digestive enzymes observations, suggesting the ability of AME to improve the functionality of diets. The growth-stimulating effects of dietary AM were also confirmed in Nile tilapia [51] and bluegill sunfish [17]. Dietary supplementation with *Astragalus* polysaccharides (ASP) also supported the growth performance of several finfish species such as *Schizothorax prenanti* [52], Nile tilapia [28,53], turbot [21], large yellow croaker [20], largemouth bass juveniles [54], crucian carp juveniles [22], and Zebrafish [23]. Differently, in a recently published paper by Sun et al. [19], AME in the diet was shown to have no discernible effect on hybrid grouper's weight increase or feed efficiency. Variations in fish species, diet, experiment design, time spent feeding, or other factors may all play a role in these disparities. The growth-promoting effects of dietary AME may be attributed to the presence of functional bioactive constituents in the AME, such as phenolic acids and flavonoids, as presented in Table 2. These important phytochemicals have already proven to increase the voluntary feeding intake and feed efficiency and improve protein retention [55]. These functional bioactive compounds could also positively enhance nutrient digestibility, which may, in turn, help improve feed utilization [56]. As reported in our current study, dietary ASP could also improve digestive enzyme activities, which subsequently helped to increase nutrient digestibility [53]. Dietary ASP has been described to be improved the gut health of fish by boosting the intestinal mucosal barrier functions [20] and the abundance of several gut-beneficial microbial communities [23,24], which were not considered in the present study, thus deserving further studies. One hundred percent survival rates across groups at the end of the feeding trial suggested that the treated fish suffered no hazardous effects from receiving AME in their diets. A possible explanation for these results involves the functional bioactive phytochemicals included in the AME, which have potent growth-promoting, hepatoprotective, antioxidant, and immunostimulant properties [55,57,58].

The whole-body composition, including moisture, CP, EE, and ash and amino acids composition, was unchanged by the test diets, suggesting that AME supplementation did not affect the assimilation of whole-body and amino acid composition. Similarly, it was found that diets supplemented with *Yucca schidigera* or *Quillaja saponaria* did not significantly affect the whole-body composition of *P. hypophthalmus* [59,60]. Moreover, dietary AME did not considerably alter CP, moisture, EE, and ash in the hybrid grouper's whole body and muscles [19]. Our findings were also in harmony with Sun et al. [21], who found that ASP-supplemented diets did not substantially change the whole-body composition of turbot. Farag et al. [28] also found that ASP-based diets did not significantly influence the whole-body proximate composition of Nile tilapia. However, our findings did not correspond to those reported by Liu et al. [20], who found that dietary supplementation with ASP at a dose rate of 0.10 or 0.15% significantly increased the whole-body CP content in large yellow croaker larvae. Several factors may be responsible for the obvious differences in the findings, as mentioned earlier, including different supplementation doses, fish species, experimental setup, feeding duration, and others. Conversely, our results also showed that none of the tested diets had influenced the essential and non-essential amino acid composition. This result is linked to non-significant changes in CP content among experimental groups. The reasons for the insignificant differences in body proximate composition and amino acid composition of groups fed AME-supplemented diets and those fed the control diet are ill-defined and require additional investigation.

It is well-documented that enhancement in the fishes' digestive enzyme activities is associated with improving the digestibility and availability of nutrients for fish [61]. A significant improvement in lipase, α-amylase, and protease enzyme production in juvenile *P. hypophthalmus* fed with different AME inclusion levels indicated that AME-based diets might have positively enhanced the nutrient digestibility and promoted nutrient absorption capacity, underpinning the growth-promoting effects of AME. This was proven by a study in which 0.1% of AST supplementation improved intestinal metabolisms by stimulating intestinal mucosal barrier function and beneficial bacteria [62]. Similarly, supplementation with *Eleutherine bulbosa*, a medicinal herb, improved the intestinal digestive enzymes in *P. hypophthalmus* [63]. Several previously published studies reported the functional ability of ASP to enhance the digestive enzymes in several other finfish species. For instance, the intestinal protease, amylase, and lipase enzyme activities are elevated in crucian carps fed ASP-based diets [22]. A similar reflection was observed in the intestinal trypsin enzymatic activity in large yellow croaker larvae [20] and intestinal protease and amylase activities in Catla when fed ASP-based diets [25]. Furthermore, dietary ASP (0.10–0.2%) increased the midgut digestive enzymes, for instance, protease and lipase activities in Asian seabass (*Lates calcarifer*) compared with those fed the reference diet [64]. However, a recent study showed no variations in α-amylase and protease enzyme activities in Nile tilapia-fed ASP-enriched diets associated with the controls [28]. These inconsistencies may be attributable to variables such as dietary supplementation dose and experiment design.

Fish liver enzymes such as AST and ALT are used as indications of liver health [65], and the increases in these enzymes in sera are considered a potential indicator of liver injury, causing leakage of these enzymes into the blood circulation from the hepatocytes [66]. Liver health was improved in *P. hypophthalmus* juveniles fed AME-based diets, manifested by a considerable decrease in liver enzymes (AST, ALT, and ALP) in fish fed diets with different AME levels compared to those provided with the reference diet. These data may suggest the hepatoprotective impacts of dietary AME, suggesting the potential application of AME as a functional additive to prevent negative impact on the liver caused by intrinsic or extrinsic factors. This could be further strengthened by the study of Jia et al. [67], who observed potential hepatoprotective effects of ASP against CCl_4-induced hepatic injury in common carp by inhibiting the elevation of AST, ALT, and lactate dehydrogenase (LDH) enzymes in the hepatocytes. Another study found a similar dietary effect of ASP on the serum ALT and AST enzyme levels in crucian carp juveniles [22]. The functional flavonoids in AME might have potential effects on the improvement in liver functions of *P. hypophthalmus* juveniles. These flavonoids conferred effective hepatoprotective functions [68]. Moreover, an earlier study also reported that *A. membranaceus* root had potent hepato-protective effects and protected hepatic cells from pathological injuries [69]. Several other reports have demonstrated the potential hepatoprotective effects of polysaccharides in *A. membranaceus* [13,67,70]. Another theory showed that AM has potent antioxidant constituents such as astragalosides, flavonoids, and polysaccharides, effectively preventing tissue injury via their antioxidant mechanisms [71]. Jin et al. have reviewed that ASP could increase the enzymatic antioxidant activities, which helps to limit and eliminate oxidative stress caused by free radicals and oxygen radicals [13].

Similar to liver enzymes, blood urea nitrogen, an indicator of the occurrence of renal damage and gill dysfunction [72], was improved by dietary AME, indicating the optimistic impacts of dietary AST supplementation on the kidney functions of the treated fish with no renal injuries. As far as we know, there were no previous reports on the effects of AME on the fish's kidney functions. *A. membranaceus* root has been described as used in treating kidney diseases in Chinese medicine [73]. It also has renal protective effects against nephropathy via the modulation of kidney function biomarkers in the blood [74]. In animal models, the positive renal protective effect of *A. membranaceus* root has been closely associated with the presence of astragalosides (astragalus saponins); they are most well-known for their ability to protect renal tubules from damage caused by free radicals [75]. However, our previously published study showed that kidney function markers such as creatinine, uric acid, and blood urea nitrogen were not significantly altered in common carp fingerlings fed diets supplemented with *Origanum vulgare* essential oil [65]. Apart from the beneficial effect of AME on enzymes associated with liver and kidney function, further research should consider the histological microstructure of the liver and kidney in relation to stress-relevant gene expression to better understand the actual modes of action of phytochemicals present in AME.

Lysozyme, a mucolytic enzyme excreted by leukocytes, can activate leukocytes and macrophages to lyse the bacterial cell walls [76,77]. Total Igs are a major part of the fish's humoral immunity playing a significant role in the fish's immune system defense and are deemed a biomarker of the fish's adaptive immune responses [44]. Improved serum immunity, including lysozyme activity and total Igs in *P. hypophthalmus* groups fed diets with different AME levels compared with the CONT, suggested the immunostimulatory roles of dietary AME. It is well documented that plant herbal extracts can stimulate the immune response of *P. hypophthalmus*. For instance, it was found that dietary supplementation with *Euphorbia hirta* extract for one month considerably increased serum lysozyme and total Igs content in *P. hypophthalmus* [9]. Correspondingly, supplementing diets with *Psidium guajava* and *Phyllanthus amarus* extracts also improved the serum lysozyme and total Igs in *P. hypophthalmus* [10]. The immunomodulatory effects of dietary AME or ASP have been described in several studies in other finfish species. Ardó et al. [15] found that dietary *A. membranaceus*-based diets significantly increased serum lysozyme and total Igs in Nile tilapia. Serum lysozyme pursuits were also increased in Nile tilapia-fed ASP-enriched diets [53]. Furthermore, ASP liposome-based diets significantly increased nitric oxide production and boosted the phagocytic activities of head kidney macrophages as well as serum lysozyme activity in large yellow croaker [78]. It was reported that polysaccharides, saponins, and water decoction of AME significantly increased phagocytic activity and serum lysozyme activities of spotted maigre (*Nibea albiflora*) [79]. It was recently reported that dietary ASP significantly increased serum lysozyme and total Igs in Nile tilapia [27]. The immunostimulatory functions of AME might be associated with the functional phytochemicals such as polyphenols, flavonoids, and phenolic acids, as mentioned above [55,80]. In addition, astragalosides present in *A. membranaceus* possess immune-boosting, anti-inflammatory, and immune-regulatory effects [81]. Further challenge trials considering the immune response at a different post-challenge time should be conducted to understand the immunostimulatory effects of AME.

The enzymatic endogenous antioxidant enzymes such as CAT, SOD, and GPx can safeguard the host against oxidative stress [82,83]. MDA is a biomarker of the lipid peroxidation process [84]. An elevation in the hepatic production of CAT, GPx, and SOD enzymes coupled with the lower production of MDA in *P. hypophthalmus* fed AME-based diets with regard to those reared in the AME0.0 group suggested the antioxidant effects of dietary AME. A study by Wu et al. [51] observed similar responses in SOD, GPx, CAT, and MDA enzyme activities to low-temperature stress in the liver of Nile tilapia when fed AME powder. In a similar pattern, dietary AME increased the enzymatic antioxidant capacity (via increased SOD, GPx, and CAT enzyme activities) of bluegill sunfish exposed to cold-water stress [17]. *Astragalus* polysaccharides also modulated the antioxidant activity of several finfish species, such as Nile tilapia [53], large yellow croaker [20], and turbot [21]. Furthermore, it was reported that ASP significantly modulated CCl4-induced oxidative stress through increasing SOD enzyme activity and total antioxidant capacity (T-AOC) and decreasing MDA concentrations in the liver of common carp [67]. The antioxidant effects of dietary AME may be associated with ASP, which has potential efficacy in increasing antioxidant enzyme activities and counteracting the negative impacts of free radicals and reactive oxygen species [85]. Notably, the HPLC analysis of AME used in the present study found many flavonoids and phenolics, compounds with strong antioxidant properties [55,86–88]. We hypothesize that these phytochemicals can positively enhance the antioxidative capacity of juvenile *P. hypophthalmus*.

5. Conclusions and Prospects

In summary, feeding AME-supplemented diets for two months improved the growth performance of juvenile *P. hypophthalmus*. This was supported by improvement in feed utilization with concurrent elevated levels of digestive enzymes. In addition, inclusion of AME in feed improved immunity and antioxidant activity without impacting liver and kidney functions. The overall improvement in fish health might be attributed to the phytochemicals (flavonoids and phenolic acids) found in the AME used in the present study. In addition, dietary supplementation of 1.0–4.5 g AME/kg diet could be regarded as a promising phyto-additive to promote the health condition and welfare of farmed *P. hypophthalmus* in addition to potentially being used as a functional additive to maintain hepatorenal functions. It may also prevent liver and kidney damage caused by intrinsic and extrinsic factors. Nevertheless, more in-depth experiments and investigations are still necessary to decipher the existent roles of the phytochemicals in AME in influencing the gut health, intestinal histomorphology, and gene expression analysis of the examined fish species.

Supplementary Materials: The following Supplementary Materials are available for download at: https://www.mdpi.com/article/10.3390/fishes7060319/s1.

Author Contributions: Conceptualization, R.M.S., H.M.R.A.-L. and M.S.; methodology, M.S. and H.A.A.; software, M.A.K. and R.M.S.; validation, H.A.A., M.S. and M.R.C.; formal analysis, M.R.C.; investigation, H.M.R.A.-L. and M.S.; resources, M.A.K. and M.S.; data curation, M.R.C.; writing—original draft preparation, H.M.R.A.-L. and M.R.C.; writing—review and editing, H.M.R.A.-L. and M.R.C.; visualization, H.M.R.A.-L.; supervision, R.M.S. and H.A.A.; project administration, H.M.R.A.-L. and R.M.S.; funding acquisition, M.A.K. and R.M.S. Each author has contributed equally to this work. All authors have read and agreed to the published version of the manuscript.

Funding: This research received no external funding.

Institutional Review Board Statement: The experimental methods, procedure, and fish rearing were approved by the Faculty of Veterinary Medicine, Kafrelsheikh University, Egypt. This study was conducted according to the guidelines of the Local Experimental Animal Care Committee, Faculty of Veterinary Medicine, Kafrelsheikh University, Egypt.

Informed Consent Statement: Not applicable.

Data Availability Statement: Data are available upon request.

Acknowledgments: All authors thank Basem Elkhayat, The Fish and Shrimp Consultant at Fish hatchery, Borg El Arab, Alexandria, Egypt, for his aid and support throughout fish rearing and experimental work.

Conflicts of Interest: The authors declare no conflict of interest.

References

1. Ali, H.; Haque, M.M.; Belton, B. Striped catfish (*Pangasianodon hypophthalmus*, Sauvage, 1878) aquaculture in Bangladesh: An overview. *Aquac. Res.* **2013**, *44*, 950–965. [CrossRef]
2. Phan, L.T.; Bui, T.M.; Nguyen, T.T.T.; Gooley, G.J.; Ingram, B.A.; Nguyen, H.V.; Nguyen, P.T.; De Silva, S.S. Current status of farming practices of striped catfish, *Pangasianodon hypophthalmus* in the Mekong Delta, Vietnam. *Aquaculture* **2009**, *296*, 227–236. [CrossRef]
3. De Silva, S.S.; Phuong, N.T. Striped catfish farming in the Mekong Delta, Vietnam: A tumultuous path to a global success. *Rev. Aquac.* **2011**, *3*, 45–73. [CrossRef]
4. Abdel-Latif, H.M.R.; Chaklader, M.R.; Shukry, M.; Ahmed, H.A.; Khallaf, M.A. A multispecies probiotic modulates growth, digestive enzymes, immunity, hepatic antioxidant activity, and disease resistance of *Pangasianodon hypophthalmus* fingerlings. *Aquaculture* **2023**, *563*, 738948. [CrossRef]
5. Dong, H.T.; Nguyen, V.V.; Phiwsaiya, K.; Gangnonngiw, W.; Withyachumnarnkul, B.; Rodkhum, C.; Senapin, S. Concurrent infections of *Flavobacterium columnare* and *Edwardsiella ictaluri* in striped catfish, *Pangasianodon hypophthalmus* in Thailand. *Aquaculture* **2015**, *448*, 142–150. [CrossRef]
6. Abdel-Latif, H.M.R.; Abdel-Daim, M.M.; Shukry, M.; Nowosad, J.; Kucharczyk, D. Benefits and applications of *Moringa oleifera* as a plant protein source in Aquafeed: A review. *Aquaculture* **2022**, *547*, 737369. [CrossRef]
7. Abdel-Latif, H.M.R.; Hendam, B.M.; Nofal, M.I.; El-Son, M.A.M. *Ginkgo biloba* leaf extract improves growth, intestinal histomorphometry, immunity, antioxidant status and modulates transcription of cytokine genes in hapa-reared *Oreochromis niloticus*. *Fish Shellfish Immunol.* **2021**, *117*, 339–349. [CrossRef]
8. Abdel-Latif, H.M.R.; Shukry, M.; Noreldin, A.E.; Ahmed, H.A.; El-Bahrawy, A.; Ghetas, H.A.; Khalifa, E. Milk thistle (*Silybum marianum*) extract improves growth, immunity, serum biochemical indices, antioxidant state, hepatic histoarchitecture, and intestinal histomorphometry of striped catfish, *Pangasianodon hypophthalmus*. *Aquaculture* **2023**, *562*, 738761. [CrossRef]
9. Nhu, T.Q.; Bich Hang, B.T.; Bach, L.T.; Buu Hue, B.T.; Quetin-Leclercq, J.; Scippo, M.-L.; Phuong, N.T.; Kestemont, P. Plant extract-based diets differently modulate immune responses and resistance to bacterial infection in striped catfish (*Pangasianodon hypophthalmus*). *Fish Shellfish Immunol.* **2019**, *92*, 913–924. [CrossRef]
10. Nhu, T.Q.; Bich Hang, B.T.; Cornet, V.; Oger, M.; Bach, L.T.; Anh Dao, N.L.; Thanh Huong, D.T.; Quetin-Leclercq, J.; Scippo, M.-L.; Phuong, N.T.; et al. Single or Combined Dietary Supply of *Psidium guajava* and *Phyllanthus amarus* Extracts Differentially Modulate Immune Responses and Liver Proteome in Striped Catfish (*Pangasianodon hyphophthalmus*). *Front. Immunol.* **2020**, *11*, 797. [CrossRef]
11. Maiti, S.; Saha, S.; Jana, P.; Chowdhury, A.; Khatua, S.; Ghosh, T.K. Effect of dietary *Andrographis paniculata* leaf extract on growth, immunity, and disease resistance against *Aeromonas hydrophila* in *Pangasianodon hypopthalmus*. *J. Appl. Aquac.* **2021**, 1–25. [CrossRef]
12. Fu, J.; Wang, Z.; Huang, L.; Zheng, S.; Wang, D.; Chen, S.; Zhang, H.; Yang, S. Review of the Botanical Characteristics, Phytochemistry, and Pharmacology of *Astragalus membranaceus* (Huangqi). *Phytother. Res.* **2014**, *28*, 1275–1283. [CrossRef] [PubMed]
13. Jin, M.; Zhao, K.; Huang, Q.; Shang, P. Structural features and biological activities of the polysaccharides from *Astragalus membranaceus*. *Int. J. Biol. Macromol.* **2014**, *64*, 257–266. [CrossRef] [PubMed]
14. Zheng, Y.; Ren, W.; Zhang, L.; Zhang, Y.; Liu, D.; Liu, Y. A review of the pharmacological action of *Astragalus* polysaccharide. *Front. Pharmacol.* **2020**, *11*, 349. [CrossRef]
15. Ardó, L.; Yin, G.; Xu, P.; Váradi, L.; Szigeti, G.; Jeney, Z.; Jeney, G. Chinese herbs (*Astragalus membranaceus* and *Lonicera japonica*) and boron enhance the non-specific immune response of Nile tilapia (*Oreochromis niloticus*) and resistance against *Aeromonas hydrophila*. *Aquaculture* **2008**, *275*, 26–33. [CrossRef]
16. Elabd, H.; Wang, H.-P.; Shaheen, A.; Yao, H.; Abbass, A. Feeding *Glycyrrhiza glabra* (liquorice) and *Astragalus membranaceus* (AM) alters innate immune and physiological responses in yellow perch (*Perca flavescens*). *Fish Shellfish Immunol.* **2016**, *54*, 374–384. [CrossRef]

17. Elabd, H.; Wang, H.-P.; Shaheen, A.; Yao, H.; Abbass, A. *Astragalus membranaceus* (AM) enhances growth performance and antioxidant stress profiles in bluegill sunfish (*Lepomis macrochirus*). *Fish Physiol. Biochem.* **2016**, *42*, 955–966. [CrossRef]
18. Angela, C.; Wang, W.; Lyu, H.; Zhou, Y.; Huang, X. The effect of dietary supplementation of *Astragalus membranaceus* and *Bupleurum chinense* on the growth performance, immune-related enzyme activities and genes expression in white shrimp, *Litopenaeus vannamei*. *Fish Shellfish Immunol.* **2020**, *107*, 379–384. [CrossRef]
19. Sun, Z.; Wei, Z.; Liu, Q.; Mai, H.; Liu, Y.; Liu, B.; Tan, X.; Ye, C. Effects of dietary *Astragalus membranaceus* (Fisch.) Bge. root extract on growth performance, plasma biochemical parameters, fish composition, liver and intestinal morphology, and genes expression in head kidney of hybrid grouper (*Epinephelus lanceolatus*♂ × *Epinephelus fuscoguttatus*♀). *Aquac. Rep.* **2022**, *22*, 100934. [CrossRef]
20. Liu, Y.; Miao, Y.; Xu, N.; Ding, T.; Cui, K.; Chen, Q.; Zhang, J.; Fang, W.; Mai, K.; Ai, Q. Effects of dietary *Astragalus* polysaccharides (APS) on survival, growth performance, activities of digestive enzyme, antioxidant responses and intestinal development of large yellow croaker (*Larimichthys crocea*) larvae. *Aquaculture* **2020**, *517*, 734752. [CrossRef]
21. Sun, Y.; Wang, X.; Zhou, H.; Mai, K.; He, G. Dietary *Astragalus* polysaccharides ameliorates the growth performance, antioxidant capacity and immune responses in turbot (*Scophthalmus maximus* L.). *Fish Shellfish Immunol.* **2020**, *99*, 603–608. [CrossRef] [PubMed]
22. Wu, S. Dietary *Astragalus membranaceus* polysaccharide ameliorates the growth performance and innate immunity of juvenile crucian carp (*Carassius auratus*). *Int. J. Biol. Macromol.* **2020**, *149*, 877–881. [CrossRef] [PubMed]
23. Li, Y.; Ran, C.; Wei, K.; Xie, Y.; Xie, M.; Zhou, W.; Yang, Y.; Zhang, Z.; Lv, H.; Ma, X.; et al. The effect of *Astragalus* polysaccharide on growth, gut and liver health, and anti-viral immunity of zebrafish. *Aquaculture* **2021**, *540*, 736677. [CrossRef]
24. Shi, F.; Lu, Z.; Yang, M.; Li, F.; Zhan, F.; Zhao, L.; Li, Y.; Li, Q.; Li, J.; Li, J.; et al. *Astragalus* polysaccharides mediate the immune response and intestinal microbiota in grass carp (*Ctenopharyngodon idellus*). *Aquaculture* **2021**, *534*, 736205. [CrossRef]
25. Harikrishnan, R.; Devi, G.; Doan, H.V.; Tapingkae, W.; Balasundaram, C.; Arockiaraj, J.; Ringø, E. Changes in immune genes expression, immune response, digestive enzymes antioxidant status, and growth of catla (*Catla catla*) fed with *Astragalus* polysaccharides against edwardsiellosis disease. *Fish Shellfish Immunol.* **2022**, *121*, 418–436. [CrossRef]
26. Pu, Y.; Wu, S. The growth performance, body composition and nonspecific immunity of white shrimps (*Litopenaeus vannamei*) affected by dietary *Astragalus membranaceus* polysaccharide. *Int. J. Biol. Macromol.* **2022**, *209*, 162–165. [CrossRef]
27. Farag, M.R.; Alagawany, M.; Khalil, S.R.; El-Hady, E.W.; Elhady, W.M.; Ismail, T.A.; Marini, C.; Di Cerbo, A.; Abdel-Latif, H.M.R. Immunosuppressive Effects of Thallium Toxicity in Nile Tilapia Fingerlings: Elucidating the Rescue Role of *Astragalus membranaceus* Polysaccharides. *Front. Vet. Sci.* **2022**, *9*, 843031. [CrossRef]
28. Farag, M.R.; Alagawany, M.; Khalil, S.R.; Moustafa, A.A.; Mahmoud, H.K.; Abdel-Latif, H.M.R. *Astragalus membranaceus* polysaccharides modulate growth, hemato-biochemical indices, hepatic antioxidants, and expression of HSP70 and apoptosis-related genes in *Oreochromis niloticus* exposed to sub-lethal thallium toxicity. *Fish Shellfish Immunol.* **2021**, *118*, 251–260. [CrossRef]
29. Abdel-Latif, H.M.R.; Dawood, M.A.O.; Menanteau-Ledouble, S.; El-Matbouli, M. The nature and consequences of co-infections in tilapia: A review. *J. Fish Dis.* **2020**, *43*, 651–664. [CrossRef]
30. Gomathi, D.; Kalaiselvi, M.; Ravikumar, G.; Devaki, K.; Uma, C. GC-MS analysis of bioactive compounds from the whole plant ethanolic extract of *Evolvulus alsinoides* (L.) L. *J. Food Sci. Technol.* **2015**, *52*, 1212–1217. [CrossRef]
31. Mattila, P.; Astola, J.; Kumpulainen, J. Determination of Flavonoids in Plant Material by HPLC with Diode-Array and Electro-Array Detections. *J. Agric. Food Chem.* **2000**, *48*, 5834–5841. [CrossRef] [PubMed]
32. Öztürk, N.; Tunçel, M.; Tunçel, N.B. Determination of Phenolic Acids by a Modified HPLC: Its Application to Various Plant Materials. *J. Liq. Chromatogr. Relat. Technol.* **2007**, *30*, 587–596. [CrossRef]
33. NRC. *Nutrient Requirements of Fish and Shrimp*; National Academy Press: Washington, DC, USA, 2011; 392p.
34. AOAC. *Official Methods of Analysis*, 13th ed.; Association of Analytical Chemists: Washington, DC, USA, 2012; 1018p.
35. Belfield, A.; Goldberg, D. Colorimetric determination of alkaline phosphatase activity. *Enzyme* **1971**, *12*, 561–568. [CrossRef]
36. Reitman, S.; Frankel, S. A colorimetric method for the determination of serum glutamic oxalacetic and glutamic pyruvic transaminases. *Am. J. Clin. Pathol.* **1957**, *28*, 56–63. [CrossRef]
37. Bernfeld, P. Amylases, α and β. In *Methods in Enzymology*; Academic Press: Cambridge, MA, USA, 1955; Volume 1, pp. 149–158.
38. Khantaphant, S.; Benjakul, S. Comparative study on the proteases from fish pyloric caeca and the use for production of gelatin hydrolysate with antioxidative activity. *Comp. Biochem. Physiol. Part B Biochem. Mol. Biol.* **2008**, *151*, 410–419. [CrossRef]
39. Shihabi, Z.K.; Bishop, C. Simplified Turbidimetric Assay for Lipase Activity. *Clin. Chem.* **1971**, *17*, 1150–1153. [CrossRef]
40. Barham, D.; Trinder, P. Enzymatic determination of uric acid. *Analyst* **1972**, *97*, 142–145. [CrossRef]
41. Coulombe, J.J.; Favreau, L. A new simple semimicro method for colorimetric determination of urea. *Clin. Chem.* **1963**, *9*, 102–108. [CrossRef]
42. Heinegård, D.; Tiderström, G. Determination of serum creatinine by a direct colorimetric method. *Clin. Chim. Acta* **1973**, *43*, 305–310. [CrossRef]
43. Ellis, A. Lysozyme assays. *Tech. Fish Immunol.* **1990**, *1*, 101–103.
44. Cuesta, A.; Meseguer, J.; Esteban, M.A. Total serum immunoglobulin M levels are affected by immunomodulators in seabream (*Sparus aurata* L.). *Vet. Immunol. Immunopathol.* **2004**, *101*, 203–210. [CrossRef] [PubMed]
45. Siwicki, A.; Anderson, D. An easy spectrophotometric assay for determining total protein and immunoglobulin levels in fish sera: Correlation to fish health. *Tech. Fish Immunol.* **1993**, *3*, 23–30.
46. Aebi, H. Catalase in vitro. In *Methods in Enzymology*; Academic Press: Cambridge, MA, USA, 1984; Volume 105, pp. 121–126.

47. McCord, J.M.; Fridovich, I. Superoxide Dismutase: An enzymic function for erythrocuprein (hemocuprein). *J. Biol. Chem.* **1969**, *244*, 6049–6055. [CrossRef]
48. Paglia, D.E.; Valentine, W.N. Studies on the quantitative and qualitative characterization of erythrocyte glutathione peroxidase. *J. Lab. Clin. Med.* **1967**, *70*, 158–169. [PubMed]
49. Ohkawa, H.; Ohishi, N.; Yagi, K. Assay for lipid peroxides in animal tissues by thiobarbituric acid reaction. *Anal. Biochem.* **1979**, *95*, 351–358. [CrossRef]
50. Uchiyama, M.; Mihara, M. Determination of malonaldehyde precursor in tissues by thiobarbituric acid test. *Anal. Biochem.* **1978**, *86*, 271–278. [CrossRef]
51. Wu, F.; Yang, C.; Wen, H.; Zhang, C.; Jiang, M.; Liu, W.; Tian, J.; Yu, L.; Lu, X. Improving low-temperature stress tolerance of tilapia, *Oreochromis niloticus*: A functional analysis of *Astragalus membranaceus*. *J. World Aquac. Soc.* **2019**, *50*, 749–762. [CrossRef]
52. Xiang, X.; Chen, J.; Zhou, X.; Wang, W. Effect of astragalus polysaccharides on growth, body composition and immune index in *Schizothorax prenanti*. *Acta Hydrobiol. Sin.* **2011**, *35*, 291–299. [CrossRef]
53. Zahran, E.; Risha, E.; AbdelHamid, F.; Mahgoub, H.A.; Ibrahim, T. Effects of dietary *Astragalus* polysaccharides (APS) on growth performance, immunological parameters, digestive enzymes, and intestinal morphology of Nile tilapia (*Oreochromis niloticus*). *Fish Shellfish Immunol.* **2014**, *38*, 149–157. [CrossRef]
54. Lin, S.-M.; Jiang, Y.; Chen, Y.-J.; Luo, L.; Doolgindachbaporn, S.; Yuangsoi, B. Effects of *Astragalus* polysaccharides (APS) and chitooligosaccharides (COS) on growth, immune response and disease resistance of juvenile largemouth bass, *Micropterus salmoides*. *Fish Shellfish Immunol.* **2017**, *70*, 40–47. [CrossRef]
55. Ahmadifar, E.; Yousefi, M.; Karimi, M.; Fadaei Raieni, R.; Dadar, M.; Yilmaz, S.; Dawood, M.A.O.; Abdel-Latif, H.M.R. Benefits of Dietary Polyphenols and Polyphenol-Rich Additives to Aquatic Animal Health: An Overview. *Rev. Fish. Sci. Aquac.* **2021**, *29*, 478–511. [CrossRef]
56. Citarasu, T. Herbal biomedicines: A new opportunity for aquaculture industry. *Aquac. Int.* **2010**, *18*, 403–414. [CrossRef]
57. Bratkov, V.M.; Shkondrov, A.M.; Zdraveva, P.K.; Krasteva, I.N. Flavonoids from the Genus Astragalus: Phytochemistry and Biological Activity. *Pharm. Rev.* **2016**, *10*, 11–32. [CrossRef]
58. Matkowski, A.; Woźniak, D.; Lamer-Zarawska, E.; Oszmiański, J.; Leszczyńska, A. Flavonoids and Phenol Carboxylic Acids in the Oriental Medicinal Plant Astragalus membranaceus Acclimated in Poland. *Z. Für Nat. C* **2003**, *58*, 602–604. [CrossRef]
59. Güroy, B.; Mantoğlu, S.; Kayalı, S.; Şahin, İ. Effect of dietary *Yucca schidigera* extract on growth, total ammonia–nitrogen excretion and haematological parameters of juvenile striped catfish *Pangasianodon hypophthalmus*. *Aquac. Res.* **2014**, *45*, 647–654. [CrossRef]
60. Güroy, B.; Mantoğlu, S.; Merrifield, D.L.; Guroy, D. Effects of dietary Nutrafito Plus on growth, haemotological parameters and total ammonia-nitrogen excretion of juvenile striped catfish *Pangasianodon hypophthalmus*. *Aquac. Res.* **2016**, *47*, 1770–1777. [CrossRef]
61. Hoseinifar, S.H.; Dadar, M.; Ringø, E. Modulation of nutrient digestibility and digestive enzyme activities in aquatic animals: The functional feed additives scenario. *Aquac. Res.* **2017**, *48*, 3987–4000. [CrossRef]
62. Shi, H.-T.; Zhao, S.-Z.; Wang, K.-L.; Fan, M.-X.; Han, Y.-Q.; Wang, H.-L. Effects of dietary *Astragalus membranaceus* supplementation on growth performance, and intestinal morphology, microbiota and metabolism in common carp (*Cyprinus carpio*). *Aquac. Rep.* **2022**, *22*, 100955. [CrossRef]
63. Rudy, A.N.; Meylianawati, O.F.A.; Yanti, P.S.; Esti, H.H. The effects of dietary *Eleutherine bulbosa* on the growth, leukocyte profile, and digestive enzyme activity of the striped catfish *Pangasianodon hypophthalmus*. *Nusant. Biosci.* **2018**, *10*, 46–51.
64. Yu, W.; Yang, Y.; Zhou, Q.; Huang, X.; Huang, Z.; Li, T.; Wu, Q.; Zhou, C.; Ma, Z.; Lin, H. Effects of dietary *Astragalus* polysaccharides on growth, health and resistance to *Vibrio harveyi* of *Lates calcarifer*. *Int. J. Biol. Macromol.* **2022**, *207*, 850–858. [CrossRef]
65. Abdel-Latif, H.M.R.; Abdel-Tawwab, M.; Khafaga, A.F.; Dawood, M.A.O. Dietary oregano essential oil improved the growth performance via enhancing the intestinal morphometry and hepato-renal functions of common carp (*Cyprinus carpio* L.) fingerlings. *Aquaculture* **2020**, *526*, 735432. [CrossRef]
66. Bruslé, J.; Anadon, G.G. The structure and function of fish liver. *Fish Morphol.* **1996**, *76*, 545–551.
67. Jia, R.; Cao, L.; Xu, P.; Jeney, G.; Yin, G. In vitro and in vivo hepatoprotective and antioxidant effects of *Astragalus* polysaccharides against carbon tetrachloride-induced hepatocyte damage in common carp (*Cyprinus carpio*). *Fish Physiol. Biochem.* **2012**, *38*, 871–881. [CrossRef] [PubMed]
68. Davila, J.C.; Lenherr, A.; Acosta, D. Protective effect of flavonoids on drug-induced hepatotoxicity in vitro. *Toxicology* **1989**, *57*, 267–286. [CrossRef]
69. Zhen-Lun, Z.; Qi-Zhen, W.; Chang-Xiao, L. Hepatoprotective effects of *Astraglus* root. *J. Ethnopharmacol.* **1990**, *30*, 145–149. [CrossRef]
70. Hamid, M.; Liu, D.; Abdulrahim, Y.; Liu, Y.; Qian, G.; Khan, A.; Gan, F.; Huang, K. Amelioration of CCl4-induced liver injury in rats by selenizing *Astragalus* polysaccharides: Role of proinflammatory cytokines, oxidative stress and hepatic stellate cells. *Res. Vet. Sci.* **2017**, *114*, 202–211. [CrossRef]
71. Shahzad, M.; Shabbir, A.; Wojcikowski, K.; Wohlmuth, H.; C Gobe, G. The antioxidant effects of Radix Astragali (Astragalus membranaceus and related species) in protecting tissues from injury and disease. *Curr. Drug Targets* **2016**, *17*, 1331–1340. [CrossRef]

72. Nelson, K.; Jones, J.; Jacobson, S.; Reimschuessel, R. Elevated Blood Urea Nitrogen (BUN) Levels in Goldfish as an Indicator of Gill Dysfunction. *J. Aquat. Anim. Health* **1999**, *11*, 52–60. [CrossRef]
73. Zhang, H.W.; Lin, Z.X.; Xu, C.; Leung, C.; Chan, L.S. *Astragalus* (a traditional Chinese medicine) for treating chronic kidney disease. *Cochrane Database Syst. Rev.* **2014**, CD008369. [CrossRef]
74. Li, M.; Wang, W.; Xue, J.; Gu, Y.; Lin, S. Meta-analysis of the clinical value of *Astragalus membranaceus* in diabetic nephropathy. *J. Ethnopharmacol.* **2011**, *133*, 412–419. [CrossRef]
75. Zhang, J.; Xie, X.; Li, C.; Fu, P. Systematic review of the renal protective effect of *Astragalus membranaceus* (root) on diabetic nephropathy in animal models. *J. Ethnopharmacol.* **2009**, *126*, 189–196. [CrossRef] [PubMed]
76. Alexander, J.B.; Ingram, G.A. Noncellular nonspecific defence mechanisms of fish. *Annu. Rev. Fish Dis.* **1992**, *2*, 249–279. [CrossRef]
77. Saurabh, S.; Sahoo, P.K. Lysozyme: An important defence molecule of fish innate immune system. *Aquac. Res.* **2008**, *39*, 223–239. [CrossRef]
78. Zhang, W.; Zhang, M.; Cheng, A.; Hao, E.; Huang, X.; Chen, X. Immunomodulatory and antioxidant effects of *Astragalus* polysaccharide liposome in large yellow croaker (*Larimichthys crocea*). *Fish Shellfish Immunol.* **2020**, *100*, 126–136. [CrossRef]
79. Wang, X.; Ding, L.; Yan, M.; Chai, X.; Lu, R.; Wang, Q.; Li, F. Polysaccharides, saponins, and water decoction of *Astragalus membranaceus* significantly enhance the non-specific immune response of spotted maigre (*Nibea albiflora*). *Isr. J. Aquac.—Bamidgeh* **2012**, *64*, 1–6. [CrossRef]
80. Chakraborty, S.B.; Hancz, C. Application of phytochemicals as immunostimulant, antipathogenic and antistress agents in finfish culture. *Rev. Aquac.* **2011**, *3*, 103–119. [CrossRef]
81. Qi, Y.; Gao, F.; Hou, L.; Wan, C. Anti-inflammatory and immunostimulatory activities of astragalosides. *Am. J. Chin. Med.* **2017**, *45*, 1157–1167. [CrossRef]
82. Chowdhury, S.; Saikia, S.K. Oxidative Stress in Fish: A Review. *J. Sci. Res.* **2020**, *12*, 145–160. [CrossRef]
83. Hoseinifar, S.H.; Yousefi, S.; Van Doan, H.; Ashouri, G.; Gioacchini, G.; Maradonna, F.; Carnevali, O. Oxidative Stress and Antioxidant Defense in Fish: The Implications of Probiotic, Prebiotic, and Synbiotics. *Rev. Fish. Sci. Aquac.* **2021**, *29*, 198–217. [CrossRef]
84. Martínez-Álvarez, R.M.; Morales, A.E.; Sanz, A. Antioxidant Defenses in Fish: Biotic and Abiotic Factors. *Rev. Fish Biol. Fish.* **2005**, *15*, 75–88. [CrossRef]
85. Li, R.; Chen, W.-C.; Wang, W.-P.; Tian, W.-Y.; Zhang, X.-G. Antioxidant activity of *Astragalus* polysaccharides and antitumour activity of the polysaccharides and siRNA. *Carbohydr. Polym.* **2010**, *82*, 240–244. [CrossRef]
86. Bors, W.; Michel, C.; Stettmaier, K. Antioxidant effects of flavonoids. *BioFactors* **1997**, *6*, 399–402. [CrossRef] [PubMed]
87. Huyut, Z.; Beydemir, Ş.; Gülçin, İ. Antioxidant and Antiradical Properties of Selected Flavonoids and Phenolic Compounds. *Biochem. Res. Int.* **2017**, *2017*, 7616791. [CrossRef]
88. Rice-Evans, C.; Miller, N.; Paganga, G. Antioxidant properties of phenolic compounds. *Trends Plant Sci.* **1997**, *2*, 152–159. [CrossRef]

fishes

MDPI

Article

Comparison of Three Artificial Diets for the Larviculture of Giant Kōkopu (*Galaxias argenteus*)

William McKay [1,*] and Andrew Jeffs [1,2]

[1] Institute of Marine Science, University of Auckland, Auckland 0985, New Zealand
[2] School of Biological Sciences, University of Auckland, Auckland 1010, New Zealand
* Correspondence: wjgmckay@gmail.com

Abstract: The selection of artificial feed is critical to the success of larviculture of fin fish and requires knowledge of the varied species-specific dietary and nutritional requirements. With the emergence of commercial aquaculture of giant kōkopu, *Galaxias argenteus*, there is a need to understand the species-specific needs for artificial feeds in larviculture. Consequently, this study compares three commercial artificial dry feeds; Otohime, Artemac and O.range on the growth of recently weaned giant kōkopu. Larvae fed with Otohime outperformed both Artemac and O.range treatments by achieving the highest wet weight after 67 days, greater by at least 47% on average than both Artemac and O.range. These differences in larval performance are likely to be due to the higher protein:energy ratio and EPA content of Otohime. High DHA and ARA in the diets in absolute terms or in relation to EPA did not result in added benefit for growth performance. This study provides an important first step in identifying the nutritional needs of larval giant kōkopu which can assist in improving their commercial aquaculture production.

Keywords: aquaculture; whitebait; larval diet; formulated feed; PUFA; HUFA

Citation: McKay, W.; Jeffs, A. Comparison of Three Artificial Diets for the Larviculture of Giant Kōkopu (*Galaxias argenteus*). *Fishes* **2022**, *7*, 310. https://doi.org/10.3390/fishes7060310

Academic Editor: Terje van der Meeren

Received: 16 September 2022
Accepted: 24 October 2022
Published: 28 October 2022

1. Introduction

Reliable larviculture to produce high quality juvenile fin fish in large quantities is one of the greatest challenges facing the aquaculture industry currently [1,2]. Low survival rates and poor growth of fin fish larvae are frequently major contributing factors to this production bottleneck [3,4]. Poor outcomes from larviculture is often the result of a combination of factors, such as a lack of knowledge of larval nutritional requirements, difficulties in supplying feeds that can be processed by rudimentary larval digestive systems, providing husbandry of small and fragile larval fish, as well as the appropriate management of the rapid and complex changes that fish undergo during this early phase of development [1–3,5–7]. The preparation and provision of appropriate feed is particularly critical for successful fish larviculture and typically accounts for 30–70% of total production costs [8], so that the optimization of larval nutrition and feeding regimes are priority targets for research in early stage aquaculture businesses [9].

Generally, feeding regimes for the larviculture of fin fish begin with several weeks of live food provision, followed by a period of co-feeding with live and artificial diets before larvae can be fully weaned to artificial diets [3,6,10]. Rotifers and *Artemia* are the most commonly used live feeds, but while these organisms have unlocked the potential for aquaculture of a multitude of fin fish species in the last few decades, they still impose limitations to successful commercial larviculture [7]. These limitations include their high costs, from both their purchase and the infrastructure and resources required to culture or prepare them [10]. Furthermore, the nutritional condition of live feeds can vary widely, with poor nutritional condition resulting in inferior survival and growth performance in cultured larvae [10,11]. As a result, significant effort is focus on identifying optimal artificial diets to replace live feeds.

Artificial diets provide several key advantages over live prey for larviculture. Most importantly, they can be customized to the species and age-specific nutritional needs of the animal in culture and are reliably uniform in composition [6,10]. This allows the reliability that is required for consistently delivering optimal survival and growth in commercial larviculture. Artificial diets are also far more practical, in that they are easily stored, require far fewer resources for preparation than live feeds, and are immediately dispensable when required [6].

Issues remain in preparing artificial larval feeds which provide for nutrient stability, as well as perceptibility, palatability, digestibility and nutritional value [5,6,12]. Of critical importance is identifying the specific needs of the species in culture, such that the optimal feed can be provided and economical production be achieved [13]. The provision of essential fatty acids are especially critical for larval marine fin fish [14–16]. The absolute and relative requirements of essential fatty acids, particularly EPA (eicosapentaenoic acid), ARA (arachidonic acid) and DHA (docosahexaenoic acid), are highly species-specific due to differences in their metabolic capabilities and nutritional requirements for development [13,15,17–21].

Galaxiid fishes are a family of freshwater fishes, found in cool-temperate southern hemisphere [22]. Most galaxiids maintain an entirely freshwater life cycle, however, amphidromy is common, such as in the case of the giant kōkopu (*Galaxias argenteus*) which is endemic to New Zealand [23,24]. The eggs of this species are deposited on river-banks, incubating terrestrially, until flooding rain events stimulate hatching, washing the larvae down rivers and out to sea [25–28]. After 3–6 months, giant kōkopu larvae, along with larvae from four other galaxiid species, undertake a mass migration back into the freshwater habitats occupied by adults [27,29]. These mass migrations occurring in confined fresh waterways provide the opportunity for targeting their harvesting to provide e highly prized "whitebait" [30,31]. Increasingly, studies show that the abundance of the species which make up this fishery are in decline due to habitat loss and predation by introduced species, with the giant kōkopu now considered threatened [31–33].

Protection of these galaxiid species largely involves increasing regulatory controls on wild fisheries activities, however, larviculture of giant kōkopu is in the early stages of commercialisation and is proposed to reduce the reliance on harvesting threatened wild populations [28,34,35]. Optimum diet selection for larvae is a significant issue impeding further commercialisation of giant kōkopu aquaculture. Recent work on this species has developed knowledge on their morphometry and energetic demands, indicating the benefits of earlier provision of larger feed particles [36,37]. Still, a significant portion of the larval production cycle (approximately 30 of 77 days) expensive live feeds are required, in large part due to the unknown nutritional of requirements of the larvae. Some early research that has been undertaken on the closely related inanga, *Galaxias maculatus*, identified that its larval diet requires high levels of alpha-linolenic fatty acid under certain salinity culture conditions [38]. However, these studies were undertaken at much lower salinities (0 and 15 ppt) than those in which giant kōkopu are reared.

This study aims to improve knowledge of the nutritional requirements of larval giant kōkopu through the comparison of the growth performance of larvae fed on three different commercially available artificial dry feeds. The results have the potential to be useful for improving the efficiency of larviculture, which is important for securing the future commercial success of the giant kōkopu aquaculture industry.

2. Materials and Methods

2.1. Experimental Animals

Gametes from 80 female and 20 male giant kōkopu were stripped and fertilized before being subjected to a 4 week incubation period in UV treated freshwater filtered to 1 µm at 4 °C. On 4 July 2016, approximately 1.2 million giant kōkopu were hatched directly into one 2500 L conical commercial larval rearing tank containing UV treated water, 35 ppt and ambient temperature < 18 °C. At 2 DAH (days after hatching) around 9000 fish were randomly selected from the commercial tank and split evenly among nine 20 L experimental tanks. This was achieved by estimating the total number of fish per liter in the transfer vessel by careful mixing and taking random 200 mL samples and then counting the number of fish in each sample to produce a mean estimate of the total number of fish.

2.2. Tank Design and Recirculation System

Experimental tanks were made from 20 L plastic (HDPE), round, blue pails, 270 mm in diameter and 380 mm in height. The water outflow pipe was set 80 mm below the rim of the pail so that each tank held 18 L. For the first 4 weeks the outflow pipe was fitted with a banjo filter using 600 μm filter mesh to prevent the escape of giant kōkopu larvae while allowing the passage of suspended particles. For the remainder of the experiment, banjo filters with 1 mm mesh were used. Surfboard wax was applied in a thick, 40 mm width strip around the inside of the tank at the water level to inhibit the climbing ability of the fish larvae.

The experimental tanks were connected to a recirculation system with the outflow from each tank being directed to a filter basket for removal of insoluble particles by a 5 μm filter mat. After passing through the filter mat, the water entered a 300 L sump containing 40 L of plastic Kaldnes-K3 media (Krüger Kaldnes AS, Norway) for biological filtration which had been preconditioned in a commercial giant kōkopu RAS system for at least six months immediately prior to experimental use. Protein was skimmed from the sump manually, as required. Each day the filter mat was changed, and 100 L of seawater was removed from the sump and replaced with natural seawater, 35 ppt, filtered to 5 μm and UV sterilized.

From the sump seawater was pumped through a UV filter and then distributed into each experimental tank using 4 mm tubing connected at the surface and bottom of the tank. Water flow to each tank was 0.28 L min^{-1} over the first 14 days of the experiment. However, inflow was suspended for 30 min during feeding events for the first 7 days. Flow rate was increased to 0.37 L min^{-1} from 15–28 days before increasing to 0.49 L min^{-1} for the following 14 days, and finally to 0.62 L min^{-1} for the remainder of the experiment.

Tanks were aerated by an air-stone at the bottom of the tank producing two medium sized (0.5 mm diameter) bubbles per second for the first 28 days of the experiment. The air-stone was then changed to provide a high number of very fine bubbles for the remaining experimental period.

Illumination of experimental tanks was provided by three 58 W fluorescent tubes, suspended 100 cm above the top edge of the tanks. Light reaching the tanks was dimmed by hanging shade cloth over the tanks for 30 min either side of the lights coming on at 0745 h and off at 1800 h.

Seawater temperature was not controlled, but was measured every 6 h with a glass thermometer (Aqua One) during the experimental period and found to vary between 14 and 18 °C, and was consistent among all tanks. Nitrate (<5 mg/L), nitrite (<0.25 mg/L), ammonia (NH_3/NH_4^{+}) (<0.25 mg/L), carbonate hardness and pH (7–8) were measured every second day using API® test kits to ensure water treatment was maintaining suitable conditions for the larvae and was within the acceptable ranges reported for the rearing of larvae [39–42]. Water quality was never found to be outside these acceptable ranges for rearing larvae from set up through to the conclusion of the experiment and was consistent among experimental tanks due to the recirculation system.

2.3. Experimental Design

Three commercially available artificial dry feeds for larval giant kōkopu were tested. Larvae were provided solely live food for the first 14 days before a prolonged weaning period after which (from 45 DAH) only artificial food was provided (Table 1).

Table 1. Experimental feeding regime for giant kōkopu larvae showing the feed provision of each feeding event and number of feeding events per day by larvae age. Instar-I *Artemia* (in–I), instar–II *Artemia* (in–II), small particle (SP), medium particle (MP), large particle (LP). * Indicates that for the period 36–44 DAH a ration of 2 g of instar–II *Artemia* were also administered with the fourth feed of the day only.

DAH	Feed	Number of Events
0–2	NIL	-
3–14	3 g in-I	4
15–21	0.1 g SP + 3 g in–II	4
22–28	0.1 g SP + 2 g in–II	4
29–35	0.25 g SP + 0.1 g MP + 2 g in–II	4
36–44 *	0.6 g MP	4
45–49	1.2 g MP	4
50–56	0.6 g MP	5
57–64	0.6 g MP + 0.3 g LP	4
64–66	0.3 g MP + 0.6 g LP	4

The first feeding treatment "OTO" used the Otohime products A (75–250 μm), B1 (250–360 μm) and B2 (360–650 μm) (Marubeni Nisshin Feed Co., Ltd., Tokyo, Japan) and are referred to as Small Particle, Medium Particle and Large Particle or "SP", "MP", and "LP", respectively (Table 2).

Table 2. Percent dry matter feed composition, energy content, and fatty acid profiles of the three larval diets tested in the present study.

	Colour	Protein (% DW)	Digestible Energy (kcal/g)	Gross Energy (kcal/g)	Est. Protein:Energy (mg/kcal)	Lipid (%)	EPA (mg/g)	DHA (mg/g)	ARA (mg/g)	Carbohydrate (%DW)	Fiber (%DW)	Ash (%DW)	Moisture (%DW)
OTO	Brick Red	53	4.07	4.66	123	9	22.9	20.3	1.1	14.5	3.5	15	6.5
ART	Mustard	57	5.01	5.50	113	9	11.2	12.7	0.9	12.0	2.0	5	7.0
ORA	Orange	56	4.39	5.01	124	13	10.0	20.0	1.2	15.1	1.0	10	5.9

(i) Data on percent dry matter was obtained from product manufacturer's specifications except for the moisture content of O.range which came from [43]. (ii) Digestible energy calculated with the equation: crude protein × 5.64 kcal/g + crude lipid × 9.44 kcal/g [44,45]. (iii) Gross energy calculated with the equation: crude protein × 5.64 kcal/g, crude lipid × 9.44 kcal/g [46] and carbohydrate (nitrogen free extract) × 4.11 kcal/g [44,45,47]. (iv) Protein energy ratio calculated with the equation: crude protein × 5.64 kcal/g × 0.877, crude lipid × 9.44 kcal/g × 0.982 [46], and carbohydrates (nitrogen free extract) × 4.11 kcal/g × 0.90 [44,45,47]. (v) EPA, DHA data obtained from manufacturers for Artemac and O.Range, and for Otohime as well as ARA for each product from [13,43,48].

The second treatment "ART" made use of the Artemac products 2 (100–200 μm), 3 (200–300 μm) and 4 (300–500 μm) (Aquafauna Bio-Marine, Inc., Hawthorne, CA, USA), again referred to as "SP", "MP", and "LP", respectively (Table 2).

The final feed treatment "ORA" used O.range products START-S (100–200 μm), WEAN-S (200–400 μm) and WEAN–L (300–500 μm) (INVE Aquaculture Inc., Salt Lake City, UT, USA) also referred to as "SP", "MP", and "LP", respectively (Table 2).

2.4. Live Food Production

Artemia cysts used to produce live feed throughout this experiment were GSL Sep-Art (INVE Aquaculture Inc., Salt Lake City, UT, USA) from the same batch.

Live feeds were administered by total wet weight, with *Artemia* being harvested and poured through a 100 μm sieve that was allowed to drip dry on a towel for 1 min and measured with electronic scales to the nearest 0.1 g.

Instar-I *Artemia* were produced by incubating cysts in natural seawater 35 ppt for 17 h at 29 °C with constant, vigorous aeration while exposed to light in 250 l *Artemia* cones. Instar-I *Artemia* were separated from unhatched cysts and husks with a magnet before rinsing in a 100 μm sieve with clean 35 ppt water. Instar-I *Artemia* were fed out immediately after harvesting from cysts.

Instar-II *Artemia* were produced and prepared for feeding under the same conditions, however, received a 27 h incubation with enrichment. After separation live animals were enriched in a 400 L tank for between 23–31 h using a proprietary enrichment formula that combines the commercially available instant algae products—Rotigrow Plus, Nanno 3600 and Tetraselmis 3600 (Reed Mariculture Inc., Salt Lake City, UT, USA). At the beginning of enrichment and 23 h later an aliquot of 60 mL of enrichment formula was added to the enrichment tank.

2.5. Sampling of Larvae

Two sampling events of larval giant kōkopu took place in this experiment. The initial sampling took place on 27 July 2016 when larvae were 23 DAH, once larvae had been exposed to SP weaning diets for the first week to establish if initial weaning performance may set a foundation for subsequent outcomes. The second sampling was at the conclusion of the experiment on 9 September 2016 when larvae were 67 DAH to determine the overall outcome of the comparative weaning treatments.

At both sampling events, randomly sampled larvae were euthanized by placing in ice water (0 °C) for 20 min, measured for total length (i.e., snout to tip of tail) and body depth (i.e., center of body at the anus across to the dorsal surface) from each tank. Measurements were conducted by placing fish on 46 μm grid plastic sheets and photographing fish under a microscope using an Olympus TG-4 camera. Images were later processed using ImageJ (ver. 1.53, National Institutes of Health) to derive measurements.

For the initial sampling event from each tank three samples of 50 fish were taken at random by gently swirling the tank, collecting fish with a small jar and pouring through 300 μm mesh. Mean wet weight (WW) of fish was determined by weighing followed by mean dry weight (DW) after freeze drying, re-weighing and dividing by the total number of fish. These lyophilized samples were then used to determine total lipid and total protein content. A further 20 fish were randomly sampled from each tank to measure total length and body depth.

For the final sampling event 20 fish were randomly sampled to total length and body depth measurements. Due to reduced numbers of fish from mortality amongst all treatments early in the experiment resulting from stress of transfer of larval into the experimental tanks, three samples of 20 fish per sample were taken from each tank to undertake WW and DW measurements. These same sampled fish were then also lyophilized and used for protein and lipid analyses. The ORA treatment was an exception where one tank had sufficient numbers for only 20, 20 and 19 fish per replicate sample and another tank with only 15 fish per replicate sample. Unfortunately, it was not possible to accurately recover and record respective mortalities of larvae throughout the experiment.

2.6. Specific Growth Rate

The mean specific growth rate (SGR) for WW was determined for each treatment tank across the duration of the 44 day experimental period between initial and final sampling events using Equation (1) [49].

$$SGR = 100(e^{g} - 1) \tag{1}$$

where: g = (lnfinal mass − lninitial mass)/(number of days between sampling events).

2.7. Lipid and Protein Composition

Lipid was extracted from larval fish samples using a modified Bligh and Dyer [50] solvent extraction method [51]. A 1.9 mL aliquot of chloroform, methanol and deionized water mixture (ratios 2:1:0.4) was added to the lyophilized samples before being vortexed for 30 s and then left to stand for 16 h. An aliquot of 0.5 mL of 0.7% sodium chloride and 0.5 mL of chloroform were added, followed by 30 s of vortexing, then centrifuging for 10 min at 1000 rpm. The chloroform-lipid layer was removed and placed in a pre-weighed glass vial. The residual layer was washed with 1 mL of chloroform, followed by 30 s vortexing, centrifuging for 10 min at 1000 rpm and the chloroform-lipid removed and added to pre-weighed glass vial. This step was repeated again using only 0.5 mL of chloroform. The glass vials were then placed in a thermal evaporator held at 39 °C under flowing nitrogen gas to remove the chloroform. The glass vials were then re-weighed to determine lipid mass which was then divided by the total larval sample dry mass and multiplied by 100 to provide lipid content as a percentage of dry weight (%DW). The total lipid (per larva) was determined by multiplying the lipid proportion (%DW) by the mean DW of individual larvae in the respective sample. For both the three replicate lipid proportion measures and the three replicate total lipid measurements per replicate tank were then used to determine the tank average, with the results from the three tanks per treatment averaged to give the treatment mean.

The protein content of larvae was measured using a bicinchoninic acid (BCA) assay (Micro BCA™ Protein Assay Kit, ThermoFisher Scientific, Auckland, New Zealand). After removing the lipid content the residual larval tissues were freeze dried and ground before the addition of sodium hydroxide and incubation in a water bath at 50 °C for 16 h. Samples were diluted and then centrifuged at 4000 rpm for 10 min at 4 °C. The resulting samples and a set of bovine serum albumin standards were placed into a 96 well-plate and reagents added to each well followed by reading absorbance at 562 nm. The protein content of the larvae was calculated using the standard curve as a percentage of dry weight (%DW). The mean total protein (per larva) was determined by multiplying the protein proportion (%DW) by the mean DW of larvae in the respective sample.

2.8. Fatty Acid Profiles

Fatty acid analyses were conducted on an aliquot of the total lipid previously extracted gravimetrically. The derivatization process was based on Lepage & Roy [52]. Laboratory controls were included during the derivatization process. This comprised a positive control containing 52 reference standards of FAs all with different concentrations and a negative control containing C19 and C23 FAs in the same concentration range as the samples. An extraction solution of 2 mL of methanol:toluene (4:1 v/v, Analytical Grade, Merck) containing internal standards (C19: nonadecanoic acid 0.083 mg mL^{-1} and C23: tridecanoic acid 0.082 mg mL^{-1}, Nu-Chek Prep., Elysian, MN, USA) was added to each sample and transferred to borosilicate tubes with Teflon-lined screw caps. Magnetic stirring bars were added to each tube. Acetyl chloride (200 mL, ECP) was added slowly, dropwise to each sample over a period of 1 min. The tubes were placed in a heating and stirring dry block at 100 °C for 1 h. After 1 h, the tubes were cooled in water and 5 mL of an aqueous solution of 6% potassium carbonate were added to each tube. The tubes were vortexed, then centrifuged at 3500× g (5 min at room temperature). The upper toluene phase was recovered and transferred to a gas chromatography (GC) vial with an insert, and a further 25% dilution was done using toluene, for analysis by GC-mass spectrometry (GC–MS) at the Auckland Science Analytical Services, at the University of Auckland. GC–MS instrument parameters were based on Kramer et al. [53]. The instrument used was an Agilent 7890B gas chromatograph coupled to a 5977C mass spectrometer with a split/splitless inlet [54]. A sample of 1 µL was injected using a CTC PAL autosampler into a glass 4 mm ID straight inlet liner packed with deactivated glass wool (Restek Sky®). The inlet temperature was 250 °C, in splitless mode, and the column flow was set at 1 mL min^{-1}, with a column head pressure of 62 kPa, giving an average linear velocity of 19 cm s^{-1}. Purge flow was set to 50 mL min^{-1} at 1 min after injection. Column selection was based on the recommendations from the official methods for the determination of trans fat (American Oil Chemists Society—[55]). The column was a fused silica Rtx-2330, which was 100 m long, 0.25 mm internal diameter, 0.2 µm highly polar stationary phase (90% biscyanopropyl and 10% cyanopropylphenyl polysiloxane, Shimadzu). Carrier gas was instrument grade helium (99.99%, BOC). The GC oven temperature programming started isothermally at 45 °C for 2 min, increased by 10 °C min^{-1} to 215 °C, held for 35 min and then increased by 40 °C min^{-1} to 250 °C and held for 10 min. The transfer line to the mass spectrometric detector (MSD) was maintained at 250 °C, the MSD source at 230 °C and the MSD quadropole at 150 °C. The detector was turned on 14.5 min into the run. The detector was run in positive-ion, electron-impact ionization mode, at 70 eV electron energy, with electron multiplier set with no additional voltage relative to the autotune value. Data were acquired at 1463 amu s^{-1} in scan mode from 41 to 420 amu, with a detection threshold of 100 ion counts. Resulting GC-MS peaks were identified on fatty acid methyl ester mass spectral library and each FA peak was quantified using an inhouse R package (RStudio, ver. 1.2.1335). The data were screened for chromatographic retention time drift, and manual correction/integration was carried out where necessary. The data set was normalized by the response of the internal standard (nonadecanoic acid), and a blank treatment was applied to correct the baseline response. The resulting normalized peak area values were used to quantify the total of each FA using linear calibration information obtained from seven calibration curve standards. The total amount of FA measured in the lipid aliquot was then adjusted for the total lipid extract to calculate the proportional contribution of each FA to the total FAMEs.

2.9. Statistical Analyses

The initial and final mean DW, WW, total length, body depth, total lipid and protein, percent lipid and protein, fatty acids, and SGR were compared among treatments using ANOVA where parametric data assumptions were satisfied. Initial and final lipid and protein concentrations as well as fatty acid profiles and SGR were arc-sine transformed prior to analysis to correct for any data distribution bias associated with percentage data [56]. Normality and equality of variance of data were tested and confirmed using the Shapiro–Wilk's and Levene's tests prior to analyses. Where data conformed to parametric assumptions a linear mixed model ANOVA was fitted to control the random effects of the tanks in each analysis. When ANOVA identified overall experimental treatment effects the differences between pairs of individual means were identified with a Tukey's test with adjustment for false discovery. Estimated mean difference and 95% confidence intervals were calculated and are presented.

Data requirements for performing parametric tests were not met for the mean final total length and depth variables for giant kōkopu larvae. Consequently, Kruskal–Wallis tests were used to compare these data and where significant differences were found, then Mann–Whitney-Wilcox post hoc comparison tests were used to compare means between treatment groups.

All statistical analyses were performed using R (RStudio, ver. 1.2.1335). All measures of variability of sampled means is reported as standard error of the mean.

3. Results

3.1. Weight

There was no difference among treatments for mean initial WW of larval fish ($F_{(2,6)} = 0.38$, $p = 0.70$); i.e., OTO 7.45 ± 0.13 mg, ART 7.52 ± 0.24 mg and ORA 7.21 ± 0.10 mg (Figure 1a). However, at the final sampling the mean WW of the fish was different among the treatments ($F_{(2,6)} = 15.99$, $p < 0.01$). Fish in the OTO treatment had a greater mean final WW (73.23 ± 3.14 mg), between 12.70 and 34.60 mg greater than ART (49.58 ± 2.15 mg, $p < 0.01$), and between 19.10 and 41.00 mg greater than ORA (43.16 ± 1.74 mg, $p < 0.01$) (Figure 1b). The mean final WW of the ART and ORA treatment groups were not different ($p = 0.30$).

The mean initial DW of the fish larvae was not different among treatments ($F_{(2,6)} = 1.03$, $p = 0.41$), OTO 1.39 ± 0.03 mg, ART 1.32 ± 0.03 mg and ORA 1.34 ± 0.02 mg (Figure 1a). However, at the final sampling there were significant differences among treatments for DW ($F_{(2,6)} = 20.84$, $p < 0.01$). OTO mean final DW (15.05 ± 0.62 mg), was between 0.26 and 0.59 times greater than ART (9.83 ± 0.42 mg, $p < 0.01$), and between 0.35 and 0.69 times greater than ORA (8.95 ± 0.37 mg, $p < 0.01$). There was no significant difference in the mean final DW between ART and ORA ($p = 0.32$) (Figure 1b).

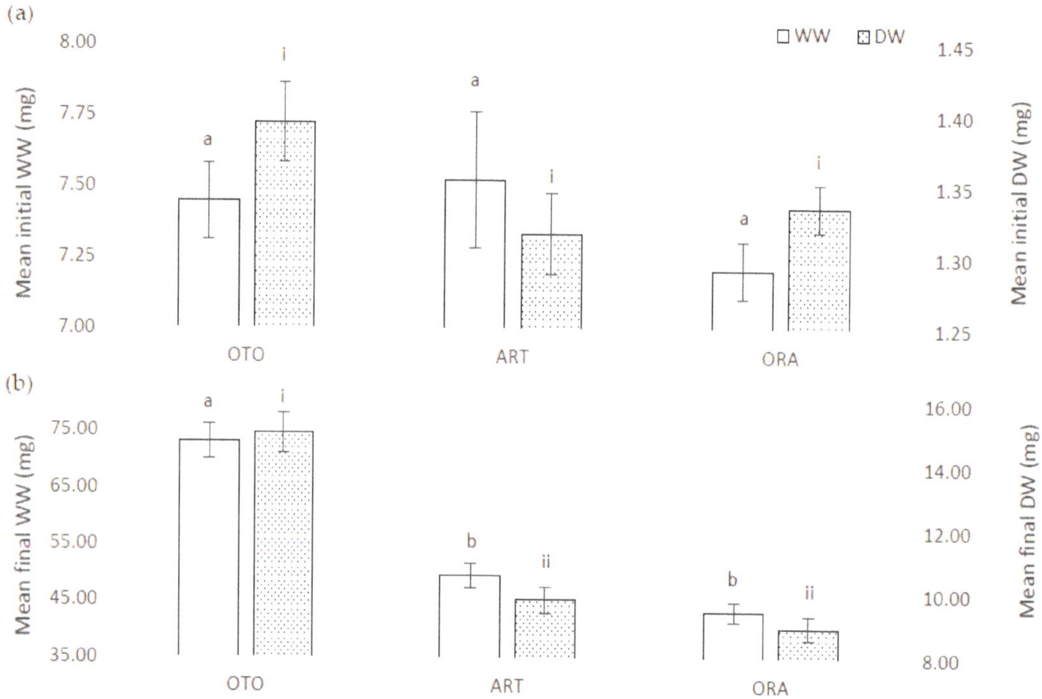

Figure 1. (**a**) Mean initial wet weight (WW) and dry weight (DW) for larval giant kōkopu from three different feed treatments; OTO (Otohime), ART (Artemac) and ORA (O.range) (mean ± SE). Means with different superscripts are significantly different among treatments for WW and among treatment for DW ($p < 0.05$). (**b**) Mean final wet weight (WW) and dry weight (DW) for larval giant kōkopu from three different feed treatments; OTO (Otohime), Artemac (ART) and ORA (O.range) (mean ± SE). Means with different superscripts are significantly different among treatments for WW and among treatment for DW ($p < 0.05$).

3.2. Length and Depth

At the initial sampling the total length of the giant kōkopu larvae was not significantly different among treatments ($F_{(2,6)} = 0.11$, $p = 0.90$); i.e., OTO 15.05 ± 0.21 mm, ART 15.02 ± 0.21 mm and ORA 14.87 ± 0.18 mg (Figure 2). However, at final sampling the mean total length of larvae was different among treatments ($X^2 = 8.52$, $p = 0.01$); i.e., OTO 27.97 ± 0.55 mm, ART 24.85 ± 0.44 mm and ORA 25.74 ± 0.58 mm (Figure 2). Mean final total length was greater in the OTO treatment than ART ($p < 0.01$), while there was no difference between OTO and ORA ($p = 0.10$), and between ART and ORA ($p = 0.13$).

The mean initial depth of the body of the larval fish was the same among treatments (i.e., 23 DAH) ($F_{(2,6)} = 0.02$, $p = 0.98$); i.e., OTO 1.09 ± 0.02 mm, ART 1.09 ± 0.02 mm and ORA 1.10 ± 0.02 mm (Figure 3). At the final sampling there was a significant difference in mean body depth among treatments ($X^2 = 22.08$, $p < 0.01$); i.e., OTO 2.55 ± 0.06 mm, ART 2.11 ± 0.06 mm and ORA 2.23 ± 0.07 mm (Figure 3). Mean final total body depth was greater in the OTO treatment than both ART ($p < 0.01$) and ORA ($p < 0.01$). There was no difference in final total body depth between ART and ORA ($p = 0.13$).

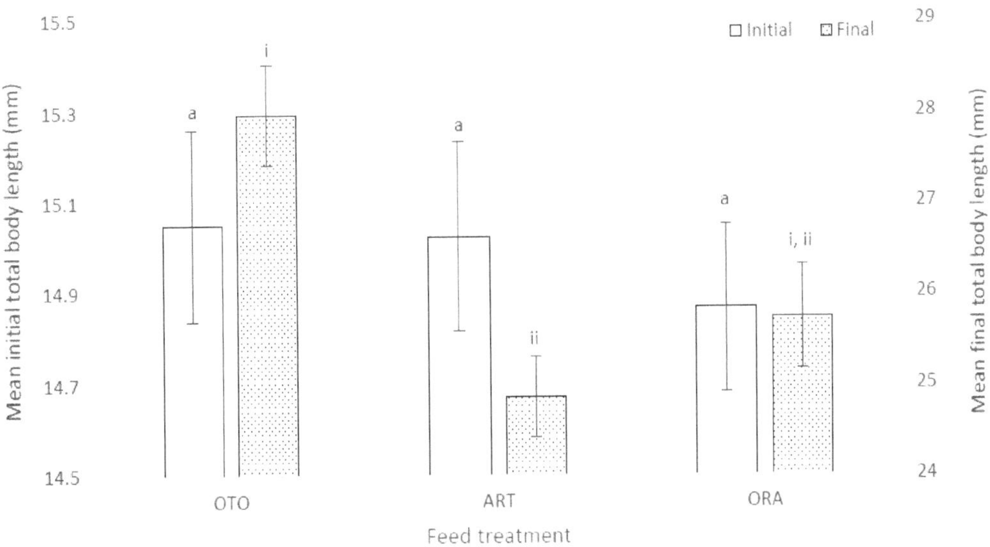

Figure 2. Mean initial and final total length of larval giant kōkopu from three different feed treatments; OTO (Otohime), ART (Artemac) and ORA (O.range) (mean ± SE). Means with different superscripts are significantly different among treatments for mean initial total length and among treatments for mean final total length ($p < 0.05$).

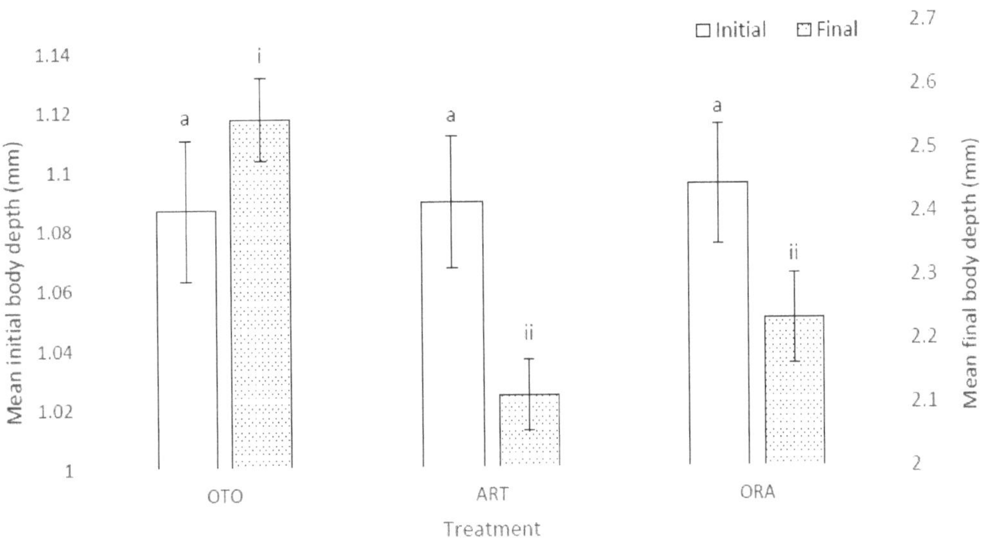

Figure 3. Mean initial and final body depth for larval giant kōkopu in three different feed treatments; OTO (Otohime), ART (Artemac) and ORA (O.range) (mean ± SE). Means with different superscripts are significantly different among treatments for mean initial body depth and among treatments for mean final body depth ($p < 0.05$).

3.3. *Specific Growth Rate*

There was a significant difference in the SGR among the three feed treatments ($F_{(2,6)} = 29.88$, $p < 0.01$). The OTO treatment achieved a greater SGR, 5.59 ± 0.10%, than both ART,

4.42 ± 0.06% ($p < 0.01$) and ORA, 4.14 ± 0.09 % ($p < 0.01$) (Figure 4). No difference was found in the SGR between ORA and ART treatments ($p = 0.20$).

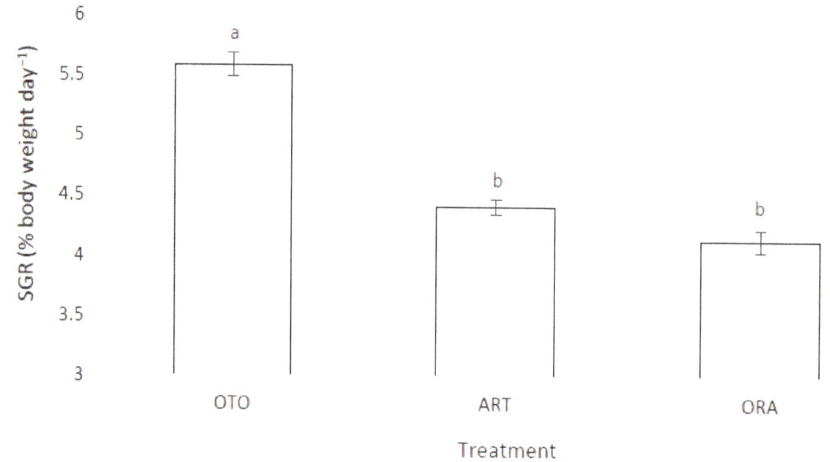

Figure 4. Mean specific growth rate (SGR) for larval giant kōkopu for three different feed treatments; OTO (Otohime), ART (Artemac) and ORA (O.range) (mean ± SE). Means with different superscripts are significantly different ($p < 0.05$).

3.4. Lipid and Protein

At the initial sampling there was no difference in proportional lipid content of the larvae among the three feed treatments ($F_{(2,6)} = 1.30$, $p = 0.34$); i.e., OTO 16.1 ± 0.2% DW, ART 15.8 ± 0.2% DW and ORA 16.0 ± 0.4% DW (Figure 5). There was no significant difference in proportional lipid content of the larvae among treatments at the final sampling ($F_{(2,6)} = 0.21$, $p = 0.82$); i.e., OTO 17.0 ± 0.6% DW, ART 15.5 ± 0.8% DW and ORA 16.7 ± 0.6% DW (Figure 5).

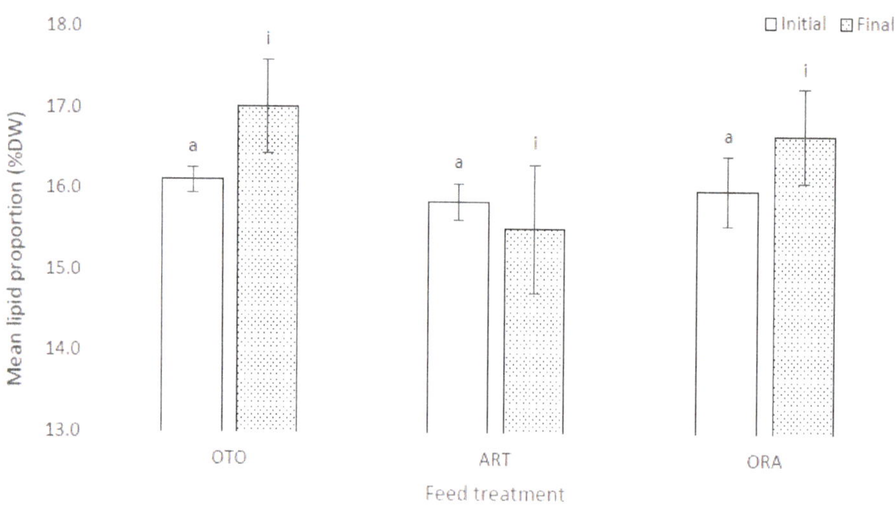

Figure 5. Mean proportional lipid content of larval giant kōkopu as a percentage of dry weight (DW) for at the initial and final sampling events for three different feed treatments; i.e., OTO (Otohime), ART (Artemac) and ORA (O.range) (mean ± SE). Means with different superscripts are significantly different within each of the set of three treatment means for each sampling event ($p < 0.05$).

There was also no difference in total lipid among treatments at the initial sampling ($F_{(2,6)} = 3.08$, $p = 0.12$); i.e., OTO 0.24 ± 0.00 mg, ART 0.20 ± 0.00 mg and ORA 0.22 ± 0.00 mg (Figure 6). However, there was a significant difference in mean total lipid among the three feed treatments at the final sampling event ($F_{(2,6)} = 17.72$, $p < 0.01$); i.e., OTO 2.43 ± 0.10 mg, ART 1.56 ± 0.05 mg and ORA 1.43 ± 0.01 mg (Figure 6). OTO accumulated between 0.51 and 1.23 mg more lipid than ART ($p < 0.01$) and between 0.64 and 1.35 mg more lipid than ORA ($p < 0.01$). There was no difference in the total lipid content of larvae between the ART and ORA treatments ($p > 0.1$).

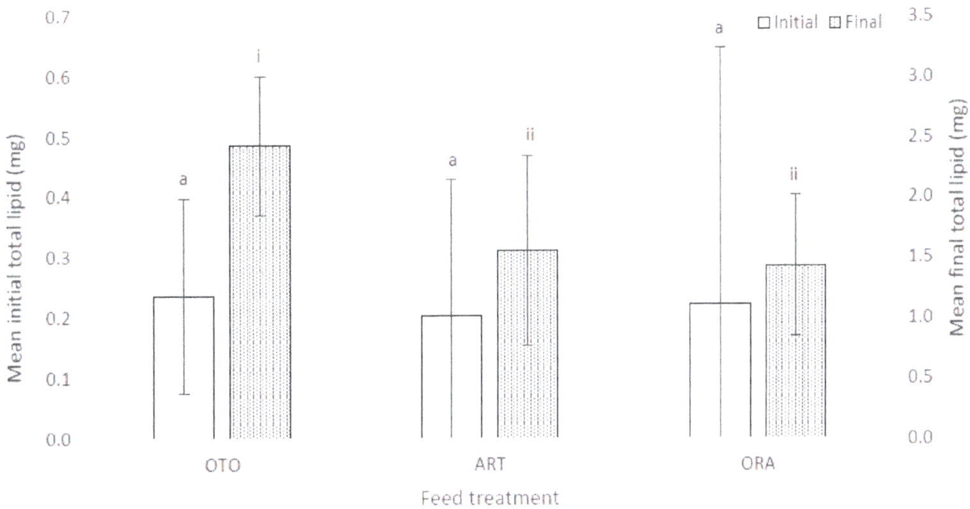

Figure 6. Mean initial and final total lipid for larval giant kōkopu in three different feed treatments; i.e., OTO (Otohime), ART (Artemac) and ORA (O.range) (mean ± SE). Means with different superscripts are significantly different within each of the set of three treatment means for each sampling event ($p < 0.05$).

At the initial sampling there was no difference in the proportional protein content of the larval giant kōkopu among the three feed treatments ($F_{(2,6)} = 2.29$, $p = 0.18$), i.e., OTO 76.8 ± 2.3% DW, ART 77.7 ± 1.5% DW and ORA 72.0 ± 1.6% DW (Figure 7). Likewise, there was no difference in the proportional protein content of larvae among the three feed treatment groups at the final sampling ($F_{(2,6)} = 0.22$, $p = 0.81$); i.e., OTO 73.5 ± 2.5% DW, ART 77.6 ± 4.7% DW and ORA 73.8 ± 1.9% DW (Figure 7).

There were no differences among treatments in the mean initial total protein of the larvae ($F_{(2,6)} = 1.17$, $p = 0.37$); i.e., OTO 1.07 ± 0.04 mg, ART 1.03 ± 0.03 mg and ORA 0.96 ± 0.03 mg (Figure 8). However, there was a significant difference in the total protein of larval fish at the final sampling ($F_{(2,6)} = 13.82$, $p < 0.01$) (Figure 8). The mean final total protein for OTO was 10.65 ± 0.67 mg, between 1.92 and 5.18 mg more than for ART (7.42 ± 0.11 mg, $p = 0.01$), and between 2.45 and 5.98 mg more than for ORA (6.77 ± 0.17 mg, $p = 0.01$) (Figure 8). There was no difference in mean final total protein between ART and ORA ($p = 0.50$) (Figure 8).

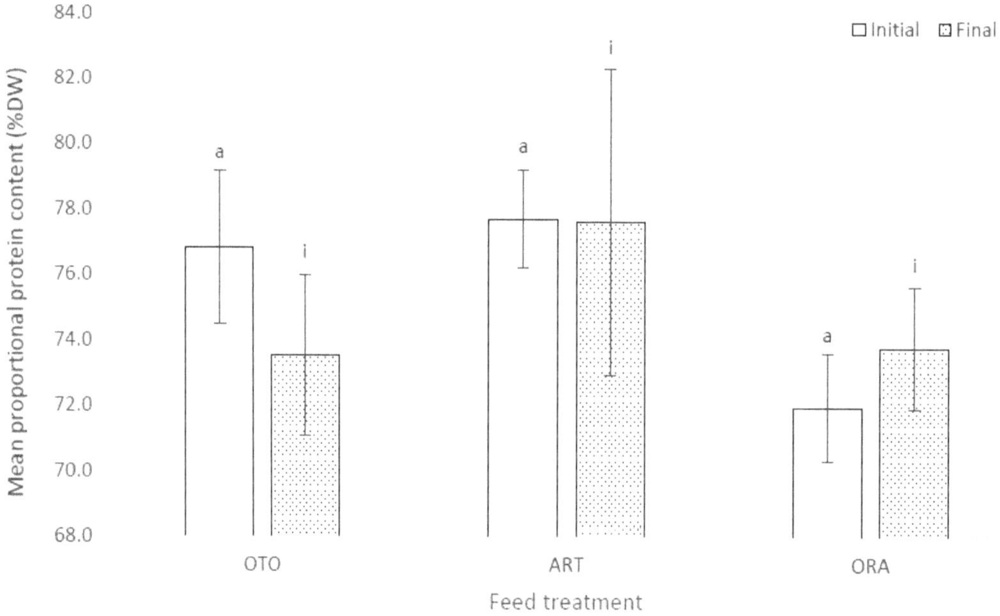

Figure 7. Mean proportional protein content measured as a percentage of dry weight (DW) for larval giant kōkopu at initial and final sampling for three feed treatments; i.e., OTO (Otohime), ART (Artemac) and ORA (O.range) (mean ± SE). Means with different superscripts are significantly different within each of the set of three treatment means for each sampling event ($p < 0.05$).

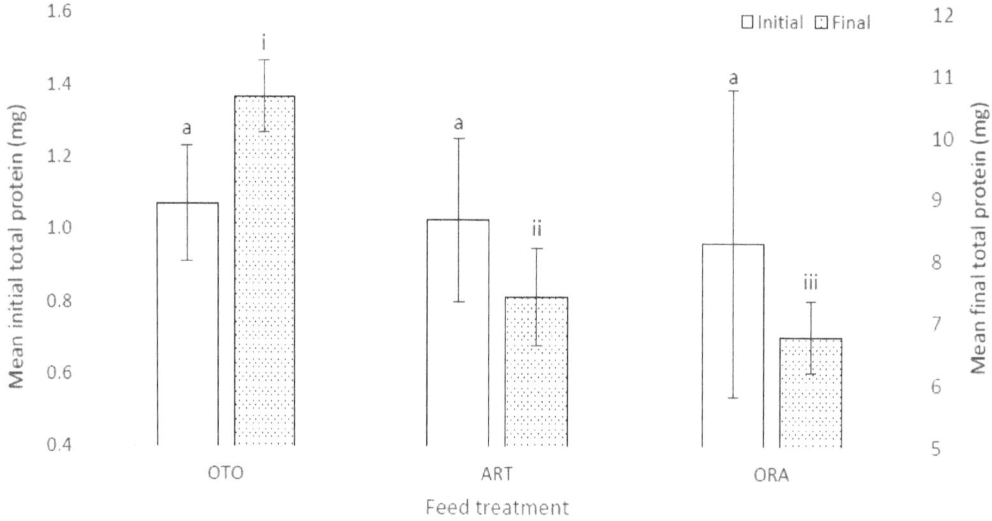

Figure 8. Mean initial and final total protein content of larval giant kōkopu for three feed treatments; i.e., OTO (Otohime), ART (Artemac) and ORA (O.range) (±SE). Means with different superscripts are significantly different within each of the set of three treatment means for each sampling event ($p < 0.05$).

3.5. Fatty Acid Profiles

Thirty one fatty acids were identified across the three treatment groups with 18 fatty acids being present as >1% of total fatty acids (Table 3).

Table 3. Mean initial and final percent fatty acid composition (±SE) of larval giant kōkopu for three feed treatments. Only fatty acids that are present at >1% are included, "Other" is the sum of all other fatty acids present at <1% (12:0, 15:0, 16:0, 16:1n-7t, 17:1-7c, 18:1n-9t, 18:3n-6c, 20:0, 20:2n-6c, 20:3n-6c, 21:0, 22:0, 24:0, 24:1n-9c). Different superscripts indicate significant differences along the row. There were no differences among initial fatty acids. ARA: arachidonic acid; DHA: docosahexaenoic acid; EPA: eicosapentaenoic acid; PUFA: polyunsaturated FA; HUFA: highly unsaturated FA.

FA	Initial Otohime	Initial Artemac	Initial O.Range	Final Otohime	Final Artemac	Final O.Range
C14:0	1.35 ± 0.06	1.33 ± 0.07	1.48 ± 0.06	7.34 ± 0.12 ^{a}	5.47 ± 0.11 ^{b}	3.56 ± 0.08 ^{c}
C16:0	25.92 ± 0.11	24.7 ± 0.14	25.48 ± 0.1	26.82 ± 0.19 ^{a}	25.80 ± 0.16 ^{a}	25.22 ± 0.09 ^{a}
C16:1n-7c	1.56 ± 0.03	1.60 ± 0.04	1.72 ± 0.02	2.22 ± 0.08 ^{b}	3.48 ± 0.04 ^{a}	1.55 ± 0.01 ^{c}
C17:0	1.22 ± 0.04	1.15 ± 0.05	1.23 ± 0.04	2.18 ± 0.10 ^{a}	1.07 ± 0.10 ^{b}	0.90 ± 0.04 ^{b}
C18:0	12.95 ± 0.12	12.15 ± 0.13	12.85 ± 0.16	6.33 ± 0.11 ^{a}	7.24 ± 0.14 ^{a}	7.73 ± 0.11 ^{a}
C18:1n-7t	3.06 ± 0.07	2.93 ± 0.07	2.88 ± 0.04	1.14 ± 0.10 ^{a}	1.22 ± 0.16 ^{a}	0.90 ± 0.06 ^{a}
C18:1n-9c	11.93 ± 0.06	12.32 ± 0.07	12.06 ± 0.08	7.97 ± 0.08 ^{b}	9.64 ± 0.11 ^{a}	7.40 ± 0.04 ^{b}
C18:2n-6c	4.07 ± 0.03	4.17 ± 0.03	4.15 ± 0.02	3.47 ± 0.03 ^{c}	4.31 ± 0.03 ^{b}	7.27 ± 0.03 ^{a}
C18:2n-6t	4.19 ± 0.03	4.30 ± 0.03	4.29 ± 0.02	3.60 ± 0.03 ^{c}	4.46 ± 0.03 ^{b}	7.43 ± 0.03 ^{a}
C18:3n-3c	17.62 ± 0.15	17.93 ± 0.14	16.69 ± 0.15	2.11 ± 0.18 ^{a}	1.85 ± 0.03 ^{a}	2.10 ± 0.03 ^{a}
C20:1n-9c	0.41 ± 0.01	0.44 ± 0.03	0.45 ± 0.01	2.04 ± 0.13 ^{a}	2.32 ± 0.20 ^{a}	0.74 ± 0.05 ^{b}
C20:4n-6c ARA	1.84 ± 0.03	2.00 ± 0.04	1.95 ± 0.04	1.11 ± 0.03 ^{b}	1.27 ± 0.05 ^{a}	1.33 ± 0.03 ^{a}
C20:5n-3 EPA	4.28 ± 0.06	4.63 ± 0.07	4.36 ± 0.06	10.08 ± 0.08 ^{a}	7.73 ± 0.05 ^{b}	7.55 ± 0.06 ^{b}
C22:1n-9c	0.34 ± 0.03	0.33 ± 0.02	0.33 ± 0.02	0.96 ± 0.13 ^{b}	2.02 ± 0.21 ^{a}	0.73 ± 0.04 ^{b}
C22:2n-6c	2.82 ± 0.04	3.07 ± 0.05	2.99 ± 0.03	5.14 ± 0.08 ^{a}	4.36 ± 0.05 ^{b}	4.19 ± 0.03 ^{b}
C22:4n-6c	1.04 ± 0.02	1.13 ± 0.03	1.09 ± 0.03	0.62 ± 0.02 ^{a}	0.48 ± 0.02 ^{b}	0.47 ± 0.01 ^{b}
C22:5n-3c	0.48 ± 0.02	0.54 ± 0.03	0.54 ± 0.03	1.77 ± 0.04 ^{b}	1.51 ± 0.03 ^{c}	2.33 ± 0.03 ^{a}
C22:6n-3c DHA	0.41 ± 0.05	0.71 ± 0.25	0.68 ± 0.16	11.07 ± 0.08 ^{b}	11.00 ± 0.15 ^{b}	14.86 ± 0.11 ^{a}
Other	4.54 ± 0.05	4.57 ± 0.05	4.78 ± 0.05	4.05 ± 0.12 ^{a}	4.77 ± 0.11 ^{a}	3.73 ± 0.05 ^{a}
Sum PUFA	37.98 ± 0.13	39.86 ± 0.18	38.18 ± 0.16	40.09 ± 0.08 ^{b}	37.80 ± 0.08 ^{b}	48.19 ± 0.12 ^{a}
Sum HUFA	8.27 ± 0.06	9.26 ± 0.15	8.88 ± 0.12	24.76 ± 0.08 ^{b}	22.11 ± 0.11 ^{c}	26.66 ± 0.12 ^{a}
EPA:ARA	2.32 ± 0.02	2.31 ± 0.01	2.23 ± 0.01	9.15 ± 0.11 ^{a}	6.13 ± 0.07 ^{b}	5.69 ± 0.03 ^{b}
DHA:EPA	0.1 ± 0.03	0.15 ± 0.10	0.15 ± 0.07	1.10 ± 0.04 ^{c}	1.43 ± 0.06 ^{b}	1.97 ± 0.03 ^{a}
DHA:ARA	0.23 ± 0.05	0.34 ± 0.16	0.34 ± 0.11	10.10 ± 0.16 ^{a}	8.79 ± 0.18 ^{a}	11.21 ± 0.10 ^{a}

For the initial sampling of larvae at the outset of the experiment, there were no differences for the proportion of any individual fatty acid detected among treatments (Table 3).

For the final sampling the proportions of some fatty acids differed among treatments (Table 3). ARA ($F_{(2,6)}$ = 6.15, p = 0.04) was higher in ORA (1.33% ± 0.03) than OTO (1.11% ± 0.03, p = 0.04) but not greater than in ART (1.27% ± 0.05, p = 0.41), while ART did not have a higher proportion of ARA than OTO (p = 0.07). EPA ($F_{(2,6)}$ = 56.21, $p < 0.01$) was higher in OTO (10.08% ± 0.08) than ART (7.73% ± 0.05, $p < 0.01$) and ORA (7.55 ± 0.06%, $p < 0.01$). DHA ($F_{(2,6)}$ = 21.37, $p < 0.01$) was proportionately higher in larvae from the ORA treatment (14.86% ± 0.11) compared to ART (11.00 ± 0.15, $p < 0.01$) and OTO (11.07% ± 0.08, $p < 0.01$).

Larvae in the ORA treatment had a higher proportion of PUFA ($F_{(2,6)}$ = 58.55, $p < 0.01$) (48.19% ± 0.12) than OTO (40.09% ± 0.08, $p < 0.01$) and ART (37.8% ± 0.08, $p < 0.01$). Likewise, HUFA ($F_{(2,6)}$ = 19.13, $p < 0.01$) were proportionately more abundant in ORA (26.66% ± 0.12), followed by OTO (24.76% ± 0.08, $p < 0.05$) and ART (22.11% ± 0.11, $p < 0.01$).

Ratios of EPA, DHA and ARA were different among treatments EPA:ARA ($F_{(2,6)}$ = 66.21, $p < 0.01$) and DHA:APA ($F_{(2,6)}$ = 62.97, $p < 0.01$), but not differ for DHA:ARA EPA ($F_{(2,6)}$ = 4.14, p = 0.07). OTO had the highest EPA:ARA ratio (9.15% ± 0.11), greater than that of ART (6.13% ± 0.07) ($p < 0.01$) and ORA (5.69% ± 0.03). ORA DHA:EPA (1.97% ± 0.03) was greater than OTO (1.10% ± 0.04) ($p < 0.01$) and ART (1.43 ± 0.06) ($p < 0.01$).

4. Discussion

The results from this experiment demonstrate the effectiveness of the artificial dry food (Otohime) for the rearing of larval giant kōkopu compared to two other commercially available artificial dry foods, ART and ORA. The OTO treatment produced giant kōkopu larvae with the greatest WW, DW, SGR, total body length and depth when compared to the ART and ORA treatments. The artificial dry food treatments did not have any influence on the proportions of either lipid or protein of the larvae at the end of the experiment. However, both the total lipid and protein content of the larvae were greater for the OTO treatment as a result of the larger overall size of the fish, than for fish provided with the ART and ORA feed treatments. At the end of the experiment there were no differences in any of the morphometric or biochemical parameters between the larvae provided with the ART and ORA treatments.

The intake of food particles is a crucial determining factor of the suitability of a feed products for larval fish and is affected by several characteristics. Perceptibility, capture/handling and palatability can impact the intake and the effectiveness of a feed item on the growth performance of larval fin fish [6]. Despite similar size and color of the different feed particles, perceptibility may have impacted on growth performances among the different dietary treatments in this experiment. The visual attractiveness of the feed, the speed at which feed particles sink through the water column of the tank and any chemical attractant can influence differences in the feeding response by the larvae [6]. Larval eyes are pigmented at hatch in giant kōkopu and relative eye size throughout the first 77 DAH of development indicate have good visual acuity [36]. However, weaning experiments show artificial feed intake is very low in the first 21 DAH indicating that these larvae are not able to recognize these feed particles [37]. Given that the current experiment included a weaning period, the non-nutritional characteristic of the artificial feed products used will have had an impact on the growth performance of larvae, potentially to the benefit of the darker colored and wider size range of Otohime. Future experiments should seek to separate out the weaning period and the artificial only period so as to determine both the best feeds by non-nutritional and nutritional characteristics and include larger feed particle items which giant kōkopu larvae have proven well adept at capturing [36,37]. The incremental increases in feed particle sizes and rates of water flow during the experiment, although kept consistent among all treatments, may have influenced food availability through affecting the period that food particles remained in suspension and available for consumption. The density and sinking rate of feed particles and their suspension via water turbulence are collectively important factors that influence feeding intake but are difficult to measure in practice [57,58].

Nutritional value is a key consideration in the selection of feed in fin fish aquaculture and is likely to have influenced the variation in growth performance amongst the feed treatments in this experiment. Formation of musculature accounts for the majority of mass increase in larval fish with dietary protein providing the amino acids required for muscle construction [59]. Dietary lipid is thought to be the primary source of energy, which is in significant demand in rapidly developing larval fish [59]. In order to realize maximal growth rates the optimum balance between the two macronutrients must be achieved [60]. The protein to energy ratio (P:E) is species-specific, with diets providing P:E ratios either side of the optimum will typically result in reduced growth. A low P:E can incur inadequate protein intake because consumption is also regulated by energetic requirements [8,44], while excessive dietary P:E lacks the energy required for catabolic and anabolic activity [8,44,61]. Despite the OTO feed treatment containing the lowest total protein (i.e., 51–51% versus 57% for ORA and 56% for ART) it is possible that the 123 mg kcal^{-1} P:E is most suitable or more readily available to meet the requirements of larval giant kōkopu as this treatment group achieved the greatest final mass and total protein. The ART diet has a lower P:E of 113 mg kcal^{-1} and corresponding with lower growth in the larvae. However, the ORA diet has a similar P:E (i.e., 124 mg kcal^{-1}) to that of OTO, but resulted in significantly lower growth performance. This outcome exemplifies the difficulty of larval fin fish diet selection due to the complex and interacting factors characteristics of aquaculture feeds where proximate analyses and ratios alone cannot be used in isolation to determine the optimum larval diet.

Generally, the fatty acid composition of a fish is a reflection of its diet [62–65]. However, digestive capability in larval fin fish is generally poor, limited by the lack of development of organs and physiological activities required to break down feed into usable nutrients [66–68]. Digestibility is fundamental to the transformation of food into utilizable nutrients and subsequently biomass and it is likely that this has had a material impact on the growth performance of larvae in this trial.

The delivery of sufficient essential PUFAs through the diet is critical during larviculture because they are utilized in tissue construction, especially for vital nervous and optic tissues [14,69]. The levels of EPA, DHA and ARA in each of the treatment diets did not always correlate directly with levels of accumulation in larval tissues. Otohime and O.Range have equally high levels of DHA in the feed particles but larvae in the ORA treatment accumulated more DHA as a proportion of total fatty acids. Although Otohime has almost twice as much DHA as Artemac, larvae from the OTO and ART treatments accumulated similar relative levels of this fatty acid. Subsequently, DHA was likely not the limiting factor to growth in these feed treatment. However, for each treatment Final DHA levels were two orders of magnitude greater than respective Initials indicating the demand and possible lack of DHA in the early diet which largely consisted with *Artemia* nauplii. This is consistent with earlier studies where DHA appears to be absent from these live feeds and a limiting factor to growth [37]. Furthermore, it has been highlighted that, current weaning protocols in the commercial hatchery have little effect on growth performance, and that weaning could be undertaken earlier as a result may in fact also reflect the improved nutritional provision of the artificial over the live feeds [36,37].

ARA content in the three larval diets is an order of magnitude lower in the diet than EPA and DHA, and was accumulated in larval tissues to a similarly low degree. ARA is highest in O.Range, of the three diets, with larvae in the ORA treatment accumulating a higher relative proportion of ARA in tissue when compared against the other treatments. Despite the importance of ARA to early development of larvae the differences in accumulated ARA within the tissues of the larvae of each treatment indicates that higher levels of ARA are not advantageous and that the levels found in the OTO diet appear to be sufficient [15,70]. This is contrary to previous findings for other species where higher levels of ARA and lower ratios of EPA:ARA were advantageous, further indicating the likely critical importance of EPA to larval giant kōkopu development [13,20,71]. Otohime has the highest EPA proportion of the three diets tested and larvae in the OTO treatment accumulated the highest levels of EPA relative to total fatty acid content. Both Artemac and O.Range have similar, low levels of EPA (roughly half that of Otohime) and EPA accumulated in larvae was equal at the end of the experiment for the latter two treatments. Furthermore, the larvae fed the O.Range diet which has an equal proportion of DHA to Otohime and slightly more ARA, performed as poorly as larvae fed Artemac which has much lower levels of each of these fatty acids. These observations are critical as they highlight the importance of high levels of EPA to giant kōkopu growth performance, as has been noted in many other species [14,21,69,72–74].

Although not measured in this study, the leaching of nutrients from feed particles (i.e., the loss of water soluble compounds out of feeds) can have negative implications for the performance of the feed for provisioning for fish growth. Upon introduction of feed particles to rearing tanks, any rapid leaching of water soluble compounds represents a subsequent loss of the nutritional value of subsequently consumed feed particles to the larvae. Due to high surface area to volume ratio of larval fish microdiets as much as 50–95% of free amino acids and protein hydrolysates can be lost through leaching within minutes of being fed into water [5,68,75–79]. Any differences in nutrient leaching of the feeds among the three treatments may have resulted in differing biomass growth rates between feed treatments with the OTO treatment potentially being more stable and therefore better able to provide nutrients to larvae after their consumption. A further study should determine the differences in the nutrient leaching among these three weaning diets.

The experimental set up may have had an impact on the performance on the artificial feeds due to stress caused due to initially transferring larvae from the commercial tanks in which they hatched. The majority of mortality occurred in the first 48 h after larvae were transferred into the experimental tanks as has been previously observed in prior experiments with this species [80]. Furthermore, recent studies have shown that the predominance of instar-I *Artemia* is not optimal as a first feed for giant kōkopu larvae, potentially compounding the post-transfer stress [80]. As such, the larvae that survived and became the subjects of this study may not best represent the performance of larvae in the commercial scale setting where the transfer in particular does not occur.

Given the scale of the experimental systems water quality was easily managed, maintaining very low ammonia levels which may not be possible at commercial scale. Closely related species *Galaxias maculatus* and *Galaxias fasciatus* have demonstrated tolerance to ammonia toxicity 1.47 and 0.80 mg/L, respectively; however, future experiments on this species should confirm the species-specific optimum ranges for giant kōkopu [42]. Furthermore, to ensure the most commercially relevant information is attained, this should focus on the early larval rearing stage [81,82]. A critical learning from this experiment is that the use of full commercial scale larviculture systems would be advantageous for future experimentation to avoid any unnecessary impact on larvae performance. As previously mentioned, larval performance may have been impacted in this study through initial handling for tank transfers, low quality live feeds as well as the small experimental scale for the tank experiment. The latter of which can impact water flow dynamics, affecting larvae behavior, as well as water quality as a result of RAS performance, all of which is likely to impact the performance of the larvae themselves [83–86]. These variations need to be removed in order to attain the most commercially relevant information possible.

This study has confirmed marked differences in performance of larval giant kōkopu fed with different commercially available feeds and that examining the differences among the nutritional status of the resulting fish and that of their diet can provide insights into basis of the observed differences. EPA appears to be more critical than DHA and ARA while a high P:E ratio is also advantageous. There are still several key aspects for future research which would lead to better understanding of the variation among these artificial diets and consequently their suitability as larval giant kōkopu diets. Nutrient leaching rate on rehydration of diets may impact whether key nutrients are still available to larvae when consumed. Furthermore, understanding differences in settling rates among diets may help to identify the characteristics of artificial diets which initiates a feeding response from larval giant kōkopu. Analyses should also be undertaken to better understand the digestive capabilities of these larvae. Furthermore, additional variables should be examined, particularly mortality rates, given the impact this has on total productivity of a larviculture system. However, the present study enables a baseline for a suitable diet from which future studies can build upon to further optimize growth performance and survival from dietary provisions. Given the paucity of information on galaxiid species in aquaculture more generally, these data should also act to inform the development of techniques for closely related species globally.

Author Contributions: Conceptualisation, W.M. and A.J.; methodology, W.M.; data analyses, W.M.; writing and editing, W.M. and A.J. All authors have read and agreed to the published version of the manuscript.

Funding: This research was funded by Callaghan Innovation under project NZPWH1501/PROP-47490-FELLOW. Logistic support was provided New Zealand Premium Whitebait Ltd.

Institutional Review Board Statement: Ethical review and approval was not required for this study under New Zealand's Animal Welfare Act 1999 because it explicitly excludes larval stages.

Data Availability Statement: The data presented in this study are available on request from the corresponding author. The data are not publicly available due to stakeholder privacy.

Acknowledgments: We would like to thank New Zealand Premium Whitebait Ltd. for the larvae and, in particular, Paul Decker and Tagried Kurwie for sharing their experiences and expertise on the larval rearing of galaxiids.

Conflicts of Interest: The authors declare no conflict of interest.

References

1. Crooks, N.; Rees, W.; Black, A.; Hide, D.; Britton, J.R.; Henshaw, A. Influence of live and dry diets on growth and survival of chub (*Leuciscus cephalus*) Larvae. *Fish. Aquac. J.* **2013**, FAJ-62. [CrossRef]
2. Nhu, V.C.; Dierckens, K.; Nguyen, H.T.; Hoang, T.M.T.; Le, T.L.; Tran, M.T.; Nys, C.; Sorgeloos, P. Effect of early co-feeding and different weaning diets on the performance of cobia (*Rachycentron canadum*) larvae and juveniles. *Aquaculture* **2010**, *305*, 52–58. [CrossRef]
3. Conceição, L.E.C.; Aragão, C.; Richard, N.; Engrola, S.; Gavaia, P.; Mira, S.; Dias, J. Novel methodologies in marine fish larval nutrition. *Fish Physiol. Biochem.* **2010**, *36*, 1–16. [CrossRef] [PubMed]
4. Shields, R.J. Larviculture of marine finfish in Europe. *Aquaculture* **2001**, *200*, 55–88. [CrossRef]
5. Hamre, K.; Yúfera, M.; Rønnestad, I.; Boglione, C.; Conceição, L.E.C.; Izquierdo, M.S. Fish larval nutrition and feed formulation: Knowledge gaps and bottlenecks for advances in larval rearing. *Rev. Aquac.* **2013**, *5*, S26–S58. [CrossRef]
6. Holt, G.J. *Larval Fish Nutrition*; Wiley-Blackwell: Oxford, UK, 2011; pp. 1–143. ISBN 978-0-470-95986-2.
7. FAO. *Manual on the Production and Use of Live Food for Aquaculture*; Lavens, P., Sorgeloos, P., Eds.; FAO Fisheries Technical Paper; Food and Agriculture Organisation: Rome, Italy, 1996; ISBN 978-92-5-103934-2.
8. Webster, C.D.; Lim, C.E. *Nutrient Requirements and Feeding of Finfish for Aquaculture*; CABI: Cambridge, UK, 2001; ISBN 978-0-85199-702-5.
9. Trushenski, J.T.; Kasper, C.S.; Kohler, C.C. Challenges and opportunities in finfish nutrition. *N. Am. J. Aquac.* **2006**, *68*, 122–140. [CrossRef]
10. Pousão-Ferreira, P.; Santos, P.; Carvalho, A.P.; Morais, S.; Narciso, L. Effect of an experimental microparticulate diet on the growth, survival and fatty acid profile of gilthead seabream (*Sparus aurata* L.) larvae. *Aquac. Int.* **2003**, *11*, 491–504. [CrossRef]
11. Kolkovski, S.; Tandler, A.; Kissil, G.W.; Gertler, A. The effect of dietary exogenous digestive enzymes on ingestion, assimilation, growth and survival of gilthead seabream (*Sparus Aurata*, Sparidae, Linnaeus) Larvae. *Fish Physiol. Biochem.* **1993**, *12*, 203–209. [CrossRef] [PubMed]
12. Jones, A.C.; Mead, A.; Kaiser, M.J.; Austen, M.C.V.; Adrian, A.W.; Auchterlonie, N.A.; Black, K.D.; Blow, L.R.; Bury, C.; Brown, J.H.; et al. Prioritization of knowledge needs for sustainable aquaculture: A national and global perspective. *Fish Fish.* **2015**, *16*, 668–683. [CrossRef]
13. Hauville, M.R.; Zambonino-Infante, J.L.; Bell, G.; Migaud, H.; Main, K.L. Impacts of three different microdiets on Florida pompano, *Trachinotus carolinus*, weaning success, growth, fatty acid incorporation and enzyme activity. *Aquaculture* **2014**, *422–423*, 268–276. [CrossRef]
14. Izquierdo, M.S. Essential fatty acid requirements of cultured marine fish larvae. *Aquac. Nutr.* **1996**, *2*, 183–191. [CrossRef]
15. Sargent, J.R.; Bell, G.; McEvoy, L.; Tocher, D.; Estevez, A. Recent developments in the essential fatty acid nutrition of fish. *Aquaculture* **1999**, *177*, 191–199. [CrossRef]
16. Watanabe, T.; Kiron, V. Prospects in larval fish dietetics. *Aquaculture* **1994**, *124*, 223–251. [CrossRef]
17. Benítez-Santana, T.; Masuda, R.; Juárez Carrillo, E.; Ganuza, E.; Valencia, A.; Hernández-Cruz, C.M.; Izquierdo, M.S. Dietary n-3 HUFA deficiency induces a reduced visual response in gilthead seabream *Sparus aurata* larvae. *Aquaculture* **2007**, *264*, 408–417. [CrossRef]
18. Koven, W. Key factors influencing juvenile quality in mariculture: A review. *Isr. J. Aquac.* **2003**, *55*, 283–297. [CrossRef]
19. Rodríguez, C.; Pérez, J.A.; Díaz, M.; Izquierdo, M.S.; Fernández-Palacios, H.; Lorenzo, A. Influence of the EPA:DHA ratio in rotifers on gilthead seabream (*Sparus aurata*) larval development. *Aquaculture* **1997**, *150*, 77–89. [CrossRef]
20. Tocher, D.R. Metabolism and functions of lipids and fatty acids in teleost fish. *Rev. Fish. Sci.* **2003**, *11*, 107–184. [CrossRef]
21. Bell, M.V.; Batty, R.S.; Dick, J.R.; Fretwell, K.; Navarro, J.C.; Sargent, J.R. Dietary Deficiency of cocosahexaenoic acid impairs vision at low light intensities in juvenile herring (*Clupea harengus* L.). *Lipids* **1995**, *30*, 443–449. [CrossRef]
22. McDowall, R.M. *New Zealand Freshwater Fishes*; Springer Netherlands: Dordrecht, 2010; ISBN 978-90-481-9270-0.
23. McDowall, R.M. *Diadromy in Fishes: Migrations between Freshwater and Marine Environments*; Timber Press: Portland, Oregon, USA, 1988; ISBN 978-0-88192-114-4.
24. McDowall, R.M. Early Hatch: A strategy for safe downstream larval transport in amphidromous gobies. *Rev. Fish Biol. Fish.* **2009**, *19*, 1–8. [CrossRef]
25. Benzie, V. Some ecological aspects of the spawning behaviour and early development of the common whitebait, *Galaxias maculatus attenuatus* (Jenyns). *Proc. N. Z. Ecol. Soc.* **1968**, *15*, 31–39.
26. Franklin, P.A.; Smith, J.; Baker, C.F.; Bartels, B.; Reeve, K. First observations on the timing and location of giant kōkopu (*Galaxias argenteus*) spawning. *N. Z. J. Mar. Freshw. Res.* **2015**, *49*, 419–426. [CrossRef]
27. McDowall, R.M.; Mitchell, C.P.; Brothers, E.B. Age at migration from the sea of juvenile galaxias in New Zealand (Pisces: Galaxiidae). *Bull. Mar. Sci.* **1994**, *54*, 385–402.

28. Mitchell, C.P. Laboratory culture of *Galaxias maculatus* and potential applications. *N. Z. J. Mar. Freshw. Res.* **1989**, *23*, 325–336. [CrossRef]
29. McDowall, R.M. *New Zealand Freshwater Fishes: A Natural History and Guide*; Revised Edition; Heinemann Reed MAF Publishing Group: Auckland, New Zealand, 1990; ISBN 978-0-7900-0022-0.
30. McDowall, R.M. *The New Zealand Whitebait Book*; Reed: Wellington, New Zealand, 1984; ISBN 978-0-589-01533-6.
31. McDowall, R.M. *Conservation and Management of the Whitebait Fishery*; Head Office, Department of Conservation: Wellington, New Zealand, 1991; ISBN 978-0-478-01317-7.
32. Main, M.R. Factors Influencing the Distribution of Kokopu and Koaro (Pisces: Galaxiidae). Ph.D. Thesis, University of Canterbury, Christchurch, New Zealand, 1988.
33. Swales, S.; West, D.W. Distribution, abundance and conservation status of native fish in some Waikato streams in the North Island of New Zealand. *J. R. Soc. N. Z.* **1991**, *21*, 281–296. [CrossRef]
34. Environment Southland. Whitebait Survey Summary of Results. 2013. Available online: http://Www.Es.Govt.Nz/Document%20Library/Research%20and%20reports/Various%20reports/Whitebait_survey_summary_of_results_2013.Pdf (accessed on 6 March 2017).
35. O'Brien, Q.; Cooper, D. Conservation breeding of shortfin eels (*Anguilla australis*) and giant kokopu (*Galaxias argenteus*) at Mahurangi Technical Institute using aquarium and aquaculture techniques. *Int. Zoo Yearb.* **2013**, *47*, 120–128. [CrossRef]
36. McKay, W.J.G.; Jeffs, A.G. Morphometric and energetic development of artificially reared giant kōkopu (*Galaxias argenteus*). *Aquaculture* **2021**, *544*, 737123. [CrossRef]
37. McKay, W.J.G.; Jeffs, A.G. Improving the weaning of larval giant kōkopu, *Galaxias argenteus*: An emerging aquaculture species. *J. World Aquac. Soc.* **2022**, jwas.12926, *Online Version of Record before inclusion in an issue*. [CrossRef]
38. Dantagnan, P.; Bórquez, A.; Pavez, C.; Hernández, A. Feeding ω-3 PUFA enriched rotifers to *Galaxias maculatus* (Jenyns, 1842) larvae reared at different salinity conditions: Effects on growth parameters, survival and fatty acids profile. *Lat. Am. J. Aquat. Res.* **2013**, *41*, 404–411. [CrossRef]
39. Brownell, C.L. Water quality requirements for first-feeding in marine fish larvae. I. ammonia, nitrite, and nitrate. *J. Exp. Mar. Biol. Ecol.* **1980**, *44*, 269–283. [CrossRef]
40. Brownell, C.L. Water quality requirements for first-feeding in marine fish larvae. II. pH, oxygen, and carbon dioxide. *J. Exp. Mar. Biol. Ecol.* **1980**, *44*, 285–298. [CrossRef]
41. West, D.W.; Boubée, J.A.T.; Barrier, R.F.G. Responses to pH of nine fishes and one shrimp native to New Zealand freshwaters. *N. Z. J. Mar. Freshw. Res.* **1997**, *31*, 461–468. [CrossRef]
42. Richardson, J. Acute Ammonia toxicity for eight New Zealand indigenous freshwater species. *N. Z. J. Mar. Freshw. Res.* **1997**, *31*, 185–190. [CrossRef]
43. Bonaldo, A.; Parma, L.; Badiani, A.; Serratore, P.; Gatta, P.P. Very early weaning of common sole (*Solea solea* L.) larvae by means of different feeding regimes and three commercial microdiets: Influence on performances, metamorphosis development and tank hygiene. *Aquaculture* **2011**, *321*, 237–244. [CrossRef]
44. Csargo, I.J.; Brown, M.L.; Ward, M.J. Evaluation of four commercial grower diets for production of largemouth bass. *J. Appl. Aquac.* **2013**, *25*, 35–49. [CrossRef]
45. Jobling, M. National Research Council (NRC): Nutrient requirements of fish and shrimp. *Aquac. Int.* **2012**, *20*, 601–602. [CrossRef]
46. Portz, L.; Cyrino, J.E.P. Digestibility of nutrients and amino acids of different protein sources in practical diets by largemouth bass *Micropterus salmoides* (Lacepéde, 1802). *Aquac. Res.* **2004**, *35*, 312–320. [CrossRef]
47. Gallagher, M.L. Apparent digestibility coefficients for some carbohydrates in diets for hybrid striped bass *Morone saxatilis* × *M. Chrysops*. *J. World Aquac. Soc.* **1997**, *28*, 429–431. [CrossRef]
48. Thériault, I.; Pernet, F. Lipid nutrition and settlement behaviour in American lobster *Homarus Americanus*. *Aquat. Biol.* **2007**, *1*, 121–133. [CrossRef]
49. Ricker, W.E. *Handbook of Computations for Biological Statistics of Fish Populations*; The Fisheries Research Board of Canada under the Control of the Honourable the Minister of Fisheries: Ottawa ON, Canada, 1958; Volume 119, pp. 1–300.
50. Bligh, E.G.; Dyer, W.J. A rapid method of total lipid extraction and purification. *Can. J. Biochem. Physiol.* **1959**, *37*, 911–917. [CrossRef]
51. Wang, M.; O'Rorke, R.; Nodder, S.D.; Jeffs, A.G.; Wang, M.; O'Rorke, R.; Nodder, S.D.; Jeffs, A.G. Nutritional composition of potential zooplankton prey of the spiny lobster phyllosoma (*Jasus edwardsii*). *Mar. Freshw. Res.* **2013**, *65*, 337–349. [CrossRef]
52. Lepage, G.; Roy, C.C. Direct transesterification of all classes of lipids in a one-step reaction. *J. Lipid Res.* **1986**, *27*, 114–120. [CrossRef]
53. Kramer, J.K.G.; Hernandez, M.; Cruz-Hernandez, C.; Kraft, J.; Dugan, M.E.R. Combining results of two GC separations partly achieves determination of all cis and trans 16:1, 18:1, 18:2 and 18:3 except CLA isomers of milk fat as demonstrated using Ag-ion SPE fractionation. *Lipids* **2008**, *43*, 259–273. [CrossRef]
54. Jeffs, A.G.; Phleger, C.F.; Nelson, M.M.; Mooney, B.D.; Nichols, P.D. Marked depletion of polar lipid and non-essential fatty acids following settlement by post-larvae of the spiny lobster *Jasus verreauxi*. *Comp. Biochem. Physiol. A. Mol. Integr. Physiol.* **2002**, *131*, 305–311. [CrossRef]
55. Mossoba, M.; Kramer, J.K.G. *Official Methods for the Determination of Trans Fat*; AOCS Press: Champaign, IL, USA, 2009.
56. Zar, J.H. *Biostatistical Analysis*, 4th ed.; Prentice Hall: Upper Saddle River, NJ, USA, 1999; ISBN 978-0-13-081542-2.

57. Rønnestad, I.; Yúfera, M.; Ueberschär, B.; Ribeiro, L.; Sæle, Ø.; Boglione, C. Feeding behaviour and digestive physiology in larval fish: Current knowledge, and gaps and bottlenecks in research. *Rev. Aquac.* **2013**, *5*, S59–S98. [CrossRef]
58. Kato, Y.; Ohshima, M.; Yamashita, Y.; Kitagawa, T.; Kimura, S. Effect of larval ontogeny, turbulence, and prey density on survival in red sea bream, *Pagrus major*, larvae and interspecies comparison of turbulence effect. *Coast. Mar. Sci.* **2012**, *35*, 262–268.
59. Qin, J.G. Larval Fish Nutrition and Rearing Technology: State of the Art and Future. In *Aquaculture Research Trends*; Schwartz, S.H., Ed.; Nova Science Publishers: New York, NY, USA, 2008; pp. 113–148.
60. Lucas, J.S.; Southgate, P.C. *Aquaculture: Farming Aquatic Animals and Plants*; Blackwell Publishing Ltd.: West Sussex, UK, 2012; ISBN 978-1-118-68793-2.
61. Goddard, S. *Feed Management in Intensive Aquaculture*; Springer US: Boston, MA, USA, 1996; ISBN 978-1-4613-1173-7.
62. Salhi, M.; Bessonart, M. Growth, survival and fatty acid composition of *Rhamdia quelen* (Quoy and Gaimard, 1824) larvae fed on artificial diet alone or in combination with *Artemia* nauplii. *Aquac. Res.* **2012**, *44*, 41–49. [CrossRef]
63. Turchini, G.M.; Ng, W.-K.; Tocher, D.R. *Fish Oil Replacement and Alternative Lipid Sources in Aquaculture Feeds*; CRC Press: Boca Raton, FL, USA, 2010; ISBN 978-1-4398-0863-4.
64. Watanabe, W.O.; Alam, S.; Ostrowski, A.D.; Montgomery, F.A.; Gabel, J.E.; Morris, J.A.; Seaton, P.J. Live prey enrichment and artificial microdiets for larviculture of Atlantic red porgy *Pagrus pagrus*. *Aquac. Rep.* **2016**, *3*, 93–107. [CrossRef]
65. Xu, H.; Turchini, G.M.; Francis, D.S.; Liang, M.; Mock, T.S.; Rombenso, A.; Ai, Q. Are fish what they eat? A fatty acid's perspective. *Prog. Lipid Res.* **2020**, *80*, 101064. [CrossRef]
66. Cousin, J.C.B.; Baudin-Laurencin, F.; Gabaudan, J. Ontogeny of enzymatic activities in fed and fasting turbot, *Scophthalmus maximus* L. *J. Fish Biol.* **1987**, *30*, 15–33. [CrossRef]
67. Kvåle, A.; Mangor-Jensen, A.; Moren, M.; Espe, M.; Hamre, K. Development and characterisation of some intestinal enzymes in Atlantic cod (*Gadus morhua* L.) and Atlantic halibut (*Hippoglossus hippoglossus* L.) larvae. *Aquaculture* **2007**, *264*, 457–468. [CrossRef]
68. Nordgreen, A.; Tonheim, S.; Hamre, K. Protein quality of larval feed with increased concentration of hydrolysed protein: Effects of heat treatment and leaching. *Aquac. Nutr.* **2009**, *15*, 525–536. [CrossRef]
69. Sargent, J.R.; Tocher, D.R.; Bell, J.G. The Lipids. In *Fish Nutrition (Third Edition)*; Halver, J.E., Hardy, R.W., Eds.; Academic Press: San Diego, CA, USA, 2003; pp. 181–257. ISBN 978-0-12-319652-1.
70. Bell, J.G.; Farndale, B.M.; Bruce, M.P.; Navas, J.M.; Carillo, M. Effects of broodstock dietary lipid on fatty acid compositions of eggs from sea bass (*Dicentrarchus labtax*). *Aquaculture* **1997**, *149*, 107–119. [CrossRef]
71. Sargent, J.R.; McEvoy, L.; Estevez, A.; Bell, G.; Bell, M.; Henderson, J.; Tocher, D. Lipid nutrition of marine fish during early development: Current status and future directions. *Aquaculture* **1999**, *179*, 217–229. [CrossRef]
72. Boglino, A.; Darias, M.J.; Ortiz-Delgado, J.B.; Özcan, F.; Estévez, A.; Andree, K.B.; Hontoria, F.; Sarasquete, C.; Gisbert, E. Commercial products for *Artemia* enrichment affect growth performance, digestive system maturation, ossification and incidence of skeletal deformities in Senegalese sole (*Solea senegalensis*) larvae. *Aquaculture* **2012**, *324–325*, 290–302. [CrossRef]
73. Izquierdo, M.S.; Koven, W. Lipids. In *Larval Fish Nutrition*; Wiley: Hoboken, NJ, USA, 2011; pp. 47–81.
74. Koven, W.; Nixon-Shtupler, O.; Lutzky, S.; Ben Atia, S.; Elkayam, A.; Tandler, A. The effect of n-3 HUFA and light intensity on hunting success in gilthead sea bream (*Sparus aurata*). *Isr. J. Aquacult. Bamidgeh.* **2012**, *64*, 1–8.
75. Baskerville-Bridges, B.; Kling, L. Development and evaluation of microparticulate diets for early weaning of Atlantic cod *Gadus morhua* larvae. *Aquac. Nutr.* **2000**, *6*, 171–182. [CrossRef]
76. Kvåle, A.; Yúfera, M.; Nygård, E.; Aursland, K.; Harboe, T.; Hamre, K. Leaching properties of three different micropaticulate diets and preference of the diets in cod (*Gadus morhua* L.) larvae. *Aquaculture* **2006**, *251*, 402–415. [CrossRef]
77. Nicklason, P.M.; Johnson, R.B. Real-time measurement of protein leaching from micro-particulate larval fish feeds. *Aquac. Res.* **2008**, *39*, 1793–1798. [CrossRef]
78. Önal, U.; Langdon, C. Development and characterization of complex particles for delivery of amino acids to early marine fish larvae. *Mar. Biol.* **2005**, *146*, 1031–1038. [CrossRef]
79. Yúfera, M.; Kolkovski, S.; Fernández-Díaz, C.; Dabrowski, K. Free Amino acid leaching from a protein-walled microencapsulated diet for fish larvae. *Aquaculture* **2002**, *214*, 273–287. [CrossRef]
80. McKay, W.J.G. Advances in Larval Development and Feed Regime Optimisation for Giant Kōkopu (*Galaxias argenteus*). Ph.D. Thesis, University of Auckland, Auckland, New Zealand, 2022.
81. Wang, W.; Wang, H.; Yu, C.; Jiang, Z. Acute toxicity of ammonia and nitrite to different ages of Pacific cod (*Gadus macrocephalus*) larvae. *Chem. Speciat. Bioavailab.* **2015**, *27*, 147–155. [CrossRef]
82. van der Meeren, T.; Mangor Jensen, A. Tolerance of Atlantic cod (*Gadus morhua* L.) larvae to acute ammonia exposure. *Aquac. Int.* **2020**, *28*, 1753–1769. [CrossRef]
83. Espmark, Å.M.; Kolarevic, J.; Åsgård, T.; Terjesen, B.F. Tank size and fish management history matters in experimental design. *Aquac. Res.* **2017**, *48*, 2876–2894. [CrossRef]
84. Davidson, J.; Summerfelt, S. Solids flushing, mixing, and water velocity profiles within large (10 and 150 m3) circular 'cornell-type' dual-drain tanks. *Aquac. Eng.* **2004**, *32*, 245–271. [CrossRef]
85. Oppedal, F.; Dempster, T.; Stien, L.H. Environmental drivers of Atlantic salmon behaviour in sea-cages: A Review. *Aquaculture* **2011**, *311*, 1–18. [CrossRef]
86. Holan, A.B.; Good, C.; Powell, M.D. 9—Health management in recirculating aquaculture aystems (RAS). In *Aquaculture Health Management*; Kibenge, F.S.B., Powell, M.D., Eds.; Academic Press: Cambridge, MA, USA, 2020; pp. 281–318, ISBN 978-0-12-813359-0.

fishes

MDPI

Article

Improving Growth, Digestive and Antioxidant Enzymes and Immune Response of Juvenile Grass Carp (*Ctenopharyngodon idella*) by Using Dietary *Spirulina platensis*

Mehwish Faheem [1,*], Rafia Jamal [1], Nadia Nazeer [1], Saba Khaliq [2], Seyed Hossein Hoseinifar [3,*], Hien Van Doan [4,5] and Marina Paolucci [6]

1 Department of Zoology, Government College University, Lahore 54000, Pakistan
2 Department of Physiology and Cell Biology, University of Health Sciences, Lahore 54600, Pakistan
3 Department of Fisheries, Gorgan University of Agricultural Sciences and Natural Resources, Gorgan 49189-43464, Iran
4 Department of Animal and Aquatic Sciences, Faculty of Agriculture, Chiang Mai University, Chiang Mai 50200, Thailand
5 Science and Technology Research Institute, Chiang Mai University, 239 Huay Keaw Rd., Suthep, Muang, Chiang Mai 50200, Thailand
6 Department of Science and Technology, University of Sannio, Via Port'Arsa 11, 82100 Benevento, Italy
* Correspondence: mehwishfaheem@gcu.edu.pk (M.F.); hoseinifar@gau.ac.ir (S.H.H.)

Abstract: The present study was designed to investigate the effect of *Spirulina platensis* (SP) supplemented diets on the growth performance, digestive enzymes, hepatic antioxidants and innate immunity biomarkers in juvenile grass carp (*Ctenopharyngodon idella*). Two hundred and forty grass carp juveniles (average weight 4.81 ± 0.13 g) were divided into four treatment groups in triplicates (20 fish/replicate) and fed with diets containing 0, 1, 5, and 10% *Spirulina* for 90 days. A significant increase in growth ($p < 0.05$) was observed in fish fed with diets having 1 and 5% *Spirulina*. Intestinal protease and lipase activities increased significantly ($p < 0.01$) in fish fed with a diet having 5% *Spirulina* while intestinal amylase activity increased significantly ($p < 0.01$) in fish fed with diets having 1, 5, and 10% *Spirulina*. Hepatic lipid peroxidation decreased significantly ($p < 0.05$) in fish fed with a 1 and 5% *Spirulina* supplemented diet. The activity of catalase, glutathione-S-transferase, and glutathione levels increased significantly ($p < 0.05$) in the livers of fish fed with 1% *Spirulina* supplemented diets while no significant difference ($p > 0.05$) was observed for hepatic superoxide dismutase levels when compared to the control. Significant increases in the skin mucus protease ($p < 0.05$), antiprotease ($p < 0.01$), lysozyme ($p < 0.001$), and peroxidase ($p < 0.05$) activities were observed in fish fed with 5% *Spirulina*-supplemented diets. Gene expression analysis of head kidney showed that fish fed with a 1% *Spirulina* diet had significantly ($p < 0.01$) higher expression of *tnf-α*, *il-8*, and *inf-γ*. In conclusion, the present study suggested that the inclusion of 5% *Spirulina platensis* in the diet of grass carp has positive effects on growth, digestive enzymes, antioxidants, and innate immunity.

Keywords: *Spirulina*; digestive enzymes; hepatic enzymes; gene expression; aquaculture; cyanobacteria

Citation: Faheem, M.; Jamal, R.; Nazeer, N.; Khaliq, S.; Hoseinifar, S.H.; Van Doan, H.; Paolucci, M. Improving Growth, Digestive and Antioxidant Enzymes and Immune Response of Juvenile Grass Carp (*Ctenopharyngodon idella*) by Using Dietary *Spirulina platensis*. *Fishes* **2022**, *7*, 237. https://doi.org/10.3390/fishes7050237

Academic Editor: Junyan Jin

Received: 16 July 2022
Accepted: 25 August 2022
Published: 5 September 2022

1. Introduction

The world population is increasing at a tremendous rate and is expected to hit 10 billion people by 2050 [1]. Food supply, especially animal-based protein, will be crucial in the coming decades. The global demand for animal-derived protein will be doubled by 2050 which is expected to intensify pressure on the need to produce more animal-based protein [2]. Fish and shellfish are the primary sources of protein for approximately 950 million people worldwide [3]. Aquaculture is expanding faster than any other food-producing sector, and intensive aquaculture has the potential to provide animal-based protein to an exponentially growing human population [4,5]. As stocking density is high in an intensive culture system, this may lead to crowding stress. Crowding stress along with poor water quality results in the emergence of disease outbreaks that inflict significant financial losses [6].

Infectious diseases have always been a great threat in intensive animal production systems [7–11]. Antibiotics are used to treat diseases and reduce pathogens and disease incidences [12] but antibiotic resistance has become a global issue in humans as well as poultry, livestock, and aquaculture [13–15]. Many researchers are now trying to discover sustainable alternatives to antibiotics [16–21]. The use of medicinal plants, macro and microalgae, herbs, and probiotics as oral immunostimulants has gained a great deal of interest throughout the globe [22,23]. Phenol, polyphenol, quinone, alkaloid, terpenoid, polypeptide compounds and lectin are present in algae, certain medicinal plants, and their by-products which promote growth, enhance the antioxidant status, and stimulate the immune system, thus providing a sustainable alternative to vaccines and antibiotics [24,25].

Algae and other plant-based feed additives are able to promote the growth of fish [26–29], protect against diseases [29,30], strengthen the immune system [22,26,31–33], stimulate hunger and enhance feed consumption [34,35], reduce stress [36–38] and improve digestion by increasing secretion of different digestive enzymes [39–41]. They also have antimicrobial and antiviral properties [39,42–44]. Algae and other plants are cheaper, eco-friendly, have minimum side effects, and are frequently used as substitutes to the costly antibiotics in fish health management. The World Health Organization (WHO) encourages the use of supplementary diets combined with algae and medicinal herbs or plants, which minimize the application of chemicals in fish diet [45].

Spirulina is a filamentous blue-green algae, which has the potential to be used in aquafeed as a growth promoter and immunostimulant [46–50]. Dried *Spirulina* powder has high protein content (up to 55–70% of dry weight). It also contains a high amount of gamma-linolenic acid (GLA), polysaccharides, phycobiliproteins, carotenoids, vitamins (especially B12), pigments such as carotenoids, and minerals. Some studies also reported the immune-stimulating effect of *Spirulina* in several fish species [5,47,51–54], but to the best of our knowledge, no study has been performed using *Ctenopharyngodon idella*, commonly known as grass carp. We hypothesized that grass carp, a herbivorous cyprinid, can show better digestive and physiological status after feeding with algal-based feed. Therefore, this study was performed to evaluate the effect of feeding *Spirulina*-based diets on the growth, physiology, and immunity of grass carp.

2. Materials and Methods

2.1. Fish Culture and Diet Preparation

A three-month feeding trial was carried out at the Animal House fish rearing facility, Department of Zoology, Government College University Lahore Campus. Fish were provided by the Himalaya Fish Hatchery, Muridke. For acclimation purposes, fish were kept in laboratory conditions for two weeks and treated with potassium permanganate to avoid any infections. During this time, fish were fed with a basal (control) diet twice a day at a rate of 3% of body weight. Temperature (20 °C), pH (7.8), electrical conductivity (618 µS/cm), and dissolved oxygen (5.7 mg/L) were measured with the help of digital meters during the experiment.

Feed was prepared by using ingredients bought from the local market. Dried *Spirulina* powder (Naturya Organic Superfoods) was purchased from a local organic store. Dried *Spirulina* was added to the fish feed and four diets were prepared with 0%, 1%, 5%, and 10% *Spirulina* supplementation. The inclusion of *Spirulina* in the diet was selected based on previously published literature [53,54]. Ingredients were ground into a fine powder and mixed together. After mixing, pellets were formed manually using a mincer. Pellets were shade dried for 48 h and were preserved in zipper bags for further use.

The diet samples were homogenized using a pestle and mortar and were analyzed by standard methods of AOAC [55]. Moisture was determined by oven-drying at 105 °C for 12 h. Crude protein (N × 6.25) was estimated by micro Kjeldahl's apparatus. Ash was determined by ignition samples at 650 °C for 12 h (Eyela-TMF 3100) to constant weight.

Crude fat was determined by the petroleum ether extraction method through the Soxtec HT2 1045 system. The carbohydrate content of the feed was measured by subtracting the % values of nutrients (moisture, protein, fat, ash) from 100. The composition and proximate analysis of *Spirulina* powder and feed are given in Table 1.

Table 1. Composition of basal and experimental diet (%).

Ingredients	Control	1% *Spirulina*	5% *Spirulina*	10% *Spirulina*
Fish meal	10	10	10	10
Wheat flour	20	19.5	18	15
Soybean meal	21.5	21	19.5	18.5
Cottonseed meal	6.5	6.5	6	6
Mustard cake	21.5	21.5	21	20
Fish oil	3	3	3	3
Soya oil	4	4	4	4
Mineral premix [a]	3	3	3	3
Vitamin premix [b]	3	3	3	3
Spirulina	0	1	5	10
cellulose	4	4	4	4
Salt	0.5	0.5	0.5	0.5
Molasses	3	3	3	3
Crude protein (%)	28	28	28.5	29
Fat	5.5	5.6	5.5	5.7
Ash	20.37	20.34	20.45	20.39
Moisture	12.89	12.77	12.6	12.79

[a] Each 1000 g of mineral premix contains copper (Cu) 0.25 g, magnesium (Mg) 25 g, calcium (Ca) 0.023 g, zinc (Zn) 2.17 g, manganese (Mn) 10 g, potassium (K) 0.5 g, selenium (Se) 0.01 g, sodium (Na) 120 g. [b] Each 1000 g of vitamin premix contains Vit A 0.8 g, Vit D_3 0.16 g, Vit E 0.38 g, Vit B_1 1 g, Vit B_2 1.25 g, Vit B_{12} 0.001 g, Vit B_3 6.25 g, Vit B_6 4 g, Pantothenic acid 54 g, folic acid 5 g.

2.2. Experimental Setup

After the acclimatization of two weeks, fish were anesthetized using clove oil and weighed individually. Fish with almost equal weight (initial weight 4.81 g $\pm$ 0.13) were stocked in a glass aquarium (20 fingerlings/group) filled with 60 L of water. Experiments were performed in triplicates; therefore, 12 aquariums were set with 20 fish in each aquarium. All aquariums were supplied with air stones to maintain dissolved oxygen. The feeding trial was started 24 h post stocking in the aquarium. Fish were fed twice a day (8–9 am and 3–4 pm) at a rate of 3% of body weight. The feeding ration was adjusted fortnightly according to the weight of the fish. The control was fed with a basal diet having 0% *Spirulina* powder, while groups II, III, and IV were fed with a diet containing 1, 5, and 10% *Spirulina*, respectively.

The feeding regime was continued for 90 days. The aquaria were siphoned every other day to remove the uneaten feed particles.

2.3. Analysis of Growth

At the end of 90 days of the feeding trial, fish from each aquarium were anesthetized using clove oil [56], and weight and length were recorded again to calculate growth. Growth parameters were recorded using the following formulas:

$$\text{Weight gain (g)} = \text{W2} - \text{W1} \tag{1}$$

where W2 is the final weight and W1 is the initial weight;

$$\text{Specific growth rate (SGR; \%g/day)} = 100\ (\text{Ln W2(g)} - \text{Ln W1(g)})/\text{T} \tag{2}$$

where W2 is the final weight (g), W1 is the initial weight (g), and T is the experimental period (day); Thermal growth coefficient = 100 (final body weight$^{1/3}$ − initial body weight$^{1/3}$/experimental days × mean daily temperature); Condition factor (K) = 100 × (body weight; g)/(body length; cm^3); Fish survival (%) = 100 (Number of fish at the end of trial/number of fish at the start of trial).

2.4. Analysis of Mucus Immunity

Mucus samples were collected at the end of the feeding trial for the determination of the innate immune response. To collect mucus samples, individual fish were added to a polythene bag having 10 mL of sterile saline solution. Fish were rubbed gently and all liquid was transferred to 50 mL falcon tubes and centrifuged. The supernatant was collected and used for the analysis of lysozyme, protease, antiprotease, and peroxidase activity as described in our previous work [57]. Briefly, lysozyme activity was measured using *Micrococcus luteus*. Change in turbulence was recorded using a spectrophotometer at 450 nm. Lysozyme activity was expressed as units/mL. For protease activity, mucus samples were incubated with azocasein, ammonium bicarbonate, and trichloroacetic acid. The reaction was stopped using NaOH. Optical density was recorded at 450 nm. Antiprotease activity was measured using trypsin as a substrate and optical density was recorded at 450 nm. For peroxidase activity, mucus samples were incubated with 3,3′,5,5′-tetramethylbenzidine (TMB) and Hank's Balanced Salt Solution (without Ca^{2+} or Mg^{2+}). Sulphuric acid was used to stop the reaction and optical density was recorded at 450 nm.

After the collection of mucus samples, fish were dissected and the intestine, liver, and kidneys were harvested and stored in pre-labelled Eppendorf's at −80 °C. The intestine was used for the estimation of digestive enzymes; liver tissue was used for antioxidant analysis, while kidney tissue was used for the expression of immune-related genes.

2.5. Oxidative Stress and Antioxidant Defense Markers Assessment

The liver was used for the assessment of antioxidative enzymes. Tissue samples were weighed and homogenized in 0.1M phosphate buffer (pH 7.4) to make 10% homogenate. A portion of the homogenate (500 µL) was used to estimate lipid peroxidation while the remaining homogenate was centrifuged at 13,000 rpm at 4 °C for 30 min. The supernatant was collected and stored in clean Eppendorf's and used for the estimation of glutathione, catalase, glutathione-S-transferase, and superoxide dismutase [56,58,59]. Protein content in the supernatant was measured using Bradford reagent.

2.6. Analysis of Digestive Enzymes

The whole intestine of ten fish from each aquarium/replicate (n = 30 for each treatment group) was extracted and kept in prelabeled, sterile falcon tubes and weighed. The entire procedure was performed on ice. A double amount of saline solution (0.86%) was added to each sample, homogenized, and centrifuged at a maximum speed of 4 °C. The supernatant was removed and stored at −80 until analysis of digestive enzymes [41].

Protease activity was measured according to [60], and the reaction mixture of 0.2 mL of supernatant and 0.2 mL of freshly prepared azocasein was taken in an Eppendorf. The sample was incubated in a water bath for 30 min at 30 °C. TCA (1.2 mL) was added and incubated for 30 min at room temperature. The reaction mixture (1.5 mL) was then centrifuged at a maximum speed for 10 min. The supernatant (1 mL) was taken in a cuvette and an equal volume of NaOH was added. Absorbance was recorded at 450 nm.

For amylase activity, 1.8 mL of 0.1M sodium phosphate buffer, 0.1 mL of 1% starch solution (substrate), and 0.1 mL of enzyme source were taken in an Eppendorf. The reaction mixture was incubated for 30 min at 37 °C. The reaction was stopped by adding 2 mL of dinitro salicylic acid (DNS). The sample was then heated in a boiling water bath for 5 min and cooled. A volume of 10 mL of water was added to the sample; a brown color appeared. Absorbance was recorded at 540 nm [61].

Lipase activity was measured according to Dar et al. [61]. Briefly, the reaction mixture consists of homogenate (1 mL), phosphate buffer (0.5 mL), and olive oil emulsion (2 mL). The mixture was shaken well and incubated at 37 °C for 24 h. The mixture was titrated using NaOH (0.0.5N) using phenolphthalein as an indicator. The milliliter equivalent of alkali consumed was taken as a measure of the activity of lipase.

2.7. Expression Analysis of Immune-Related Genes

Head kidney (100 mg) from 5 fish per replicate (n = 15 per treatment group) was used to extract RNA with Trizol reagent. Quality and quantity of RNA were checked on nanodrop (Thermo Fisher Scientific). First-strand complementary DNA was synthesized using a commercially available RevertAid First Strand cDNA synthesis kit (Thermo Fisher Scientific). Resultant cDNA was diluted 10 times and used to study relative gene expression. Primer sequence, accession number, and annealing temperature of genes are listed in Table 2. Primers were validated using conventional PCR. Real-time PCR was used to analyze mRNA levels of selected immune-related genes and housekeeping genes. The reaction mixture (20 µL) contained syber green master mix (10 µL), cDNA (1 µL), and 1 µL forward and reverse primers. The expression of *tumor necrosis factor-α*, *interleukin-8*, and *interferon-γ* was estimated by the $2^{-\Delta\Delta ct}$ method [62]. *Actin-β* was used as a reference control. Each sample was analyzed in triplicate.

Table 2. Primer sequence, annealing temperature, product size, and accession number of selected genes.

Genes	Primer Sequence	Annealing Temperature (°C)	Product Size	Accession Number
tnf-α	GGTGCATACGACCCTGAAGT TTTTGCCTCCATAGGAATCG	60	244	JQ040498.1
il-8	ATGAGTCTTAGAGGTCTGGGTG ACAGTGAGGGCTAGGAGGG	60	118	JN663841
inf-γ	TGCATGTAGGCGGATATCAA GAGGGCGCATAAGTCTGAAG	60	192	FJ695520.1
actin-β	ACCCACACTGTGCCCATCTACG ATTTCCCTCTCGGCTGTGGTGG	60	146	JQ991014.1

2.8. Statistical Analysis

All the data of growth and biochemical parameters are represented as mean ± S.E.M. ANOVA followed by Tukey's test was used to report statistical differences among groups. Statistical difference $p < 0.05$ is represented with a single asterisk (*), while double (**) and triple (***) asterisks represent statistical differences of $p < 0.01$ and $p < 0.001$, respectively. All the data were analyzed using GraphPad Prism.

3. Results

3.1. Growth Performance

A significant increase in weight gain was recorded in fish fed with 1 and 5% *Spirulina* powder. However, fish fed with a diet supplemented with 10% *Spirulina* did not differ statistically from the control. Percentage weight gain was highest in the fish fed with a diet supplemented with 5% *Spirulina* powder; the polynomial regression equation is $y = -2.8202x^2 + 32.476x + 54.179$; $R^2 = 0.9259$. The survival rate was 100% in all the treatment groups throughout the study period (Table 3). Polynomial regression analysis of dietary inclusion of *Spirulina* and specific growth rate is $y = -0.0102x^2 + 0.1117x + 0.3354$; $R^2 = 0.8693$ while the polynomial regression equation for feed conversion is $y = -0.0061x^2 + 0.0658x + 0.1781$; $R^2 = 0.8677$.

Table 3. Growth performance of grass carp fed with 0, 1, 5, and 10% *Spirulina* for 90 days.

Parameters	Control	1% *Spirulina*	5% *Spirulina*	10% *Spirulina*
Initial body weight (g)	4.9 ± 0.125	4.975 ± 0.182	4.65 ± 0.175	4.75 ± 0.153
Final body weight (g)	7.393 ± 0.4362	10.19 ± 0.7686 *	11.11 ± 0.7644 ***	7.948 ± 0.4511
%-weight gain	43.67 ± 8.310	92.43 ± 17.92	117.8 ± 16.32 **	89.09 ± 13.07
SGR	0.2874 ± 0.036	0.5035 ± 0.056	0.6154 ± 0.037	0.4401 ± 0.054
Thermal condition factor	0.1498 ± 0.020	0.2772 ± 0.034 *	0.3397 ± 0.024 ***	0.2261 ± 0.028
Survival rate (100%)	100	100	100	100

Statistical difference $p < 0.05$ is represented with a single asterisk (*), while double (**) and triple (***) asterisks represent statistical differences of $p < 0.01$ and $p < 0.001$, respectively.

3.2. Mucus Immunity

After 90 days of feeding a *Spirulina*-supplemented diet, lysozyme, and anti-protease activity increased significantly in groups fed with 5 and 10% *Spirulina*. Mucus protease and peroxidase activity increased significantly only in the group fed with 5% *Spirulina* (Table 4).

Table 4. Skin mucus enzyme activity of grass carp fed with 0, 1, 5, 10% *Spirulina* for 90 days.

Parameters	Control	1%-*Spirulina*	5%-*Spirulina*	10%-*Spirulina*
Protease (%)	51.91 ± 6.14	61.86 ± 6.480	80.53 ± 6.57 *	63.17 ± 7.26
Antiprotease (%)	66.78 ± 14.4	120.2 ± 13.84	152.2 ± 15.36 **	130.6 ± 16.77 *
Lysozyme (U/L)	2.056 ± 0.318	4.5 ± 0.606	7.00 ± 1.069 ***	5.33 ± 0.991 *
Peroxidase (U/mL)	0.2611 ± 0.030	0.2951 ± 0.028	0.4390 ± 0.057 *	0.248 ± 0.049

Statistical difference $p < 0.05$ is represented with a single asterisk (*), while double (**) and triple (***) asterisks represent statistical differences of $p < 0.01$ and $p < 0.001$, respectively.

3.3. Digestive Enzyme Activity

The activity of digestive enzymes *viz* protease, lipase, and amylase are shown in Figure 1a–c. A significant increase was reported for intestinal protease activity in fish fed with 5% ($p < 0.001$) and 10% ($p < 0.05$) *Spirulina* powder. Amylase activity increased significantly ($p < 0.001$) in all treatment groups compared to the control. Intestinal lipase activity increased only in fish fed with 1% ($p < 0.05$) and 5% ($p < 0.001$) *Spirulina* powder (Figure 1c).

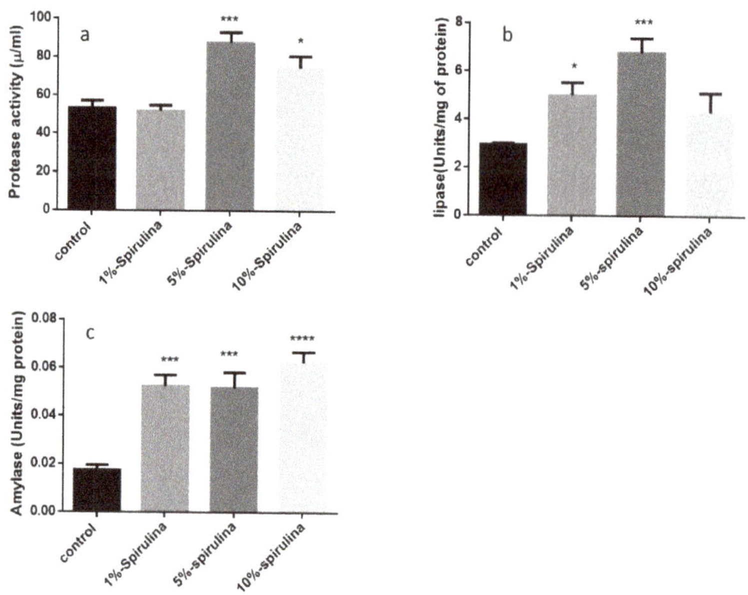

Figure 1. Activity of protease (**a**), lipase (**b**) and amylase (**c**) in the intestine of grass carp fed with various levels of *Spirulina* supplemented diet for 90 days. Statistical difference $p < 0.05$ is represented with a single asterisk (*), while triple (***) asterisks represent statistical difference of $p < 0.001$, (****) represent statisticall difference of $p < 0.0001$.

3.4. Hepatic Anti-Oxidants

Hepatic lipid peroxidation decreased significantly in all groups, and a significant decrease was recorded in fish fed with 1 and 5% *Spirulina* powder (Figure 2a). Hepatic superoxide dismutase increased in all groups fed with *Spirulina*, however, a significant increase was only observed in fish fed with 10% *Spirulina* powder (Figure 2b). The activity of reduced glutathione increased significantly in the liver of fish fed with 1 and 10%

Spirulina powder (Figure 2c). A significant increase in glutathione-S-transferase activity was recorded in fish fed with 1 and 5% *Spirulina* powder (Figure 2d). Catalase activity was increased only in groups fed with the lowest level of *Spirulina* (2e) while a non-significant decrease was observed in the liver of fish fed with 5 and 10% *Spirulina* powder for 90 days.

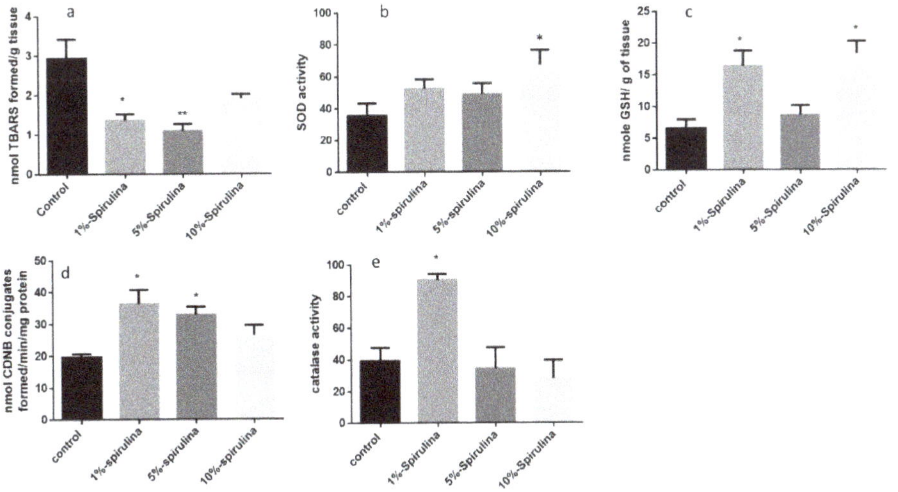

Figure 2. Hepatic antioxidant activity of grass carp fed with 0,1,5 and 10% *Spirulina* for 90 days. Data expressed as mean ± S.E.M *—represent significant difference $p < 0.05$. **—represent statistical differences of $p < 0.01$. (**a**) Lipid peroxidation is expressed as nmol of thiobarbituric acid reactive substance (TBARS) formed. (**b**) Superoxide dismutase activity. (**c**) Reduced glutathione level. (**d**). Glutathione-S-transferase activity is expressed as nmol of 1-chloro-2,4-dinitrobenzene (CDNB) conjugates formed. (**e**) Catalase activity.

3.5. Gene Expression

Expression of immune-related genes was recorded in kidney tissue of grass carp after 90 days of feeding with different levels of *Spirulina*. The mRNA expression of *tumor necrosis factor-α* increased significantly in the group fed with 1 and 10% *Spirulina* powder (Figure 3a). A decrease in the mRNA level of *interleukin-8* was recorded in the head kidney of fish fed with 1% *Spirulina*, while no significant difference was observed in groups fed with 5% *Spirulina* powder, however, the inclusion of 10% *Spirulina* resulted in a significant increase in mRNA expression of *il-8* (Figure 3b). Expression of *interferon-γ* increased significantly in the head kidney of fish fed with 1 and 5% *Spirulina* powder (Figure 3c).

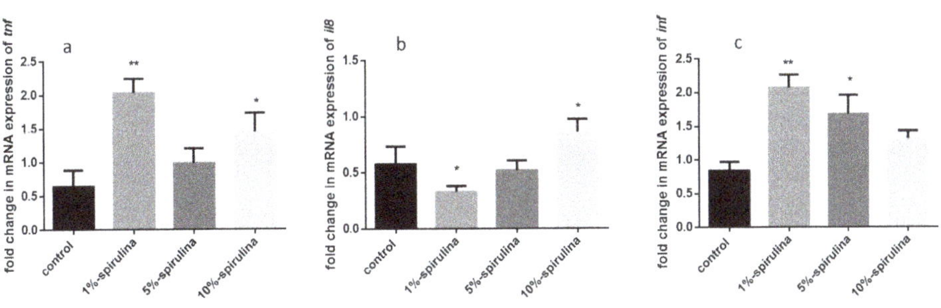

Figure 3. Relative mRNA expression of *tnf-α* (**a**), *il-8* (**b**) and *ifn-γ* (**c**) in head kidney of grass carp fed with 0, 1, 5 and 10% *Spirulina* for 90 days. Values are presented as mean ± S.E.M. * = $p < 0.05$; ** = $p < 0.01$.

4. Discussion

Many algal species, plants, and plant extracts (aqueous, methanolic, ethanolic) are now being successfully used in the aquaculture industry and had a positive effect on fish growth, antioxidant status, immunity, and resistance against water-borne pathogens [26,27,57,63–68].

Evaluation of growth performance is one of the important parameters to assess the efficacy of feed additives [36]. In the present work, weight gain, %WG, SGR, and TCR was highest in fish fed with a 1 and 5% *S. platensis* supplement diet for 90 days. A similar growth-promoting effect was observed in other fish species [49,53,62] fed with *Spirulina*-supplemented diets (2.5–10%). Food components are also assessed by their ability to be digested and absorbed in the gut, thus promoting growth [69]. The degradation of food in the digestive tract is performed with the assistance of enzymes. Assessment of intestinal enzymes (proteases, lipases, amylases, trypsin, and chymotrypsins) provides information about the physiological status of the gut [70]. The intestinal enzyme activity of grass carp was enhanced after the inclusion of *Spirulina* in the diet. Previous studies also reported that the inclusion of *Spirulina* in the diet of aquatic organisms significantly enhanced the digestion and absorption process [47]. An increase in digestive enzymes leads to better nutrient absorption that can explain the growth-promoting effect of *Spirulina* [47,71] as observed in the present study.

Unfavorable water quality parameters, high stocking density, and other environmental factors can induce oxidative stress in fish. Several antioxidant compounds (glutathione) and enzymes (catalase, superoxide dismutase, glutathione peroxidase, glutathione reductase, etc.) make up the body's anti-oxidant defense system that detoxifies reactive oxygen and nitrogen species through a series of reaction cascades. Recently, a great deal of research focused on supplementing the fish diet with additives that can enhance the natural anti-oxidant level and alleviate oxidative stress. In the present study, fish were not challenged with any stress but it is generally accepted that an enhanced and better anti-oxidant system will provide better resistance against oxidative stress. In the present study, the level of lipid peroxidation decreased in the fish fed with 1 and 5% *Spirulina* diets. The activity of catalase, glutathione-S-transferase, and glutathione levels increased significantly ($p < 0.05$) in fish livers fed with 1% *Spirulina*-supplemented diets. *Spirulina* is a rich source of bioactive compounds such as catechins, phycobiliproteins, allophycocyanin, and phycocyanins [72]. Catechins have the ability to chelate metal ions, scavenge reactive oxygen species, and produce antioxidant enzymes [73]. Similarly, phycocyanin and allophycocyanin have the properties of antioxidants [74]. The presence of these bioactive components in *Spirulina* may be responsible for the improved antioxidant status of grass carp.

The innate immune response provides protection against pathogens. Fish skin mucus is the first line of defense against pathogens. Feeding fish with a diet containing various additives that can enhance the immunity of a mucosal surface [1,5,24,27,33,67,71] has become an active area of research in the last decade. Innate immune enzymes and molecules present in fish skin mucus are lysozymes, esterases, proteases, antiproteases, anti-microbial peptides, and complement proteins [75]. These mucosal enzymes and molecules have strong anti-microbial activity against both Gram-positive and -negative bacteria, thus, improved status of these molecules in mucus may protect the fish against water-borne pathogens.

The grass carp fingerlings fed with 5% *Spirulina* diets had significantly increased mucosal lysozyme, protease, antiprotease, and peroxidase activities. A similar increase in skin mucosal innate immune biomarkers was reported in other fish species when fed with diets containing *Spirulina* and our results are consistent with previous studies on other fish species [47,52,76,77]. Many plants and algae are rich in antimicrobial peptides, essential oils, polysaccharides, saponins, and phenolic compounds that are effective against infections. These secondary metabolites modulate the active sites of enzymes and also modulate the receptor sites hence, enhancing immunity [78].

Our gene expression analysis revealed that fish fed with a diet containing 1 and 5% *Spirulina* powder had higher expression of *tnf-α* and *inf-γ*. Tumor necrosis factor-α is a pro-inflammatory cytokine and is used as a biomarker of innate immune status in fish.

Tnf-α actively recruits lymphocytes to fight infection and stimulate the cellular and humoral immune response [79]. Similar up-regulation of *tnf-α* was reported in Nile tilapia, common carp, and rainbow trout when fed with diets supplemented with *Spirulina* [5,47,51,79,80]. Mast cells are considered an important source for the synthesis and release of cytokines, such as *tnf-α* [81]. Active compounds such as C-phycocyanin present in *Spirulina* [82] can influence mast cells which may be responsible for increased expression of *tnf-α* as observed in the present study. Tumor necrosis factor-α along with interferon-γ and interleukin-8 activate natural killer cells, macrophages, and cytotoxic-T tells and also augment phagocytosis that leads to the inactivation of viruses and eradication of pathogens [23]. An in vitro study revealed that *Spirulina* induced the secretion of interferon-gamma in the peripheral blood mononuclear cells [83]. Polysaccharides found in many microalgae can increase the expression of pro-inflammatory cytokines, thus acting as immunostimulating agents in aquaculture [84].

In conclusion, the results of the study displayed that the inclusion of *Spirulina* (up to 5%) is effective for improvement in growth performance, antioxidant, digestive enzymes, and innate immune biomarkers in grass carp.

Author Contributions: M.F and S.H.H. designed the experiment. R.J. and N.N. performed experiments and collected samples under the supervision of M.F. S.K. supervised gene expression studies. M.F. and S.K. prepared the manuscript. S.H.H.; H.V.D. and M.P. reviewed the final draft. All authors have read and agreed to the published version of the manuscript.

Funding: This research did not receive any funding.

Institutional Review Board Statement: The study was approved by the Board of Studies, Department of Zoology GC University, Lahore.

Data Availability Statement: The data that support the findings of this study are available from the corresponding author upon reasonable request.

Conflicts of Interest: The authors declare no conflict of interest.

References

1. Hoseinifar, S.H.; Sun, Y.-Z.; Zhou, Z.; Van Doan, H.; Davies, S.J.; Harikrishnan, R. Boosting immune function and disease bio-control through environment- friendly and sustainable approaches in finfish aquaculture: Herbal therapy scenarios. *Rev. Fish. Sci. Aquac.* **2020**, *28*, 303–321. [CrossRef]
2. Henchion, M.; Hayes, M.; Mullen, A.M.; Fenelon, M.; Tiwari, B. Future protein supply and demand: Strategies and factors influencing a sustainable equilibrium. *Foods* **2017**, *6*, 53.
3. Pradeepkiran, J.A. Aquaculture role in global food security with nutritional value: A review. *Transl. Anim. Sci.* **2019**, *3*, 903–910. [PubMed]
4. FAO—Food and Agricultural Organization. *The State of World Fisheries and Aquaculture 2018-Meeting the Sustainable Development Goals*; FAO: Rome, Italy, 2018.
5. Al-Deriny, S.H.; Dawood, M.A.O.; Abou Zaid, A.; El-Trasa, W.F.; Paray, B.A.; Van Doan, H.; Mohamed, R.A. The synergistic effects of *Spirulina platensis* and *Bacillus amyloliquefaciens* on the growth performance, intestinal histomorphology, and immune response of Nile tilapia (*Oreochromis niloticus*). *Aquac. Rep.* **2020**, *17*, 100390.
6. Holmes, A.H.; Moore, L.S.; Sundsfjord, A.; Steinbakk, M.; Regmi, S.; Karkey, A.; Piddock, L.J. Understanding the mechanisms and drivers of antimicrobial resistance. *Lancet* **2016**, *387*, 176–187. [PubMed]
7. Bosila, M.A.; Mekky, H.M.; Fedawy, H.S.; Elbayomi, K.M.; Amer, M.M. Histopathological lesion of arthritis in *Mycoplasma synoviae* naturally infected breeder chicken in Egypt. *Int. J. Vet. Sci* **2021**, *10*, 72–74.
8. Saeed, N.M. Sequence analysis and comparison of infectious bursal disease virus affecting indigenous Kurdish breed and broiler chickens in Sulaymaniyah, Kurdistan Region of Iraq. *Pak. Vet. J.* **2021**, *41*, 249–253.
9. Tahir, A.; Khan, M.A.; Bibi, K.; Bibi, S.; Rauf, F.; Ayaz, F. Prevalence of Colibacillosis in Young Broiler Chicks and Antibiogram of *Escherichia coli* in Different Areas of Hazara Region. *Adv. Life Sci.* **2021**, *8*, 238–240.
10. Abu, S.M.; Monira, N.; Asmaul, H.; E-Elahi ATM Mahbub, A.T.M.; Mustafa, K.A.H.; Masudur, R.M. Seroprevalence of Newcastle disease in layer chickens and pathology in clinically affected chickens at Gazipur, Bangladesh. *Cont. Vet. J.* **2022**, *2*, 35–41.
11. Ul-Rahman, A.; Shabbir, M.A.B.; Mehmood, A.; Shabbir, M.Z. Genotypic and subgenotypic diversity of avian paramyxoviruses 2, 4 and 6. *Pak. Vet. J.* **2021**, *41*, 156–159.
12. Elazab, S.T.; Elshater, N.S.; Hashem, Y.H.; Abdelaziz, A.S. Pharmacokinetics of tildipirosin in healthy and *Mycoplasma gallisepticum* infected chickens. *Int. J. Vet. Sci.* **2021**, *10*, 119–123.

13. Iraqi, M.; Nasef, S.A.; El-Enbaawy, M. Phenotypic and genotypic characteristics of antimicrobial and disinfectant resistance gram-negative bacteria involved in early broiler chick mortality. *Int. J. Vet. Sci.* **2021**, *10*, 129–134.
14. Sharif, M.; Tunio, S.A.; Bano, S. Synergistic effects of Zinc oxide nanoparticles and conventional antibiotics against methicillin resistant *Staphylococcus aureus*. *Adv. Life Sci.* **2021**, *8*, 167–171.
15. Betelhem, T.; Shubisa, A.L.; Bari, F.D. Isolation, identification and antimicrobial resistance of *Staphylococcus aureus* isolates from mastitis cases of lactating dairy cows found in Sululta and Holleta Towns, Oromia, Ethiopia. *Agrobiol. Rec.* **2022**, *8*, 27–34.
16. Al-Sarraj, F.M.B. A Review on the impacts of *Azadirachta indica* on Multi-drug Resistant Extended Spectrum Beta Lactamase positive of *Escherichia coli* and *Klebsiella pneumonia*. *Adv. Life Sci.* **2021**, *8*, 228–232.
17. Mohamed, H.F.; Atta, A.H.; Darwish, A.S.; Atef, M. Effect of probiotics on the pharmacokinetic aspects and tissue residues of difloxacin in broiler chickens. *Pak. Vet. J.* **2021**, *41*, 269–273.
18. Mohsin, M.; Li, L.; Huang, X.; Aleem, M.T.; Habib, Y.J.; Shehata, A.I.; Afzal, M.Z.; Abbas, R.Z.; Abbas, A.; Yin, G. Immunogenicity and protective efficacy of probiotics with EtIMP1C against *Eimeria tenella* challenge. *Pak. Vet. J.* **2021**, *41*, 274–278.
19. Rafay, M.; Ghaffar, M.U.; Abid, M.; Malik, Z.; Madnee, M. Phytochemicals analysis and antimicrobial activities of *chinops echinatus* from Cholistan desert, Pakistan. *Agrobiol. Rec.* **2021**, *5*, 21–27.
20. Rani, Z.; Abbas, R.Z.; Abbas, A.; Saeed, Z.; Rehman, T.; Mehmood, K.; Rehman, A.; Hussain, K. In vitro and in vivo anticoccidial effects of butyric acid and its impact on blood and serum chemistry of broiler chickens. *Kafkas. Univ. Vet. Fak. Derg.* **2021**, *27*, 583–588.
21. Rehman, K.; Hamayun, M.; Khan, S.S.; Ahmad, W.S. Efficiency of virgin's mantle (*Fagonia cretica* L.) as an antibacterial and antifungal agent. *Adv. Life Sci.* **2021**, *8*, 233–237.
22. Ma, K.; Bao, Q.; Wu, Y.; Chen, S.; Zhao, S.; Wu, H.; Fan, J. Evaluation of Microalgae as Immunostimulants and Recombinant Vaccines for Diseases Prevention and Control in Aquaculture. *Front. Bioeng. Biotechnol.* **2020**, *8*, 590431. [PubMed]
23. Ahmadifar, E.; Fallah, H.P.; Yousefi, M.; Dawood, M.A.O.; Hoseinifar, S.H.; Adineh, H.; Yilmaz, S.; Paolucci, M.; Van Doan, V. The gene regulatory roles of herbal extracts on the growth, immune system, and reproduction of fish. *Animals* **2021**, *11*, 2167.
24. Harikrishnan, R.; Balasundaram, C.; Heo, M.S. Impact of plant products on innate and adaptive immune system of cultured finfish and shellfish. *Aquaculture* **2011**, *317*, 1–15.
25. Harikrishnan, R.; Kim, J.S.; Kim, M.C.; Balasundaram, C.; Heo, M.S. *Lactuca indica* extract as feed additive enhances immunological parameters and disease resistance in *Epinephelus bruneus* to *Streptococcus iniae*. *Aquaculture* **2011**, *318*, 43–47.
26. Van Doan, H.; Hoseinifar, S.H.; Chitmanat, C.; Jaturasitha, S.; Paolucci, M.; Ashouri, G.; Dawood, M.A.O.; Esteban, M.A. The effects of Thai ginseng, *Boesenbergia rotunda* powder on mucosal and serum immunity, disease resistance, and growth performance of Nile tilapia (*Oreochromis niloticus*) fingerlings. *Aquaculture* **2019**, *513*, 734388.
27. Van Doan, H.; Lumsangkul, C.; Hoseinifar, S.H.; Hung, T.Q.; Stejskal, V.; Ringø, E.; Dawood, M.A.O.; Esteban, M.A. Administration of watermelon rind powder to Nile tilapia (*Oreochromis niloticus*) culture under biofloc system: Effect on growth performance, innate immune response, and disease resistance. *Aquaculture* **2020**, *528*, 735574.
28. Mohammadi, G.; Rafiee, G.; El Basuini, M.F.; Van Doan, H.; Ahmed, H.A.; Dawood, M.A.O.; Abdel-Latif, H.M.R. Oregano (*Origanum vulgare*), St John's-wort (*Hypericum perforatum*), and lemon balm (*Melissa officinalis*) extracts improved the growth rate, antioxidative, and immunological responses in Nile tilapia (*Oreochromis niloticus*) infected with *Aeromonas hydrophila*. *Aquac. Rep.* **2020**, *18*, 100445.
29. Abdel-Tawwab, M.; El-Araby, D.A. Immune and antioxidative effects of dietary licorice (*Glycyrrhiza glabra* L.) on performance of Nile tilapia, *Oreochromis niloticus* (L.) and its susceptibility to *Aeromonas hydrophila* infection. *Aquaculture* **2021**, *530*, 735828.
30. Stratev, D.; Zhelyazkov, G.; Noundou, X.S.; Krause, R.W.M. Beneficial effects of medicinal plants in fish diseases. *Aquac. Int.* **2018**, *26*, 289–308.
31. Bilen, S.; Altunoglu, Y.C.; Ulu, F.; Biswas, G. Innate immune and growth promoting responses to caper (*Capparis spinosa*) extract in rainbow trout (*Oncorhynchus mykiss*). *Fish Shellfish Immunol.* **2016**, *57*, 206–212.
32. Van Doan, H.; Hoseinifar, S.H.; Naraballobh, W.; Jaturasitha, S.; Tongsiri, S.; Chitmanat, C.; Ringø, E. Dietary inclusion of orange peels derived pectin and *Lactobacillus plantarum* for Nile tilapia (*Oreochromis niloticus*) cultured under indoor biofloc systems. *Aquaculture* **2019**, *508*, 98–105.
33. Srichaiyo, N.; Tongsiri, S.; Hoseinifar, S.H.; Dawood, M.A.O.; Jaturasitha, S.; Esteban, M.A.; Ringø, E.; Van Doan, H. The effects gotu kola (*Centella asiatica*) powder on growth performance, skin mucus, and serum immunity of Nile tilapia (*Oreochromis niloticus*) fingerlings. *Aquac. Rep.* **2020**, *16*, 100239.
34. Gabriel, N.N.; Qiang, J.; Ma, X.Y.; He, J.; Xu, P.; Liu, K. Dietary *Aloe vera* improves plasma lipid profile, antioxidant, and hepatoprotective enzyme activities in GIFT-tilapia (*Oreochromis niloticus*) after *Streptococcus iniae* challenge. *Fish Physiol. Biochem.* **2015**, *41*, 1321–1332. [PubMed]
35. Kuebutornye, F.K.; Abarike, E.D. The contribution of medicinal plants to tilapia aquaculture: A review. *Aquac. Int.* **2020**, *28*, 965–983.
36. Sonmez, A.Y.; Bilen, S.; Alak, G.; Hisar, O.; Yanık, T.; Biswas, G. Growth performance and antioxidant enzyme activities in rainbow trout (*Oncorhynchus mykiss*) juveniles fed diets supplemented with sage, mint and thyme oils. *Fish Physiol. Biochem.* **2015**, *41*, 165–175. [PubMed]

37. Awed, E.M.; Sadek, K.M.; Soliman, M.K.; Khalil, R.H.; Younis, E.M.; Abdel-Warith, A.A.-W.; Van Doan, H.; Dawood, M.A.O.; Abdel-Latif, H.M.R. *Spirulina platensis* Malleviated the oxidative damage in the gills, liver, and kidney organs of Nile tilapia intoxicated with sodium sulphate. *Animals* **2020**, *10*, 2423.
38. Yousefi, M.; Abtahi, B.; Adineh, H.; Hoseinifar, S.H.; Mirghaed, A.T.; Paolucci, M.; Van Doan, H. Effects of dietary arginine supplementation on cytokine- and antioxidant-related gene expressions in common carp (*Cyprinus carpio*) fingerling during ammonia toxicity. *Aquac. Res.* **2021**, *52*, 2751–2758.
39. Citarasu, T. Herbal biomedicines: A new opportunity for aquaculture industry. *Aquac. Int.* **2010**, *18*, 403–414.
40. Lee, J.Y.; Gao, Y. Review of the application of garlic, Allium sativum, in aquaculture. *J. World Aquac. Soc.* **2012**, *43*, 447–458.
41. Xu, A.; Shang-Guan, J.; Li, Z.; Gao, Z.; Huang, Y.C.; Chen, Q. Effects of dietary Chinese herbal medicines mixture on feeding attraction activity, growth performance, nonspecific immunity and digestive enzyme activity of Japanese seabass (*Lateolabrax japonicus*). *Aquac. Rep.* **2020**, *17*, 100304.
42. Kiron, V. Fish immune system and its nutritional modulation for preventive health care. *Anim. Feed Sci. Technol.* **2012**, *173*, 111–133.
43. Egamberdieva, D.; Wirth, S.; Behrendt, U.; Ahmad, P.; Berg, G. Antimicrobial activity of medicinal plants correlates with the proportion of antagonistic endophytes. *Front. Microbiol.* **2017**, *8*, 199. [CrossRef] [PubMed]
44. Khan, M.F.; Tang, H.; Lyles, J.T.; Pineau, R.; Mashwani, Z.-U.-R.; Quave, C.L. Antibacterial properties of medicinal plants from Pakistan against multidrug-resistant ESKAPE pathogens. *Front. Pharmacol.* **2018**, *9*, 815. [CrossRef] [PubMed]
45. Dada, A.A. Improvement of tilapia (*Oreochromis niloticus* Linnaeus, 1758) growth performance fed three commercial feed additives in diets. *J. Aquac. Res. Dev.* **2015**, *6*, 2.
46. Radhakrishnan, S.; Bhavan, P.S.; Seenivasan, C.; Shanthi, R.; Muralisankar, T. Replacement of fishmeal with *Spirulina platensis*, *Chlorella vulgaris* and *Azolla pinnata* on non-enzymatic and enzymatic antioxidant activities of *Macrobrachium rosenbergii*. *J. Basic Appl. Zool.* **2014**, *67*, 25–33.
47. Sheikhzadeh, N.; Mousavi, S.; Oushani, A.K.; Firouzamandi, M.; Mardani, K. *Spirulina platensis* in rainbow trout (*Oncorhynchus mykiss*) feed: Effects on growth, fillet composition, and tissue antioxidant mechanisms. *Aquac. Int.* **2019**, *27*, 1613–1623.
48. Raji, A.A.; Jimoh, W.A.; Abu Bakar, N.H.; Mohd Taufek, N.H.; Muin, H.; Alias, Z.; Milow, P.; Abdul Razak, S. Dietary use of *Spirulina* (Arthrospira) and *Chlorella* instead of fish meal on growth and digestibility of nutrients, amino acids and fatty acids by African catfish. *J. Appl. Phycol.* **2020**, *32*, 1763–1770.
49. Xia, Y.; Liu, C.; Fei, S.; Liu, H.; Han, D.; Jin, J.; Yang, Y.; Zhu, X.; Xie, S. *Arthrospira platensis* additive enhances the growth performance and antioxidant response in hybrid yellow catfish (*Pelteobagrus fulvidraco*♀ × *Pelteobagrus vachelli*♂). *Aquac. Rep.* **2021**, *20*, 100721.
50. Ragap, H.M.; Khalil, R.H.; Mutawie, H. Immunostimulant effects of dietary *Spirulina platensis* on tilapia *Oreochromis niloticus*. *J. App. Pharm. Sci.* **2012**, *2*, 26–31.
51. Watanuki, H.; Ota, K.; Tassakka, A.C.M.A.; Kato, T.; Sakai, M. Immunostimulant effects of dietary *Spirulina platensis* on carp, *Cyprinus carpio*. *Aquaculture* **2006**, *258*, 157–163.
52. Adel, M.; Yeganeh, S.; Dadar, M.; Sakai, M.; Dawood, M.A.O. Effects of dietary *Spirulina platensis* on growth performance, humoral and mucosal immune responses and disease resistance in juvenile great sturgeon (*Huso huso* Linnaeus, 1754). *Fish Shellfish Immuno.* **2016**, *56*, 436–444.
53. Teimouri, M.; Yeganeh, S.; Mianji, G.R.; Najafi, M.; Mahjoub, S. The effect of *Spirulina platensis* meal on antioxidant gene expression, total antioxidant capacity, and lipid peroxidation of rainbow trout (*Oncorhynchus mykiss*). *Fish Physiol. Biochem.* **2019**, *45*, 977–986. [CrossRef]
54. Yu, W.; Wen, G.; Lin, H.; Yang, Y.; Huang, X.; Zhou, C.; Zhang, Z.; Duan, Y.; Huang, Z.; Li, T. Effects of dietary *Spirulina platensis* on growth performance, hematological and serum biochemical parameters, hepatic antioxidant status, immune responses and disease resistance of Coral trout *Plectropomus leopardus* (Lacepede, 1802). *Fish Shellfish Immunol.* **2018**, *74*, 649–655. [PubMed]
55. Association of Official Analytical Chemists. *Official Methods of Analysis*, 16th ed.; Association of Official Analytical Chemists: Washington, DC, USA, 1995.
56. Latif, M.; Faheem, M.; Asmatullah. Study of oxidative stress and histo-biochemical biomarkers of diethyl phthalate induced toxicity in a culturable fish, *Labeo rohita*. *Pak. Vet. J.* **2020**, *40*, 202–208. [CrossRef]
57. Faheem, M.; Khaliq, S.; Mustafa, N.; Rani, S.; Lone, K.P. Dietary *Moringa oleferia* leaf meal induce growth, innate immunity and cytokine expression in grass carp, *Ctenopharyngodon Idella*. *Aquac. Nut.* **2020**, *26*, 1164–1172.
58. Faheem, M.; Lone., K.P. Oxidative stress and histopathologic biomarkers of exposure to bisphenol-A in the freshwater fish, *Ctenopharyngodon idella*. *Braz. J. Pharm. Sci.* **2017**, *53*, e17003.
59. Faheem, M.; Zahid, Z.; Ferreira, N.G.C. Toxicity assessment of dibutyl phthalate in grass carp: An integrated biomarker approach. *Pak. Vet. J.* **2021**, *41*, 365–371.
60. Allameh, S.K.; Ringø, E.; Yusoff, F.M.; Daud, H.M.; Ideris, A. Dietary supplement of *Enterococcus faecalis* on digestive enzyme activities, short-chain fatty acid production, immune system response and disease resistance of Javanese carp (*Puntius gonionotus*, Bleeker 1850). *Aquac. Nutr.* **2017**, *23*, 331–338.
61. Dar, S.A.; Srivastava, P.P.; Nazir, M.I.; Jahan, I.; Varghese, T.; Hajam, I.A.; Gupta, S. Modulations of digestive and metabolic enzymes profiles during restriction feeding in *Labeo rohita* (Hamilton, 1822) fingerlings. *N. Am. J. Aquac.* **2021**, *83*, 58–66.

62. Livak, K.J.; Schmittgen, T.D. Analysis of relative gene expression data using real-time quantitative PCR and the $2-\Delta\Delta CT$ method. *Methods* **2001**, *25*, 402–408. [CrossRef]
63. Hoseinifar, S.H.; Yousefi, S.; Capillo, G.; Paknejad, H.; Khalil, M.; Tabarraei, A.; Van Doan, V.H.; Span, N.; Faggio, C. Mucosal immune parameters, immune and antioxidant defence related genes expression and growth performance of zebrafish (*Danio rerio*) fed on *Gracilaria gracilis* powder. *Fish Shellfish Immunol.* **2018**, *83*, 232–237. [PubMed]
64. Van Doan, H.; Hoseinifar, S.H.; Sringarm, K.; Jaturasitha, S.; Khamlor, T.; Dawood, M.A.O.; Esteban, M.A.; Soltani, M.; Mustha, M.S. Effects of elephant's foot (*Elephantopus scaber*) extract on growth performance, immune response, and disease resistance of Nile tilapia (*Oreochromis niloticus*) fingerlings. *Fish Shellfish Immunol.* **2019**, *93*, 328–335.
65. Harikrishnan, R.; Devi, G.; Van Doan, H.V.; Balasundaram, C.; Thamizharasan, S.; Hoseinifar, S.H.; Abdel-Tawwab, M. Effect of diet enriched with *Agaricus bisporus* polysaccharides (ABPs) on antioxidant property, innate-adaptive immune response and pro-anti inflammatory genes expression in *Ctenopharyngodon idella* against *Aeromonas hydrophila*. *Fish Shellfish Immunol.* **2021**, *114*, 238–252. [PubMed]
66. Ghafarifarsani, H.; Rashidian, G.; Sheikhlar, A.; Farsani, M.N.; Hoseinifar, S.H.; Van Doan, H.V. The use of dietary oak acorn extract to improve haematological parameters, mucosal and serum immunity, skin mucus bactericidal activity, and disease resistance in rainbow trout (*Oncorhynchus mykiss*). *Aquac. Res.* **2021**, *52*, 2518–2527.
67. Rufchaei, R.; Nedaei, S.; Hoseinifar, S.H.; Hassanpour, S.; Golshan, M.; Bourani, M.S. Improved growth performance, serum and mucosal immunity, haematology and antioxidant capacity in pikeperch (*Sander lucioperca*) using dietary water hyacinth (*Eichhornia crassipes*) leaf powder. *Aquac. Res.* **2021**, *52*, 2194–2204.
68. Latif, M.; Faheem, M.; Asmatullah; Hoseinifar, S.H.; Van Doan, H. Protective efficacy of *Nigella sativa* seeds against diethyl phthalate induced growth retardation, oxidative stress and histo-biochemical damages in *Labeo rohita*. *Aquaculture* **2021**, *533*, 736065.
69. Akintunde, E.A. Digestive enzymes in the digestive enzymes in the gut of *Sarotherodon galilaeus* (syn. *Tilapia galilaea*, family cichlidae) of the Lake Kainji, Nigeria. *Niger. J. Sci.* **1985**, *18*, 22–25.
70. Guzman, C.; Gaxiola, G.; Rosa, C.; Torre-Blanco, A. The effect of dietary protein and total energy content on digestive enzyme activities, growth and survival of *Litopenaeus setiferus* post larvae. *Aquac. Nut.* **2005**, *7*, 113–122.
71. Hoseinifar, S.H.; Dadar, M.; Khalil, M.; Cerezuela, R.; Esteban, M.Á. Effect of dietary supplementation of palm fruit extracts on the transcriptomes of growth, antioxidant enzyme and immune-related genes in common carp (*Cyprinus carpio*) fingerlings. *Aquac. Res.* **2017**, *48*, 3684–3692.
72. Takeuchi, T.; Lu, J.; Yoshizaki, G.; Satoh, S. Effect on the growth and body composition of juvenile tilapia Oreochromis niloticus fed raw *Spirulina*. *Fish Sci.* **2002**, *68*, 34–40.
73. Bernatoniene, J.; Kopustinskiene, D.M. The Role of Catechins in Cellular Responses to Oxidative Stress. *Molecules* **2018**, *23*, 965.
74. Esteban, M.A. An overview of the immunological defenses in fish skin. *ISRN Immunol.* **2012**, *2012*, 853470. [CrossRef]
75. Renugadevi, K.; Valli Nachiyar, C.; Sowmiya, P.; Sunkar, S. Antioxidant activity of phycocyanin pigment extracted from marine filamentous cyanobacteria Geitlerinema sp TRV57. *Biocatal. Agric. Biotechnol.* **2018**, *16*, 237–242.
76. Promya, J.; Chitmanat, C. The effects of *Spirulina* platensis and Cladophora algae on the growth performance, meat quality and immunity stimulating capacity of the African sharptooth catfish (*Clarias gariepinus*). *Int. J. Agric. Biol.* **2011**, *13*, 77–82.
77. Mahmoud, M.M.; El-Lamie, M.M.; Kilany, O.E.; Dessouki, A.A. *Spirulina* (*Arthrospira platensis*) supplementation improves growth performance, feed utilization, immune response, and relieves oxidative stress in Nile tilapia (*Oreochromis niloticus*) challenged with Pseudomonas fluorescens. *Fish Shellfish Immunol.* **2018**, *72*, 291–300. [PubMed]
78. Wink, M. Evolution of toxins and antinutritional factors in plants with special emphasis on Leguminosae. In *Poisonous Plants and Related Toxins*; CABI Publishing: Oxford, UK, 2004; pp. 1–25.
79. Galindo-Villegas, J.; Mulero, I.; García-Alcazar, A.; Muñoz, I.; Peñalver-Mellado, M.; Streitenberger, S.; Scapigliati, G.; Meseguer, J.; Mulero, V. Recombinant TNFα as oral vaccine adjuvant protects European sea bass against vibriosis: Insights into the role of the CCL25/CCR9 axis. *Fish Shellfish Immunol.* **2013**, *35*, 1260–1271.
80. Abdellatief, S.A.; Abdel Rahman, A.N.; Abdallah, F.D. Evaluation of Immunostimulant Activity of *Spirulina* platensis (*Arthrospira platensis*) and Sage (*Salvia officinalis*) in Nile Tilapia (*Oreochromis niloticus*). *Zagazig Vet. J.* **2018**, *46*, 25–36.
81. Shih, C.M.; Cheng, S.N.; Wong, C.S.; Kuo, Y.L.; Chou, T.C. Antiinflammatory and antihyperalgesic activity of C-phycocyanin. *Anesth. Analg.* **2009**, *108*, 1303–1310.
82. Liu, Q.; Huang, Y.; Zhang, R.; Cai, T.; Cai, Y. Medical Application of *Spirulina platensis* Derived C-Phycocyanin. *Evid. Based Complement. Alternat. Med.* **2016**, 7803846.
83. Mao, T.K.; Van De Water, J.; Gershwin, M.E. Effect of *Spirulina* on the secretion of cytokines from peripheral blood mononuclear cells. *J. Med. Food.* **2000**, *3*, 135–140.
84. Mohan, k.; Ravichandran, S.; Muralisankar, T.; Uthayakumar, V.; Chandirasekar, R.; Seedevi, P.; Rajan, D.K. Potential uses of fungal polysaccharides as immunostimulants in fish and shrimp aquaculture: A review. *Aquaculture* **2019**, *500*, 250–263.

fishes

MDPI

Article

Passionfruit (*Passiflora edulis*) Peel Powder Stimulates the Immune and Antioxidant Defense System in Nile Tilapia, *Oreochromis niloticus*, Cultivated in a Biofloc System

Piyatida Outama [1], Nguyen Vu Linh [1], Chinh Le Xuan [1], Supreya Wannavijit [1], Sudaporn Tongsiri [2], Chanagun Chitmanat [2], Napatsorn Montha [1] and Hien Van Doan [1,3,*]

1 Department of Animal and Aquatic Sciences, Faculty of Agriculture, Chiang Mai University, Chiang Mai 50200, Thailand
2 Faculty of Fisheries Technology and Aquatic Resources, Maejo University, Chiang Mai 50290, Thailand
3 Innovative Agriculture Research Center, Faculty of Agriculture, Chiang Mai University, Chiang Mai 50200, Thailand
* Correspondence: hien.d@cmu.ac.th

Abstract: This study aimed to assess the impacts of dietary supplementation with passionfruit (*Passiflora edulis*) peel powder (PSPP) on the growth, immune response, and expression of immune and antioxidant-related genes in Nile tilapia (*Oreochromis niloticus*) maintained in a biofloc system. Fish were fed basal diets supplemented with different doses of PSPP at 10 g kg^{-1} (PSPP10), 20 g kg^{-1} (PSPP20), 40 g kg^{-1} (PSPP40), and 80 g kg^{-1} (PSPP80). The basal diet, without PSPP-supplementation, was used as a control at 0 g kg^{-1} (PSPP0). We observed that the dietary supplementation groups fed different levels of PSPP exhibited no substantial difference or only slight increases in growth performance and immunological response in Nile tilapia ($p > 0.05$), whereas fish fed diets supplemented with PSPP at concentrations of 10 g kg^{-1}, 20 g kg^{-1}, and 40 g kg^{-1} had significantly higher mRNA transcripts (approximately 1.5–4.5 fold) of immune (*il-1*, *il-8*, and *lbp*) and antioxidant (*gst-α*, *gpx*, and *gsr*) gene expressions than fish in the control treatment group (0 g kg^{-1}). These findings suggest that dietary supplementation with PSPP may effectively stimulate the immune and antioxidant defense system and may function as feed additives in Nile tilapia cultured in a biofloc system.

Keywords: feed additives; immune gene expressions; antioxidant defensive system

Citation: Outama, P.; Linh, N.V.; Xuan, C.L.; Wannavijit, S.; Tongsiri, S.; Chitmanat, C.; Montha, N.; Van Doan, H. Passionfruit (*Passiflora edulis*) Peel Powder Stimulates the Immune and Antioxidant Defense System in Nile Tilapia, *Oreochromis niloticus*, Cultivated in a Biofloc System. *Fishes* **2022**, *7*, 233. https://doi.org/10.3390/fishes7050233

Academic Editor: Lluís Tort

Received: 1 August 2022
Accepted: 23 August 2022
Published: 2 September 2022

1. Introduction

Nile tilapia (*Oreochromis niloticus*) has been extensively produced in more than 100 nations globally, generating around 7.3 million tons in 2021, because of its flexibility, high growth, resistance to stress and disease, and great economic value [1–4]. Nonetheless, like with any intensively cultured fish, tilapia farming imposes significant strains on the water quality for fish farming and increases the occurrence of pathogenic infections—especially bacterial diseases [5,6]—resulting in a high mortality rate (up to 95%) and massive economic losses [7,8]. Antibiotics and chemotherapeutics have been commonly used by farmers all over the world to control disease outbreaks in fish farming [9]. However, these activities have caused the outgrowth of antibiotic-resistant bacteria [10,11]. Due to restrictions on the use of antibiotics in aquaculture, the development of innovative ways to supply appropriate feed additives and develop cost-effective methods of disease prevention and treatment for fish has become a top concern [12,13]. Consequently, natural immunostimulants (such as prebiotics, probiotics, and synbiotics) are promising alternatives for modifying the bacterial population and attempting to control infectious disease outbreaks in aquaculture by enhancing dietary intake, nutritional absorption, and immune defense systems in aquatic animals [14–17]. In this respect, fruit by-products have been identified as potential supplements in the diets of aquatic species [18–20]. Fruit by-products used as feed additives have the potential to minimize waste, reduce aquafeed costs, and offer raw materials for the nutritional sectors [21]. Additionally, utilizing these by- and co-products would also have a positive influence on the environment and provide additional advantages to farmers [22,23].

Passionfruit (*Passiflora edulis*) is a species of the Passifloraceae family, which has more than 500 species [24]. It is found mostly in North America, but also in tropical and subtropical Southeast Asia, Australia, and New Zealand due to its economic and medicinal properties [25,26]. The passionfruit extract market is projected to reach USD 1028.6 million by 2029 [27]. Peels are created in great amounts during the processing of passionfruit to produce passionfruit juice [28]. Moreover, passionfruit peel is a by-product of the fruit processing industry that makes up around 50% of the weight of the fruit [29], which is typically thrown away as waste [30,31]. The passionfruit peel includes a variety of bioactive components, including phenolic compounds, flavonoids, cyanogenic chemicals, anthocyanin, minerals, polysaccharides, and vitamins [28,32–35]. Numerous investigations using passionfruit by-products as feed additives have been conducted on sheep [36], swine [37,38], quail [39,40], and poultry [41,42]. For fish farming, the incorporation of passionfruit seed meal (including its oil residue) in diets for tambaqui (*Colossoma macropomum*) [43,44] and passionfruit juice in tilapia has been investigated [45]. However, there have been few studies on the influences of passionfruit peel powder (PSPP) on the growth and overall wellbeing of common commercial fish species—particularly Nile tilapia.

Biofloc technology (BFT) is an alternative approach that mixes aggregates of algae, protozoa, or bacteria with particulate organic substances to improve water quality, waste treatment, and disease prevention in intensive aquaculture systems. It has been proposed as a cost-effective alternative to intensive systems since it improves water quality without requiring water exchange and produces microbial protein for aquatic species. Biofloc is a microbial community composed mostly of prokaryotic and eukaryotic microorganisms and different organic particulates [46–48]. Biofloc functions as a nutrition supply for aquatic creatures in this system, assisting in growth enhancement, pathogen reduction, and zero-water exchange. Additionally, BFT has a beneficial impact on the host immune system, increasing resistance to diseases and infections [49,50]. Therefore, this study aimed to assess the influence of dietary supplementation with powdered passionfruit peel on growth, immunological responses, and the expression of key immune-antioxidant-related genes in Nile tilapia raised in a biofloc system.

2. Materials and Methods

2.1. Preparation of Powdered Passionfruit Peel and Experimental Diets

Passionfruit was obtained from local markets at Chiang Mai (Thailand). Passionfruit peels were oven-dried at 60 °C for 48 h. The dried peel was then ground into a powder using a mill and stored at 4 °C until used in the fishes' diets. To prepare the dough, the feedstuffs were blended and then soybean oil and distilled water were added. The dough was then pelleted (2 mm in diameter) in a pelleting machine. Pellets were dried at 50 °C to attain 10% moisture and stored in sealed polyethylene bags at 4 °C until used. The proximate composition of the diets was determined using the techniques recommended by AOAC [51], and the crude fat content was measured using an extractable matter machine (Soxtec^{TM} 8000, Hilleroed, Denmark). The basal diets were modified according to the descriptions reported previously [52], which proved their suitability for Nile tilapia by their different levels of PSPP. The ingredients and elemental composition of the basal diets are shown in Table 1.

Table 1. Formulation and proximate composition of the experimental diets (g kg^{-1}).

	PSPP	PSPP0	PSPP10	PSPP20	PSPP40	PSPP80
Fish meal	-	150	150	150	150	150
Corn meal	-	200	200	200	200	200
Soybean meal	-	390	387	384	383	380
Wheat flour	-	70	70	70	70	70
Rice bran	-	150	150	150	135	100
PSPP	-	0	10	20	40	80
Cellulose	-	20	13	6	2	0
Soybean oil	-	5	5	5	5	5
Premix	-	10	10	10	10	10
Vitamin C (98%)	-	5	5	5	5	5
Total	-	1000	1000	1000	1000	1000
Proximate composition of the experimental diets (%)						
Crude protein	10.1	32.80	35.2	34.5	33.6	32.3
Crude lipid	1.01	2.85	3.45	4.18	3.62	3.60
Fiber	25.73	3.68	4.36	5.45	5.21	6.44
Ash	8.02	7.59	8.38	7.96	7.87	7.99
Dry matter	95.13	99.16	97.05	96.16	98	97.69
Gross Energy (cal/g)	2731	4273	4278	4203	4185	4085

2.2. Culture Conditions

Three weeks before starting the experiment, floc inoculants were generated in each tank (150 L) by adding sea salt (400 g), dolomite (5 g), wheat flour (2 g), and molasses (5 g). After formation, the floc quantity was kept constant throughout the experiment at a level of approximately 8.21 ± 0.15 mL per tank, by maintaining the NH_3 concentration at 0.11 ± 0.005 mg L^{-1} and modifying the C:N ratio (15:1) by adding molasses and probiotics (PondPlus, Bayer Thai Co., Ltd., Bangkok, Thailand) [53]. The C:N ratio was calculated using a diagrammatic representation of residual nitrogen grades and food intake [54].

2.3. Experimental Design

Nile tilapia fingerlings were purchased from a tilapia hatchery in Chiang Mai, Thailand. Fish were acclimatized and fed a commercial meal CP-9950 (Charoen Pokphand Foods Public Co., Ltd., Bangkok, Thailand) for two weeks. Prior to conducting additional experiments, twenty randomly selected fish were subjected to bacterial and parasite examinations to guarantee their health. A total of 300 Nile tilapia (14.22 ± 0.05 g) were randomly assigned into five dietary treatment groups with PSPP supplemented as follows: control-PSPP0 (0 g kg^{-1}), PSPP10 (10 g kg^{-1}), PSPP20 (20 g kg^{-1}), PSPP40 (40 g kg^{-1}), and PSPP80 (80 g kg^{-1}). Fish were maintained in 150 L glass tanks. The experimental trials were conducted in triplicate with 20 fish per tank. Fish were fed twice daily at 8:30 a.m. and 4:30 p.m. at 3% body weight for eight weeks. Temperature, pH, and dissolved oxygen were maintained at 25–29 °C, 7.5–7.9, and 5 mg L^{-1}, respectively.

2.4. Growth Rates

Initial weight (IW), weight gain (WG), survival rate (SR), specific growth rate (SGR), and feed conversion rate (FCR) in Nile tilapia were determined after four and eight weeks of feeding trial as follows [55]:

$$\text{WG (g)} = \text{FW} - \text{IW}$$

$$\text{SGR}\left(\%\ \text{day}^{-1}\right) = 100 \times \frac{\text{FW} - \text{IW}}{60\ \text{d}}$$

$$\text{FCR (g)} = \frac{\text{total feed given}}{\text{weight gain}}$$

$$\text{SR (\%)} = \frac{\text{final number}}{\text{initial number}} \times 100$$

Weights were measured using a balance with an accuracy of 0.1 g. Additionally, any dead fish were tallied and the mortality rate computed during the experiment.

2.5. Immune Response Analysis

2.5.1. Sample Collection

To examine immunological activities, skin mucus and blood samples (6 fish/treatment) were collected. Before collecting samples, clove oils (5 mL L^{-1}) were used to anesthetize fish. For skin mucus sample collection, the experimental fish were gently rubbed for 2 min in a plastic bag containing 10 mL of 50 mM NaCl (Merck, Germany). The mixture was immediately transferred into new sterile tubes and centrifuged at 1500× *g* at 4 °C for 10 min. The mucus samples (1 mL) were then kept at −80 °C until used. For blood sampling, blood was collected according to the protocols previously reported [56]. Briefly, approximately 1 mL of fish blood was obtained using a 1 mL syringe at the caudal vein. Blood samples were promptly withdrawn and placed into sterile tubes (without adding anticoagulant). The samples were kept for an hour at room temperature and at 4 °C for a further hour. The samples were then centrifuged, and the serum samples were collected and stored at −80 °C for further analysis.

2.5.2. Lysozyme and Peroxidase Assay

Lysozyme assays were conducted following the procedures reported by Parry, et al. [57], with the minor modifications in Van Doan et al. [55]. The equivalent unit of activity for each sample was calculated in accordance with the standard curve, which was constructed by plotting the decrease in the optical density value against the concentration, ranging from 0–20 μL mL^{-1} of hen egg-white lysozyme (Sigma-Aldrich, Inc., USA) and represented as μg mL^{-1} of serum.

Peroxidase activity was determined according to the protocols described in Quade and Roth [58] and Cordero, et al. [59], with minor modifications in Van Doan et al. [55]. The peroxidase assay was represented in units (U) per mg of skin mucus or serum proteins, where a unit was defined as the quantity of serum or mucus proteins that produced a change in absorbance equal to one.

2.6. Total RNA Extraction and Real-Time PCR (qPCR) Analysis

A total of 40–50 mg of fish tissues (gills and liver) was sampled for RNA extraction using TRIzol and an RNA extraction kit (Invitrogen, Waltham, MA, USA). The quality and quantity were measured using spectrophotometers (Thermo Fisher Scientific, Waltham, MA, USA) at wavelengths of 260 and 280 nm. The first-strand complementary DNA (cDNA) was synthesized with 1 μg total RNA using the iScript™ cDNA Synthesis Kit (BIO-RAD, Hercules, CA, USA). The analysis was conducted in triplicate using 100 ng of cDNA and iTaq Universal SYBR Green on a CFX Connect™ real-time PCR (BIO-RAD, Hercules, CA, USA). The qPCR was conducted at 95 °C for 30 s, 40 cycles of 95 °C for 15 s, and 60 °C for 30 s and followed by 95 °C for 15 s, 60 °C for 60 s, and 95 °C for 15 s. Expression levels was analyzed according to the $2^{-\Delta\Delta Ct}$ method [60] The qPCR results were normalized to the *18S rRNA* gene. The primers used for the qPCR analysis in this study are presented in Table 2.

Table 2. Primers used for the qPCR analysis in this study.

Target Genes	Sequence (5′-3′)	Tm (°C)	Product Size (bp)	Ref.
18S rRNA	GTGCATGGCCGTTCTTAGTT CTCAATCTCGTGTGGCTGAA	60	150	[61]
il-1	GTCTGTCAAGGATAAGCGCTG ACTCTGGAGCTGGATGTTGA	59	200	[61]
il-8	CTGTGAAGGCATGGGTGTG GATCACTTTCTTCACCCAGGG	59	196	[61]

Table 2. *Cont.*

Target Genes	Sequence (5′-3′)	Tm (°C)	Product Size (bp)	Ref.
lbp	ACCAGAAACTGCGAGAAGGA GATTGGTGGTCGGAGGTTTG	59	200	[61]
gst-α	ACTGCACACTCATGGGAACA TTAAAAGCCAGCGGATTGAC	60	190	[61]
gpx	GGTGGATGTGAATGGAAAGG CTTGTAAGGTTCCCCGTCAG	60	190	[61]
gsr	CTGCACCAAAGAACTGCAAA CCAGAGAAGGCAGTCCACTC	60	172	[61]

Note: F: Forward, R: Reverse, bp: base pair.

2.7. Statistical Analysis

The Kolmogorov–Smirnov test was used to evaluate the normality of the data. Means were compared using Duncan's multiple range test. Growth rates, immunological responses, and gene expression levels were analyzed using ANOVA analysis. SAS v9.4 statistical software (Cary, NC, USA) was used for all the statistical analyses [62]. $p < 0.05$ was denoted as a significant difference.

3. Results

3.1. Growth Performance Analysis

In this study, the growth parameters observed in different dietary treatment groups are presented in Table 3. There was no significant difference in final weight (FW), weight gain (WG), feed conversion ratio (FCR), or specific growth rate (SGR) between fish fed PSPP-supplemented diets and those fed only a basal diet (0 g kg^{-1} PSPP) after four and eight weeks of the experimental trial. The survival rate (SR) of all treatment groups exceeded 95% at the conclusion of the feeding studies. No significant difference in any groups of the dietary PSPP-supplemented diets were detected ($p > 0.05$).

Table 3. Growth performances and feed utilization in Nile tilapia with different levels of PSPP-supplemented diets after four and eight weeks of the feeding trial. Different letters in the same row indicate statistically significant differences ($p < 0.05$). Data are presented as mean ± SE.

	PSPP 0	PSPP 10	PSPP 20	PSPP 40	PSPP 80
IW (g)	14.22 ± 0.04 ^{a}	14.23 ± 0.04 ^{a}	14.25 ± 0.05 ^{a}	14.15 ± 0.03 ^{a}	14.23 ± 0.03 ^{a}
FW (g)					
4 weeks	28.60 ± 1.64 ^{a}	29.05 ± 1.11 ^{a}	28.69 ± 0.64 ^{a}	29.73 ± 1.23 ^{a}	29.67 ± 0.37 ^{a}
8 weeks	55.03 ± 0.64 ^{a}	55.01 ± 0.91 ^{a}	53.84 ± 1.35 ^{a}	55.93 ± 1.36 ^{a}	54.05 ± 0.79 ^{a}
SGR (%)					
4 weeks	2.32 ± 0.18 ^{a}	2.37 ± 0.13 ^{a}	2.33 ± 0.06 ^{a}	2.47 ± 0.13 ^{a}	2.45 ± 0.05 ^{a}
8 weeks	2.26 ± 0.02 ^{a}	2.25 ± 0.03 ^{a}	2.21 ± 0.04 ^{a}	2.29 ± 0.04 ^{a}	2.22 ± 0.03 ^{a}
WG (g)					
4 weeks	14.38 ± 1.61 ^{a}	14.82 ± 1.13 ^{a}	14.44 ± 0.59 ^{a}	15.58 ± 1.21 ^{a}	15.44 ± 0.39 ^{a}
8 weeks	40.81 ± 0.64 ^{a}	40.78 ± 0.90 ^{a}	39.59 ± 1.33 ^{a}	41.78 ± 1.34 ^{a}	39.82 ± 0.82 ^{a}
FCR					
4 weeks	1.00 ± 0.07 ^{a}	1.03 ± 0.03 ^{a}	1.11 ± 0.08 ^{a}	0.98 ± 0.06 ^{a}	1.01 ± 0.00 ^{a}
8 weeks	0.96 ± 0.03 ^{a}	1.01 ± 0.03 ^{a}	1.01 ± 0.02 ^{a}	0.98 ± 0.02 ^{a}	1.01 ± 0.02 ^{a}
SR (%)					
4 weeks	100	98	97	100	98
8 weeks	98	95	95	98	98

Note: IW: Initial weight; FW: Final weight; SGR: Specific growth rate; WG: Weight gain; FCR: Feed conversion ratio; SR: Survival rate.

3.2. Analysis of Skin Mucus Immune Responses

Lysozyme and peroxidase activities in skin mucus in Nile tilapia after four and eight weeks of feeding are presented in Figure 1. No significant difference ($p > 0.05$) in lysozyme activity was observed between fish fed PSPP-supplemented diets and those fed only a basal diet (0 g kg^{-1} PSPP) after four and eight weeks of feeding. Only fish fed the 20 g kg^{-1}

PSPP (PSPP20) diet had significantly ($p < 0.05$) higher peroxidase activity than those with other treatments after four weeks of feeding trial (Figure 1B). No significant difference was found in peroxidase activity after eight weeks of feeding ($p > 0.05$).

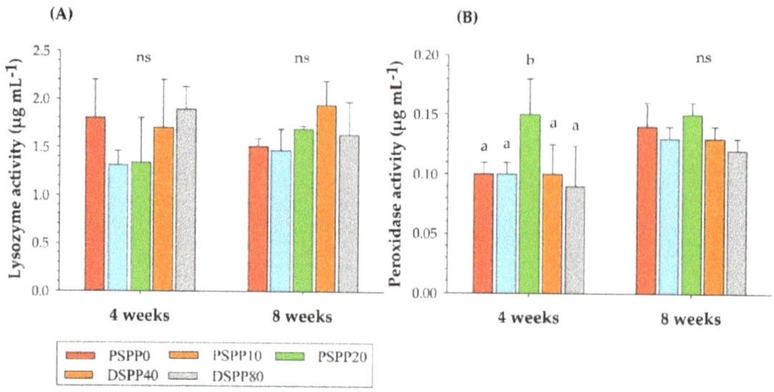

Figure 1. Lysozyme (**A**) and peroxidase (**B**) activity in the skin mucus of Nile tilapia after four and eight weeks of feeding with experimental diets: 0 g kg^{-1} (PSPP0) control, 10 g kg^{-1} (PSPP10), 20 g kg^{-1} (PSPP20), 40 g kg^{-1} (PSPP40), and 80 g kg^{-1} (PSPP80) cultivated for eight weeks. Data are presented as mean ± SE. Significantly different levels are denoted by superscript letters ($p < 0.05$). "ns" denotes no significant difference.

3.3. Analysis of Serum Immune Responses

Serum immunological activities (lysozyme and peroxidase) were determined in this study using serum samples obtained after four and eight weeks of feeding. Figure 2 summarizes the impact of the experimental diets on serum immunological activity. Peroxidase and lysozyme activities in serum showed substantially higher levels in fish fed dietary PSPP-supplemented diets than those fed the basal diet without PSPP supplementation after four and eight weeks of the experimental trial. No statistically significant differences in lysozyme activity were detected in any of the dietary supplementation treatments ($p > 0.05$) after four or eight weeks of feeding. On the other hand, the PSPP10 diet substantially enhanced serum peroxidase activity compared to the control PSPP0 group ($p < 0.05$). At eight weeks post-feeding, a statistically significant change in the activity of peroxidase was detected between the PSPP-supplemented treatments (PSPP20 and PSPP80) and the PSPP0 control treatment ($p < 0.05$).

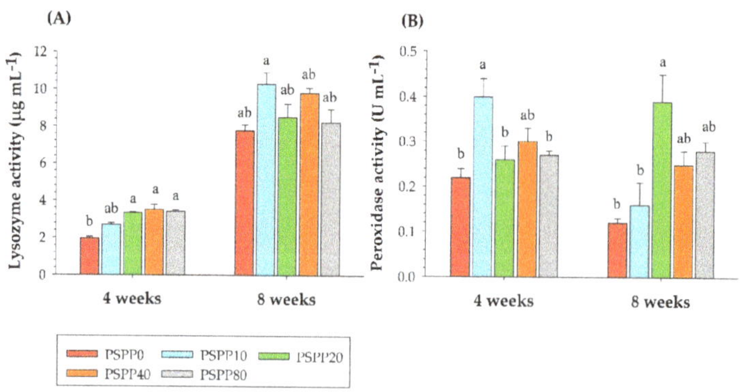

Figure 2. Lysozyme (**A**) and peroxidase (**B**) activity in the serum of Nile tilapia after four and eight weeks of feeding with experimental diets: 0 g kg^{-1} (PSPP0) control, 10 g kg^{-1} (PSPP10), 20 g kg^{-1} (PSPP20), 40 g kg^{-1} (PSPP40), and 80 g kg^{-1} (PSPP80) cultivated for eight weeks. Data are presented as mean ± SE. Significantly different levels are denoted by superscript letters ($p < 0.05$).

3.4. Analysis of Immune and Antioxidant-Related Gene Expression

Fish tissues (gills and liver) were collected to investigate relative immune and antioxidant gene expressions by qPCR. Three relative immune genes (*lbp*, *il-1*, and *il-8*) and three antioxidant genes (*gsr*, *gst-α*, and *gpx*) were investigated in this study. A significant upregulation (approximately 2–2.7 fold) was found in the gill tissues of fish fed with dietary supplementation with PSPP compared to those fed the basal diet without PSPP supplementation (the control group, PSPP0). The greatest level of mRNA transcripts was observed in the PSPP20 diet groups (approximately 3.1–3.7 fold) in *lbp*, *gst-α*, and *gpx*, whereas *il-1* and *il-8* had the highest levels in the dietary PSPP40 and PSPP10 treatments, respectively (Figure 3).

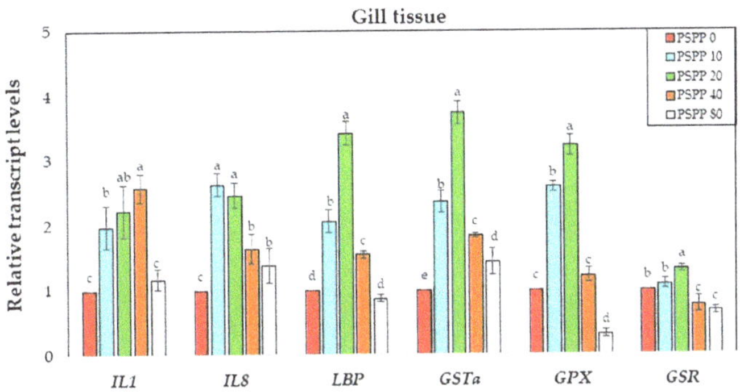

Figure 3. Expression transcript levels of interleukin-1 (*il-1*), interleukin-8 (*il-8*), lipopolysaccharide-binding protein (*lbp*), glutathione S-transferase-α (*GST-α*), glutathione peroxidase (*gpx*), and glutathione reductase (*gsr*) in gill tissues of Nile tilapia ($n = 5$) fed diets supplemented 0 g kg^{-1} PSPP, 10 g kg^{-1} PSPP, 20 g kg^{-1} PSPP, 40 g kg^{-1} PSPP, and 80 g kg^{-1} PSPP after eight weeks of feeding. The mRNA transcript levels of immune and antioxidant genes were normalized by *18S rRNA*. The mRNA transcript level of the 0 g kg^{-1} PSPP control group was set at 1. Data are presented as mean ± SE. Significantly different levels are denoted by superscript letters ($p < 0.05$).

A substantial difference in the mRNA transcripts of the examined genes was identified in liver tissues (Figure 4). Fish given the dietary PSPP20 expressed the greatest levels of mRNA transcripts in all target genes (except *lbp*) compared to the other dietary treatment and the control group (approximately 2–4.3 fold). *Lbp* expression was considerably upregulated in fish fed with dietary PSPP40 (approximately 2.5-fold), followed by PSPP20 (approximately 2.0-fold) and PSPP10 (approximately 1.8-fold).

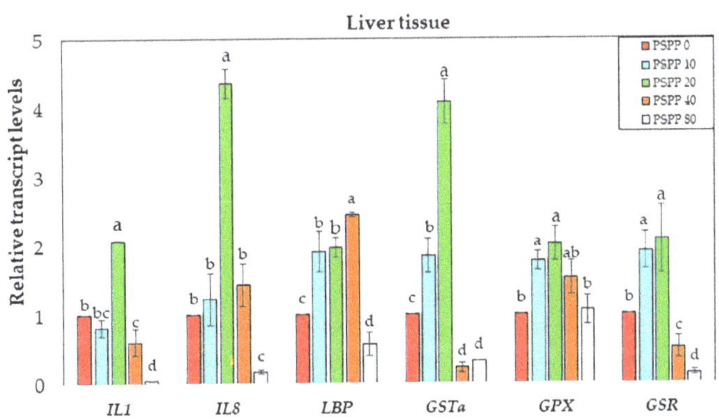

Figure 4. Expression transcript levels of interleukin-1 (*il-1*), interleukin-8 (*il-8*), lipopolysaccharide-binding protein (*lbp*), glutathione S-transferase-α (*gst-α*), glutathione peroxidase (*gpx*), and glutathione

reductase (*gsr*) in liver tissues of Nile tilapia (n = 5) fed diets supplemented 0 g kg^{-1} PSPP, 10 g kg^{-1} PSPP, 20 g kg^{-1} PSPP, 40 g kg^{-1} PSPP, and 80 g kg^{-1} PSPP after eight weeks of feeding. The mRNA transcript levels of immune and antioxidant genes were normalized by *18S rRNA*. The mRNA transcript level of the 0 g kg^{-1} PSPP control group was set at 1. Data are presented as mean ± SE. Significantly different levels are denoted by superscript letters ($p < 0.05$).

4. Discussion

The use of fruit by-products as feed additives in aquaculture is an efficient approach to conserve the environment and generate more income for farmers. By-products, such as peels and seeds, contain many substances with health-promoting effects [63–65].

After eight weeks of feeding, we observed that growth performance and feed consumption were unaffected by the PSPP supplement diets. The findings of this study were consistent with those of earlier studies on Jaraqui (*Semaprochilodus insignis*) and tambaqui fed passionfruit seed cake [66]; silver catfish (*Schilbe intermedius*) fed grape, orange, guava, and fig residues [67]; Nile tilapia and African catfish (*Clarias gariepinus*) fed dehydrated lemon peels [68]; Nile tilapia fed passionfruit seed oil and pomegranate peel [69,70]; rainbow trout (*Oncorhynchus mykiss*) fed dehydrate lemon peel [71]; Asian sea bass (*Lates calcarifer*) fed fermented lemon peel [72,73]; and giant freshwater prawn (*Macrobrachium rosenbergii*) fed biogas sludge meal [74], suggesting that the PSPP may not promote the production of digestive enzymes or intestinal absorption due to the large levels of soluble and insoluble fiber in PSPP [75]. Dietary fiber in PSPP has been shown by Vuolo, et al. [76] to decrease glucose and lipid absorption, resulting in less energy storage and increased lipid and glucose excretion. There was no discernible difference between the dietary PSPP-supplemented groups in this study. According to Ramli, et al. [33], PSPP has various valuable active components, including phenolic compounds, flavonoids, cyanogenic chemicals, anthocyanin, minerals, polysaccharides, and vitamins—which may account for its beneficial impact on growth performance [28,34,77]. Additionally, the PSPP contains considerable amounts of vitamin C that can be fortified into fish feed [78]. More investigations are needed to clarify the impact of these extracts on the growth performance of Nile tilapia cultivated in biofloc systems.

Skin mucus plays an important role in fish immune responses [79]. Lysozyme and peroxidase are important indicators of the immune defense system of fish; it has lytic action against bacteria and participates in phagocytic activity, neutrophil activation, and the complement system [80,81]. Lysozyme and peroxidase activities were greater in fish fed PSPP diets than in the control group after four and eight weeks of the feeding experiment. The addition of fruit by-products or extracts to diets, especially powdered passionfruit peel, has a beneficial effect on the immunological activity of Nile tilapia, striped catfish (*Pangasianodon hypophthalmus*), black rockfish (*Sebastes schlegelii*), and gilthead seabream (*Sparus aurata*) [61,82–84].

il-1 and *il-8* are responsible for regulating inflammatory processes in the innate immune system to stimulate phagocytes to engulf microorganisms [85]. Antioxidant-related genes are involved in the glutathione protection mechanism, responsible for hydrogen peroxide removal (H_2O_2). A phase II xenobiotic metabolic enzyme, glutathione S-transferase (GST), combines with electrophilic chemicals to produce bigger endogenic complexes known as glutathione S-conjugates, which are then expelled from the body [86]. GPX transforms H_2O_2 into H_2O via the oxidation of glutathione (GSH) to glutathione disulfide (GSSG). GSH is revived by GSR after it has been oxidized by the oxidative reduction of NADPH [87]. Increased immune responses and gene expression levels in fish fed powdered passionfruit peel are likely to be the result of an overall improvement in health and wellbeing due to a combination of several health benefits associated with dietary PSPP. These include (i) greater immunity against pathogens, indicated by elevated lysozyme and peroxidase levels in skin mucus and serum, and by elevated *il-1*, *il-8*, and *lbp* mRNA transcript levels in the gills and liver tissues; (ii) enhanced antioxidant activity, indicated by elevated mRNA transcript levels of *gst-α*, *gpx* and *gsr*; and (iii) PSPP may stimulate the immune defense system in fish, thereby improving survival rates and disease resistance in fish.

The successful application of biofloc in aquaculture depends on the presence of both prebiotics and probiotics. The addition of PSPP in a biofloc aquaculture system may be involved in several processes, such as stimulating the proliferation of favorable bacteria, inhibiting the growth of pathogenic microorganisms, and improving the gastrointestinal condition of fish [61,88–90]. On the other hand, the recycling of nitrogen via its conversion to microbial biomass in biofloc increases the populations of favorable bacteria, enhancing host immunity [91].

5. Conclusions

Diets containing powdered passionfruit peel at concentrations of 10 to 20 g kg^{-1} improved expression levels of innate immune and antioxidant-related genes in Nile tilapia cultured in a biofloc system. However, fish fed PSPP-supplemented feed had no significantly differences in growth performance; further studies should explore this issue to gain a better understanding of the impacts of PSPP in Nile tilapia.

Author Contributions: P.O.; investigation, N.V.L.; writing—original draft preparation, C.L.X.; formal analysis, S.W.; formal analysis, S.T.; resources, C.C.; data curation, N.M.; visualization, H.V.D.; conceptualization, project administration, and writing—review and editing. All authors have read and agreed to the published version of the manuscript.

Funding: This research work was partially supported by Chiang Mai University (CoE2565).

Institutional Review Board Statement: This study was conducted in accordance with international guidelines and approved by the Chiang Mai University Committee (No. RAGIACUC002/2565).

Data Availability Statement: The data presented in this study are available on request from the corresponding author.

Acknowledgments: The authors take this opportunity to thank the National Research Council of Thailand for supporting and helping with the study. This research work was partially supported by Chiang Mai University.

Conflicts of Interest: The authors declare no conflict of interest.

References

1. Magbanua, T.O.; Ragaza, J.A. Selected dietary plant-based proteins for growth and health response of Nile tilapia Oreochromis niloticus. *Aquac. Fish.* **2022**. [CrossRef]
2. Fialho, N.S.; Valenti, W.C.; David, F.S.; Godoy, E.M.; Proença, D.C.; Roubach, R.; Wolff Bueno, G. Environmental sustainability of Nile tilapia net-cage culture in a neotropical region. *Ecol. Indic.* **2021**, *129*, 108008. [CrossRef]
3. Mai, T.T.; Kayansamruaj, P.; Soontara, C.; Kerddee, P.; Nguyen, D.H.; Senapin, S.; Costa, J.Z.; Del-Pozo, J.; Thompson, K.D.; Rodkhum, C.; et al. Immunization of Nile Tilapia (Oreochromis niloticus) Broodstock with Tilapia Lake Virus (TiLV) Inactivated Vaccines Elicits Protective Antibody and Passive Maternal Antibody Transfer. *Vaccines* **2022**, *10*, 167. [CrossRef] [PubMed]
4. Munguti, J.M.; Nairuti, R.; Iteba, J.O.; Obiero, K.O.; Kyule, D.; Opiyo, M.A.; Abwao, J.; Kirimi, J.G.; Outa, N.; Muthoka, M.; et al. Nile tilapia (Oreochromis niloticus Linnaeus, 1758) culture in Kenya: Emerging production technologies and socio-economic impacts on local livelihoods. *Aquac. Fish Fish.* **2022**, *2*, 265–276. [CrossRef]
5. Paredes-Trujillo, A.; Mendoza-Carranza, M. A systematic review and meta-analysis of the relationship between farm management, water quality and pathogen outbreaks in tilapia culture. *J. Fish Dis.* **2022**, *00*, 1–20. [CrossRef]
6. Alazab, A.; Sadat, A.; Younis, G. Prevalence, antimicrobial susceptibility, and genotyping of Streptococcus agalactiae in Tilapia fish (Oreochromis niloticus) in Egypt. *J. Adv. Vet. Anim Res.* **2022**, *9*, 95–103. [CrossRef]
7. Chen, S.-W.; Liu, C.-H.; Hu, S.-Y. Dietary administration of probiotic Paenibacillus ehimensis NPUST1 with bacteriocin-like activity improves growth performance and immunity against Aeromonas hydrophila and Streptococcus iniae in Nile tilapia (*Oreochromis niloticus*). *Fish Shellfish. Immunol.* **2019**, *84*, 695–703. [CrossRef]
8. Liu, G.; Zhu, J.; Chen, K.; Gao, T.; Yao, H.; Liu, Y.; Zhang, W.; Lu, C. Development of Streptococcus agalactiae vaccines for tilapia. *Dis. Aquat. Org.* **2016**, *122*, 163–170. [CrossRef]
9. Payne, C.J.; Turnbull, J.F.; MacKenzie, S.; Crumlish, M. Investigating the Effect of an Oxytetracycline Treatment on the Gut Microbiome and Antimicrobial Resistance Gene Dynamics in Nile Tilapia (*Oreochromis niloticus*). *Antibiotics* **2021**, *10*, 1213. [CrossRef]

10. Sun, B.-Y.; He, W.; Yang, H.-X.; Tian, D.-Y.; Jian, P.-Y.; Wu, K.; Yang, C.-G.; Song, X.-H. Increased susceptibility to Aeromonas hydrophila infection in grass carp with antibiotic-induced intestinal dysbiosis. *Aquaculture* **2022**, *552*, 737969. [CrossRef]
11. Schar, D.; Zhao, C.; Wang, Y.; Larsson, D.G.J.; Gilbert, M.; van Boeckel, T.P. Twenty-year trends in antimicrobial resistance from aquaculture and fisheries in Asia. *Nat. Commun.* **2021**, *12*, 5384. [CrossRef] [PubMed]
12. Kraemer, S.A.; Ramachandran, A.; Perron, G.G. Antibiotic Pollution in the Environment: From Microbial Ecology to Public Policy. *Microorganisms* **2019**, *7*, 180. [CrossRef] [PubMed]
13. Thornber, K.; Verner-Jeffreys, D.; Hinchliffe, S.; Rahman, M.M.; Bass, D.; Tyler, C.R. Evaluating antimicrobial resistance in the global shrimp industry. *Rev. Aquac.* **2020**, *12*, 966–986. [CrossRef] [PubMed]
14. Abdel-Latif, H.M.R.; Yilmaz, E.; Dawood, M.A.O.; Ringø, E.; Ahmadifar, E.; Yilmaz, S. Shrimp vibriosis and possible control measures using probiotics, postbiotics, prebiotics, and synbiotics: A review. *Aquaculture* **2022**, *551*, 737951. [CrossRef]
15. Yilmaz, S.; Yilmaz, E.; Dawood, M.A.O.; Ringø, E.; Ahmadifar, E.; Abdel-Latif, H.M.R. Probiotics, prebiotics, and synbiotics used to control vibriosis in fish: A review. *Aquaculture* **2022**, *547*, 737514. [CrossRef]
16. Iwashita, M.K.P.; Addo, S.; Terhune, J.S. 9—Use of pre- and probiotics in finfish aquaculture. In *Feed and Feeding Practices in Aquaculture*, 2nd ed.; Davis, D.A., Ed.; Woodhead Publishing: Oxford, UK, 2022; pp. 269–289.
17. Mugwanya, M.; Dawood, M.A.O.; Kimera, F.; Sewilam, H. Updating the Role of Probiotics, Prebiotics, and Synbiotics for Tilapia Aquaculture as Leading Candidates for Food Sustainability: A Review. *Probiotics Antimicrob. Proteins* **2022**, *14*, 130–157. [CrossRef]
18. Buruiana, C.-T.; Gómez, B.; Vizireanu, C.; Garrote, G. Manufacture and evaluation of xylooligosaccharides from corn stover as emerging prebiotic candidates for human health. *LWT* **2017**, *77*, 449–459. [CrossRef]
19. Yao, W.; Gong, Y.; Li, L.; Hu, X.; You, L. The effects of dietary fibers from rice bran and wheat bran on gut microbiota: An overview. *Food Chem. X* **2022**, *13*, 100252. [CrossRef]
20. Deehan, E.C.; Zhang, Z.; Riva, A.; Armet, A.M.; Perez-Muñoz, M.E.; Nguyen, N.K.; Krysa, J.A.; Seethaler, B.; Zhao, Y.-Y.; Cole, J.; et al. Elucidating the role of the gut microbiota in the physiological effects of dietary fiber. *Microbiome* **2022**, *10*, 77. [CrossRef]
21. Gómez-García, R.; Campos, D.A.; Aguilar, C.N.; Madureira, A.R.; Pintado, M. Valorization of melon fruit (*Cucumis melo* L.) by-products: Phytochemical and Biofunctional properties with Emphasis on Recent Trends and Advances. *Trends Food Sci. Technol.* **2020**, *99*, 507–519. [CrossRef]
22. Donner, M.; Verniquet, A.; Broeze, J.; Kayser, K.; de Vries, H. Critical success and risk factors for circular business models valorising agricultural waste and by-products. *Resour. Conserv. Recycl.* **2021**, *165*, 105236. [CrossRef]
23. Thakur, N.; Nigam, M.; Tewary, R.; Rajvanshi, K.; Kumar, M.; Shukla, S.K.; Mahmoud, G.A.-E.; Gupta, S. Drivers for the behavioural receptiveness and non-receptiveness of farmers towards organic cultivation system. *J. King Saud Univ. Sci.* **2022**, *34*, 102107. [CrossRef]
24. Yan, C.; Muhammad Rizwan, H.; Liang, D.; Reichelt, M.; Mithöfer, A.; Scholz, S.S.; Oelmüller, R.; Chen, F. The effect of the root-colonizing Piriformospora indica on passion fruit (*Passiflora edulis*) development: Initial defense shifts to fitness benefits and higher fruit quality. *Food Chem.* **2021**, *359*, 129671. [CrossRef] [PubMed]
25. He, X.; Luan, F.; Yang, Y.; Wang, Z.; Zhao, Z.; Fang, J.; Wang, M.; Zuo, M.; Li, Y. *Passiflora edulis*: An Insight Into Current Researches on Phytochemistry and Pharmacology. *Front. Pharmacol.* **2020**, *11*, 617. [CrossRef]
26. Hu, Y.; Jiao, L.; Jiang, M.-H.; Yin, S.; Dong, P.; Zhao, Z.-M.; Yang, D.-P.; Ho, P.-T.; Wang, D.-M. A new C-glycosyl flavone and a new neolignan glycoside from Passiflora edulis Sims peel. *Nat. Prod. Res.* **2018**, *32*, 2312–2318. [CrossRef] [PubMed]
27. FMI. *Passion Fruit Extract Market projected to Reach USD 1,028.6 Million by 2029—Comprehensive Research Report by FMI*; FMI: Maharashtra, India, 2022.
28. Chutia, H.; Mahanta, C.L. Green ultrasound and microwave extraction of carotenoids from passion fruit peel using vegetable oils as a solvent: Optimization, comparison, kinetics, and thermodynamic studies. *Innov. Food Sci. Emerg. Technol.* **2021**, *67*, 102547. [CrossRef]
29. Abboud, K.Y.; Iacomini, M.; Simas, F.F.; Cordeiro, L.M.C. High methoxyl pectin from the soluble dietary fiber of passion fruit peel forms weak gel without the requirement of sugar addition. *Carbohydr. Polym.* **2020**, *246*, 116616. [CrossRef]
30. Mohd Basri, M.S.; Abdul Karim Shah, N.N.; Sulaiman, A.; Mohamed Amin Tawakkal, I.S.; Mohd Nor, M.Z.; Ariffin, S.H.; Abdul Ghani, N.H.; Mohd Salleh, F.S. Progress in the Valorization of Fruit and Vegetable Wastes: Active Packaging, Biocomposites, By-Products, and Innovative Technologies Used for Bioactive Compound Extraction. *Polymers* **2021**, *13*, 3503. [CrossRef]
31. Ghada, B.; Pereira, E.; Pinela, J.; Prieto, M.A.; Pereira, C.; Calhelha, R.C.; Stojković, D.; Soković, M.; Zaghdoudi, K.; Barros, L.; et al. Recovery of Anthocyanins from Passion Fruit Epicarp for Food Colorants: Extraction Process Optimization and Evaluation of Bioactive Properties. *Molecules* **2020**, *25*, 3203. [CrossRef]
32. Ramli, A.N.M.; Manap, N.W.A.; Bhuyar, P.; Azelee, N.I.W. Passion fruit (Passiflora edulis) peel powder extract and its application towards antibacterial and antioxidant activity on the preserved meat products. *SN Appl. Sci.* **2020**, *2*, 1–11. [CrossRef]
33. Cao, Q.; Teng, J.; Wei, B.; Huang, L.; Xia, N. Phenolic compounds, bioactivity, and bioaccessibility of ethanol extracts from passion fruit peel based on simulated gastrointestinal digestion. *Food Chem.* **2021**, *356*, 129682. [CrossRef] [PubMed]
34. Ferreira, M.d.P.; Teixeira Tarley, C.R. Bioaccessibility estimation of metallic macro and micronutrients Ca, Mg, Zn, Fe, Cu and Mn in flours of oat and passion fruit peel. *LWT* **2021**, *150*, 111880. [CrossRef]
35. Júnior, J.E.L.; da Costa, J.M.; Neiva, J.N.M.; Rodriguez, N.M. Physical-chemical characterization of tropical fruit by-products for use in animal feed. *Rev. Ciência Agronômica* **2006**, *37*, 70.

36. Fachinello, M.R.; Pozza, P.C.; Moreira, I.; Carvalho, P.L.O.; Castilha, L.D.; Pasquetti, T.J.; Esteves, L.A.C.; Huepa, L.M.D. Effect of passion fruit seed meal on growth performance, carcass, and blood characteristics in starter pigs. *Trop. Anim. Health Prod.* **2015**, *47*, 1397–1403. [CrossRef]
37. Perondi, D.; Moreira, I.; Pozza, P.C.; Carvalho, P.L.d.O.; Pasquetti, T.J.; Huepa, L.M.D. Passion fruit seed meal at growing and finishing pig (30-90 kg) feeding. *Ciência Agrotecnologia* **2014**, *38*, 390–400. [CrossRef]
38. Wadhwa, M.; Bakshi, M.P.; Makkar, H.P. Waste to worth: fruit wastes and by-products as animal feed. *CAB Rev.* **2015**, *10*, 1–26. [CrossRef]
39. Fachinello, M.R.; Pozza, P.C.; Furlan, A.C.; Paula, V.R.C.d.; Bonagurio, L.P.; Marcato, S.M.; Leal, I.F.; Huepa, L.M.D. Nutritional evaluation of passion fruit seed meal for meat quails. *Rev. Bras. De Saúde Produção Anim.* **2016**, *17*, 202–213. [CrossRef]
40. Togashi, C.K.; Fonseca, J.B.; Soares, R.d.T.R.N.; da Costa, A.P.D.; da Silveira, K.F.; Detmann, E. Subprodutos do maracujá em dietas para frangos de corte. *Acta Sci. Anim. Sci.* **2008**, *30*, 395–400. [CrossRef]
41. Togashi, C.K.; Fonseca, J.B.; Soares, R.d.T.R.N.; Gaspar, A.; Detmann, E. Composição em ácidos graxos dos tecidos de frangos de corte alimentados com subprodutos de maracujá. *Rev. Bras. De Zootec.* **2007**, *36*, 2063–2068. [CrossRef]
42. De Souza, A.M.; Campeche, D.F.B.; Moraes, G.; de Melo, F.; da Cruz Neto, M.A.; Melo, J.F.B. Replacing cornmeal with mango meal in diets for juvenile tambaqui Colossoma macropomum: Growth and metabolic parameters. *Bol. Inst. Pesca* **2018**, *44*, e248. [CrossRef]
43. Wegbecher, F.X. *Bactérias Celulolíticas e o Uso de Resíduo de Maracujá (Passiflora edulis) em Rações Extrusadas Para Juvenis de Tambaqui (Colossoma macropomum)*; Instituto Nacional de Pesquisas da Amazônia: Petrópolis, Brazil, 2010.
44. dos Santos, E.A.; Ribeiro, A.E.C.; Barcellos, T.T.; Monteiro, M.L.G.; Mársico, E.T.; Caliari, M.; Júnior, M.S.S. Exploitation of byproducts from the passion fruit juice and tilapia filleting industries to obtain a functional meat product. *Food Biosci.* **2021**, *41*, 101084. [CrossRef]
45. Ahmad, I.; Babitha Rani, A.; Verma, A.; Maqsood, M. Biofloc technology: An emerging avenue in aquatic animal healthcare and nutrition. *Aquac. Int.* **2017**, *25*, 1215–1226. [CrossRef]
46. Durigon, E.G.; Lazzari, R.; Uczay, J.; de Alcântara Lopes, D.L.; Jerônimo, G.T.; Sgnaulin, T.; Emerenciano, M.G.C. Biofloc technology (BFT): Adjusting the levels of digestible protein and digestible energy in diets of Nile tilapia juveniles raised in brackish water. *Aquac. Fish.* **2020**, *5*, 42–51. [CrossRef]
47. Wei, G.; Shan, D.; Li, G.; Li, X.; Tian, R.; He, J.; Shao, Z. Prokaryotic communities vary with floc size in a biofloc-technology based aquaculture system. *Aquaculture* **2020**, *529*, 735632. [CrossRef]
48. Khanjani, M.H.; Sharifinia, M.; Hajirezaee, S. Recent progress towards the application of biofloc technology for tilapia farming. *Aquaculture* **2022**, *552*, 738021. [CrossRef]
49. Abakari, G.; Wu, X.; He, X.; Fan, L.; Luo, G. Bacteria in biofloc technology aquaculture systems: Roles and mediating factors. *Rev. Aquac.* **2022**, *14*, 1260–1284. [CrossRef]
50. AOAC. *Official Methods of Analysis of Association of Official Analytical Chemists*, 18th ed.; Association of Officiating Analytical Chemists: Washington, DC, USA, 2005.
51. Van Doan, H.; Hoseinifar, S.H.; Harikrishnan, R.; Khamlor, T.; Punyatong, M.; Tapingkae, W.; Yousefi, M.; Palma, J.; El-Haroun, E. Impacts of pineapple peel powder on growth performance, innate immunity, disease resistance, and relative immune gene expression of *Nile tilapia, Oreochromis niloticus*. *Fish Shellfish. Immunol.* **2021**, *114*, 311–319. [CrossRef]
52. Avnimelech, Y. *Biofloc Technology: A Practical Guide Book*; World Aquaculture Society: Baton Rouge. LA, USA, 2015.
53. Panigrahi, A.; Saranya, C.; Sundaram, M.; Vinoth Kannan, S.R.; Das, R.R.; Satish Kumar, R.; Rajesh, P.; Otta, S.K. Carbon: Nitrogen (C:N) ratio level variation influences microbial community of the system and growth as well as immunity of shrimp (*Litopenaeus vannamei*) in biofloc based culture system. *Fish Shellfish. Immunol.* **2018**, *81*, 329–337. [CrossRef]
54. Van Doan, H.; Lumsangkul, C.; Sringarm, K.; Hoseinifar, S.H.; Dawood, M.A.O.; El-Haroun, E.; Harikrishnan, R.; Jaturasitha, S.; Paolucci, M. Impacts of Amla (*Phyllanthus emblica*) fruit extract on growth, skin mucosal and serum immunities, and disease resistance of Nile tilapia (*Oreochromis niloticus*) raised under biofloc system. *Aquac. Rep.* **2022**, *22*, 100953. [CrossRef]
55. Xuan, C.L.; Wannavijit, S.; Outama, P.; Lumsangkul, C.; Tongsiri, S.; Chitmanat, C.; Doan, H.V. Dietary inclusion of rambutan (*Nephelium lappaceum* L.) seed to Nile tilapia (*Oreochromis niloticus*) reared in biofloc system: Impacts on growth, immunity, and immune-antioxidant gene expression. *Fish Shellfish. Immunol.* **2022**, *122*, 215–224. [CrossRef] [PubMed]
56. Parry, R.M., Jr.; Chandan, R.C.; Shahani, K.M. A rapid and sensitive assay of muramidase. *Proc. Soc. Exp. Biol. Med.* **1965**, *119*, 384–386. [CrossRef] [PubMed]
57. Quade, M.J.; Roth, J.A. A rapid, direct assay to measure degranulation of bovine neutrophil primary granules. *Vet. Immunol. Immunopathol.* **1997**, *58*, 239–248. [CrossRef]
58. Cordero, H.; Cuesta, A.; Meseguer, J.; Esteban, M.Á. Changes in the levels of humoral immune activities after storage of gilthead seabream (Sparus aurata) skin mucus. *Fish Shellfish. Immunol.* **2016**, *58*, 500–507. [CrossRef]
59. Livak, K.J.; Schmittgen, T.D. Analysis of Relative Gene Expression Data Using Real-Time Quantitative PCR and the $2^{-\Delta\Delta CT}$ Method. *Methods* **2001**, *25*, 402–408. [CrossRef] [PubMed]
60. Le Xuan, C.; Wannavijit, S.; Outama, P.; Montha, N.; Lumsangkul, C.; Tongsiri, S.; Chitmanat, C.; Hoseinifar, S.H.; van Doan, H. Effects of dietary rambutan (*Nephelium lappaceum* L.) peel powder on growth performance, immune response and immune-related gene expressions of striped catfish (Pangasianodon hypophthalmus) raised in biofloc system. *Fish Shellfish. Immunol.* **2022**, *124*, 134–141. [CrossRef] [PubMed]

61. Wakai, K.; Hamajima, N.; Okada, R.; Naito, M.; Morita, E.; Hishida, A.; Kawai, S.; Nishio, K.; Yin, G.; Asai, Y. SAS/STAT 9.1 user's guide SAS/STAT 9.1 user's guide, 2004. *J. Epidemiol.* **2009**, *19*, 72–80. [CrossRef]
62. Ahmed, M.M.; Badawy, M.T.; Ahmed, F.K.; Kalia, A.; Abd-Elsalam, K.A. Chapter 10—Fruit peel waste-to-wealth: Bionanomaterials production and their applications in agroecosystems. In *Agri-Waste and Microbes for Production of Sustainable Nanomaterials*; Abd-Elsalam, K.A., Periakaruppan, R., Rajeshkumar, S., Eds.; Elsevier: Amsterdam, The Netherlands, 2022; pp. 231–257.
63. Samtiya, M.; Aluko, R.E.; Dhewa, T.; Moreno-Rojas, J.M. Potential Health Benefits of Plant Food-Derived Bioactive Components: An Overview. *Foods* **2021**, *10*, 839. [CrossRef]
64. Tlais, A.Z.A.; Fiorino, G.M.; Polo, A.; Filannino, P.; Di Cagno, R. High-Value Compounds in Fruit, Vegetable and Cereal Byproducts: An Overview of Potential Sustainable Reuse and Exploitation. *Molecules* **2020**, *25*, 2987. [CrossRef]
65. Perar, K.; Fonseca, F.; Affonso, E.; Nobre, A. Passion Fruit (*Passiflora edulis*) Seed Cake as a Feed Ingredient for Jaraqui (*Semaprochilodus insignis*) and Tambaqui (*Colossoma macropomum*). *J. Aquac. Mar. Biol.* **2017**, *6*, 173. [CrossRef]
66. Lazzari, R.; Uczay, J.; Henriques, J.; Durigon, E.; Kunz, D.; Peixoto, N.; Fronza, D. Growth and digestive enzymes of silver catfish fed diets containing fruit residue. *Arq. Bras. De Med. Veterinária E Zootec.* **2019**, *71*, 323–330. [CrossRef]
67. Abdel Rahman, A.N.; ElHady, M.; Shalaby, S.I. Efficacy of the dehydrated lemon peels on the immunity, enzymatic antioxidant capacity and growth of Nile tilapia (*Oreochromis niloticus*) and African catfish (*Clarias gariepinus*). *Aquaculture* **2019**, *505*, 92–97. [CrossRef]
68. Conrado, A.; Iunes, R.; Neyrão, I. Performance of Nile tilapia (*Oreochromis niloticus*) juveniles fed diets with different levels of passion fruit seed oil (*Passiflora edulis*). *Livest. Res. Rural. Dev.* **2020**, *32*, 8.
69. Hamed, H.S.; Abdel-Tawwab, M. Dietary pomegranate (*Punica granatum*) peel mitigated the adverse effects of silver nanoparticles on the performance, haemato-biochemical, antioxidant, and immune responses of Nile tilapia fingerlings. *Aquaculture* **2021**, *540*, 736742. [CrossRef]
70. Chekani, R.; Akrami, R.; Ghiasvand, Z.; Chitsaz, H.; Jorjani, S. Effect of dietary dehydrated lemon peel (*Citrus limon*) supplementation on growth, hemato-immunolological and antioxidant status of rainbow trout (*Oncorhynchus mykiss*) under exposure to crowding stress. *Aquaculture* **2021**, *539*, 736597. [CrossRef]
71. Zhuo, L.-C.; Chen, C.-F.; Lin, Y.-H. Dietary supplementation of fermented lemon peel enhances lysozyme activity and susceptibility to Photobacterium damselae for orange-spotted grouper, Epinephelus coioides. *Fish Shellfish. Immunol.* **2021**, *117*, 248–252. [CrossRef] [PubMed]
72. Zhuo, L.-C.; Abang Zamhari, D.N.J.b.; Yong, A.S.K.; Shapawi, R.; Lin, Y.-H. Effects of fermented lemon peel supplementation in diet on growth, immune responses, and intestinal morphology of Asian sea bass, Lates calcarifer. *Aquac. Rep.* **2021**, *21*, 100801. [CrossRef]
73. Whangchai, N.; Yaibouathong, D.; Junluthin, P.; Balakrishnan, D.; Unpaprom, Y.; Ramaraj, R.; Pimpimol, T. Effect of biogas sludge meal supplement in feed on growth performance molting period and production cost of giant freshwater prawn culture. *Chemosphere* **2022**, *301*, 134638. [CrossRef]
74. Abboud, K.Y.; da Luz, B.B.; Dallazen, J.L.; Werner, M.F.d.P.; Cazarin, C.B.B.; Maróstica Junior, M.R.; Iacomini, M.; Cordeiro, L.M.C. Gastroprotective effect of soluble dietary fibres from yellow passion fruit (*Passiflora edulis f. flavicarpa*) peel against ethanol-induced ulcer in rats. *J. Funct. Foods* **2019**, *54*, 552–558. [CrossRef]
75. Vuolo, M.M.; Lima, G.C.; Batista, Â.G.; Carazin, C.B.B.; Cintra, D.E.; Prado, M.A.; Júnior, M.R.M. Passion fruit peel intake decreases inflammatory response and reverts lipid peroxidation and adiposity in diet-induced obese rats. *Nutr. Res.* **2020**, *76*, 106–117. [CrossRef]
76. Pertuzatti, P.B.; Sganzerla, M.; Jacques, A.C.; Barcia, M.T.; Zambiazi, R.C. Carotenoids, tocopherols and ascorbic acid content in yellow passion fruit (*Passiflora edulis*) grown under different cultivation systems. *LWT Food Sci. Technol.* **2015**, *64*, 259–263. [CrossRef]
77. Shabir, U.; Dar, J.S.; Bhat, A.H.; Ganai, B.A.; Khan, I.A. Isolation and characterization of β-defensin-like protein 1 from epidermal mucus of fungal infected fish (*Cyprinus carpio*) and assessment of its antimicrobial potencies. *Aquac. Rep.* **2022**, *23*, 101056. [CrossRef]
78. Alvarez-Pellitero, P. Fish immunity and parasite infections: From innate immunity to immunoprophylactic prospects. *Vet. Immunol. Immunopathol.* **2008**, *126*, 171–198. [CrossRef] [PubMed]
79. Yu, Y.Y.; Ding, L.G.; Huang, Z.Y.; Xu, H.Y.; Xu, Z. Commensal bacteria-immunity crosstalk shapes mucosal homeostasis in teleost fish. *Rev. Aquac.* **2021**, *13*, 2322–2343. [CrossRef]
80. Wannavijit, S.; Outama, P.; Le Xuan, C.; Lumsangkul, C.; Lengkidworraphiphat, P.; Tongsiri, S.; Chitmanat, C.; Doan, H.V. Modulatory effects of longan seed powder on growth performance, immune response, and immune-antioxidant related gene expression in Nile tilapia (*Oreochromis niloticus*) raised under biofloc system. *Fish Shellfish. Immunol.* **2022**, *123*, 460–468. [CrossRef] [PubMed]
81. Lee, D.-Y.; Lee, C.-H.; Kim, K.-D.; Lim, H.J.; Kim, H.S. Effects of diet supplementation with plant juice processing by-products on juvenile black rockfish (*Sebastes schlegelii*) growth performance, feed utilization, non-specific immunity, and disease resistance against Vibrio harveyi. *Aquac. Rep.* **2021**, *21*, 100831. [CrossRef]
82. Silva-Brito, F.; Cardoso, A.; Machado, M.; Ramos-Pinto, L.; Hinzmann, M.; Abreu, H.; Costas, B.; Magnoni, L. Dietary supplementation with Gracilaria gracilis by-products modulates the immune status and oxidative stress response of gilthead seabream (Sparus aurata) stimulated with Photobacterium damselae subsp. piscicida. *Fish Shellfish. Immunol.* **2022**, *126*, 164–177. [CrossRef]

83. Oliva-Teles, A. Nutrition and health of aquaculture fish. *J. Fish Dis.* **2012**, *35*, 83–108. [CrossRef]
84. Lall, S.P.; Dumas, A. 3—Nutritional requirements of cultured fish: Formulating nutritionally adequate feeds. In *Feed and Feeding Practices in Aquaculture*, 2nd ed.; Davis, D.A., Ed.; Woodhead Publishing: Oxford, UK, 2022; pp. 65–132.
85. Cruvinel, W.d.M.; Mesquita Júnior, D.; Araújo, J.A.P.; Catelan, T.T.T.; Souza, A.W.S.d.; Silva, N.P.d.; Andrade, L.E.C. Immune system: Part I. Fundamentals of innate immunity with emphasis on molecular and cellular mechanisms of inflammatory response. *Rev. Bras. De Reumatol.* **2010**, *50*, 434–447. [CrossRef]
86. Ahlf, W.; Heise, S. Sediment Toxicity Assessment: Rationale for effect classes (5 pp). *J. Soils Sediments* **2005**, *5*, 16–20. [CrossRef]
87. Imai, H.; Nakagawa, Y. Biological significance of phospholipid hydroperoxide glutathione peroxidase (PHGPx, GPx4) in mammalian cells. *Free. Radic. Biol. Med.* **2003**, *34*, 145–169. [CrossRef]
88. Panigrahi, A.; Esakkiraj, P.; Das, R.R.; Saranya, C.; Vinay, T.; Otta, S.K.; Shekhar, M.S. Bioaugmentation of biofloc system with enzymatic bacterial strains for high health and production performance of Penaeus indicus. *Sci. Rep.* **2021**, *11*, 1–13. [CrossRef] [PubMed]
89. Van Doan, H.; Lumsangkul, C.; Hoseinifar, S.H.; Harikrishnan, R.; Balasundaram, C.; Jaturasitha, S. Effects of coffee silverskin on growth performance, immune response, and disease resistance of Nile tilapia culture under biofloc system. *Aquaculture* **2021**, *543*, 736995. [CrossRef]
90. Van Doan, H.; Lumsangkul, C.; Jaturasitha, S.; Meidong, R.; Hoseinifar, S.H.; Dawood, M.A. Modulation of growth, skin mucus and serum immunities, and disease resistance of Nile tilapia fed host-associated probiotic (*Lactobacillus paracasei* l61-27b). *Aquac. Nutr.* **2021**, *27*, 3–12. [CrossRef]
91. Mohammadi, G.; Rafiee, G.; Tavabe, K.R.; Abdel-Latif, H.M.; Dawood, M.A. The enrichment of diet with beneficial bacteria (single-or multi-strain) in biofloc system enhanced the water quality, growth performance, immune responses, and disease resistance of Nile tilapia (*Oreochromis niloticus*). *Aquaculture* **2021**, *539*, 736640. [CrossRef]

fishes

MDPI

Article

Effects of *Macleaya cordata* Extract on Growth Performance, Serum Biochemical Parameters, and Intestinal Health of Juvenile American Eel (*Anguilla rostrata*)

Runan Chen, Linxin Huang and Shaowei Zhai *

Engineering Research Center of the Modern Industry Technology for Eel, Ministry of Education, Fisheries College of Jimei University, Xiamen 361021, China
* Correspondence: zhaisw@jmu.edu.cn; Tel.: +86-592-6181420

Abstract: The present trial was conducted to evaluate the supplementation effects of *Macleaya cordata* extract (MCE) on growth performance, serum biochemical parameters, and intestinal health of the juvenile American eel (*Anguilla rostrata*). The 480 juvenile American eels (10.93 ± 0.06 g) were randomly divided into four groups. They were fed on diets supplemented with MCE levels of 0, 25 mg/kg, 50 mg/kg, and 100 mg/kg for ten weeks, respectively. The 50 mg/kg or 100 mg/kg MCE could significantly improve growth performance, and increase the activities of acid phosphatase and alkaline phosphatase, as well as the level of high-density lipoprotein cholesterol. These levels of MCE also decreased the levels of D-lactate acid, triglyceride, and total cholesterol and the activities of aspartate aminotransferase, alanine aminotransferase, and diamine oxidase. The antioxidant ability, muscular thickness, and fold height of the intestine were enhanced by 50 mg/kg or 100 mg/kg MCE. There was no significant difference in the above parameters of groups fed with 50 mg/kg or 100 mg/kg of MCE. The beneficial effects on the intestinal microbiota were demonstrated in the group fed with 50 mg/kg MCE. In conclusion, the 50 mg/kg MCE could be used in the diet to improve the growth performance and health status of the juvenile American eels.

Keywords: *Macleaya cordata* extract; growth performance; serum lipid; antioxidant ability; intestinal microbiota; intestinal health; *Anguilla rostrata*

Citation: Chen, R.; Huang, L.; Zhai, S. Effects of *Macleaya cordata* Extract on Growth Performance, Serum Biochemical Parameters, and Intestinal Health of Juvenile American Eel (*Anguilla rostrata*). *Fishes* **2022**, *7*, 229. https://doi.org/10.3390/fishes7050229

Academic Editors: Marina Paolucci and Shunsuke Koshio

Received: 9 August 2022
Accepted: 29 August 2022
Published: 1 September 2022

1. Introduction

Global aquaculture development increased rapidly in recent years, and the sector has become one of the fastest-growing industries in the animal-derived food production system. However, the intensification and commercialization of aquaculture have been associated with frequent outbreaks of infectious diseases [1,2]. The use of antibiotics and chemotherapeutics is the main strategy in commercial aquaculture for controlling diseases. However, the development of drug-resistant pathogens, drug residues, and environmental pollution have generated great concerns [2]. Recently, growing interest has arisen in the utilization of natural products of the medical plant as an eco-friendly measure to ensure the sustainability of aquaculture. These extracts from the natural plant usually possess multiple biological activities such as antimicrobial, anti-inflammatory, immunostimulant, and antioxidant properties [2,3]. *Macleaya cordata* is a traditional herb primarily distributed in China, North America, and Europe. The main active ingredients of *Macleaya cordata* extract (MCE) are isoquinoline alkaloids that are mainly composed of sanguinarine, chelerythrine, some minority of protopine, berberine, coptisine, allocryptopine, etc. [3,4]. Based on the nucleophilic character of the iminium moiety and the polycyclic planar structure, MCE exhibits a broad spectrum of biological activities, such as anti-inflammatory [5,6], antioxidant [7,8], anti-bacterial [9], and anti-parasitic effects [10]. In 2004, compounds containing sanguinarine and chelerythrine from MCE were registered as feed additives in the European Union, and MCE is employed as a feed additive for swine and poultry widely [3,11]. In aquatic animals, previous studies revealed that dietary supplementation with MCE containing 1.5% sanguinarine and 0.75% chelerythrine could promote the growth performance of Caspian roach fry (*Rutilus rutilus*) and red tilapia (*Oreochromis niloticus*) [12,13]. Similarly, it was reported that the MCE containing 1.5% sanguinarine could increase weight gain and the average daily gain of common carp (*Cyprinus carpio*) [14]. However, dietary MCE containing 0.15% sanguinarine supplementation could not improve the growth performance of grass carp (*Ctenopharyngodon idellus*) fed high cottonseed and rapeseed meal diets [5]. These results indicated that the growth-promotion effects of MCE might vary in different fish species or diets. Little information is available regarding the growth-promotion effects of dietary MCE on other farmed fish species.

Being one of the most common freshwater cultured fish in the world, the eel has made important contributions to the development of the Chinese fisheries economy. With the natural stocks of European eel (*Anguilla anguilla*) and Japanese eel (*Anguilla japonica*) declining sharply, the American eel (*Anguilla rostrata*) has become one of the most popular farmed species in southeastern China [15,16]. Although dietary supplementation with MCE appears to exert beneficial effects on multiple species of fish, no research is available about the application of MCE in the diet of the American eel. The present study is aimed to evaluate the effects of dietary MCE supplementation on the growth performance, serum biochemical parameters, and intestinal health of the American eel.

2. Materials and Methods

2.1. Feeding Trial

One thousand juvenile American eels with similar body weights were obtained from Fujian Jinjiangzhiman Aquatic Technology Co., Ltd., Zhangzhou, China. Before the formal trial, the fish were acclimatized in two PVC tanks with bottom center drains (110 cm diameter, 80 cm height) and about 800 L water volume supplied with 5 L/min of degassed and dechlorinated municipal water. All the eels were fed on a commercial powder feed (Fuzhou Sea Horse Feed Co., Ltd., Fuzhou, Fujian, China) two times daily (6:00 and 18:00). The powder diet was mixed with 1:1.1 volume water to form a dough shape, and then the dough feed was placed on a feeding table for fish. The uneaten feed was siphoned out 20–25 min after feeding. The commercial diet was mainly composed of white fish meal, brown fish meal, pre-gelatinized starch, yeast powder, extruded soybean, and compound premix. The proximate composition of the commercial diet was crude protein 46.58%, crude fat 6.70%, ash 12.35%, and moisture 7.36%. During the acclimation period, the water quality parameters were maintained at 24–26 °C, pH 7.0–7.5, dissolved oxygen 7.0–9.0 mg/L, total ammonia nitrogen 0.2–0.6 mg/L, nitrite 0.03–0.06 mg/L.

After four weeks of acclimation, 480 American eels with similar body weights (10.93 ± 0.06 g/fish) were selected and randomly distributed into 16 tanks. The 16 tanks were randomly divided into four treatment groups fed the diets with MCE levels being 0, 25 mg/kg, 50 mg/kg, and 100 mg/kg, respectively. The four treatment groups were the control group, MCE25 group, MCE50 group, and MCE100 group, respectively. There were four replicates in each treatment group with 30 fish per replicate. The trial period was ten weeks.

MCE (An orange powder product containing 1.5% sanguinarine and 0.75% chelerythrine) was manufactured by Hunan Micolta Bioresource Co., Ltd., Changsha, China. The batch number of the MCE product was 2,103,291. The basal diet for the control group was the commercial feed with no MCE supplementation during the acclimation period, and the diets for the other MCE groups were prepared with the inclusion of MCE at 25 mg/kg, 50 mg/kg, and 100 mg/kg, respectively. The fish in the formal trial were cultured in 16 circular PVC tanks (320 L water) with a water recirculation system and an automatic temperature control device. The fish management and the water quality parameters during the formal trial period were maintained the same as those in the acclimation period. The consumption of diet in each tank was recorded daily.

2.2. Sample Collection

At the end of the feeding trial, all fish in each tank was deprived for 24 h. The fish of each tank were anesthetized with 0.1 mg/L eugenol, weighed, and counted to calculate the growth performance parameters. The blood of nine fish from each tank was sampled, treated, and mixed as one sample before analysis of serum biochemical parameters according to the procedure of Zhang et al. [15]. After the four fish of each tank was dissected, the intestine tissue was sampled and immersed in Bouin's solution (which consisted of 75 mL saturated picric acid aqueous solution, 25 mL formaldehyde, and 5 mL glacial acetic acid) for the observation of intestinal morphology. The remaining intestine samples were also collected and placed into 1.5 mL sterilization freeze tubes. Intestine samples were homogenized with 10 times the volume (volume/weight) of precooled normal saline (0.86%). Ground the mixture with a grinder (Tissuelyser-24, Shanghai Jingxin Industrial Development Co., Ltd., Shanghai, China), then the homogenate was centrifuged at 4 °C 3000 r/min for 10 min, collected the supernatant in centrifugal tubes, frozen at 80 °C before the analysis of antioxidant parameters. In addition, two intestine samples were selected from each tank for the analysis of the microbial diversity and community structure.

2.3. Growth Performance Calculation

At the end of the trial, the following growth parameters were calculated.

Weight gain rate (WGR, %) = 100 × [final fish weight of each tank (g) − initial fish weight of each tank(g)]/initial fish weight of each tank(g).

Specific growth rate (SGR, %/d) = [Ln final weight of each tank (g) − Ln initial weight of each tank (g) × 100]/trial days (d).

Feeding rate (FR, %) = 100 × feed consumption of each tank (g)/[((initial fish weight of each tank (g) + final fish weight of each tank(g))/2]/trial days (d).

Feed efficiency (FE, %) = 100 × [final fish weight of each tank (g) − the initial fish weight of each tank (g)]/feed intake of each tank (g).

Survival rate (SR, %) = 100 × (final fish number of each tank)/(initial fish number of each tank).

2.4. Measurement of Serum Biochemical Parameters

Total cholesterol (TC), triglyceride (TG), high-density lipoprotein cholesterol (HDL-C), low-density lipoprotein cholesterol (LDL-C), glutamic-oxaloacetic transaminase (GOT), glutamic-pyruvic transaminase (GPT), diamine oxidase (DAO), D-lactate (D-lac) in the serum were measured using commercial kits (Nanjing Jiancheng Bioengineering Co., Ltd., Nanjing, China) according to the manufacturer's instructions manual.

2.5. Measurement of Intestinal Antioxidant Parameters

The activities of superoxide dismutase (SOD), catalase (CAT), and glutathione peroxidase (GSH-PX) and the levels of total antioxidant capacity(T-AOC) and malondialdehyde (MDA) in the intestine were measured by assay kits (Nanjing Jiancheng Biotechnology Co., Ltd., Nanjing, China). The detailed determination steps were referred to in the manufacturer's instructions manual.

2.6. Observation and Measurement of Intestinal Morphology

The intestinal issues were dehydrated through 75% ethanol, cleared with xylene, and embedded in paraffin. The 5-μm histological section was cut by a microtome (KD-TS3A, Beijing Century Kexin Scientific Instrument Co., Ltd., Beijing, China), then stained with hematoxylin and eosin (H & E). Intestinal slices were observed and photographed with a light microscope (Olympus BX80-JPA) and analyzed using Image-Pro Plus 6.0 (Media Cybernetics, Silver Spring, MD, USA).

2.7. Intestinal Microbiota Profiling

The extraction and quality detection of intestinal total DNA, the design of primers for PCR amplification of bacterial 16S rDNA V3-V4 region, and data analysis were the same as the description by Shi et al. [17]. The high-throughput sequencing analysis was performed on the platform of Illumina Miseq PE300 with the assistance of Beijing Allwegene Tech. Co., Ltd. (Beijing, China). The data of high-throughput sequencing analysis was analyzed using QIIME (version 1.8.0) and Mothur (version 1.31.2) according to the description in the study of Shi et al. [17].

2.8. Statistical Analysis

The results of this trial were presented as means ± SD ($n = 4$). The data of growth performance, serum biochemical parameters, antioxidant parameters of the intestine, and the parameters of intestinal morphology from the present trial were subjected to Duncan's multiple comparisons in a one-way ANOVA model to estimate the statistical significance ($p < 0.05$) by SPSS 17.0 statistical software (SPSS Inc., Chicago, IL, USA). The data expressed as percentages were subjected to square arcsine transformation before statistical analysis. Kruskal Wallis method was used to conduct a non-parametric test for intestinal differential bacteria analysis ($p < 0.05$).

3. Results

3.1. Growth Performance

As shown in Table 1, the FBW, WGR, and SGR in the MCE50 group and MCE100 group were significantly higher than those of the control group and MCE25 group ($p < 0.05$). Compared with the other groups, the FR was increased significantly in the MCE50 group ($p < 0.05$). The FE in the MCE100 group was significantly higher than those of the other groups ($p < 0.05$). No significant differences in those growth parameters were found between the control group and the MCE25 group ($p > 0.05$).

Table 1. Effects of dietary MCE supplementation on growth performance of juvenile American eel.

Items	Control Group	MCE25 Group	Mce50 Group	Mce100 Group
FBW (g/fish)	19.86 ± 0.49^{a}	19.77 ± 0.42^{a}	$22..93 \pm 0.57^{b}$	23.30 ± 0.49^{b}
WGR (%)	81.82 ± 4.59^{a}	79.76 ± 2.54^{a}	109.88 ± 4.76^{b}	113.93 ± 3.97^{b}
SGR (%/d)	0.81 ± 0.03^{a}	0.79 ± 0.02^{a}	1.00 ± 0.03^{b}	1.03 ± 0.02^{b}
FR (%)	1.02 ± 0.03^{a}	1.01 ± 0.01^{a}	1.17 ± 0.02^{b}	1.05 ± 0.04^{a}
FE (%)	76.69 ± 4.10^{a}	76.00 ± 2.39^{a}	82.80 ± 3.41^{ab}	93.28 ± 2.89^{b}
SR (%)	100	97.5 ± 1.60	99.17 ± 0.83	100

Values are means ± SD ($n = 4$). ^{a,b} values with different superscripts in the same row are significantly different ($p < 0.05$). FBW = final body weight; WGR = weight gain rate; SGR = specific growth rate; FR = feeding rate; FE = feed efficiency; SR = survival rate.

3.2. Serum Biochemical Parameters

As shown in Table 2, the levels of TC and TG in the MCE100 group were significantly lower than those of the other groups ($p < 0.05$), and there was no significant difference in TC and TG levels among the other three groups ($p > 0.05$). The HDL-C level in the MCE50 group and MCE100 group were significantly higher than those of the other two groups ($p < 0.05$). The activities of GOT and GPT only in the MCE50 group were significantly lower than those of the control group ($p < 0.05$). Compared with the control group, the DAO activity and D-lac level in MCE groups were significantly decreased ($p < 0.05$), and no significant differences in the DAO activity and D-lac level (except for the MCE100 group) were observed among the MCE groups ($p > 0.05$).

Table 2. Effects of dietary MCE supplementation on the serum biochemical parameters of the juvenile American eel.

Items	Control Group	MCE25 Group	Mce50 Group	Mce100 Group
TC (mmol/L)	32.85 ± 0.77 [b]	31.74 ± 0.44 [b]	31.51 ± 1.19 [b]	28.14 ± 0.59 [a]
TG (mmol/L)	6.00 ± 0.11 [b]	5.89 ± 0.07 [b]	5.61 ± 0.08 [b]	4.18 ± 0.14 [a]
HDL-C mmol/L)	4.61 ± 0.18 [a]	4.71 ± 0.12 [a]	5.59 ± 0.18 [b]	5.61 ± 0.17 [b]
LDL-C (mmol/L)	21.77 ± 0.89 [a]	21.79 ± 1.23 [a]	21.47 ± 0.35 [a]	21.81 ± 0.69 [a]
GOT (IU/L)	13.58 ± 0.57 [b]	12.98 ± 0.35 [ab]	11.99 ± 0.47 [a]	12.67 ± 0.34 [ab]
GPT (IU/L)	8.77 ± 0.37 [b]	8.42 ± 0.26 [ab]	7.41 ± 0.38 [a]	8.24 ± 0.21 [ab]
DAO (U/L)	29.22 ± 0.71 [b]	23.75 ± 1.01 [a]	21.88 ± 1.03 [a]	22.00 ± 1.21 [a]
D-lac (mmol/L)	1.41 ± 0.02 [c]	1.02 ± 0.01 [b]	0.98 ± 0.03 [b]	0.83 ± 0.05 [a]

Values are means ± SD (n = 4). [a,b,c] values with different superscripts in the same row are significantly different ($p < 0.05$). GPT = glutamic-pyruvic transaminase; GOT = glutamic-oxaloacetic transaminase; TC = total cholesterol; TG = triglyceride; HDL-C = high-density lipoprotein cholesterol; LDL-C = low-density lipoprotein cholesterol; DAO = diamine oxidase; D-lac = D-lactate.

3.3. Intestinal Antioxidant Parameters

The antioxidant parameters of juvenile American eels in different treatment groups were shown in Table 3. Compared with the control group, the T-AOC level and the GSH-PX activity in the MCE groups were significantly increased ($p < 0.05$). The SOD activity was significantly increased only in the MCE100 group ($p < 0.05$), and no significant difference was presented between the control group and MCE groups ($p > 0.05$). The CAT activity in the MCE50 group and MCE100 group were significantly higher than those in the control group and MCE25 group ($p < 0.05$), and there was no significant difference between the MCE50 group and MCE100 group ($p > 0.05$). The MDA levels in the MCE groups were significantly lower than that in the control group ($p < 0.05$), and the MDA level of the MCE50 group was significantly lower than that of the MCE25 group ($p < 0.05$).

Table 3. Effects of dietary MCE supplementation on intestinal antioxidant parameters of the juvenile American eel.

Items	Control Group	MCE25 Group	MCE50 Group	MCE100 Group
T-AOC (U/mg prot)	0.13 ± 0.01 [a]	0.18 ± 0.01 [b]	0.25 ± 0.01 [c]	0.23 ± 0.02 [c]
SOD (U/mg prot)	57.99 ± 1.76 [a]	58.73 ± 0.12 [a]	59.72 ± 0.49 [a]	67.71 ± 0.94 [b]
CAT (U/mg prot)	4.31 ± 0.15 [a]	4.34 ± 0.01 [a]	5.62 ± 0.37 [b]	6.18 ± 0.07 [b]
GSH-PX (U/mg prot)	15.75 ± 0.88 [a]	19.06 ± 0.09 [b]	21.94 ± 0.33 [c]	20.68 ± 0.58 [bc]
MDA (nmol/mg prot)	30.66 ± 0.12 [c]	22.16 ± 0.64 [b]	19.19 ± 0.98 [a]	21.12 ± 0.66 [ab]

Values are means ± SD (n = 4). [a,b,c] values with different superscripts in the same row are significantly different ($p < 0.05$). T-AOC = total antioxidant capacity; SOD = superoxide dismutase; CAT = catalase; GSH-PX = glutathione peroxidase; MDA = malondialdehyde.

3.4. Intestinal Morphology

The results of the intestinal morphology of juvenile American eel were shown in Figure 1 and Table 4.

Compared with the control group, the MT and FH of the intestine in the MCE50 group (Figure 1C) and MCE100 group (Figure 1D) were significantly increased ($p < 0.05$), and the MT in the MCE100 group was significantly higher than that of MCE50 group ($p < 0.05$). No significant difference was observed between the control group (Figure 1A) and the MCE25 group (Figure 1B) in parameters of intestinal morphology ($p > 0.05$).

3.5. Intestinal Microbiota

As shown in Table 5, there was no obvious difference in indexes of Shannon and Chao1 between the MCE50 group and the control group. The number of OTUs in the MCE50 group was higher than that in the control group, and the Chao 1 and Shannon indexes were similar between the MCE50 group and the control group. The values of coverage rate were above 98%, suggesting that the majority of intestinal bacteria might be identified.

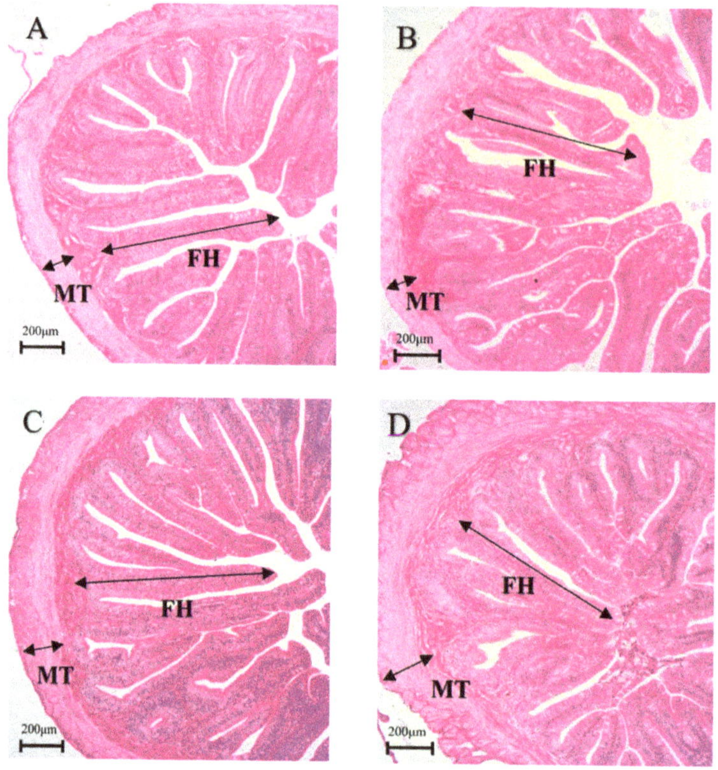

Figure 1. Effects of dietary *Macleaya cordata* extract (MCE) supplementation on the intestinal morphology of juvenile American eel. (**A**) Control group, (**B**) MCE25 group, (**C**) MCE50 group, and (**D**) MCE100 group (Magnification × 100). FH = fold height; MT = muscular thickness.

Table 4. Effects of dietary MCE supplementation on the muscular thickness and fold height of intestine of the juvenile American eel.

Items	Control Group	MCE25 Group	MCE50 Group	MCE100 Group
MT (μm)	131.11 ± 1.20 ^{a}	137.59 ± 2.74 ^{a}	146.10 ± 1.63 ^{b}	165.48 ± 2.64 ^{c}
FH (μm)	530.69 ± 8.89 ^{a}	559.52 ± 5.05 ^{a}	619.47 ± 6.50 ^{b}	599.97 ± 8.36 ^{b}

Values are means ± SD (n = 4). ^{a,b,c} values with different superscripts in the same row are significantly different ($p < 0.05$). MT = muscular thickness; FH = fold height.

Table 5. Comparison of alpha diversity of intestinal microbiota of juvenile American eel between the control group and MCE50 group.

Items	Control Group	MCE50 Group
OTUs	201	600
Chao 1	148.96 ± 21.20	167.22 ± 14.58
Shannon	0.73 ± 0.34	0.85 ± 0.16
Coverage rate (%)	99	98

Values are means ± SD (n = 4). values with no different superscripts in the same row are significantly different ($p > 0.05$).

As shown in Figure 2, the predominant bacteria at the phylum level in the intestine of the MCE50 group and the control group were Tenericutes, Fusobacteria, and Firmicutes. Compared with the control group, there was an increasing trend of relative abundance of Fusobacteria and a decreasing trend of the relative abundances of Tenericutes and Firmicutes in the MCE50 group.

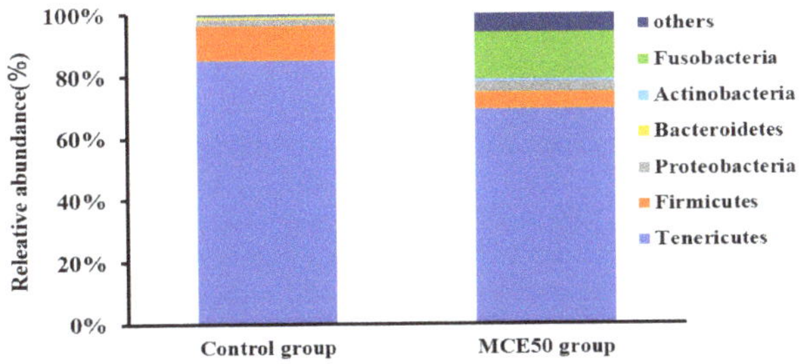

Figure 2. Comparison of intestinal microbiota at phylum level of juvenile American eels between the control group and MCE50 group.

The differential bacteria at genus level in the intestine of the control group and MCE50 group were presented in Figure 3. Compared with the control group, the lower relative abundance of *Acinetobacter* and the higher relative abundances of *Saccharopolyspora*, *Thermoactinomyces*, and *Cronobacter* were in the MCE50 group ($p < 0.05$).

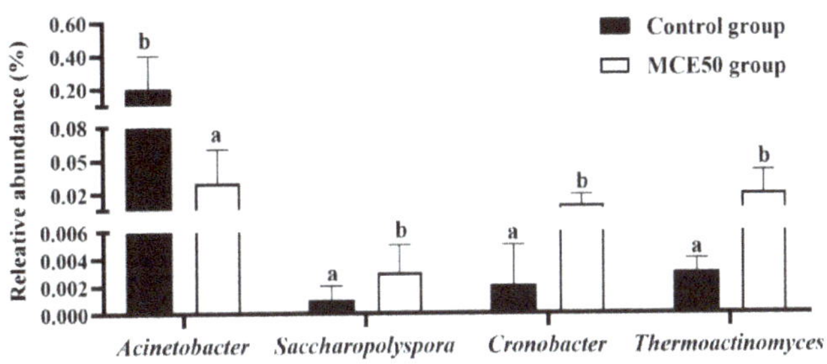

Figure 3. Comparison of differential bacteria at genus level in the intestine of juvenile American eels between the control group and MCE50 group. ^{a,b} Different superscripts on the bar are significantly different ($p < 0.05$).

4. Discussion

The results of the present trial revealed that dietary MCE supplementation at 50 mg/kg or 100 mg/kg could improve the growth performance of juvenile American eels. Similarly, dietary supplementation with MCE at 500 mg/kg could increase the final weight, weight gain, and specific growth rate of Caspian roach fry [12], as well as weight gain and average daily gain of common carp [14]; dietary supplementation with MCE (Sangrovit®) at 25 mg/kg might improve weight gain and SGR of red tilapia [13]. However, the WGR and other growth parameters were not significantly improved by 0.2–0.8 g/kg MCE supplementation in the grass carp fed high cottonseed and rapeseed meal diets [5]. Those inconsistent results of MCE supplementation in the diets might be caused by the differences in nutrients level, fish species, and feed composition.

The levels of TG, TC, HDL-C, and LDL-C in serum are important parameters of lipid metabolism in animals, and the lipid-lowering effect in serum is considered to be beneficial to health [18]. In this study, the levels of TC and TG were reduced by MCE supplementation. And the HDL-C level was increased by 50 mg/kg or 100 mg/kg MCE supplementation. Those results suggested that MCE might have a hypolipidemic effect in juvenile American eels. Similar results were observed in the previous study of Caspian roach fed diets with

MCE supplementation [12]. The lipid-lowering effect of MCE might attribute to altering bile acid metabolism and activating FXR signaling [19].

The GPT and GOT are important biomarkers for liver health and are mainly distributed in the mitochondria of liver cells. The elevated activities of GPT and GOT in the serum mean that there is tissue damage in the liver [20]. In this study, only dietary supplementation of MCE at 50 mg/kg could decrease the activities of GOT and GPT, it indicated that an appropriate level of MCE supplementation might improve the liver health status of the juvenile American eel. This phenomenon was supported by the effect of 375 or 750 μg/kg MCE (in the form of sanguinarine purity > 95%) on alleviating elevation of GOT and GPT activities in serum induced by H_2O_2 or lipopolysaccharide in rice field eels (*Monopterus albus*) [6,21]. However, Rawling et al. [13] reported that there was no significant difference in GOT and GPT activities of red tilapia with dietary MCE supplementation.

It was found that the mucosal damage in the intestine might lead to increased permeability and enable a large quantity of D-lac and DAO to get into the peripheral circulation [11,22]. The D-lac and DAO are usually regarded as indicators of damage to the intestinal barrier system [22]. In this study, dietary MCE supplementation decreased the D-lac level and the DAO activity in the serum. This point was confirmed in the study of grass carp fed high cottonseed and rapeseed meal diets on restoring the barrier function of the tightly connected gut by MCE supplementation [5]. In addition, the same results were also found in growing piglets fed diets with MCE supplementation [11]. These results suggested that dietary MCE supplementation could improve the barrier function of the intestine. Further work should be conducted to clarify the specific mechanism for the amelioration of the barrier function.

T-AOC is an index generally for assessing the ability of the nonenzymatic antioxidant defense system of fish [23]. The SOD, GSH-PX, and CAT are the main parameters to assess antioxidant ability in the enzymatic system, and they work together to catalyze free radicals and produce non-toxic compounds [24]. MDA is the product of lipid-peroxidation, which contributes to the production of reactive oxygen radicals. It could be used as an indicator of cellular oxidative damage indirectly [25]. In this study, the higher level of T-AOC and the activities of GSH-PX and CAT were found in certain MCE groups with decreasing MDA levels. The results from this trial were consistent with those of MCE supplemented in grass carp [5], koi carp (*Cryprinus carpiod*) [25], and pacific white shrimp (*Litopenaeus vannamei*) [26]. The improvement of antioxidant ability in the intestine might be related to MCE inhibiting phorbol myristate-induced oxidative burst and the NADPH oxidase complex and decreasing the production of reactive oxygen species [7].

The intestinal FH and MT are the well-defined parameters for the health status in the intestine of fish [27]. The absorption of nutrients in the intestine is determined by the villi length of the intestinal fold [28]. In the present study, the FH and MT of the intestine of the American eel were significantly increased in the MCE50 group and MCE100 group. Similarly, the FH and MT of the intestine were also increased in the grass carp fed the high cottonseed and rapeseed meal by dietary MCE supplementation [5]. The improvement of intestinal morphology suggested dietary MCE supplementation might enhance the ability to digest and absorb nutrients, and this might be one of the important factors to improve the growth performance of juvenile American eel. The amelioration of intestinal morphology might be related to MCE inhibiting the colonization of pathogenic bacteria and increasing the colonization of probiotic bacteria in the intestine [29].

Generally, the intestinal microbiota plays the important role in animal growth and nutrition metabolism [30]. The beneficial effects of medical plant-derived extracts on intestinal microbiota are increasingly reported in aquaculture [5,31,32]. The indexes of OTUs, Chao 1, and Shannon are usually used to evaluate the species abundance and richness of the intestine microbiota [27,33]. A high microbiota diversity is generally considered beneficial for intestinal health [34]. In our study, the OTUs number in the MCE50 group was higher than that in the control group, which indicated that dietary MCE supplementation might improve the richness of the intestinal microbiota of American eel. Similar results were observed in koi carp and *Trionyx sinensis* with dietary MCE supplementation [25,35]. The values of coverage indices were above 98%, which suggested that the majority of intestinal bacteria in this study might be identified [17].

In the present trial, the relative abundances of both Tenericutes and Firmicutes were lowered in the MCE50 group. The decreasing proportion of Tenericutes in the intestine might be beneficial to gut health because the lower abundance of Tenericutes was found in the intestine of the European eel of the fast-growth group in comparison with the stunted-growth group [17,34]. The lower proportion of Firmicutes in the intestine was also found in the American eel fed diet with oligomeric proanthocyanidins supplementation [36]. It was reported that some strains belonging to Firmicutes in the intestine might be associated with the decreasing growth rate of the European eel [17]. The relative abundance of Fusobacteria was increased in the MCE50 group. The Fusobacteria might produce more butyrate and synthesize numerous vitamins [37], which might play essential roles in preventing the development of intestinal inflammation and inflammatory bowel disease [38].

The results of the Lefse analysis in the present study was shown that there was a lower relative abundance of *Acinetobacter* and higher relative abundances of *Saccharopolyspora*, *Thermoactinomyces*, and *Cronobacter* in the intestine of the MCE50 group. *Acinetobacter* is a potential pathogen in aquaculture [39]. *Saccharopolyspora* sp. is reported as the major producer of secondary metabolites that have a broad spectrum of biological activities including anti-microbial, antioxidant, and inhibitory effects of some enzymes to indirectly improve the health status of the animal intestines [40,41]. The *Thermoactinomyces* species are heat-resistant spore-forming bacteria, and they are capable of producing proteases [42]. Although the *Cronobacter* turicensis might cause lethal infection in zebrafish larvae [43], some studies indicated that the beneficial or detrimental effects of *Cronobacter* in aquatic animals might be due to the differences in fish species, individual sizes, living environments, and other factors [44]. The role of *Cronobacter* in the healthy intestine of American eel fed with MCE is needed to clarify in future research. From the changes in intestinal microbiota, it was obvious that the MCE might have the ability to optimize the bacteria composition in the intestine by promoting the multiplication of some potentially beneficial bacteria and decreasing the relative abundances of the certain potential pathogen. This phenomenon caused by MCE was also found in red tilapia [13], common carp [14], koi carp [25], Pacific white shrimp [26], loach (*Misgurnus anguillicaudatus*) [45], and even grass carp fed with a higher level of cottonseed and rapeseed meal [5]. The determinant for MCE exerting antibacterial activity is the presence of reactive iminium bond in the molecule of sanguinarine and chelerythrine [46], and the possible mechanisms might be involved in inhibiting the formation of bacteria biofilm or interfering with the bacterial cytokinesis [9,47].

In general, both 50 mg/kg and 100 mg/kg MCE supplementation could have beneficial effects on growth performance, serum biochemical parameters, intestinal antioxidant ability, and intestinal morphology of juvenile American eel. Most of those parameters of growth performance and health status were not improved by 100 mg/kg MCE supplementation. Considering the changes in all the parameters in the present trial and the cost of MCE, the 50 mg/kg MCE should be supplemented in the diet of the juvenile American eel.

5. Conclusions

In conclusion, the appropriate level of dietary MCE supplementation could exert beneficial effects on the growth performance, serum biochemical parameters, antioxidant ability, morphology, and microbiota in the intestine of the juvenile American eel. The 50 mg/kg MCE in the diet was recommended to improve the growth performance and health status of the juvenile American eel. The application effects of MCE under the practical culture conditions should be confirmed in future studies.

Author Contributions: Methodology, L.H. and S.Z.; validation, L.H., R.C. and S.Z.; formal analysis, L.H. and R.C.; investigation, L.H., R.C. and S.Z.; data curation, S.Z.; writing—original draft preparation, R.C.; writing—review and editing, R.C. and S.Z.; project administration, S.Z.; funding acquisition, S.Z. All authors have read and agreed to the published version of the manuscript.

Funding: This research was funded by the Outward Cooperation Project in the Science and Technology Plan Program of Fujian Province (2020I0020), the earmarked fund for China Agriculture Research System of MOF and MARA (CARS-46), and the Open Fund of Engineering Research Center of the Modern Industry Technology for Eel, Ministry of Education of China (RE202010).

Institutional Review Board Statement: The study was conducted according to the guidelines of the Declaration of Helsinki and was approved by the Animal Care Advisory Committee of Jimei University (Approval No. 2019-0906-003).

Data Availability Statement: The data used during the current study are available from the corresponding author on reasonable request.

Acknowledgments: The authors thank Keding Xia for providing the trial fish.

Conflicts of Interest: The authors declare that they have no conflict of interest.

References

1. Naylor, R.L.; Hardy, R.W.; Buschmann, A.H.; Bush, S.R.; Cao, L.; Klinger, D.H.; Little, D.C.; Lubchenco, J.; Shumway, S.E.; Troell, M. A 20-year retrospective review of global aquaculture. *Nature* **2021**, *591*, 551–563. [CrossRef] [PubMed]
2. Tadese, D.A.; Song, C.; Sun, C.; Liu, B.; Liu, B.; Zhou, Q.; Xu, P.; Ge, X.; Liu, M.; Xu, X.; et al. The role of currently used medicinal plants in aquaculture and their action mechanisms: A review. *Rev. Aquac.* **2022**, *14*, 816–847. [CrossRef]
3. Hou, B.; Zeng, J. Biological activities of sanguinarine and application of *Macleaya cordata* extract in animal production. *Chin. J. Anim. Nutr.* **2018**, *30*, 413–420. [CrossRef]
4. Lin, L.; Liu, Y.C.; Huang, J.L.; Liu, X.B.; Qing, Z.X.; Zeng, J.G.; Liu, Z.Y. Medicinal plants of the genus Macleaya (*Macleaya cordata*, *Macleaya microcarpa*): A review of their phytochemistry, pharmacology, and toxicology. *Phytother. Res.* **2017**, *32*, 19–48. [CrossRef] [PubMed]
5. Liu, Y.L.; Zhong, L.; Chen, T.; Shi, Y.; Xu, S.D. Dietary sanguinarine supplementation on the growth performance, immunity and intestinal health of grass carp (*Ctenopharyngodon idellus*) fed cottonseed and rapeseed meal diets. *Aquaculture* **2020**, *528*, 735521. [CrossRef]
6. Shi, Y.; Zhong, L.; Chen, K.J.; Fan, Y.D.; Xie, K.; Zhang, J.Z.; Dai, J.; Hu, Y. Sanguinarine attenuates hydrogen peroxide-induced toxicity in liver of *Monopterus albus*: Role of oxidative stress, inflammation and apoptosis. *Fish Shellfish Immun.* **2022**, *125*, 190–199. [CrossRef]
7. Qin, F.; Patel, R.; Yan, C.; Liu, W. NADPH oxidase is involved in angiotensin II induced apoptosis in H9C2 cardiac muscle cells: Effects of apocynin. *Free Radical Bio. Med.* **2006**, *40*, 236–246. [CrossRef]
8. Guan, G.; Ding, S.; Yin, Y.; Duraipandiyan, V.; AlDhabi, N.A.; Liu, G. *Macleaya cordata* extract alleviated oxidative stress and altered innate immune response in mice challenged with enterotoxigenic *Escherichia coli*. *Sci. China Life Sci.* **2019**, *62*, 1019–1027. [CrossRef]
9. Cushnie, T.P.; Cushnie, B.; Lamb, A.J. Alkaloids: An overview of their antibacterial, antibiotic-enhancing and anti-virulence activities. *Int. J. Antimicrob. Ag.* **2014**, *44*, 377–386. [CrossRef]
10. Zhou, S.; Li, W.X.; Wang, Y.Q.; Zou, H.; Wu, S.G.; Wang, G.T. Anthelmintic efficacies of three common disinfectants and extracts of four traditional Chinese medicinal plants against *Gyrodactylus kobayashii* (Monogenea) in goldfish (*Carassius auratus*). *Aquaculture* **2017**, *466*, 72–77. [CrossRef]
11. Liu, G.; Guan, G.P.; Fang, J.; Martinez, Y.; Chen, S.; Bin, P.; Duraipandiyan, V.; Gong, T.; Tossou, M.C.; Al-Dhabi, N.A.; et al. *Macleaya cordata* extract decreased diarrhea score and enhanced intestinal barrier function in growing piglets. *Biomed Res. Int.* **2016**, *2016*, 1069585. [CrossRef] [PubMed]
12. Imanpoor, M.R.; Roohi, Z. Effects of Sangrovitt®-supplemented diet on growth performance, blood biochemical parameters, survival and stress resistance to salinity in the Caspian roach (*Rutilus rutilus*) fry. *Aquac. Res.* **2015**, *47*, 2874–2880. [CrossRef]
13. Rawling, M.D.; Merrifield, D.L.; Davies, S.J. Preliminary assessment of dietary supplementation of Sangrovit® on red tilapia (*Oreochromis niloticus*) growth performance and health. *Aquaculture* **2009**, *294*, 118–122. [CrossRef]
14. Abdelnaby, E.A.; Mohamed, M.F.; Gammazh, A.K. Pharmacological studies of feed additives (Sanguinarine and Saccharomyces cerevisiae) on growth performance, hematological and intestinal bacterial count with challenge test by *Aeromonas hydrophila* in *Cyprinus carpio*. *Global. Anim. Sci. J.* **2013**, *11*, 1154–1172. [CrossRef]
15. Zhang, M.L.; Wang, Y.; Zhai, S.W. Effects of dietary compound acidifiers supplementation on growth performance and intestinal health of juvenile American eels (*Anguilla rostrata*) cultured in cement tanks. *Isr. J. Aquacult-Bamid.* **2021**, *73*, 1520998. [CrossRef]
16. Yuan, Y.; Yuan, Y.M.; Dai, Y.Y.; Gong, Y.C.; Yuan, Y.Q. Development status and trends in the eel farming industry in Asia. *N. Am. J. Aquacult.* **2021**, *84*, 3–17. [CrossRef]

17. Shi, Y.; Ma, D.Y.; Zhai, S.W. Revealing the difference of intestinal microbiota composition of cultured European eels (*Anguilla anguilla*) with different growth rates. *Isr. J. Aquacult-Bamid.* **2020**, *72*, 959575. [CrossRef]
18. Guo, D.; Xie, M.; Xiao, H.; Xu, L.; Zhang, S.; Chen, X.; Wu, Z. *Bacillus subtilis* supplementation in a high-fat diet modulates the gut microbiota and ameliorates hepatic lipid accumulation in grass carp (*Ctenopharyngodon idella*). *Fishes* **2022**, *7*, 94. [CrossRef]
19. Tian, Y.; Cai, J.; Gui, W.; Nichols, R.G.; Koo, I.; Zhang, J.; Patterson, A.D. Berberine directly impacts the gut microbiota to promote intestinal farnesoid X receptor activation. *Drug Metab. Dispos.* **2019**, *47*, 86–93. [CrossRef]
20. Farris, N.W.; Kim, D.; Hamidoghli, A.; Won, S.; Lee, S.; Bae, J. Dietary α-Tocopheryl acetate and arachidonic acid synergistically improves superoxide dismutase activity in female Japanese eel broodstock, *Anguilla japonica*. *Aquaculture* **2020**, *522*, 735100. [CrossRef]
21. Shi, Y.; Hu, Y.; Liu, Y.Y.; Che, C.B. Effects of sanguinarine on immune and intestinal inflammation related to gene expression in rice field eels (*Monopterus albus*) induced by LPS. *J. Fish. Sci. China* **2020**, *27*, 125–137. [CrossRef]
22. Fukudome, I.; Kobayashi, M.; Dabanaka, K.; Maeda, H.; Okamoto, K.; Okabayashi, T.; Hanazaki, K. Diamine oxidase as a marker of intestinal mucosal injury and the effect of soluble dietary fiber on gastrointestinal tract toxicity after intravenous 5-fluorouracil treatment in rats. *Med. Mol. Morphol.* **2013**, *47*, 100–107. [CrossRef] [PubMed]
23. Mourente, G.; Tocher, D.R.; Diaz, E.; Grau, A.; Pastor, E. Relationships between antioxidants, antioxidant enzyme activities and lipid peroxidation products during early development in *dentex dentex* eggs and larvae. *Aquaculture* **1999**, *179*, 309–324. [CrossRef]
24. Shi, Q.; Xiong, X.; Wen, Z.; Qin, C.; Li, R.; Zhang, Z.; Gong, Q.; Wu, X. Cu/Zn superoxide dismutase and catalase of Yangtze sturgeon, *Acipenser dabryanus*: Molecular cloning, tissue distribution and response to fasting and refeeding. *Fishes* **2022**, *7*, 35. [CrossRef]
25. Zhang, R.; Wang, X.W.; Zhu, J.Y.; Liu, L.L.; Liu, Y.C.; Zhu, H. Dietary sanguinarine affected immune response, digestive enzyme activity and intestinal microbiota of koi carp (*Cryprinus carpiod*). *Aquaculture* **2019**, *502*, 72–79. [CrossRef]
26. Bussabong, P.; Rairat, T.; Chuchird, N.; Keetanon, A.; Phansawat, P.; Cherdkeattipol, K.; Pichitkul, P.; Kraitavin, W. Effects of isoquinoline alkaloids from *Macleaya cordata* on growth performance, survival, immune response, and resistance to *Vibrio parahaemolyticus* infection of Pacific white shrimp (*Litopenaeus vannamei*). *PLoS ONE* **2021**, *16*, e0251343. [CrossRef]
27. Su, X.; Ji, D.; Yao, J.; Zou, Y.; Yan, M. Comparative analysis of intestinal characteristics of Largemouth Bass (*Micropterus salmoides*) and intestinal flora with different growth rates. *Fishes* **2022**, *7*, 65. [CrossRef]
28. Saravanan, K.; Sivaramakrishnan, T.; Praveenraj, J.; Kiruba, S.R.; Haridas Harsha, K.S.; Varghese, B. Effects of single and multi-strain probiotics on the growth, hemato-immunological, enzymatic activity, gut morphology and disease resistance in Rohu, *Labeo rohita*. *Aquaculture* **2021**, *540*, 736749. [CrossRef]
29. Xu, Y.; Li, Y.; Xue, M.; Xiao, Z.; Fan, Y.; Zeng, L.; Zhou, Y. Effects of dietary *Enterococcus faecalis* YFI-G720 on the growth, immunity, serum biochemical, intestinal morphology, intestinal microbiota, and disease resistance of Crucian Carp (*Carassius auratus*). *Fishes* **2022**, *7*, 18. [CrossRef]
30. Ingerslev, H.C.; Jørgensen, L.V.G.; Strube, M.L.; Larsen, N.; Dalsgaard, I.; Boye, M. The development of the gut microbiota in rainbow trout (*Oncorhynchus mykiss*) is affected by first feeding and diet type. *Aquaculture* **2014**, *424*, 24–34. [CrossRef]
31. Feher, M.; Fauszt, P.; Tolnai, E.; Fidler, G.; Pesti, A.G.; Stagel, A.; Szucs, I.; Biro, S.; Remenyik, J.; Paholcsek, M.; et al. Effects of phytonutrient-supplemented diets on the intestinal microbiota of *Cyprinus carpio*. *PLoS ONE* **2021**, *16*, e0248537. [CrossRef] [PubMed]
32. Zhu, W.J.; Xiao, S.Y.; Chen, S.Y.; Xu, Q.; Yang, Z.; Liu, J.; Wang, H.Q.; Lan, S.L. Effects of fermented mulberry leaves on growth, serum antioxidant capacity, digestive enzyme activities and microbial compositions of the intestine in Crucian (*Carassius carassius*). *Aquac. Res.* **2021**, *52*, 6356–6366. [CrossRef]
33. Burtseva, O.; Kublanovskaya, A.; Fedorenko, T.; Lobakova, E.; Chekanov, K. Gut microbiome of the white sea fish revealed by 16s rRNA metabarcoding. *Aquaculture* **2020**, *533*, 736175. [CrossRef]
34. Lin, M.; Zeng, C.X.; Jia, X.Q.; Zhai, S.W.; Li, Z.Q.; Ma, Y. The composition and structure of the intestinal microflora of *Anguilla marmorata* at different growth rates: A deep sequencing study. *J. Appl. Microbiol.* **2019**, *126*, 1340–1352. [CrossRef]
35. Chen, Z.N.; Wang, X.Q.; Luo, L.T.; Wang, P.; Tu, K.F.; Yang, T.; Hu, Y.Z.; Xiong, G. High throughput sequencing analysis of the effects of sanguinarine on *Trionyx sinensis* intestinal microbiota. *Prog. Fish. Sci.* **2021**, *42*, 177–185. [CrossRef]
36. Zhai, S.; Wang, Y.; He, Y.; Chen, X.H. Oligomeric proanthocyanidins counteracts the negative effects of high level of dietary histamine on American Eel (*Anguilla rostrata*). *Front. Mar. Sci.* **2020**, *7*, 549145. [CrossRef]
37. Kostic, A.D.; Gevers, D.; Pedamallu, C.S.; Michaud, M.; Duke, F.; Ear, A.M.; Ojesina, A.I.; Jung, J.; Bass, A.J.; Tabernero, J.; et al. Genomic analysis identifies association of *Fusobacterium* with colorectal carcinoma. *Genome Res.* **2012**, *22*, 292–298. [CrossRef]
38. Lobionda, S.; Sittipo, P.; Kwon, H.Y.; Lee, Y.K. The role of gut microbiota in intestinal inflammation with respect to diet and extrinsic stressors. *Microorganisms* **2019**, *7*, 271. [CrossRef]
39. Jiang, Y.; Liu, Z.; Liu, X.; Xu, Y.; Bussabong, B.; Wang, B. Structural characteristics and succession of intestinal microbiota for *Paralichthys olivaceus* during the early life stage. *Aquac. Res.* **2019**, *50*, 529–540. [CrossRef]
40. Azman, A.S.; Mawang, C.I.; Khairat, J.E.; Abubakar, S. Actinobacteria-a promising natural source of anti-biofilm agents. *Int. Microbiol.* **2019**, *22*, 403–409. [CrossRef]
41. Krishnamoorthy, M.; Dharmaraj, D.; Rajendran, K.; Karuppiah, K.; Ethiraj, K. Pharmacological activities of coral reef associated *actinomycetes*, *Saccharopolyspora* sp. IMA1. *Biocatal. Agr. Biotechnol.* **2020**, *28*, 101748. [CrossRef]

42. Gupta, T.B.; Risson, A.N.; Brightwell, G.; Maclean, P.; Jauregui, R. Draft genome sequence of *Thermoactinomyces* vulgaris strain AGRTWHS02, isolated from pasture soil of a sheep dairy farm in New Zealand. *Microbiol. Resour. Announc.* **2022**, *11*, e0007622. [CrossRef] [PubMed]
43. Fehr, A.; Eshwar, A.K.; Neuhauss, S.C.; Ruetten, M.; Vaughan, L. Evaluation of zebrafish as a model to study the pathogenesis of the opportunistic pathogen *Cronobacter* turicensis. *Emerg. Microbes. Infec.* **2015**, *4*, e29. [CrossRef] [PubMed]
44. Das, S.K.; Kumar, S.H.; Nayak, B.B.; Lekshmi, M. Isolation and identification of *Cronobacter* spp. from fish and shellfish sold in retail markets. *Curr. Microbiol.* **2021**, *78*, 1973–1980. [CrossRef] [PubMed]
45. Xia, X.H.; Liu Gao, P.W.; Xiao, L.C.; Shuang, S.Y.; Cong, H.; Du, Q.Y.; Zhang, X.W. Effects of *Macleaya cordata* extract on TLR20 and the proinflammatory cytokines in acute spleen injury of loach (*Misgurnus anguillicaudatus*) against *Aeromonas hydrophila* infection. *Aquaculture* **2021**, *544*, 737105. [CrossRef]
46. Miao, F.; Yang, X.J.; Zhou, L.; Hu, H.J.; Zheng, F.; Ding, X.D.; Sun, W. Structural modification of sanguinarine and chelerythrine and their antibacterial activity. *Natural Product Res.* **2011**, *25*, 863–875. [CrossRef]
47. Beuria, T.K.; Santra, M.K.; Panda, D. Sanguinarine blocks cytokinesis in bacteria by inhibiting FtsZ assembly and bundling. *Biochemistry* **2005**, *44*, 16584–16593. [CrossRef]

fishes

Article

Dietary Protein Modifies Hepatic Glycolipid Metabolism, Intestinal Immune Response, and Resistance to *Streptococcus agalactiae* of Genetically Improved Farmed Tilapia (GIFT: *Oreochromis niloticus*) Exposed to High Temperature

Dongyu Huang [1], Hualiang Liang [2], Jian Zhu [1,2], Mingchun Ren [1,2,*] and Xianping Ge [1,2]

1 Wuxi Fisheries College, Nanjing Agricultural University, Wuxi 214081, China
2 Key Laboratory of Integrated Rice-Fish Farming Ecology, Ministry of Agriculture and Rural Affairs, Freshwater Fisheries Research Center, Chinese Academy of Fishery Sciences, Wuxi 214081, China
* Correspondence: renmc@ffrc.cn

Abstract: The present study investigates the effects of dietary protein levels on glucolipid metabolism, immune function, and resistance to *Streptococcus agalactiae* of genetically improved farmed tilapia (GIFT) exposed to high temperature. Six practical diets were prepared to feed 360 fish (initial weight 43.78 ± 0.12 g) with graded protein levels (26.45%, 29.28%, 31.69%, 33.68%, 36.18%, and 38.75% dry matter). The results showed that 26.45% dietary protein significantly improved glycolysis by increasing PK mRNA levels, while the 29.28% and 31.69% dietary protein levels promoted gluconeogenesis by increasing PEPCK and G6Pase mRNA levels. For lipid metabolism, 26.45% dietary protein enhanced lipid synthesis by increasing PPAR-γ, SREBP1c, and FAS mRNA levels, while 31.69% dietary protein enhanced the level of lipolysis by increasing the PPAR-α and CPT1 mRNA levels. The highest plasma TG and TC contents were observed in the 29.28% and 31.69% dietary protein groups, respectively. In terms of antioxidants and immunity, the 31.69% dietary protein level activated the expression levels of HSP90 mRNA, thus increasing the expression levels of antioxidant-related genes (CAT, SOD, and GPx), and upregulating the anti-inflammatory factor IL-10 mRNA levels. In addition, regarding the antioxidant enzymes, the highest GSH content was found in the 29.28% dietary protein group, while the 31.69% dietary protein group had the maximum GSH-Px activity. The lowest plasma ALT and AST activities were observed in the 31.69% dietary protein group. Ultimately, the survival rate of juvenile GIFT fed 31.69% dietary protein was highest after a *Streptococcus agalactiae* challenge. Overall, 29.28–31.69% dietary protein was recommended in the diet of GIFT in a high-temperature environment.

Keywords: dietary protein; glycolipid metabolism; immune response; *Streptococcus agalactiae*; high temperature

Citation: Huang, D.; Liang, H.; Zhu, J.; Ren, M.; Ge, X. Dietary Protein Modifies Hepatic Glycolipid Metabolism, Intestinal Immune Response, and Resistance to *Streptococcus agalactiae* of Genetically Improved Farmed Tilapia (GIFT: *Oreochromis niloticus*) Exposed to High Temperature. *Fishes* **2022**, *7*, 202. https://doi.org/10.3390/fishes7040202

Academic Editors: Marina Paolucci and Shunsuke Koshio

Received: 6 July 2022
Accepted: 10 August 2022
Published: 12 August 2022

1. Introduction

As one of the three major nutrients, protein is the main component of the organism, and affects the metabolic and immune processes of fish. According to previous research on grass carp (*Ctenopharyngodon idellus*) [1] and top-mouth culter (*Erythroculter ilishaeformis*) [2], optimum dietary protein levels could effectively regulate glycolipid metabolism, including decreasing lipid synthesis and enhancing glycolytic capacity. Furthermore, previous studies on Japanese sea bass (*Lateolabrax japonicus*) [3] and leopard coral grouper (*Plectropomus leopardus*) [4] have shown that the appropriate dietary protein levels have a significant enhancing effect on immunity. Conversely, low-protein feeds with high levels of carbohydrates can lead to the accumulation of body fat in animals [5,6], and deficient dietary protein significantly reduces antioxidant enzyme activities and resistance to bacterial infection [7]. Moreover, excessive dietary protein levels will cause immunosuppression to some extent in fish [8,9], and excess protein will be broken down for energy consumption; then, ammonia nitrogen and oxidation products are produced in the process, which can be toxic to the body [10].

In aquaculture, high-temperature stress will cause a stress response, resulting in a series of changes, including changes in mRNA and metabolites [11–13], which directly affect biological functions, resulting in metabolic disorders and reduced disease resistance in fish. A previous study showed that the stress response of cells under stressful conditions is closely related to alterations in energy metabolism (glucose and lipid metabolism), with high temperatures significantly altering lipid and carbohydrate metabolism [14]. Furthermore, Zhao et al. [15] reported that when turbot (*Scophthalmus maximus*) is exposed to high temperatures, lipid metabolism has an important regulatory role in stress resistance. In addition, it was also found in grass carp that high-temperature stress could decrease fatty acid synthesis and weaken carbohydrate metabolism [16]. In addition, high temperature also affects the immune system of fish, reducing nonspecific and specific immunity and leading to a decrease in resistance to pathogenic bacteria [17]. Cheng et al. [18] showed that the immunity of spotted grouper (*Epinephelus coioides*) was suppressed during heat stress and that sustained high temperatures would result in a much lower immunity against *Vibrio alginolyticus* than that at optimal temperatures. Wang et al. [19] reported that the antioxidant system of scallops (*Chlamys farreri*) is challenged by high-temperature stress and is unable to fully repair the oxidative damage caused by the stress of high temperatures combined with a bacterial infection. Thus, understanding the changes in the metabolic and immune mechanisms of the body under a high temperature environment can help reduce the potential negative effects of heat stress.

Tilapia, *Oreochromis niloticus*, is the most exported farmed fish in China, with a total production of 1.65 million tons of tilapia farmed in 2020 [20]. Tilapia was once known for its ease of culture and strong disease resistance, but in recent years, with the expansion of culture scale and the increase in culture density, diseases occur frequently and have become increasingly serious, especially *Streptococcus agalactiae* disease, which seriously threatens the healthy development of the tilapia farming industry. The optimal water temperature for tilapia growth is 29–31 °C [21]. In our previous study, it was found that the protein requirement of tilapia at high temperatures was lower than those at suitable temperatures in terms of growth performance [22]. When the water temperature is higher than 32 °C, tilapia are more likely to be infected with *Streptococcus agalactiae* [23]. As Kayansamruaj et al. [24] showed, inflammation-related genes were significantly upregulated at high temperatures, causing massive inflammatory responses and acute fish mortality. This result indicated that the resistance of tilapia to the pathogenic bacteria, *Streptococcus agalactiae*, is reduced under high temperatures, for example, due to suppression of the immune system.

To date, no studies have reported the effects of protein levels in high temperature environments on the glycolipid metabolism and immune system of tilapia, as well as the effects on antimicrobial capacity. Thus, in this study, genetically improved farmed tilapia (GIFT), one of the tilapia strains, was chosen for this experiment. Our purpose was to examine the mechanisms of dietary protein levels on the glycolipid metabolism and immune capacity under high temperature in tilapia, to improve the antimicrobial capacity of tilapia under high-temperature stress through nutritional strategies.

2. Materials and Methods

2.1. Experimental Diets

Diets with six different protein levels (26.45%, 29.28%, 31.69%, 33.68%, 36.18%, and 38.75% dry matter) were designed (Table 1). The main protein sources were the fish meal, soybean meal, rapeseed meal, cottonseed meal, and wheat flour. The lipid source is fish oil. All raw materials were first crushed and sieved through 60 mesh and then mixed with water and oil. Then, the mixture was pelletized into 2 mm-diameter pellets through a pelletizer (F-26 [II], South China University of Technology, China), air-dried at room temperature, and maintained at −20 °C until further use.

Table 1. Formulation and proximate composition (% dry matter) of experimental diets.

Ingredients	Diet 1	Diet 2	Diet 3	Diet 4	Diet 5	Diet 6
Fish meal [a]	2.00	2.00	2.00	2.00	2.00	2.00
Rapeseed meal [a]	25.00	25.00	25.00	25.00	25.00	25.00
Soybean meal [a]	2.00	10.00	18.00	26.00	34.00	43.00
Cottonseed meal [a]	9.00	9.00	9.00	9.00	9.00	9.00
Wheat flour [a]	35.00	29.30	23.60	17.90	12.20	6.00
Soybean oil	2.50	2.50	2.50	2.50	2.50	2.50
Choline chloride	0.50	0.50	0.50	0.50	0.50	0.50
Vitamin C (35%)	0.05	0.05	0.05	0.05	0.05	0.05
Vitamins premix [b]	2.00	2.00	2.00	2.00	2.00	2.00
Mineral premix [c]	2.00	2.00	2.00	2.00	2.00	2.00
Calcium dihydrogen phosphate	2.50	2.50	2.50	2.50	2.50	2.50
Rice bran	10.00	8.00	6.50	5.00	3.50	2.00
Microcrystalline cellulose	4.62	4.35	3.71	3.08	2.26	0.96
Ethoxy quinoline	0.01	0.01	0.01	0.01	0.01	0.01
Bentonite	2.00	2.00	2.00	2.00	2.00	2.00
Lysine [d]	0.32	0.26	0.14	0.00	0.00	0.00
Methionine [d]	0.33	0.38	0.40	0.42	0.47	0.48
Threonine [d]	0.17	0.15	0.09	0.04	0.02	0.00
Analyzed proximate composition						
Dry matter (%)	94.16	93.29	93.29	93.82	92.79	92.21
Crude protein (%)	26.45	29.28	31.69	33.68	36.18	38.75
Crude lipid (%)	4.56	4.35	4.65	4.39	4.60	4.31
Crude ash (%)	10.56	10.54	10.94	11.22	11.25	11.74
Crude fiber (%)	6.17	6.32	6.57	6.81	7.05	7.36
NFE [e]	46.42	42.80	39.44	37.72	33.71	30.05
Gross energy (KJ/g)	17.96	18.03	18.00	18.03	18.06	18.14

Note: The feed formulation referred to our previous study [22]. [a] Fish meal, crude protein 65.8%, crude lipid 9.5%; Rapeseed meal, crude protein 41.3%, crude lipid 6.1%; Soybean meal, crude protein 50.8%, crude lipid 4.3%; Cottonseed meal, crude protein 53.7%, crude lipid 1.4%; Wheat flour, crude protein 13.1%, crude lipid 4.0%. They were obtained from Wuxi Tongwei feedstuffs Co., Ltd., Wuxi, China. [b] Vitamins premix were obtained from HANOVE Animal Health Products Co. LTD (IU, mg/kg of diet): Vitamin A, 900,000 IU; Vitamin D, 250,000 IU; Vitamin E, 4500 mg; Vitamin K3, 220 mg; Vitamin B1, 320 mg; Vitamin B2, 1090 mg; Vitamin B5, 2000 mg; Vitamin B6, 5000 mg; Vitamin B12, 116 mg; Pantothenate, 1000 mg; Folic acid, 165 mg; Choline, 60,000 mg; Biotin, 50 mg Niacin acid, 2500 mg. [c] Mineral premix was obtained from HANOVE Animal Health Products Co. LTD (g/kg of diet): calcium biphosphate, 20 g; sodium chloride, 2.6 g; potassium chloride, 5 g; magnesium sulphate, 2 g; ferrous sulphate, 0.9 g; zinc sulphate, 0.06 g; cupric sulphate, 0.02 g; manganese sulphate, 0.03 g; sodium selenate, 0.02 g; cobalt chloride, 0.05 g; potassium iodide, 0.004 g. [d] Lysine, methionine and threonine, obtained from Feeer Co., LTD (Shanghai, China). [e] NFE (nitrogen free extract, %) = dry matter (%)—(crude protein (%) + crude lipid (%) + crude ash (%) + crude fiber (%)).

The chemical compositions of the dried experimental diets were assessed based on the established methods of the AOAC [25]. The protein content was determined by the Kjeldahl method (Auto Kjeldahl apparatus: Hanon K1100 (Jinan Hanon Instruments Co., Ltd., Jinan, China)). The lipid content was determined by the Soxhlet method (Auto fat analyzer: Hanon SOX606 (Jinan Hanon Instruments Co., Ltd., Jinan, China)). The ash content was determined by the combustion method (Muffle: XL-2A (Hangzhou Zhuochi Instrument Co., Ltd., Hangzhou, China)). The fiber content was determined by the FiberCap method (Fiber analysis system (FiberCap™ 2021, FOSS, USA)). The gross energy was determined by the combustion method (Oxygen bomb calorimeter: IKA C6000 (IKA WORKS GUANGZHOU, Guangzhou, China)).

2.2. Experimental Fish and Feeding Management

Juvenile GIFT were obtained from the breeding farm of the Freshwater Fisheries Research Center of the Chinese Academy of Fishery Sciences. A total of 360 healthy GIFT (average initial weight 43.78 ± 0.12 g) were evenly distributed in 18 floating cages. The floating cages with a square shape (1 m × 1 m × 1 m) were hung on the floating frame, with the buoyancy of the floating frame adjusted to allow the cages to remain in the upper layer of water. The netting of the cages is made of mesh sewn together, and the mesh area of the netting is 1 cm×1 cm, which can ensure good water exchange and prevent fish from escaping. Fish were fed to satiation three times daily for four weeks. The natural water temperature ranged from 32 °C to 36 °C, and the water quality parameters during the trial were as follows: dissolved oxygen > 6.0 mg/L, and pH was kept at 7.5–8.0.

2.3. *Sampling Procedure*

After 4 weeks, three fish in the cage were selected randomly to take blood samples, intestine tissues, and liver tissues. Blood samples were obtained from the caudal vein and then immediately centrifuged at 3000 rpm for 10 min at 4 °C. The separated plasma samples were stored at −20 °C until they were analyzed, and the tissues were frozen in a −80 °C freezer for later analysis.

2.4. *Plasma Biochemical Analysis*

Plasma biochemical parameters (TG: triglyceride, TC: total cholesterol, GLU: glucose, ALT: alanine transaminase, and AST: aspartic transaminase) were measured by a BS-400 automatic biochemical analyzer (Mindray, Shenzhen, China) using the corresponding Mindray Kits.

2.5. *Analysis of Intestinal Antioxidant Indices*

Intestinal antioxidant indices, including the activities of intestinal total catalase (CAT), total superoxide dismutase (T-SOD), glutathione peroxidase (GSH-Px), and the levels of glutathione (GSH) and malondialdehyde (MDA), were assayed by relevant assay kits (Jian Cheng Bioengineering Institute, Nanjing, China).

2.6. *Total RNA Extraction and Real-Time RT–PCR Analysis*

First, total RNA was extracted, and the quality and quantity of the RNA were evaluated with a spectrophotometer. Finally, the reaction system was set up and run on a real-time PCR machine. The reagents and machine models used in the above process were the same as those used in our previous study [16]. Moreover, β-actin was used as the internal reference gene, and the specific primers for the target genes used are shown in Table 2. The mRNA expression levels were determined according to Pfaffl's mathematical model [26].

2.7. *Streptococcus agalactiae Challenge Test*

After sampling, 10 fish (average body weight 109.73 g) from each cage were challenged with *Streptococcus agalactiae* (*S. agalactiae*) in indoor recirculating culture barrels (180 L), the water temperature was controlled at 33–35 °C, the pH value ranged from 7.6 ± 0.2, and dissolved oxygen levels were maintained at 6–7 mg/L. Before the challenge experiment, the pre-experiment was carried out to determine the half-lethal concentration (1 × 10^6 CFU/mL) of *S. agalactiae* using a bacterial turbidimeter (SGZ-6AXJ, Yue Feng Instrument Co. Ltd., Shanghai, China). The specific method is described in our previous study [27]. Then, the fish were challenged by intraperitoneal injection with 1 mL/100 g (1% of body weight). The mortality rate within 144 h was recorded.

2.8. *Data Analysis*

The data were subjected to normality and homogeneity tests. The experimental data (means ± SEM) were analyzed using SPSS 24.0 statistical software for one-way analysis of variance (ANOVA). When the difference was significant ($p < 0.05$), Duncan's multiple comparisons were performed.

Table 2. Primer sequence for qRT-PCR.

Gene	Forward Primer (5′-3′)	Reverse Primer (5′-3′)
CAT [a]	GGAAGAGGATGACGAAGAG	GTTACGGCGAGATGATGT
CPT1 [b]	TCAACACCACACGCATTCCT	AAAGTAGCGCCCTTTGTGGT
FAS [c]	TCATCCAGCAGTTCACTGGCATT	TGATTAGGTCCACGGCCACA
G6Pase [d]	AGCGCGAGCCTGAAGAAGTACT	ATGGTCCACAGCAGGTCCACAT
GK [e]	GACATGAGGACATTGACAAGGGAA	CTTGATGGCGTCTCTGAGTAAACC
GPx [f]	CCAAGAGAACTGCAAGAACGA	CAGGACACGTCATTCCTACAC
HO-1 [g]	CTTGCCCGTGTGGAATCACT	AGATCACCGAGGTAGCGAGT
HSP90 [h]	ATCATCAATGTCCAGCATCA	CATCTTCGCAGCATACCA
IFN-γ [i]	ATGGCTACCACAGTGAGGGCAG	AACTCTGGGGCGACCTTTAGC
IL-8 [j]	CTGTGAAGGCATGGGTGTGGAG	TCGCAGTGGGAGTTGGGAAGAA
IL-10 [k]	CTGCTAGATCAGTCCGTCGAA	GCAGAACCGTGTCCAGGTAA
PEPCK [l]	CTGCGCAAGTACAGCAACTG	TCATGGCTTTGTCCCACTCC
PK [m]	GCACTCCTCAGCTGGTTAAT	GCAAGCACTAGAGCAGGATTT
PPAR-α [n]	TCCAAAAGAAGAACCGAAACA	TTCCACCTCTTTCTCAACCAT
PPAR-γ [o]	TTTACCCATCAAACTGACCAC	GAGGAAATGGAGGCGTAGT
SOD [p]	ACAGAAGAGAAGTATCAGGAG	CACCGTAACAGCAGACAT
SREBP1c [q]	TGCAGCAGAGAGACTGTATCCGA	ACTGCCCTGAATGTGTTCAGACA
TNF-α [r]	AAGCCAAGGCAGCCATCCAT	TTGACCATTCCTCCACTCCAGA
β-actin	CCACACAGTGCCCATCTACGA	CCACGCTCTGTCAGGATCTTCA

[a] CAT, catalase. [b] CPT1, carnitine palmitoyl transterase-1. [c] FAS, fatty acid synthase. [d] G6Pase, glucose-6-phosphatase. [e] GK, glucokinase. [f] GPx, glutathione peroxidase. [g] HO-1, heme oxygenase-1. [h] HSP90, heat shock protein 90. [i] IFN-γ, interferon γ. [j] IL-8, interleukin 8. [k] IL-10, interleukin 10. [l] PEPCK, phosphoenolpyruvate carboxykinase. [m] PK, pyruvate kinase. [n] PPAR-α, peroxisome proliferators-activated receptor-α. [o] PPAR-γ, peroxisome proliferators-activated receptor-γ. [p] SOD, superoxide dismutase. [q] SREBP1c, sterol-regulatory element binding protein 1c. [r] TNF-α, tumor necrosis factor α.

3. Results

3.1. Plasma Biochemical Composition

Table 3 presents the plasma biochemical variables. Among the plasma variables, the plasma glucose (GLU) contents were insignificant across the treatments ($p > 0.05$). The plasma triglyceride (TG) content showed an increasing tendency with increasing dietary protein levels up to 29.28% ($p < 0.05$), and decreased thereafter. The highest plasma total cholesterol (TC) content was observed in the 31.69% dietary protein group ($p < 0.05$). The group with 31.69% dietary protein in the feed had significantly decreased plasma alanine transaminase (ALT) and aspartate aminotransferase (AST) activities ($p < 0.05$).

Table 3. Plasma biochemical parameters.

Dietary Protein (%)	GLU (mmol/L)	TG (mmol/L)	TC (mmol/L)	ALT (U/L)	AST (U/L)
26.45	16.63 ± 1.59	42.32 ± 0.86 ^{b,c}	3.24 ± 0.08 ^{a,b}	22.16 ± 2.35 ^{a}	118.04 ± 11.00 ^{a,b}
29.28	18.61 ± 0.79	45.64 ± 1.86 ^{c}	3.22 ± 0.07 ^{a,b}	25.60 ± 1.89 ^{a,b}	117.57 ± 13.99 ^{a,b}
31.69	15.35 ± 0.89	45.39 ± 2.90 ^{c}	3.47 ± 0.08 ^{b}	21.25 ± 3.68 ^{a}	75.60 ± 16.75 ^{a}
33.68	17.98 ± 1.14	39.60 ± 2.37 ^{a,b}	3.15 ± 0.13 ^{a}	22.72 ± 4.05 ^{a}	93.88 ± 14.74 ^{a,b}
36.18	18.57 ± 0.63	34.61 ± 1.18 ^{a}	3.15 ± 0.06 ^{a}	36.59 ± 3.32 ^{b,c}	132.18 ± 18.08 ^{b,c}
38.75	16.35 ± 1.12	35.44 ± 1.44 ^{a}	3.16 ± 0.08 ^{a}	46.10 ± 5.75 ^{c}	170.29 ± 20.01 ^{c}

Data are expressed as means with SEM (n = 9). Means in the same column with different superscripts ^{a,b,c} are significantly different ($p < 0.05$).

3.2. Intestinal Enzyme and Antioxidant Status

Table 4 shows that the GSH contents and GSH-Px activities were affected by dietary protein levels ($p < 0.05$). The highest GSH content was observed in the 29.28% dietary protein group, while the 31.69% dietary protein level yielded the largest GSH-Px activity. In addition, the other intestinal antioxidant enzyme activity indices (CAT, T-SOD, and MDA) were not affected ($p > 0.05$).

Table 4. Antioxidant enzyme activities of GIFT fed with diets containing six levels of dietary protein under high temperature.

Dietary Protein (%)	CAT (U/mg Protein)	T-SOD (U/mg Protein)	MDA (nmol/mg Protein)	GSH (μmol/g Protein)	GSH-Px (U/mg Protein)
26.45	1.68 ± 0.14	0.83 ± 0.06	0.25 ± 0.04	21.64 ± 1.64 [a,b]	6.59 ± 0.67 [b]
29.28	1.61 ± 0.12	0.73 ± 0.05	0.26 ± 0.06	24.99 ± 1.37 [b]	6.70 ± 0.61 [b]
31.69	1.60 ± 0.08	0.74 ± 0.06	0.21 ± 0.04	22.64 ± 2.29 [a,b]	7.42 ± 0.77 [b]
33.68	1.51 ± 0.14	0.70 ± 0.03	0.16 ± 0.02	18.27 ± 1.71 [a]	6.17 ± 0.64 [a,b]
36.18	1.46 ± 0.09	0.71 ± 0.03	0.31 ± 0.03	17.71 ± 1.33 [a]	5.56 ± 0.53 [a,b]
38.75	1.53 ± 0.09	0.68 ± 0.03	0.27 ± 0.05	17.27 ± 1.76 [a]	4.52 ± 0.44 [a]

Data are expressed as means with SEM (n = 9). Means in the same column with different superscripts [a,b] are significantly different ($p < 0.05$).

3.3. Gene Expression Analysis of Glucose Metabolism

Figure 1 shows the results of the gene expression analysis of the glucose metabolism-related genes. The mRNA expression levels of PK were significantly upregulated by 26.25% dietary protein ($p < 0.05$, Figure 1B). The GK mRNA showed the same phenomenon as PK, but the difference was not significant ($p > 0.05$, Figure 1A). The highest PEPCK and G6Pase mRNA levels were observed in the 29.28% and 31.69% dietary protein groups, respectively ($p < 0.05$, Figure 1C,D).

3.4. Gene Expression Analysis of Lipid Metabolism

Figure 2 shows the results of the gene expression analysis of the lipid metabolism-related genes. Dietary protein (31.69%) significantly upregulated PPAR-α mRNA expression levels ($p < 0.05$, Figure 2A), and CPT1 mRNA showed the same phenomenon ($p < 0.05$, Figure 2B). The PPAR-γ, SREBP1c and FAS mRNA expression levels decreased with increasing dietary protein levels, and the 26.45% dietary protein group produced the peak values ($p < 0.05$, Figure 2C–E).

3.5. Gene Expression Analysis of HSP90 and Antioxidant Status

Figure 3 shows the results of the gene expression analysis of HSP90 and antioxidant-related genes. The HSP90 mRNA expression levels increased with increasing dietary protein levels up to 31.69% ($p < 0.05$, Figure 3A) and decreased thereafter.

The expression levels of the antioxidant-related genes CAT, SOD, and GPx increased with increasing dietary protein levels up to 31.69%, and then decreased. The maximum levels were found in the 31.69% dietary protein group ($p < 0.05$, Figure 3C–E). However, HO-1 mRNA was not affected by the protein treatments ($p > 0.05$, Figure 3B).

3.6. Gene Expression Analysis of Immunity

Figure 4 shows the results of the gene expression analysis of immunity-related genes. Dietary protein levels did not markedly affect the TNF-α, IFN-γ, and IL-8 mRNA expression levels ($p > 0.05$, Figure 4A–C). In addition, the mRNA expression level of IL-10 reached a maximum value of 31.69% dietary protein ($p < 0.05$, Figure 4D).

3.7. *Streptococcus agalactiae* Challenge Test

Figure 5 shows the survival rate of the GIFT fed with different dietary protein levels with the *Streptococcus agalactiae* challenge after 144 h. The highest survival rate of GIFT was observed among fish given food with 31.69% dietary protein at 144 h ($p < 0.05$).

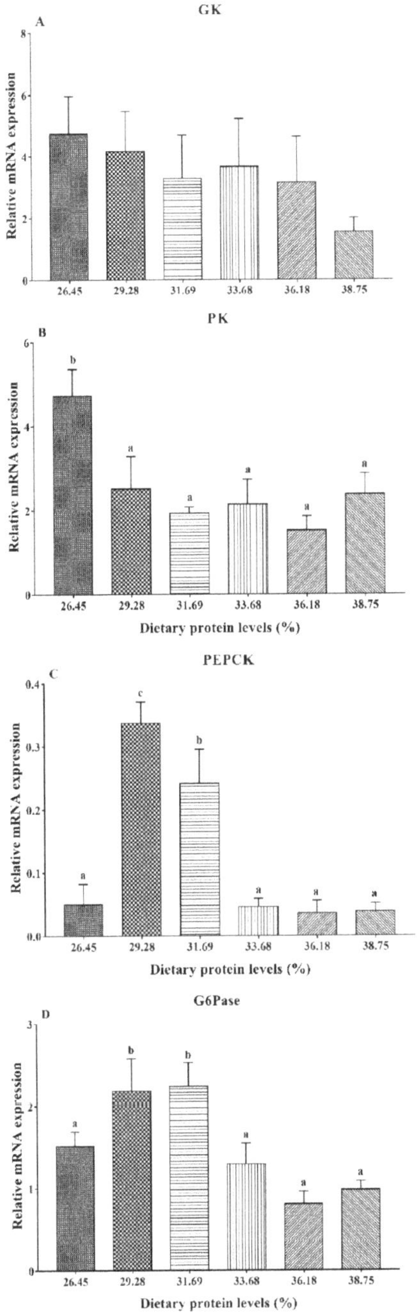

Figure 1. Relative mRNA expressions of glucose metabolism-related genes with dietary protein levels. (**A**) GK; (**B**) PK; (**C**) PEPCK; (**D**) G6Pase. Data are expressed as means with SEM (n = 9). Values with different superscripts are significantly different ($p < 0.05$).

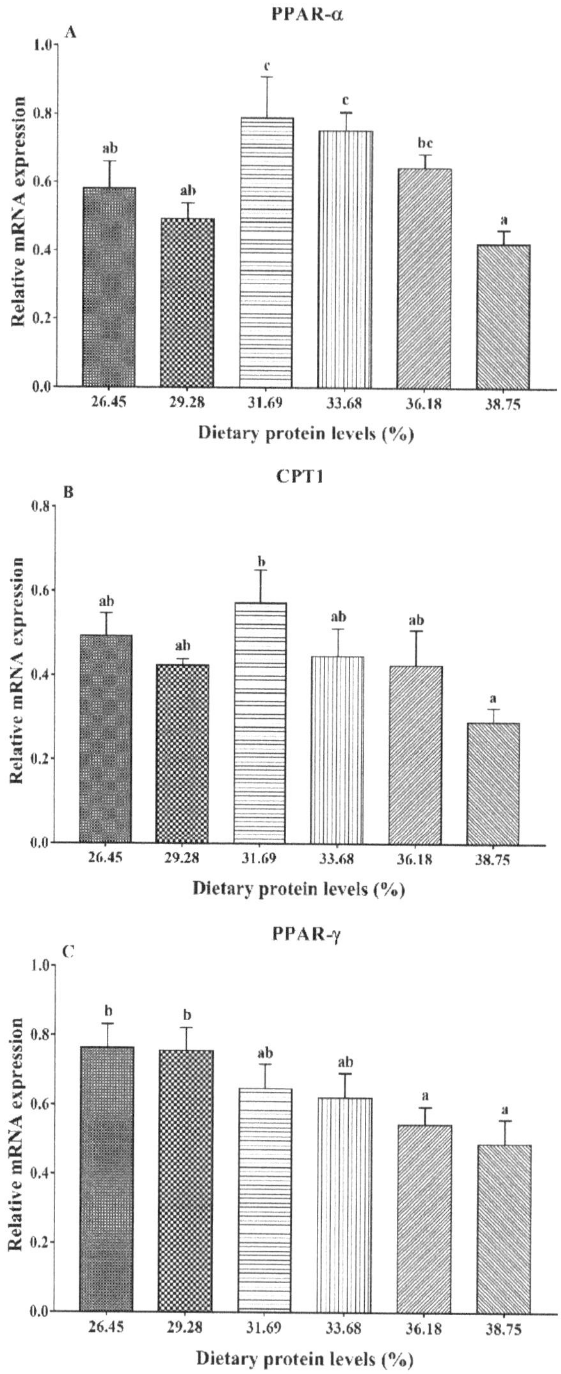

Figure 2. *Cont.*

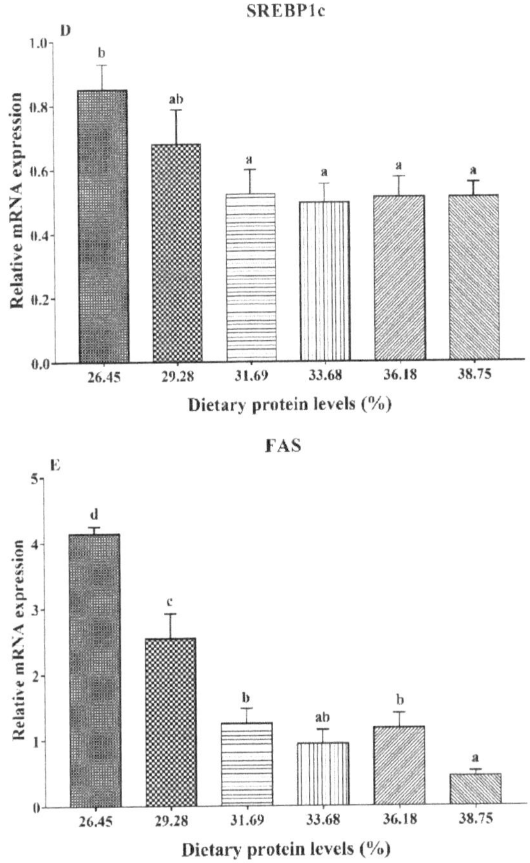

Figure 2. Relative mRNA expressions of lipid metabolism-related genes with dietary protein levels. (**A**) PPAR-α; (**B**) CPT1; (**C**) PPAR-γ; (**D**) SREBP1c; (**E**) FAS. Data are expressed as means with SEM (n = 9). Values with different superscripts are significantly different ($p < 0.05$).

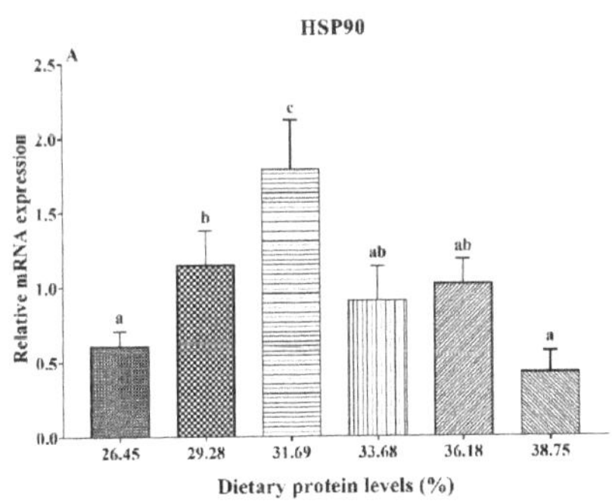

Figure 3. *Cont.*

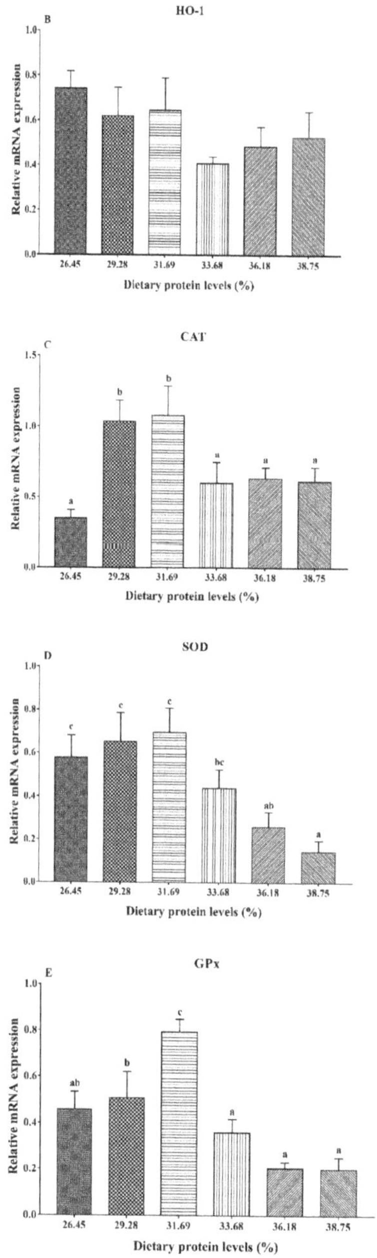

Figure 3. Relative mRNA expressions of HSP90 and antioxidant-related genes with dietary protein levels. (**A**) HSP90; (**B**) HO-1; (**C**) CAT; (**D**) SOD; (**E**) GPx. Data are expressed as means with SEM (n = 9). Values with different superscripts are significantly different ($p < 0.05$).

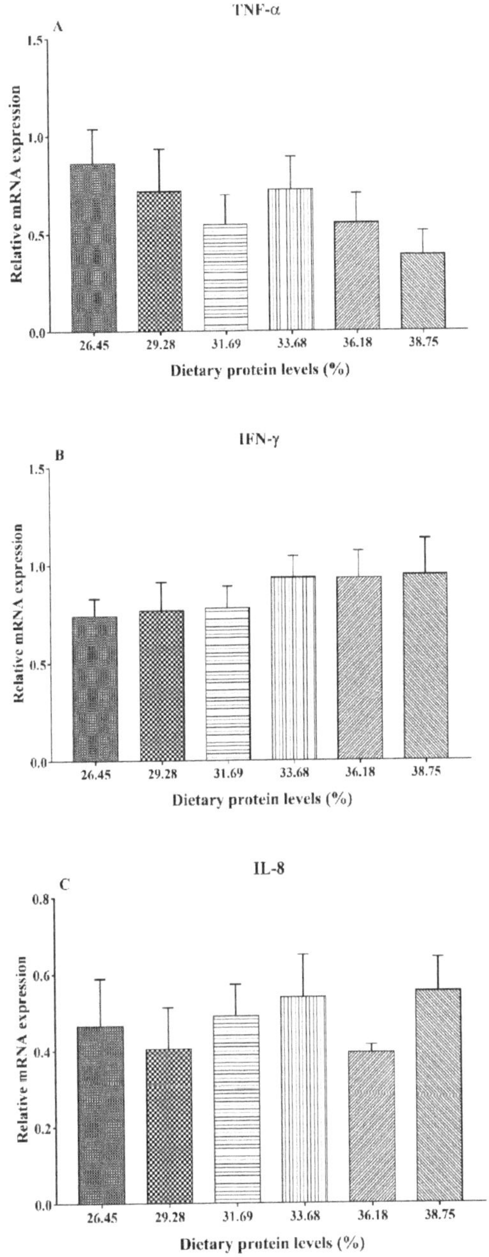

Figure 4. *Cont.*

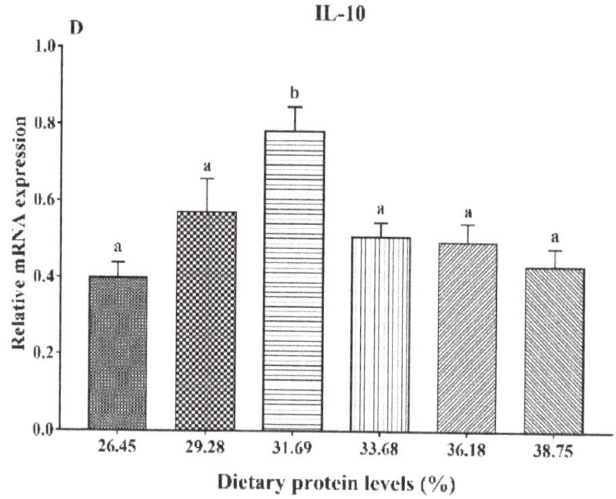

Figure 4. Relative mRNA expressions of antioxidant and immune-related genes with dietary protein levels. (**A**) TNF-α; (**B**) IFN-γ; (**C**) IL-8; (**D**) IL-10. Data are expressed as means with SEM (n = 9). Values with different superscripts are significantly different ($p < 0.05$).

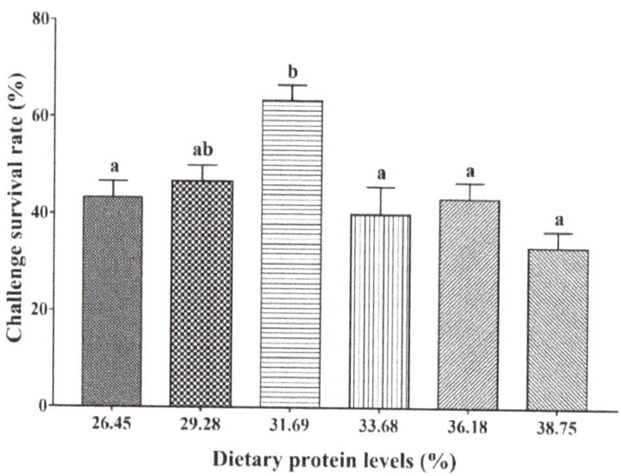

Figure 5. The survival rate of GIFTs fed with different dietary protein levels with *Streptococcus agalactiae* challenge after 144h. Data are expressed as means with SEM (n = 3). Values with different superscripts are significantly different ($p < 0.05$).

4. Discussion

This study demonstrated that dietary protein had significant effects on the glucose and lipid metabolism of GIFT under high temperatures. In this study, low-protein diets increased the expression of glycolysis-related genes, and 26.45% dietary protein resulted in the highest PK mRNA expression levels. It was found that increasing the carbohydrate levels in the feed induces an increase in glycolytic enzyme activities in the liver, which was consistent with a previous study on rainbow trout (*Oncorhynchus mykiss*) and common carp (*Cyprinus carpio*) [28]. In addition, the gluconeogenesis-related genes PEPCK and G6Pase were activated in the 29.28% and 31.69% dietary protein groups, respectively, which were higher than the levels in the 26.45% dietary protein group. This finding indicated that metabolic regulation occurred in GIFT when fed low-protein diets (26.45%, 29.28%, and 31.69%) with high carbohydrate levels, which means that low-protein diets enable rapid adaptation of hepatic glucose metabolism and increased enzyme activity under high temperatures, thus increasing the availability of glucose [29]. Cai et al. [30] reported that a low-protein diet may be more appropriate at higher temperatures due to alterations in liver metabolism. Plasma GLU is an important energy supplier in the fish body and can directly provide energy for various life activities of the fish [31]. In our current study, the plasma GLU contents did not differ significantly among all groups, which was different from other reports that blood glucose levels tend to increase with carbohydrate contents [32,33]. Our current study suggested that dietary protein levels did not influence the energy homeostasis of juvenile GIFT, which may be caused by external temperature factors. However, studies investigating the effect of dietary protein on the plasma parameters of fish under high-temperature stress are still limited, and further investigation is needed.

Regarding lipid metabolism, the PPAR signaling pathway plays an important role in regulating the transcription of genes [34]. PPAR-α and PPAR-γ regulate lipid catabolism and synthesis in lipid metabolism homeostasis, respectively [35]. In this experiment, the expression levels of PPAR-γ mRNA presented a decreasing trend with increasing dietary protein levels. In addition, the downstream factor FAS also showed the same tendency: the highest mRNA level was found in the lowest protein diet (26.45%), which was higher than that in the other groups. This result was in line with our previous study [1] on grass carp, which indicated that low-protein diets could cause an accumulation of lipids in the liver. Furthermore, SREBP1c has a positive correlation with lipid synthesis-related genes [36], which also presented the same phenomenon as PPAR-γ. In addition, the lipolysis-related gene PPAR-α showed a trend of increasing and then decreasing, and the 31.69% dietary protein group achieved the peak value. As a downstream signaling molecule, CPT1 also regulates fatty acid β-oxidation [37], and it exhibited the same trend in response to dietary protein treatments. Our current study indicated that appropriate dietary protein (31.69%) could promote lipolysis and release more energy. As the main sources of blood lipids, the plasma TG and TC contents in the low-protein diets (26.45%, 29.28%, and 31.69%) were higher than those in the high-protein diets (33.68%, 36.18%, and 38.75%), and the highest levels were observed in the 29.28% and 31.69% dietary protein groups. The reason for this result may be the high carbohydrate contents in the low-protein diets, and, thus, the fish can synthesize fat from carbohydrates [38].

High-temperature stress also suppresses the immune system of the fish body while facilitating the growth and reproduction of pathogenic bacteria and reducing the resistance of tilapia to pathogenic bacteria [39]. In the current study, dietary protein levels also had significant effects on the antioxidant status and immune response of GIFT under high temperatures. Under stressful conditions, HSP90 acts as a regulatory enzyme to prevent irreversible protein aggregation and to improve cellular tolerance to stress [40]. Our experimental results showed that an appropriate dietary protein level (31.69%) could activate the expression of HSP90, which was higher than the levels with other dietary treatments. This result indicated that an appropriate dietary protein level (31.69%) enhances the ability to scavenge free radicals and improves the immunity of GIFT under high-temperature stress. Rokutan et al. [41] revealed that increased oxygen radicals can act as a stressor to induce the production of HSP, while HSP can increase the level of peroxidase, inhibit the key enzymes that produce oxygen radicals, and ultimately scavenge oxygen radicals. In addition, HSP significantly attenuates protein exudation during inflammation and inhibits the inflammatory response [42].

High-temperature stress can cause oxidative stress in fish, leading to oxidative damage. The body responds by regulating gene expression levels and regulating the key antioxidant enzymes for oxidative stress [43]. According to previous studies, high expression of antioxidant enzymes can prevent oxidative stress in fish [44]. Numerous studies have shown that the activity of antioxidant enzymes decreases under high-temperature stress, indicating that fish are unable to eliminate the damage produced by peroxides [45,46].

Interestingly, appropriate dietary protein levels could improve the antioxidant enzymes in some fish species [47,48], which supports our findings. In the current study, the highest GSH content was found in the 29.28% dietary protein group, while the 31.69% dietary protein group yielded the maximum GSH-Px activity. In addition, the corresponding antioxidant enzyme genes (CAT, SOD, and GPx) increased continuously with increasing dietary protein levels; however, they started to decrease when the dietary protein level exceeded 31.69%. As downstream regulators of HSPs [49], CAT, SOD, and GPx mRNA showed the same tendency as HSP90, which suggested that the 31.69% dietary protein treatment could significantly improve the intestinal antioxidant capacity of GIFT.

On the other hand, high levels of HSP90 produce a strong stimulus to the body and enhanced immune function [50]. According to previous studies, heat shock proteins reduce damage to the body from inflammatory responses by inhibiting reactive oxygen species and cytokines [42,51]. In this experiment, the pro-inflammatory factors TNF-α, IFN-γ, and IL-8 were not affected by dietary protein treatments under high temperature, while the anti-inflammatory factor IL-10 showed a similar tendency to HSP90, and the highest expression level was found in the 31.69% dietary protein group. Li et al. [52] reported that simultaneous activation of HSPs and anti-inflammatory factors helped to improve the immunity of common carp (*Cyprinus carpio* L.), which was consistent with our findings in tilapia. In addition, plasma ALT and AST activities can be used as indicators of fish health [53]. In this study, the 31.69% dietary protein group had the lowest ALT and AST activities, while the highest levels were observed with the highest protein diet (38.75%), which suggested that high-protein diets are not good for fish health under high-temperature stress.

In addition, tilapia are more susceptible to *Streptococcus agalactiae* infection in high-temperature environments. Thus, it is important to select the appropriate dietary protein level to improve the antibacterial ability. In this study, the survival rate of juvenile GIFT fed 31.69% dietary protein was significantly higher than that of other groups (except 29.28% dietary protein) after a *Streptococcus agalactiae* challenge, which suggested that 31.69% dietary protein level could also enhance the immunity and antioxidant capacity to resist pathogenic bacterial infection.

A previous report has shown that suitable water temperatures for tilapia growth range from 29 to 31 °C [21]. Under appropriate temperature, numerous studies on tilapia have shown that the optimal protein levels of tilapia range from 33% to 35% [54–56], which is higher than the optimal protein requirement (29.28–31.69%) under high temperature in this study. The differences could be explained by high temperatures beyond the optimal growth temperatures inducing retarded growth; meanwhile, high-protein feeds are more detrimental to protein utilization [57]. Combined with the immunization results of this experiment, it can be concluded that adequate reduction of dietary protein levels at high temperatures is more beneficial for tilapia to improve its disease resistance than under suitable temperature conditions.

5. Conclusions

Under a high-temperature environment, low-protein feed (26.45%) enhanced the level of glycolysis and lipid synthesis, and supplementation with appropriate protein (29.28% and 31.69%) enhanced the level of gluconeogenesis and lipolysis. In addition, appropriate dietary protein (29.28% and 31.69%) can effectively improve the antioxidant capacity, enhance immune function, and strengthen the antibacterial capacity of GIFT at high temperature (Figure 6). Overall, 29.28–31.69% dietary protein was recommended in the diet of GIFT in a high-temperature environment.

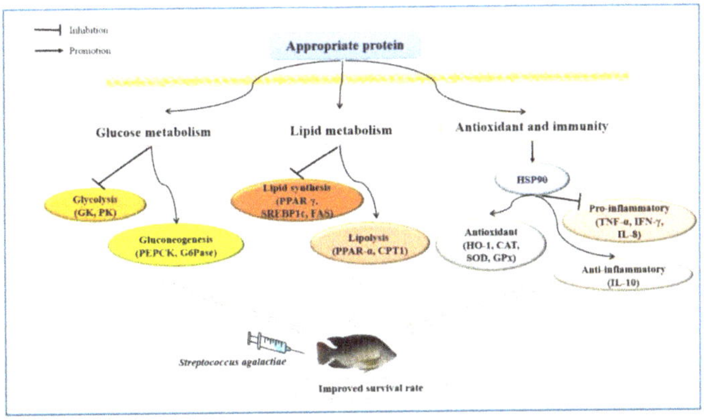

Figure 6. Regulation mechanism of improving health status by appropriate dietary protein levels in GIFT under high-temperature environment.

Author Contributions: Conceptualization, M.R.; Data curation, D.H.; Funding acquisition, M.R.; Investigation, D.H., H.L. and J.Z.; Methodology, D.H., H.L. and J.Z.; Project administration, M.R.; Supervision, M.R. and X.G.; Writing—original draft, D.H.; Writing—review & editing, M.R. and X.G. All authors have read and agreed to the published version of the manuscript.

Funding: This study was financially supported by the National Key Research and Development Program of China (2018YFD0900400), the earmarked fund for CARS (CARS-46).

Institutional Review Board Statement: The study was conducted according to Standardization Administration of China protocols and guidelines (GB/T 35892-2018), and approved by the Institutional Animal Care and Ethics Committee of Nanjing Agricultural University (Permit number: SYXK (Su) 2011–0036).

Data Availability Statement: The authors confirm that the data supporting the findings of this study are available within the manuscript, tables and figures.

Conflicts of Interest: The authors declare no conflict of interest.

References

1. Yu, H.; Liang, H.; Ren, M.; Ge, X.; Ji, K.; Huang, D.; Pan, L.; Xia, D. A study to explore the effects of low dietary protein levels on the growth performance and nutritional metabolism of grass carp (Ctenopharyngodon idella) fry. *Aquaculture* **2022**, *546*, 737324. [CrossRef]
2. Lu, X.; Wei, H.; Yang, C.; Li, Q.; Li, P.; Chen, J.; Sun, Y.; Wen, H.; Jiang, M.; Wang, G. Effects of dietary protein levels on growth performance and liver transcriptome changes in juvenile top-mouth culter Erythroculter ilishaeformis. *Aquac. Rep.* **2021**, *21*, 100964. [CrossRef]
3. Liang, M.; Wang, J.; Chang, Q.; Mai, K. Effects of different levels of fish protein hydrolysate in the diet on the nonspecific immunity of Japanese sea bass, Lateolabrax japonicus (Cuvieret Valenciennes, 1828). *Aquac. Res.* **2005**, *37*, 102–106. [CrossRef]
4. Xia, S.; Sun, J.; Li, M.; Zhao, W.; Zhang, D.; You, H.; Rajkumar, M.; Wu, B. Influence of dietary protein level on growth performance, digestibility and activity of immunity-related enzymes of leopard coral grouper, Plectropomus leopardus (Lacépède, 1802). *Aquac. Nutr.* **2020**, *26*, 242–247. [CrossRef]
5. Ojaveer, H.; Morris, P.C.; Davies, S.J.; Russell, P. The response of thick-lipped grey mullet, Chelon labrosus (Risso), to diets of varied protein-to-energy ratio. *Aquac. Res.* **1996**, *27*, 603–612. [CrossRef]
6. Doran, O.; Moule, S.K.; Teye, G.A.; Whittington, F.M.; Hallett, K.G.; Wood, J.D. A reduced protein diet induces stearoyl-CoA desaturase protein expression in pig muscle but not in subcutaneous adipose tissue: Relationship with intramuscular lipid formation. *Br. J. Nutr.* **2006**, *95*, 609–617. [CrossRef]
7. Pascual, C.; Zenteno, E.; Cuzon, G.; Suárez, J.; Sánchez, A.; Gaxiola, G.; Taboada, G.; Maldonado, T.; Rosas, C. Litopenaeus vannamei juveniles energetic balance and immunological response to dietary proteins. *Aquaculture* **2004**, *239*, 375–395. [CrossRef]
8. Kiron, V.; Watanabe, T.; Fukuda, H.; Okamoto, N.; Takeuchi, T. Protein nutrition and defence mechanisms in rainbow trout Oncorhynchus mykiss. *Comp. Biochem. Physiol. Part A Physiol.* **1995**, *111*, 351–359. [CrossRef]

9. Ma, S.; Guo, Y.; Sun, L.; Fan, W.; Liu, Y.; Liu, D.; Huang, D.; Li, X.; Zhang, W.; Mai, K. Over high or low dietary protein levels depressed the growth, TOR signaling, apoptosis, immune and anti-stress of abalone Haliotis discus hannai. *Fish Shellfish Immunol.* **2020**, *106*, 241–251. [CrossRef]
10. Sun, L.; Chen, H.; Huang, L. Growth, faecal production, nitrogenous excretion and energy budget of juvenile yellow grouper (Epinephelus awoara) relative to ration level. *Aquaculture* **2007**, *264*, 228–235. [CrossRef]
11. Hu, Y.-C.; Kang, C.-K.; Tang, C.-H.; Lee, T.-H. Transcriptomic Analysis of Metabolic Pathways in Milkfish That Respond to Salinity and Temperature Changes. *PLoS ONE* **2015**, *10*, e0134959. [CrossRef]
12. Callaghan, N.I.; Tunnah, L.; Currie, S.; MacCormack, T.J. Metabolic Adjustments to Short-Term Diurnal Temperature Fluctuation in the Rainbow Trout (*Oncorhynchus mykiss*). *Physiol. Biochem. Zool.* **2016**, *89*, 498–510. [CrossRef]
13. Forgati, M.; Kandalski, P.K.; Herrerias, T.; Zaleski, T.; Machado, C.; Souza, M.R.D.P.; Donatti, L. Effects of heat stress on the renal and branchial carbohydrate metabolism and antioxidant system of Antarctic fish. *J. Comp. Physiol. B* **2017**, *187*, 1137–1154. [CrossRef]
14. Fernandez, M.V.S.; Johnson, J.S.; Abuajamieh, M.; Stoakes, S.K.; Seibert, J.T.; Cox, L.; Kahl, S.; Elsasser, T.H.; Ross, J.W.; Isom, S.C.; et al. Effects of heat stress on carbohydrate and lipid metabolism in growing pigs. *Physiol. Rep.* **2015**, *3*, e12315. [CrossRef]
15. Zhao, T.; Ma, A.; Huang, Z.; Liu, Z.; Sun, Z.; Zhu, C.; Yang, J.; Li, Y.; Wang, Q.; Qiao, X.; et al. Transcriptome analysis reveals that high temperatures alter modes of lipid metabolism in juvenile turbot (Scophthalmus maximus) liver. *Comp. Biochem. Physiol. Part D Genom. Proteom.* **2021**, *40*, 100887. [CrossRef]
16. Huang, D.; Ren, M.; Liang, H.; Ge, X.; Xu, H.; Wu, L. Transcriptome analysis of the effect of high-temperature on nutrient metabolism in juvenile grass carp (Ctenopharyngodon idellus). *Gene* **2021**, *809*, 146035. [CrossRef]
17. Aepahi, A.; Heidarieh, M.; Mirvaghefi, A.; Rafiee, G.R.; Farid, M.; Sheikhzadeh, N. Effects of water temperature on the susceptibility of rainbow trout to streptococcus agalactiae. *Acta Sci. Vet.* **2013**, *41*, 1–5.
18. Cheng, A.-C.; Cheng, S.-A.; Chen, Y.-Y.; Chen, J.-C. Effects of temperature change on the innate cellular and humoral immune responses of orange-spotted grouper Epinephelus coioides and its susceptibility to Vibrio alginolyticus. *Fish Shellfish Immunol.* **2009**, *26*, 768–772. [CrossRef]
19. Wang, X.; Wang, L.; Zhang, H.; Ji, Q.; Song, L.; Qiu, L.; Zhou, Z.; Wang, M.; Wang, L. Immune response and energy metabolism of Chlamys farreri under Vibrio anguillarum challenge and high temperature exposure. *Fish Shellfish Immunol.* **2012**, *33*, 1016–1026. [CrossRef]
20. Fisheries and Aquaculture Software. FishStatJ-Software for Fishery and Aquaculture Statistical Time Series. In FAO Fisheries and Aquaculture Department, Food and Agriculture Organization; Rome, Italy. 2020. Available online: https://www.fao.org/home/en/ (accessed on 28 December 2020).
21. Amal, M.N.A.; Zamri-Saad, M. Streptococcosis in tilapia (Oreochromis niloticus): A review. *Pertanika J. Trop. Agric. Sci.* **2011**, *34*, 195–206.
22. Huang, D.; Liang, H.; Ren, M.; Ge, X.; Zhang, Q.; Gu, J. The optimum dietary protein requirement of the genetically improved farmed tilapia (GIFT: *Oreochromis niloticus*): Effects on growth performance and protein metabolism via GH-IGF axis and TOR signalling pathway at different seasonal growth stages. *Aquac. Res.* **2022**, 1–15. [CrossRef]
23. Rodkhum, C.; Kayansamruaj, P.; Pirarat, N.; Zhou, W.; Liu, Y.; Chen, G.H. Effect of water temperature on susceptibility to Streptococcus agalactiae serotype Ia infection in Nile tilapia (Oreochromis niloticus). *Thai J. Vet. Med.* **2011**, *41*, 309.
24. Kayansamruaj, P.; Pirarat, N.; Hirono, I.; Rodkhum, C. Increasing of temperature induces pathogenicity of Streptococcus agalactiae and the up-regulation of inflammatory related genes in infected Nile tilapia (Oreochromis niloticus). *Veter-Microbiol.* **2014**, *172*, 265–271. [CrossRef] [PubMed]
25. Association of Official Analytical Chemists (AOAC). *Official Methods of Analysis of the Association of Official Analytical Chemists*, 15th ed.; Association of Official Analytical Chemists Inc.: Arlington, TX, USA, 2003.
26. Pfaffl, M.W. A new mathematical model for relative quantification in real-time RT-PCR. *Nucleic Acids Res.* **2001**, *29*, 2002–2007. [CrossRef] [PubMed]
27. Liang, H.; Ji, K.; Ge, X.; Ren, M.; Liu, B.; Xi, B.; Pan, L. Effects of dietary arginine on antioxidant status and immunity involved in AMPK-NO signaling pathway in juvenile blunt snout bream. *Fish Shellfish Immunol.* **2018**, *78*, 69–78. [CrossRef]
28. Panserat, S.; Médale, F.; Blin, C.; Brèque, J.; Vachot, C.; Plagnes-Juan, E.; Gomes, E.; Krishnamoorthy, R.; Kaushik, S. Hepatic glucokinase is induced by dietary carbohydrates in rainbow trout, gilthead seabream, and common carp. *Am. J. Physiol. Integr. Comp. Physiol.* **2000**, *278*, R1164–R1170. [CrossRef] [PubMed]
29. Kumar, S.; Sahu, N.P.; Pal, A.K.; Subramanian, S.; Priyadarshi, H.; Kumar, V. High dietary protein combats the stress of Labeo rohita fingerlings exposed to heat shock. *Fish Physiol. Biochem.* **2011**, *37*, 1005–1019. [CrossRef]
30. Cai, L.-S.; Wang, L.; Song, K.; Lu, K.-L.; Zhang, C.-X.; Rahimnejad, S. Evaluation of protein requirement of spotted seabass (Lateolabrax maculatus) under two temperatures, and the liver transcriptome response to thermal stress. *Aquaculture* **2020**, *516*, 734615. [CrossRef]
31. Atli, G.; Ariyurek, S.Y.; Kanak, E.G.; Canli, M. Alterations in the serum biomarkers belonging to different metabolic systems of fish (Oreochromis niloticus) after Cd and Pb exposures. *Environ. Toxicol. Pharmacol.* **2015**, *40*, 508–515. [CrossRef]
32. Yu, H.; Ai, Q.; Mai, K.; Ma, H.; Cahu, C.L.; Infante, J.L.Z. Effects of dietary protein levels on the growth, survival, amylase and trypsin activities in large yellow croaker, Pseudosciaena Crocea R., larvae. *Aquac. Res.* **2012**, *43*, 178–186. [CrossRef]

33. Ye, C.; Wu, Y.; Sun, Z.; Wang, A. Dietary protein requirement of juvenile obscure puffer, *Takifugu obscurus*. *Aquac. Res.* **2017**, *48*, 2064–2073. [CrossRef]
34. Ibabe, A.; Bilbao, E.; Cajaraville, M.P. Expression of peroxisome proliferator-activated receptors in zebrafish (Danio rerio) depending on gender and developmental stage. *Histochem. Cell Biol.* **2004**, *123*, 75–87. [CrossRef]
35. Kota, B.P.; Huang, T.H.-W.; Roufogalis, B.D. An overview on biological mechanisms of PPARs. *Pharmacol. Res.* **2005**, *51*, 85–94. [CrossRef]
36. Gou, N.; Chang, Z.; Deng, W.; Ji, H.; Zhou, J. Effects of dietary lipid levels on growth, fatty acid composition, antioxidant status and lipid metabolism in juvenile *Onychostoma macrolepis*. *Aquac. Res.* **2019**, *50*, 3369–3381. [CrossRef]
37. Kerner, J.; Hoppel, C. Fatty acid import into mitochondria. *Biochim. Biophys. Acta—Mol. Cell Biol. Lipids* **2000**, *1486*, 1–17. [CrossRef]
38. Wu, F.N.; Wen, H.; Jiang, M.; Liu, W.; Liu, B.; Tian, J.; Yang, C.G. Effects of different dietary carbohydrate levels on growth performance and blood biochemical parameters of juvenile GIFT tilapia (Oreochromis niloticus). *J. Northwest A F Univ.-Nat. Sci. Ed.* **2012**, *40*, 8–14.
39. Wongsathein, D. *Factors Affecting Experimental Streptococcus Agalactiae Infection in Tilapia, Oreochromis Niloticus*; University of Stirling: Stirling, UK, 2012; Available online: https://hdl.handle.net/1893/10375 (accessed on 27 September 2012).
40. Miyata, Y.; Yahara, I. The 90-kDa heat shock protein, HSP90, binds and protects casein kinase II from self-aggregation and enhances its kinase activity. *J. Biol. Chem.* **1992**, *267*, 7042–7047. [CrossRef]
41. Rokutan, K.; Hirakawa, T.; Teshima, S.; Nakano, Y.; Miyoshi, M.; Kawai, T.; Konda, E.; Morinaga, H.; Nikawa, T.; Kishi, K. Implications of heat shock/stress proteins for medicine and disease. *J. Med. Investig.* **1998**, *44*, 137–147.
42. Weiss, Y.G.; Maloyan, A.; Tazelaar, J.; Raj, N.; Deutschman, C.S. Adenoviral transfer of HSP-70 into pulmonary epithelium ameliorates experimental acute respiratory distress syndrome. *J. Clin. Investig.* **2002**, *110*, 801–806. [CrossRef]
43. Liang, H.; Mokrani, A.; Ji, K.; Ge, X.; Ren, M.; Pan, L.; Sun, A. Effects of dietary arginine on intestinal antioxidant status and immunity involved in Nrf2 and NF-κB signaling pathway in juvenile blunt snout bream, Megalobrama amblycephala. *Fish Shellfish Immunol.* **2018**, *82*, 243–249. [CrossRef]
44. Liu, J.; Mai, K.; Xu, W.; Zhang, Y.; Zhou, H.; Ai, Q. Effects of dietary glutamine on survival, growth performance, activities of digestive enzyme, antioxidant status and hypoxia stress resistance of half-smooth tongue sole (Cynoglossus semilaevis Günther) post larvae. *Aquaculture* **2015**, *446*, 48–56. [CrossRef]
45. Liu, B.; Xie, J.; Ge, X.; Xu, P.; Wang, A.; He, Y.; Zhou, Q.; Pan, L.; Chen, R. Effects of anthraquinone extract from Rheum officinale Bail on the growth performance and physiological responses of Macrobrachium rosenbergii under high temperature stress. *Fish Shellfish Immunol.* **2010**, *29*, 49–57. [CrossRef]
46. Yu, M.-F.; Zhao, X.-M.; Cai, H.; Yi, J.-M.; Hua, G.-H. Dihydropyridine Enhances the Antioxidant Capacities of Lactating Dairy Cows under Heat Stress Condition. *Animals* **2020**, *10*, 1812. [CrossRef]
47. Yan, J.; Li, Y.; Liang, X.; Zhang, Y.; Dawood, M.A.; Matuli'C, D.; Gao, J. Effects of dietary protein and lipid levels on growth performance, fatty acid composition and antioxidant-related gene expressions in juvenile loach *Misgurnus anguillicaudatus*. *Aquac. Res.* **2017**, *48*, 5385–5393. [CrossRef]
48. Xu, J.; Feng, L.; Jiang, W.-D.; Wu, P.; Liu, Y.; Jiang, J.; Kuang, S.-Y.; Tang, L.; Zhou, X.-Q. Different dietary protein levels affect flesh quality, fatty acids and alter gene expression of Nrf2-mediated antioxidant enzymes in the muscle of grass carp (Ctenopharyngodon idella). *Aquaculture* **2018**, *493*, 272–282. [CrossRef]
49. Yamamoto, Y.; Kume, M.; Yamaoka, Y. Implications of heat shock proteins during liver surgery and liver perfusion. *Isol. Liver Perfus. Hepatic Tumors* **1998**, *147*, 157–172. [CrossRef]
50. Lauber, K.; Brix, N.; Ernst, A.; Hennel, R.; Krombach, J.; Anders, H.; Belka, C. Targeting the heat shock response in combination with radiotherapy: Sensitizing cancer cells to irradiation-induced cell death and heating up their immunogenicity. *Cancer Lett.* **2015**, *368*, 209–229. [CrossRef]
51. Wong, H.R. Heat shock proteins. Facts, thoughts, and dreams. *Shock* **1999**, *11*, 323–325. [CrossRef]
52. Li, Y.; Ding, W.; Li, X. Acute exposure of glyphosate-based herbicide induced damages on common carp organs via heat shock proteins-related immune response and oxidative stress. *Toxin Rev.* **2019**, *40*, 1071–1083. [CrossRef]
53. Sheikh, Z.A.; Ahmed, I. Impact of environmental changes on plasma biochemistry and hematological parameters of Himalayan snow trout, Schizothorax plagiostomus. *Comp. Clin. Pathol.* **2019**, *28*, 793–804. [CrossRef]
54. Wang, K.W.; Takeuchi, T.; Watanabe, T. Effect of dietary protein levels on growth of Tilapia nilotica. *Bull. Jpn. Soc. Sci. Fish.* **1985**, *51*, 133–140. [CrossRef]
55. Sayed, A.N. Optimum Crude Protein Requirement of the Fingerlings Nile Tilapia (Oreochromis niloticus). *Biol. Sci.* **2018**, *2*, 1–8. [CrossRef]
56. Kpundeh, M.D.; Qiang, J.; He, J.; Yang, H.; Xu, P. Effects of dietary protein levels on growth performance and haemato-immunological parameters of juvenile genetically improved farmed tilapia (GIFT), Oreochromis niloticus. *Aquac. Int.* **2015**, *23*, 1189–1201. [CrossRef]
57. Lu, K.L.; Cai, L.S.; Wang, L.; Song, K.; Zhang, C.X.; Rahimnejad, S. Effects of dietary protein/energy ratio and water temperature on growth performance, digestive enzymes activity and non-specific immune response of spotted seabass (*Lateolabrax macu-latus*). *Aquac. Nutr.* **2020**, *26*, 2023–2031. [CrossRef]

fishes

MDPI

Article

Effects of Non-Heated and Heat Processed Krill and Squid Meal-Based Diet on Growth Performance and Biochemical Composition in Juvenile Pacific Bluefin Tuna *Thunnus orientalis*

Jeong-Hyeon Cho [1,†], Takayuki Kurimoto [1,‡], Yutaka Haga [1,*], Yuji Kamimura [2], Akira Itoh [3] and Shuichi Satoh [1,§]

1 Graduate School of Marine Science and Technology, Tokyo University of Marine Science and Technology, Konan, Minato 4-5-7, Tokyo 108-8477, Japan; cjh0123@korea.kr (J.-H.C.); taka127coco@gmail.com (T.K.); ssatoh@fpu.ac.jp (S.S.)
2 Amami Fish Farm Co., Shinokawa Branch, Oshima, Kagoshima 894-1742, Japan; y-kamimura@maruha-nichiro.co.jp
3 Maruha Nichiro Co., Koto, Tokyo 135-8608, Japan; a-itoh@maruha-nichiro.co.jp
* Correspondence: haga@kaiyodai.ac.jp; Tel.: +81-3-5463-0555
† Current address: Jeju Fisheries Research Institute, National Institute of Fisheries Science, Jeju City 63610, Korea.
‡ Current address: The Nisshin OilliO Group, Ltd., Yokohama 235-8558, Japan.
§ Current address: Faculty of Marine Science and Technology, Fukui Prefectural University, Fukui 917-0003, Japan.

Citation: Cho, J.-H.; Kurimoto, T.; Haga, Y.; Kamimura, Y.; Itoh, A.; Satoh, S. Effects of Non-Heated and Heat Processed Krill and Squid Meal-Based Diet on Growth Performance and Biochemical Composition in Juvenile Pacific Bluefin Tuna *Thunnus orientalis*. *Fishes* **2022**, 7, 83. https://doi.org/10.3390/fishes7020083

Academic Editor: Geneviève Corraze

Received: 8 March 2022
Accepted: 30 March 2022
Published: 5 April 2022

Abstract: This study investigated the effects of krill and squid meal and their heat processing on the growth performance and biochemical composition of juvenile Pacific bluefin tuna (PBT) *Thunnus orientalis*. An experiment using a 2 × 2 factorial design examined the effects of two dietary protein sources (squid and krill meal) and heat treatment (heated and non-heated). Prey fish were provided to a reference group. Fish with an initial mean weight of 74.1 mg were fed one of the five diets. After six days of the feeding trial, the fish fed with krill meal and non-heated diets showed improved growth compared to those fed with the squid meal and heated diets. Fish fed the non-heated diets showed significantly higher whole-body crude protein and crude lipid contents than fish fed the heated diets. These results suggest that nutrient availability could be improved by using krill meal and the non-heated treatment to improve the growth performance of juvenile PBT.

Keywords: Pacific bluefin tuna; squid meal; krill meal; heat treatment; growth performance

1. Introduction

Tuna aquaculture is one of the most important aquacultures worldwide. In 2017, the global production of tuna species was approximately 7.89 million tonnes [1]. Japan is not only one of the countries with the highest consumption of tuna species in the world, but total tuna production was as high as 195,200 metric tons in 2018 [2]. In particular, the total production of Pacific bluefin tuna (PBT) in Japan gradually increased from 2,000 metric tons in 2000 to 17,600 metric tons in 2018 [2,3]. This species is a top predator, grows quickly, and requires high dietary protein. Aquafeed is typically based on fish meal which is made from forage fish, such as sardine *Sardinops melanostictus* and anchovy *Engraulis ringens*; the major producers and exporters are Peru and Chile [4]. However, the rapid rise in global demand for aquafeed has reflected soaring prices [5]. In addition, fish meal production relies on natural fish resources. The dependence of the aquaculture industry on high dietary fish meal consumption is a serious concern regarding its sustainability. To overcome these limitations, considerable research efforts are being made to reduce the dependency of aquafeed manufacturers on fish meal [6]. In the last few decades, several studies have focused on plant proteins that can be utilized as fish meal alternatives [7–10]. However, plant protein sources have relatively imbalanced amino acid profiles, low palatability, low nutrient digestibility, and anti-nutritional factors are present [6,9,11–15].

In terms of marine resources, Antarctic krill *Euphausia superba* is one of the most promising resources because of large biomass levels unparalleled anywhere else in the world's oceans. Furthermore, it has been commercially harvested since the 1960s, and today it is targeted by active fisheries of several nations [16,17]. Compared to conventional animal proteins, krill has several advantages such as similar amino acid and fatty acid composition with those of marine farmed fish [18,19] and higher contents of carotenoids, chitin, nucleotides, phospholipids, taurine, and vitamins [20–23]. Squid has been another popular aquafeed ingredient since the 1980s. Squid meal has high protein and strong palatability, is rich in taurine and polar lipids, and excellent feed ingredient [24–27]. Furthermore, improvement of egg quality and fish growth by squid meal based-diet were reported [24,28]. Although krill and squid meals are expensive as the main protein source due to their high unit price, they are ideal as supplemental proteins for low fishmeal diet that can effectively provide the nutrients that are lacking in plant ingredients.

Free amino acids and peptides are considered solubilized dietary proteins because they are water soluble and easily absorbed [29]. Most of the raw protein sources are denatured during heat cooking process, which reduces utility of amino acid and peptide from it. The krill and squid meals available in the market are heat processed, and therefore, they have low nutritive value because of decreased levels of free amino acids and peptides. Therefore, if squid and krill meal without heat processing are available, they may be an effective protein source for tuna. Previously, we showed that higher feed consumption and growth of red sea bream *Pagrus major* fed non-heated krill and squid meal-based diets [28]. However, there are few studies investigate utility of squid and krill meal without heat processing as dietary ingredients for tuna.

Therefore, this study aimed to evaluate the effect of krill and squid-based diets and their heat treatment on the biochemical composition and growth performance of PBT juveniles.

2. Materials and Methods

2.1. Experimental Diets

The test diets were formulated with 51.4% of four different animal protein sources (heated squid meal, HS; non-heated squid meal, NHS; heated krill meal, HK; non-heated krill meal, NHK). Raw squid and krill were pulverized using a centrifugal milling devise equipped with sieve (2.5 mm diameter, ZM500, Retsch Co, Clifton, NJ, USA). The resultant pulverized was kept under −30 °C. The frozen samples were then lyophilized by an automatic vacuum freeze drying devise (RLE-206II, Kyowa Vacuum Eng., Co, Saitama, Japan). The freeze-dried squid or krill meal were sieved using a 500 μm sieve. To make the HK and HS, the freeze-dried meals were kept for 12 h under 105 °C. Porcine blood meal, defatted horse mackerel meal, albumin from hen egg, DHA70E, fish oil, vitamin E, α-starch, taurine, calcium phosphate, choline chloride, sodium ascorbyl phosphate, mineral and vitamin premix, and bonito peptide were included in the test diets (Table 1). These ingredients were mixed and pelletized by fluidized bed granulation which is suspending particles in an air stream and liquid binder such as carboxy methyl cellulose is sprayed from the top of the system down onto the fluidized bed. The resultant pellets (ca. Ø 750 μm) were sieved. The moisture of diets (sinking pellets) was removed in the freeze dryer for 12 h and then kept under −30 °C until use. Spangled emperor *Lethrinus nebulosus* larvae which is popular prey for PBT larvae, was used as control (prey fish, PF). Fertilized eggs obtained from spangled emperor were introduced and hatched in 200 L aquaria.

Table 1. Formula, proximate composition, and water-soluble and -insoluble protein content in the test diets.

Ingredients (%) [2]	Treatment [1]				
	HS	**NHS**	**HK**	**NHK**	**PF**
Heated squid meal	51.4	-	-	-	-
Non-heated squid meal	-	51.4	-	-	-
Heated krill meal	-	-	51.4	-	-
Non-heated krill meal	-	-	-	51.4	-
Porcine blood meal	5.0	5.0	5.0	5.0	-
Defatted horse mackerel meal	12.0	12.0	12.0	12.0	-
Chicken egg albumin	6.6	6.6	6.6	6.6	
Chicken egg lecithin	3.2	3.2	3.2	3.2	-
DHA70E [3]	1.3	1.3	1.3	1.3	-
Taurine	1.0	1.0	1.0	1.0	-
Fish oil [4]	6.7	6.7	8.1	8.1	-
α-Starch	2.0	2.0	2.0	2.0	-
Monobasic calcium phosphate	1.9	1.9	1.9	1.9	-
Vitamin E (50%)	0.1	0.1	0.1	0.1	-
Choline chloride	0.8	0.8	0.8	0.8	-
Sodium ascorbyl phosphate	0.1	0.1	0.1	0.1	-
Mineral mixture [5]	1.5	1.5	1.5	1.5	-
Vitamin mixture [6]	3.0	3.0	3.0	3.0	-
Bonito peptide [7]	2.0	2.0	2.0	2.0	-
Carboxymethyl cellulose	2.0	2.0	2.0	2.0	-
Cellulose	1.4	1.4	-	-	-
Proximate composition (%, dry-weight)					
Moisture	7.6	6.6	6.3	7.6	91.8
Crude protein	57.2	58.4	56.8	56.2	63.7
Water-soluble protein	4.1	7.8	3.3	4.9	-
Water-insoluble protein	48.9	48.6	51.5	47.6	-
Crude lipid	22.4	24.8	21.2	21.8	22.4
Crude ash	8.3	8.0	13.2	14.1	12.3

[1] HS: heated squid meal; NHS: non-heated squid meal; HK: heated krill meal; NHK: non-heated krill meal; PF: prey fish, Spangled emperor larvae *Lethrinus nebulosus*. [2] Squid meal (CP: 67.6%, CL: 13.5%); Krill meal (CP: 67.7%, CL: 10.7%); Porcine blood meal (CP: 71.9%, CL: 1.3%); Defatted horse mackerel meal (CP: 78.6%, CL: 2.9%); Egg albumin (CP: 82.1%, CL: 0.1%). [3] DHA70E (Harima Foods Co., Osaka, Japan). [4] Cod liver oil (Kanematsu Shintoa Foods Co., Tokyo, Japan). [5] Mineral mixture (mg/kg diet): Na (as NaCl) 197; Mg (as $MgSO_4 \cdot 7H_2O$) 735; Fe (as $FeC_6H_5O_7 \cdot 5H_2O$) 258; Zn (as $ZnSO_4 \cdot 7H_2O$) 40; Mn (as $MnSO_4 \cdot 5H_2O$) 18; Cu (as $CuSO_4 \cdot 5H_2O$) 3.9; Al (as $AlCl_3 \cdot 6H_2O$) 0.56; Co (as $CoCl_2 \cdot 6H_2O$) 0.15; I (as KIO_3) 0.89; α-cellulose carrier. [6] Vitamin mixture (amount/kg diet): thiamine hydrochloride, 60 mg; riboflavin, 100 mg; pyridoxine hydrochloride, 40 mg; cyanocobalamin, 0.1 mg; ascorbic acid, 5000 mg; niacin, 400 mg; calcium pantothenate, 100 mg; inositol, 2000 mg; biotin, 6 mg; folic acid 15 mg; *p*-aminobenzoic acid, 50 mg; vitamin K_3, 50 mg; vitamin A acetate, 9000 IU; vitamin D3, 9000 IU. [7] Feeding stimulants. Values of proximate composition are presented as means of triplication.

Moisture content in the formulated diets ranged from 6.3–7.6%, and in the PF group moisture content was 91.8%. The formulated diets were isonitrogenous (57%) and isolipidic (22%) (Table 1). Crude protein and lipid levels in the control were 63.7 and 22.4%, respectively. The NHS and NHK diets contained higher water-soluble protein than the heated meal diets (the protein contents of NHS and NHK were 7.8% and 4.9% compared to 4.1% and 3.3% for HS and HK, respectively). Total and free amino acid content in the non-heated meal diets were greater than those of the heated meal diets (Tables 2 and 3). The most abundantly observed free amino acids in the krill-based diets (HK and NHK) were arginine and glycine, whereas the NHS diet contained a significant portion of proline. The control diet PF had the highest gross free amino acid level (5.69 g/100 g) compared to the other diets (2.55~4.57 g/100 g) (Table 3). Docosahexaenoic acid (DHA) level of essential fatty acids was highest in the PF group (25.0% of total fatty acid), followed by the squid-based diets (15.0~18.3%) and the krill-based diets (11.5~11.7%), whereas eicosapentaenoic acid (EPA) was highest in the krill-based diets (11.5~11.7%), followed by the squid-based diets (8.2~10.7%) and the PF group (6.3%) (Table 4).

Table 2. Total amino acid content of the test diets (g/100 g, dry-weight).

	HS	NHS	HK	NHK	PF
		Essential amino acids			
Arginine	2.25	2.91	2.68	2.83	3.06
Lysine	2.83	3.12	2.67	3.55	4.77
Histidine	1.86	1.00	0.92	1.15	2.71
Phenylalanine	1.93	2.11	1.99	1.99	2.27
Leucine	3.19	3.32	3.20	3.42	4.08
Isoleucine	1.27	1.42	1.32	1.30	1.39
Methionine	0.97	1.25	1.06	1.16	1.61
Valine	1.48	1.74	1.62	1.53	1.82
Threonine	1.99	2.02	1.87	1.82	2.82
Tryptophan	0.41	0.47	0.37	0.61	0.55
		Non-essential amino acids			
Taurine	1.73	2.02	1.64	1.86	0.76
Alanine	2.86	2.89	2.83	2.68	2.96
Tyrosine	1.44	1.57	1.56	1.59	2.09
Cystine	0.43	0.50	0.33	0.39	N/D
Cystathionine	0.13	0.07	0.09	0.12	0.07
Glycine	2.18	2.40	2.76	2.91	2.14
Glutamic acid	6.09	6.82	6.40	6.33	6.69
Serine	2.25	2.37	2.22	2.22	3.67
Aspartic acid	4.42	4.70	4.73	4.67	4.26
Proline	1.79	2.56	1.73	2.29	2.35
Total	41.49	45.25	42.00	44.42	50.09

Values are presented as means of triplication. HS, heated squid meal; NHS, non-heated squid meal; HK, heated krill meal; NHK, non-heated krill meal; PF, prey fish; Spangled emperor larvae, *Lethrinus nebulosus*. N/D: not detected (detection limit: 0.01 g/100 g, dry-weight).

Table 3. Free amino acid content of the test diets (g/100 g, dry-weight).

	HS	NHS	HK	NHK	PF
		Essential amino acids			
Arginine	0.07	0.14	0.17	0.36	0.58
Lysine	0.05	0.12	0.06	0.22	0.65
Histidine	0.06	0.06	0.04	0.01	1.08
Phenylalanine	0.01	0.08	0.02	0.07	0.33
Leucine	0.03	0.19	0.04	0.14	0.57
Isoleucine	0.02	0.08	0.03	0.07	0.16
Methionine	N/D	0.03	N/D	0.03	0.19
Valine	0.03	0.09	0.04	0.10	0.20
Threonine	0.03	0.08	0.03	0.06	0.13
Tryptophan	N/D	0.01	N/D	0.01	0.13
		Non-essential amino acids			
Taurine	1.29	1.62	1.20	1.46	0.53
Alanine	0.15	0.28	0.11	0.24	0.31
Tyrosine	0.02	0.10	0.02	0.08	0.27
Cystine	0.03	0.04	N/D	N/D	N/D
Cystathionine	N/D	0.01	0.01	0.01	0.04
Glycine	0.04	0.11	0.48	0.74	0.08
Glutamic acid	0.01	0.11	0.01	0.03	0.21
Serine	0.03	0.07	0.03	0.05	0.09
Aspartic acid	0.06	0.11	0.06	0.05	0.10
Proline	0.62	1.23	0.42	0.58	0.04
Total	2.55	4.57	2.76	4.31	5.69

Values are presented as means of triplication. N/D: not detected (detection limit: 0.01 g/100 g, dry-weight). HS, heated squid meal; NHS, non-heated squid meal; HK, heated krill meal; NHK, non-heated krill meal; PF, prey fish; Spangled emperor larvae, *Lethrinus nebulosus*.

Table 4. Fatty acid composition (area% of total lipid) of the test diets.

Fatty Acids	Experimental Diets				
	HS	NHS	HK	NHK	PF
14:0	3.7	3.8	5.5	6.2	4.4
16:0	18.0	17.5	19.2	17.5	19.7
16:1n−7	5.5	5.0	5.3	6.0	4.2
18:0	4.6	4.1	3.4	2.9	5.2
18:1n−9	11.7	10.3	12.0	11.3	9.7
18:1n−7	3.5	3.3	3.7	3.8	2.0
18:2n−6	2.8	2.2	4.1	3.4	2.7
18:3n−3	0.4	0.5	0.5	0.5	0.7
18:4n−3	1.0	1.0	1.4	1.4	1.5
20:1n−9	3.2	2.8	1.7	1.9	0.5
20:1n−11	2.7	2.5	1.4	1.6	0.3
20:4n−6	1.5	1.6	1.5	1.3	1.7
20:4n−3	0.4	0.4	0.3	0.3	0.5
20:5n−3	8.2	10.7	11.7	11.5	6.3
22:1n−11	4.0	3.1	2.8	2.8	0.3
22:1n−13	0.7	0.6	0.7	0.6	0.1
22:5n−3	0.9	0.8	0.5	0.5	1.1
22:6n−3	15.0	18.3	11.5	11.7	25.0
22:6n−3/20:5n−3	1.8	1.7	1.0	1.0	4.0
Others	12.1	11.5	12.8	15.0	13.6
Σn−3 LC−PUFA [1]	25.5	31.2	25.4	25.4	34.4

Values are presented as means of triplication. HS, heated squid meal; NHS, non-heated squid meal; HK, heated krill meal; NHK, non-heated krill meal; PF, prey fish; Spangled emperor larvae, *Lethrinus nebulosus*. [1] LC−PUFA, long chain poly unsaturated fatty acids; Σn−3 LC−PUFA: 18:4n−3, 20:4n−3, 20:5n−3, 22:5n−3, 22:6n−3.

2.2. Feeding Experiment and Sampling Schedule

Fertilized eggs were collected natural spawning of Pacific bluefin tuna (PBT) reared at Amami Fish Farm Co, Kagoshima, Japan. The rotifers *Brachionus rotundiformis* and *Artemia* nauplii were offered to larval PBT before they became 10 mm of total length prior to start of the feeding experiment. The larval PBT were fed larvae of spangled emperor *Lethrinus nebulosus* with yolk-sac until the PBT reached 20 mm in total length. Pacific bluefin tuna larvae with total length of 20.5 ± 0.2 mm (19 days post hatching, dph) were introduced into ten 500-L circular polycarbonate aquaria. The aquaria had walls carrying black tape in a checkered pattern to prevent collision of fish against aquarium walls [30] (Table 5).

Table 5. Rearing conditions for feeding trial of Pacific bluefin tuna juveniles.

Average Total Length of Initial Fish (mm) [1]	20.5 ± 0.2
Average body weight of initial fish (mg) [1]	74.1 ± 2.7
Age of initial fish body (day post hatching)	19
Tank volume (L)	500
Number of fish (ind./tank)	240
Rearing period (days)	Weaning period 3 Sole feeding of test diet period 6
Water temperature (°C) [2]	27.8 ± 0.6
pH[2]	8.1 ± 0.0
Dissolved oxygen (mg/L) [2]	11.0 ± 1.7
Photoperiod	11L (07:00–18:00):13D
Exchange rate of sea water (% tank volume/day)	1200
Aeration (mL/min)	800

[1] Mean ± standard deviation (SD) (n = 100). [2] Mean ± SD (n = 10).

The initial three days were designated as acclimation period for the formulated diet (19–21 dph), and the following six days were allocated for feeding with only the test diets for the juveniles (22–28 dph). A commercial diet for marine fish (CP: 58.1%, CL: 19.4%, dry weight) was provided to all treatment groups hourly between 7:00 and 19:00 during the weaning period. Moreover, PF was offered to all groups (three times a day at 7:00, 13:00, and 19:00). The PF offered was reduced continuously; PF: PBT larvae (ind.: ind.) = 150:1, 120:1, and 90:1, so as to promote acclimation of PBT larvae to the experimental diets. PBT juveniles were fed only one of the experimental diets after a three-day weaning period. Fish in the PF group were subjected to satiation feeding of PF for whole period of the feeding trial. Dietary treatment was carried out in duplicate. Rearing of fish was finished after the sixth day from the beginning of the trial to have enough fish for chemical composition analysis. The daily seawater exchange rate was 1200% of the capacity of an aquarium. The average of dissolved oxygen concentration and water temperature were 11.0 ± 1.7 mg/L and 27.8 ± 0.6 °C, respectively (Table 5).

Prior to each sampling, fish were starved for 12 h. In this case, 50 fish were collected just before the beginning of the feeding trial and kept at −80 °C. PBT were also collected at 4, 8, and 10 days after feeding to evaluate fish growth. Ten fish were measured at four and eight days, and 120 fish were measured at ten days, for fork and total length. The whole body was weighed to obtain wet weight for calculation of the condition factor (CF). The carcasses were kept under −80 °C.

The bottom of aquaria was cleaned every day with a syphon, and survival of fish was estimated by counting the number of dead fish. Number of dead fish was subtracted from the fish number introduced at the start of the feeding trial for calculating survival rate. CF, specific growth rate (SGR), thermal growth coefficient (TGC), and weight gain (WG) were calculated by the formulae:

$$\text{WG (\%)} = [\text{wet weight of the final fish (g)} - \text{wet weight of the initial fish (g)}]/\text{wet weight of the initial fish (g)} \times 100$$

$$\text{SGR (\%)} = [\ln \text{wet weight of the final fish (g)} - \ln \text{wet weight of the initial fish (g)}] \times 100/\text{time (days)}$$

$$\text{TGC} = (\text{wet weight of the final fish (g)}^{1/3} - \text{wet weight of the initial fish (g)}^{1/3}) \times (\text{sum day-degrees Celsius})^{-1} \times 1000$$

$$\text{CF} = \text{wet weight (g)}/[\text{fork length (cm)}]^3 \times 100$$

2.3. Chemical Analyses

Diets and carcass were analyzed using standard methods for dry and wet matter, crude protein, and ash [31]. Chemical analysis was conducted in triplicate for each sample at the Laboratory of Fish Nutrition, Tokyo University of Marine Science and Technology, Minato, Tokyo, Japan, and the data was averaged. Moisture content was determined gravimetrically by drying the sample in a dry oven at 105 °C until achieving a constant weight. The samples were incinerated under 650 °C for 8 h by a muffle furnace (FO200, Yamato Co., Tokyo, Japan) for ash content determination. The Kjeldahl method with using an automatic titlator (Kjeltec 2400, FOSS Japan Co, Tokyo, Japan) and conversion index of 6.25 were employed for crude protein analysis. Dietary water-soluble protein content was determined according to de Schrijyer and Ollevier (2000) [32]. Chloroform and methanol mixture (2:1) was used for crude lipids extraction [33]. Preparation of fatty acid methyl ester (FAME) was followed the previously reported methods [34]. FAME was analyzed by a gas chromatograph (GC-2025, Shimadzu, Tokyo, Japan) installed with a fiberglass capillary column (30 m × 0.32 mm, i. d., SUPERCO-WAX10, Sigma-Aldrich Co, St. Louis, MO, USA) and peak area of FAME was measured by a recorder (C-R8A Chromatopac; Shimadzu). Temperature of the column oven was elevated from 170 °C at 2 °C/min for 40 min and kept under 250 °C. An automatic amino acid analyzer (JLC-500/v; JEOL Co., Tokyo, Japan) was used for the total and free amino acid analysis as previously described [35]. Samples were hydrolyzed with 10 mL 2% sulfosalicylic acid (*w/v*, Wako Fujifilm Co, Tokyo, Japan), or 4M methanesulfonic acid (Sigma-Aldridge Co, St. Louis, MO, USA) for free and total amino acid analysis, respectively. The digesta was homogenized for 1.5 min and centrifuged at 1610× *g* under 4 °C for 15 min twice by a devise (SRX-201; Tommy Co., Tokyo, Japan); the upper layers were pooled and filtered with a membrane filter (0.45 µm, Millipore Co,

Darmstadt, Germany). The filtrate was analyzed for amino acid. Constitutional amino acid content was calculated by subtracting the free amino acid from the total amino acid.

2.4. Statistical Analyses

Data on fish growth, initial and final carcass composition were analyzed by a one-way analysis of variance (ANOVA) followed by Tukey's multiple range tests. The main effects of dietary protein source and the heat processing were tested using a two-way ANOVA. Statistical significance was accepted when probability was below 95%. IBM SPSS 19 (SPSS Inc., Chicago, IL, USA) was used for statistical analysis.

3. Results

3.1. Survival and Growth of PBT Juveniles

The survival of juvenile PBT fed the treatment and control diets during the test period is presented in Figure 1. There were no significant differences in survival during the period from one to five days (including the weaning period of one to three days) in all treatment groups. However, significant differences in survival were observed thereafter until the final day. The highest survival of the final fish was found in the PF group (75.3%), followed by the NHK, HK, NHS, and HS groups (52.6%, 51.9%, 45.9%, and 25.8%, respectively). There were no significant differences among the HK, NHK, and NHS groups on the final day, but these groups differed significantly from the HS group. Survival in the krill-based diet groups (HK and NHK) was similar during the rearing period (Figure 1). Survival was not affected by the heat treatment. Effects were significant for protein sources during the rearing period; however, no interactive effects of two factors were detected (Table 6).

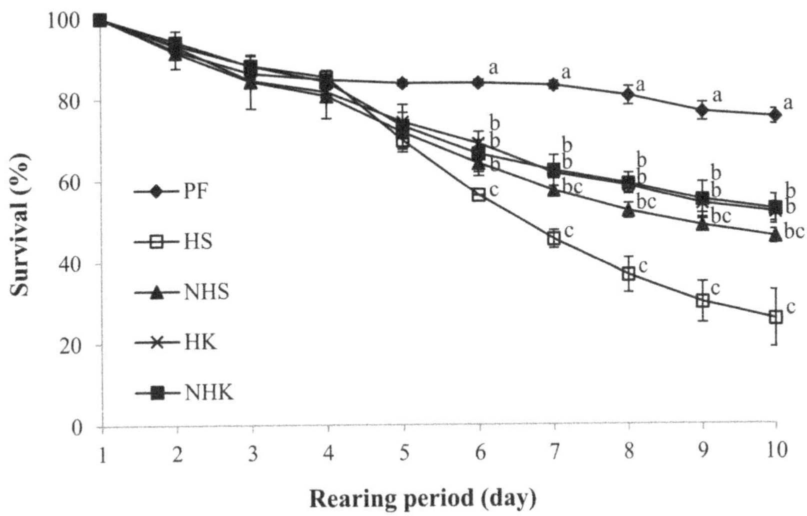

Figure 1. Survival of Pacific bluefin tuna juveniles fed different test diets during the rearing period. Different superscript letters indicate significant difference among the dietary groups (Tukey's test, $p < 0.05$). HS, heated squid meal; NHS, non-heated squid meal; HK, heated krill meal; NHK, non-heated krill meal; PF, prey fish; Spangled emperor larvae, *Lethrinus nebulosus*.

Table 6. Growth performance, proximate composition and biological indices of Pacific bluefin tuna fed the test diets [1] for 9 days.

	Initial	Final (28 dph)					*p*		
	(19 dph)	HS	NHS	HK	NHK	PF	I	H	I × H
Growth performance									
Total length (mm)	20.5 ± 0.2	33.2 ± 0.2^{d}	35.7 ± 0.3^{c}	34.6 ± 0.3^{c}	39.0 ± 0.3^{b}	47.7 ± 0.4^{a}	*	*	*
Body depth (mm)	4.2 ± 0.0	6.4 ± 0.0^{e}	7.1 ± 0.1^{c}	6.7 ± 0.1^{d}	7.4 ± 0.1^{b}	8.6 ± 0.1^{a}	*	*	ns
Body weight (mg)	74.1 ± 2.7	251.0 ± 5.6^{d}	355.3 ± 10.7^{c}	318.6 ± 8.3^{c}	480.4 ± 12.1^{b}	837.4 ± 21.3^{a}	*	*	*
CF	0.8 ± 0.1	0.7 ± 0.1^{c}	0.8 ± 0.1^{b}	0.7 ± 0.1^{b}	0.8 ± 0.1^{a}	0.8 ± 0.1^{b}	*	*	*
WG (%)		238.7 ± 3.1^{c}	379.6 ± 16.1^{c}	330.0 ± 20.9^{c}	548.3 ± 3.1^{b}	1030.1 ± 76.4^{a}	*	*	ns
SGR (%)		13.6 ± 0.1^{d}	17.5 ± 0.0^{c}	16.2 ± 0.8^{c}	20.5 ± 0.2^{b}	27.4 ± 0.5^{a}	*	*	ns
TGC		0.2 ± 0.0^{d}	0.4 ± 0.0^{c}	0.3 ± 0.0^{cd}	0.5 ± 0.0^{b}	1.1 ± 0.1^{a}	*	*	ns
Survival rate (%)		25.8 ± 9.9^{c}	45.9 ± 2.4^{bc}	51.9 ± 3.3^{b}	52.6 ± 5.2^{b}	75.3 ± 2.5^{a}	*	ns	ns
Proximate composition (%, wet-weight)									
Moisture	83.5 ± 0.2	83.7 ± 0.2^{ab}	82.9 ± 0.1^{c}	84.1 ± 0.3^{a}	83.3 ± 0.0^{bc}	83.4 ± 0.2^{bc}	*	*	ns
Crude protein	12.4 ± 0.0	11.2 ± 0.3^{c}	11.9 ± 0.1^{b}	11.6 ± 0.1^{bc}	12.6 ± 0.0^{a}	12.9 ± 0.1^{a}	*	*	ns
Crude lipid	2.0 ± 0.1	2.0 ± 0.0^{c}	2.6 ± 0.1^{a}	1.8 ± 0.1^{c}	2.2 ± 0.0^{b}	1.6 ± 0.0^{d}	*	*	*
Crude ash	2.6 ± 0.1	3.2 ± 0.2	2.7 ± 0.2	2.9 ± 0.2	2.7 ± 0.1	2.8 ± 0.1	ns	*	ns

Values of total length, body depth, body weight and CF (condition factor) are means ± SD of 100 (19 dph) or 120 (28 dph) fish. Values are means ± SD of 2 groups of fish ($n = 2$; WG, SGR, TGC and survival rate) or 3 groups of fish ($n = 3$; proximate composition). Values in a same row with different letter are significantly different (Tukey's test, $p < 0.05$). I, ingredients (squid meal and krill meal); H, heated and non-heated treatment; ns, no significant difference (two-way ANOVA, $p > 0.05$); *, $p < 0.05$. WG, weight gain; SGR, specific growth rate; TGC, thermal growth coefficient. [1] HS, heated squid meal; NHS, non-heated squid meal; HK, heated krill meal; NHK, non-heated krill meal; PF, prey fish; Spangled emperor larvae, *Lethrinus nebulosus*.

The growth performances of juvenile PBT during all test periods are shown in Table 6, Figures 2–4. The best growth was found in the PF group, and its difference was significant among all groups ($p < 0.05$). The PBT juveniles in non-heated diet groups grew significantly better than those in the heated treatment diet groups ($p < 0.05$), and the fish in krill meal-based diet groups grew significantly better than those in the squid meal-based diet groups ($p < 0.05$) (Figures 2 and 3). The WG, SGR, and TGC of fish were affected by both protein sources and heat treatments, but there were no interactive effects of either factor (Figure 4).

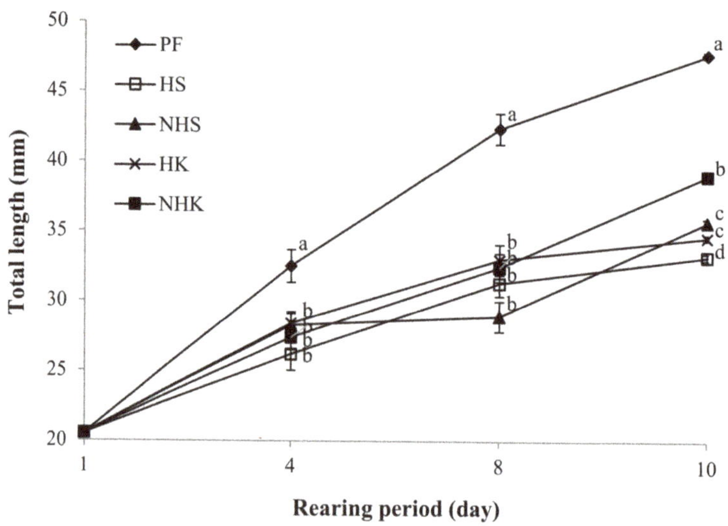

Figure 2. Final total length of Pacific bluefin tuna juveniles fed different test diets during the rearing period. Different superscript letters indicate significant difference among the dietary groups (Tukey's test, $p < 0.05$; 1st day, $n = 50$; 4th and 8th day, $n = 10$; 10th day, $n = 120$). HS, heated squid meal; NHS, non-heated squid meal; HK, heated krill meal; NHK, non-heated krill meal; PF, prey fish; Spangled emperor larvae, *Lethrinus nebulosus*.

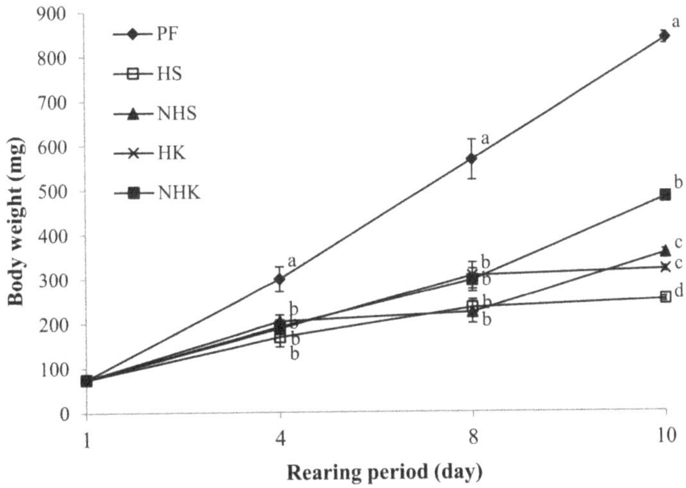

Figure 3. Final body weight of Pacific bluefin tuna juveniles fed different diets during the rearing period. Different superscript letters indicate significant difference among the dietary groups (Tukey's test, $p < 0.05$; 1st day, $n = 50$; 4th and 8th day, $n = 10$; 10th day, $n = 120$). HS, heated squid meal; NHS, non-heated squid meal; HK, heated krill meal; NHK, non-heated krill meal; PF, prey fish; Spangled emperor larvae, *Lethrinus nebulosus*.

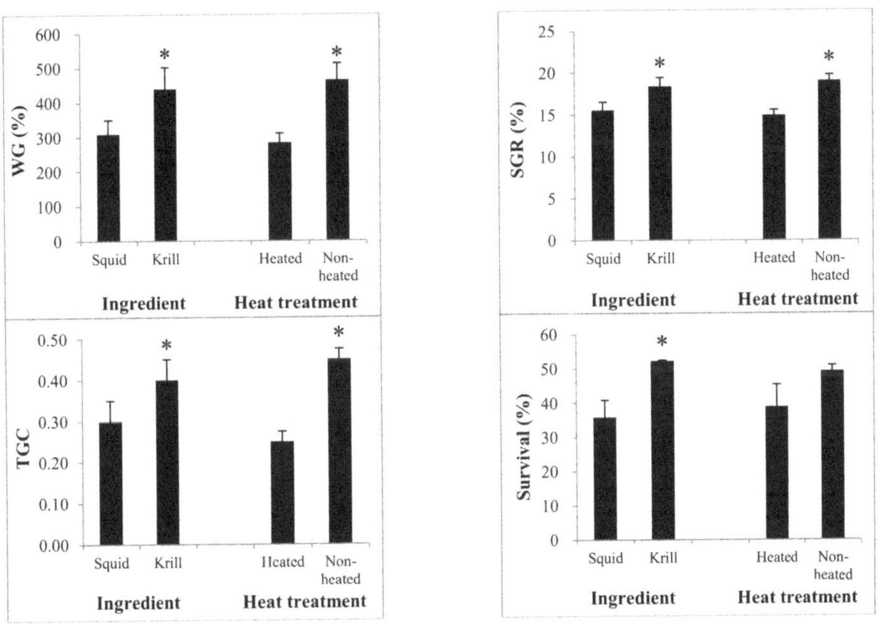

Figure 4. Means of WG (g, **upper-left**), SGR (%, **upper-right**), TGC (**lower-left**), and survival (%, **lower-right**) of juvenile Pacific bluefin tuna after rearing period for the main effects of ingredient and heat treatment. Data was expressed as means ± SE. WG, weight gain; SGR, specific growth rate; TGC, thermal growth coefficient. The asterisks indicate significant differences by two-way ANOVA analysis ($p < 0.05$).

3.2. Chemical Property of PBT

The highest moisture content was noted in the HK group, followed by the HS, PF, NHK, and NHS groups ($p < 0.05$). The crude protein contents in the PF and NHK groups were significantly higher than those in the other groups, whereas the crude lipid content in the NHS group was significantly higher than in the other groups ($p < 0.05$). No significant differences were observed in the crude ash among all the groups ($p > 0.05$). The moisture, crude protein, and crude lipid of the carcasses were affected by both protein source and heat treatment, but an interactive effect between both factors was detected only for crude lipids. By the final day, the CF observed in the NHK group was significantly higher than those of the PF, NHS, and HK groups, which were significantly higher than that of the HS group ($p < 0.05$).

Significant differences between the PF and the other groups were observed in all constitutional amino acid levels except valine and tryptophan. Specifically, most of the amino acid levels were higher in the PF group than in the other treatment groups. Constitutional amino acids were affected by both the protein source and heat treatment, while the free amino acids tended to be affected by the interaction. Only free histidine was affected by both factors (Table 7).

Table 7. Constitutional and free amino acid content of Pacific bluefin tuna fed the test diets [1] for 9 days (g/100 g, dry-weight).

	Constitutional Amino Acid								
	Initial	Final (28 dph)					p		
	(19 dph)	HS	NHS	HK	NHK	PF	I	H	I × H
Essential amino acid									
Arginine	2.56 ± 0.27	1.67 ± 0.47 ^{c}	2.31 ± 0.16 ^{bc}	2.53 ± 0.16 ^{ab}	2.65 ± 0.09 ^{ab}	3.14 ± 0.07 ^{a}	*	*	ns
Lysine	3.36 ± 0.38	2.21 ± 0.48 ^{c}	3.06 ± 0.22 ^{b}	3.30 ± 0.11 b	3.52 ± 0.08 ^{ab}	4.14 ± 0.04 ^{a}	*	*	ns
Histidine	0.77 ± 0.15	0.53 ± 0.15 ^{c}	0.76 ± 0.04 ^{b}	0.82 ± 0.03 ^{ab}	0.88 ± 0.02 ^{ab}	1.01 ± 0.03 ^{a}	*	*	ns
Phenylalanine	1.89 ± 0.25	1.42 ± 0.37 ^{b}	1.84 ± 0.13 ^{ab}	2.05 ± 0.04 ^{a}	2.05 ± 0.10 ^{a}	2.34 ± 0.07 ^{a}	*	*	ns
Leucine	3.18 ± 0.40	2.35 ± 0.50 ^{c}	3.08 ± 0.24 ^{b}	3.34 ± 0.11 ^{ab}	3.48 ± 0.09 ^{ab}	3.82 ± 0.11 ^{a}	*	*	ns
Isoleucine	1.25 ± 0.16	0.93 ± 0.19 ^{b}	1.11 ± 0.21 ^{ab}	1.21 ± 0.10 ^{ab}	1.31 ± 0.05 ^{a}	1.45 ± 0.03 ^{a}	*	ns	ns
Methionine	0.92 ± 0.12	0.84 ± 0.29 ^{b}	0.93 ± 0.14 ^{b}	1.21 ± 0.08 ^{ab}	1.18 ± 0.13 ^{ab}	1.45 ± 0.10 ^{a}	*	ns	ns
Valine	1.28 ± 0.18	0.98 ± 0.49	1.14 ± 0.20	1.28 ± 0.11	1.31 ± 0.03	1.53 ± 0.04	ns	ns	ns
Threonine	2.24 ± 0.25	1.66 ± 0.28 ^{b}	2.14 ± 0.08 ^{a}	2.21 ± 0.04 ^{a}	2.39 ± 0.08 ^{a}	2.50 ± 0.04 ^{a}	*	*	ns
Tryptophan	0.39 ± 0.08	0.26 ± 0.13	0.41 ± 0.05	0.47 ± 0.07	0.36 ± 0.06	0.50 ± 0.11	ns	ns	*
	Free amino acid							p	
Essential amino acid									
Arginine	0.39 ± 0.04	0.79 ± 0.20 ^{a}	0.35 ± 0.25 ^{bc}	0.29 ± 0.01 ^{bc}	0.57 ± 0.02 ^{ab}	0.15 ± 0.00 ^{c}	ns	ns	*
Lysine	0.49 ± 0.01	0.84 ± 0.16 ^{a}	0.43 ± 0.26 ^{bc}	0.36 ± 0.01 ^{bc}	0.61 ± 0.02 ^{ab}	0.18 ± 0.00 ^{c}	ns	ns	*
Histidine	0.69 ± 0.01	0.29 ± 0.04 ^{d}	0.43 ± 0.05 ^{c}	0.33 ± 0.00 ^{d}	0.52 ± 0.01 ^{b}	0.78 ± 0.01 ^{a}	*	*	ns
Phenylalanine	0.32 ± 0.00	0.50 ± 0.16 ^{a}	0.21 ± 0.17 ^{b}	0.17 ± 0.01 ^{b}	0.38 ± 0.01 ^{ab}	0.11 ± 0.00 ^{b}	ns	ns	*
Leucine	0.46 ± 0.01	0.75 ± 0.22 ^{a}	0.31 ± 0.26 ^{b}	0.26 ± 0.01 ^{b}	0.52 ± 0.02 ^{ab}	0.17 ± 0.00 ^{b}	ns	ns	*
Isoleucine	0.23 ± 0.00	0.35 ± 0.09 ^{a}	0.16 ± 0.13 ^{b}	0.13 ± 0.00 ^{b}	0.26 ± 0.01 ^{ab}	0.10 ± 0.00 ^{b}	ns	ns	*
Methionine	0.22 ± 0.00	0.32 ± 0.09 ^{a}	0.15 ± 0.11 ^{ab}	0.13 ± 0.00 ^{b}	0.25 ± 0.01 ^{ab}	0.08 ± 0.01 ^{b}	ns	ns	*
Valine	0.37 ± 0.01	0.52 ± 0.12 ^{a}	0.23 ± 0.19 ^{b}	0.19 ± 0.00 ^{b}	0.36 ± 0.01 ^{ab}	0.17 ± 0.00 ^{b}	ns	ns	*
Threonine	0.25 ± 0.00	0.35 ± 0.07 ^{a}	0.19 ± 0.12 ^{b}	0.15 ± 0.00 ^{b}	0.25 ± 0.01 ^{ab}	0.14 ± 0.00 ^{b}	ns	ns	*
Tryptophan	0.05 ± 0.02	0.15 ± 0.07 ^{a}	0.04 ± 0.04 ^{b}	0.04 ± 0.00 ^{b}	0.09 ± 0.01 ^{ab}	0.02 ± 0.00 ^{b}	ns	ns	*

Values are presented as means of triplication. Values in a same row with different superscript letters are significantly different (Tukey's test, $p < 0.05$). I, ingredients (squid meal and krill meal); H, heated and non-heated treatment; ns, no significant difference (two-way ANOVA, $p > 0.05$); *, $p < 0.05$. [1] HS, heated squid meal; NHS, non-heated squid meal; HK, heated krill meal; NHK, non-heated krill meal; PF, prey fish; Spangled emperor larvae, *Lethrinus nebulosus*.

The major fatty acids in the carcass were 16:0 (palmitic), 18:0 (stearic), 18:1n−9 (oleic), 20:5n−3 (eicosapentaenoic, EPA), and 22:6n−3 (docosahexaenoic, DHA) acids (Table 8). The sum of these major fatty acids accounted for more than 66.2%, 70.0%, 69.9%, 68.4%, and 77.4% of the total fatty acids in the HS, NHS, HK, NHS, and PF treatment groups, respectively. The major saturated fatty acid (SAFA) was 16:0 in all groups. Regarding polyunsaturated fatty acids (PUFAs), bluefin tuna is considered to be a good source of n−3 fatty acids, particularly DHA, which exhibited the highest levels in the PF group. DHA occurred in a higher proportion than EPA in all treatment groups. The sum of DHA and EPA reached 28.9%, 34.6%, 32.7%, 30.5%, and 35.5% for HS, NHS, HK, NHK, and PF, respectively.

Table 8. Fatty acid composition (area% of total lipid) of whole fish body of Pacific bluefin tuna for 9 days.

	Initial	Final (28 dph)					*p*		
	(19 dph)	HS	NHS	HK	NHK	PF	I	H	I × H
14:0	3.1 ± 0.6	1.6 ± 0.1	1.4 ± 0.2	1.6 ± 0.5	1.6 ± 0.1	0.9 ± 0.2	ns	ns	ns
16:0	25.4 ± 1.2	18.0 ± 0.7 [b]	18.6 ± 0.8 [b]	19.8 ± 0.9 [b]	18.8 ± 2.2 [b]	23.7 ± 0.4 [a]	ns	ns	ns
16:1n−7	3.2 ± 0.5	2.5 ± 0.1 [bc]	3.4 ± 0.2 [ab]	3.2 ± 0.5 [ab]	4.1 ± 0.2 [a]	1.8 ± 0.3 [c]	*	*	ns
18:0	11.3 ± 2.1	11.0 ± 0.9 [a]	6.2 ± 0.3 [c]	7.5 ± 0.2 [bc]	6.1 ± 0.5 [c]	8.9 ± 0.5 [b]	*	*	*
18:1n−9	8.1 ± 1.0	8.3 ± 0.2 [c]	10.6 ± 0.4 [b]	9.9 ± 0.2 [bc]	13.0 ± 1.3 [a]	9.3 ± 0.6 [bc]	*	*	ns
18:1n−7	2.0 ± 0.2	2.2 ± 0.0 [c]	3.0 ± 0.1 [b]	2.8 ± 0.1 [bc]	3.7 ± 0.4 [a]	2.2 ± 0.1 [c]	*	*	ns
18:2n−6	2.2 ± 0.2	1.9 ± 0.2 [c]	2.6 ± 0.1 [b]	3.9 ± 0.3 [a]	3.9 ± 0.2 [a]	1.9 ± 0.1 [c]	*	ns	*
18:3n−3	0.4 ± 0.1	0.2 ± 0.0	0.3 ± 0.0	0.2 ± 0.0	0.3 ± 0.0	0.2 ± 0.0	ns	*	ns
18:4n−3	0.7 ± 0.1	0.8 ± 0.3	0.6 ± 0.0	0.6 ± 0.2	0.8 ± 0.1	0.3 ± 0.1	ns	ns	ns
20:1n−9	0.5 ± 0.1	1.8 ± 0.2 [a]	1.8 ± 0.1 [a]	1.2 ± 0.3 [b]	1.5 ± 0.1 [ab]	0.3 ± 0.0 [c]	*	ns	ns
20:1n−11	0.4 ± 0.1	1.4 ± 0.1 [ab]	1.5 ± 0.1 [a]	0.9 ± 0.2 [c]	1.2 ± 0.1 [bc]	0.3 ± 0.0 [d]	*	*	ns
20:2n−6	0.1 ± 0.0	0.2 ± 0.0 [b]	0.2 ± 0.0 [a]	0.1 ± 0.0 [c]	0.1 ± 0.0 [c]	0.1 ± 0.0 [c]	*	*	*
20:4n−6	3.0 ± 0.4	2.7 ± 0.4 [b]	3.3 ± 0.2 [b]	4.4 ± 0.6 [a]	3.5 ± 0.2 [ab]	3.7 ± 0.1 [ab]	*	ns	*
20:4n−3	0.3 ± 0.1	0.2 ± 0.0 [b]	0.3 ± 0.0 [a]	0.2 ± 0.0 [ab]	0.2 ± 0.0 [ab]	0.3 ± 0.0 [ab]	ns	*	ns
20:5n−3	5.6 ± 1.0	7.1 ± 0.4 [b]	8.6 ± 0.4 [ab]	8.2 ± 0.3 [ab]	9.4 ± 0.8 [a]	5.0 ± 0.3 [c]	*	*	ns
22:5n−6	0.9 ± 0.1	0.6 ± 0.1 [b]	0.9 ± 0.1 [ab]	0.8 ± 0.1 [ab]	0.9 ± 0.0 [ab]	1.1 ± 0.2 [a]	ns	*	ns
22:6n−3	22.8 ± 3.5	21.8 ± 3.9 [ab]	26.0 ± 2.2 [ab]	24.5 ± 3.5 [ab]	21.1 ± 2.0 [b]	30.5 ± 0.2 [a]	ns	ns	ns
22:6n−3/20:5n−3	4.2 ± 0.6	3.1 ± 0.5 [b]	3.0 ± 0.2 [b]	3.0 ± 0.5 [b]	2.3 ± 0.1 [b]	6.1 ± 0.4 [a]	ns	ns	ns
Others	9.8 ± 1.5	16.3 ± 5.4	8.8 ± 4.3	8.3 ± 2.1	7.7 ± 2.0	9.3 ± 1.8	ns	ns	ns
Σn−3 LC−PUFA [1]	30.7 ± 4.5	30.7 ± 4.0	36.6 ± 2.5	34.6 ± 3.2	32.7 ± 2.9	37.4 ± 0.2	ns	ns	ns

[1] LC−PUFA, long chain poly unsaturated fatty acids; Σn−3 LC−PUFA: 18:4n−3, 20:4n−3, 20:5n−3, 22:5n−3, 22:6n−3. Values are presented as means of triplication. Values in a same row with different superscript letters are significantly different (Tukey's test, $p < 0.05$). I, ingredients (squid meal and krill meal); H, heated and non-heated treatment; ns, no significant difference (two-way ANOVA, $p > 0.05$); *, $p < 0.05$. HS, heated squid meal; NHS, non-heated squid meal; HK, heated krill meal; NHK, non-heated krill meal; PF, prey fish; Spangled emperor larvae, *Lethrinus nebulosus*.

4. Discussion

Kvåle et al. (2009) [36] reported that different feeding practices greatly affect fish performance, and the weaning period and method of introducing formulated diets are important. In the present study, based on the findings of Haga et al. (2010) [37], Cho et al. (2016) [38] and Cho et al. (2022) [39], we fed yolk-sac larvae of spangled emperor and a commercial diet on the first day, but the frequency of feeding yolk-sac larvae was gradually reduced over the subsequent two days due to acclimation to the formulated diet. According to Cho et al. (2016) [38], early larvae of PBT is able to weaned to formulated diets once they successfully accepted and ingested a suitable formulated diet. Similarly, in the present study, no significant differences were observed in fish growth and survival immediately after the weaning period, but we were able to observe significant differences in growth performance after changing their diet (Figures 1–3).

In the present study, there were significant differences in the contents of water-soluble proteins and free amino acids between the heated and non-heated diets (Tables 1 and 3). Considering an extruded pelleting process which is subjected to heat treatment, protein denaturation of the non-heated meal could be expected. To avoid protein denaturation of the non-heated meal, an extruder was not used to prepare the test diet in the present study. According to Cho et al. (2018) [28], in a study comparing the effects of heat treatment of ingredients on the growth performance of juvenile red sea bream *Pagrus major*, the growth of fish fed a non-heated diet was significantly better than that of fish fed a heated diet. Cho et al. attributed this difference in growth performance to the high water-soluble protein and free amino acid content in the feed. Watanabe (1982) [40] reported the incorporation of water-soluble proteins by pinocytosis in rectal epithelium cells in larval and juvenile teleosts. In the present study, it is suggested that the highly water-soluble protein contained in the non-heated diets has been effectively absorbed and assimilated in the early stages of

PBT. Furthermore, higher utilization of free amino acids was demonstrated in the larvae and juvenile of fish [41,42]. Current result showed that approximately 1.7 times higher gross free amino acid contained in the non-heated diets than the heated treatment diets, suggesting that they were utilized more effectively for growth.

Satoh (2005) [43] reported the in vitro digestibility of fish meal and raw fish by pepsin and trypsin from yellowtail *Seriola quinqueradiata*. Both pepsin and trypsin activity of fish fed raw fish were much higher than in those fed fish meal, and these enzymes worked better on protein from raw fish compared to that from fish meal. Furthermore, Seoka et al. (2010) [27] reported that the growth and survival (at 30 days post hatching) of PBT juvenile fed non-heated Toyama squid *Watasenia scintillans* meal-based diet were better than that fed a commercial diet. Yellowtail and tuna species feed on live fish in natural ecosystems; therefore, it is thought that the proteins of non-heated treatment without denaturation are highly digestible; therefore, using a non-heated protein source in an aquaculture setting is likely to be superior in terms of digestibility. It was reported that PBT juveniles exhibited higher gustatory responses to alanine, leucine, valine, methionine, isoleucine, and proline [44]. Here we demonstrated that the gross levels of these amino acids in non-heated diets (1.90 and 1.16 g/100 g) were higher than those in the heated diets (0.85 and 0.64 g/100 g). These results suggest that a diet containing high levels of water-soluble proteinous component such as free amino acids that are effective for fish growth can be produced without heating the ingredients and such a diet will promote the feed intake of PBT juveniles. Furthermore, Marubeni Nisshin Feed Co, Tokyo, Japan, developed a commercial formulated diet (Magokoro) based on enzyme-treated Chilean fish meal called BioCP from LANDES Co., Talcahuano, Chile, for PBT juveniles. Higher water solubility after enzyme treatment has been well documented for fish protein [45], and 60% of protein in BioCP accounts for soluble protein. However, the production of BioCP is expected to decline in the near future. When fish meal is in short supply, non-heated proteins can be more competitive depending on the availability of soluble proteins in the market.

The effects of feed ingredient on nutritional status were investigated in the present study. As a result, nutritional status indices, including growth performance, whole body composition, constitutional amino acid, and fatty acid altered. Fish fed krill meal-based diet (HK and NHK) had a positive WG, SGR, TGC, and survival. It was suggested that krill meal diet promotes digestive ability of yellowtail by regulating the digestive enzyme secretion in intestine [43].

Compared to squid meal, krill meal is considered to be a more effective ingredient for marine fish, including for the early life stage of PBT. The crude protein in krill contains 40% of the extractable nitrogen-containing components that are not contained in the protein itself [46]. Since the extractable component contains free amino acids and peptides and does not require decomposition by proteases, it might be easily absorbed. In the present study, the content of water-soluble proteins in the krill meal was not high. This could be because the extract component was dissolved and lost when it thawed. The essential amino acid contents, except tryptophan and valine, of the whole fish body were affected by the feed ingredients. This finding suggests that the amino acid balance of krill meal makes it more suitable than squid meal for PBT juveniles.

Krill is rich in astaxanthin, which has antioxidant activity, whereas squid meal is low in astaxanthin [47]. Although the content of astaxanthin was not measured in the present study, it is expected that the krill meal-based diet contains astaxanthin, which is considered to be effective in preventing the oxidation of long-chain polyunsaturated fatty acids. The fatty acid 18:1n−9 was identified as the major monounsaturated fatty acid (MUFA) and was significantly higher in the heated treatment (HS and HK) and PF groups. The most abundant saturated fatty acid (SAFA) in the present study was 16:0, which is considered to be a predominant source of potential metabolic energy in fish during growth [48], and it is the predominant SAFA in the main live feed source of cultured tuna [48,49]. The ratio of DHA/EPA in raw feed for PBT was reported to exceed 2.0 [50], and the DHA/EPA ratio of

PF used in the present study was 4.0. Further, Seoka et al. (2008) [51] reported that PBT juveniles fed yolk-sac larvae of striped beakfish *Oplegnathus fasciatus* with a DHA/EPA ratio of 3.6 show better growth performance than those fed a formulated diet with a DHA/EPA ratio of 1.7. It has also been considered that the dietary DHA/EPA ratio affects the growth performance of marine fish larvae and juveniles, and a ratio of at least 1.0 is appropriate for normal growth [52,53]. Although the DHA/EPA ratio of the test diet used in the present study exceeded 1.0, the growth in the test group was lower than that of the PF group (4.0).

Although the present study clearly demonstrated that non-heated meal is a suitable protein source for formulated diet for PBT juveniles, overall survival of the PBT fed the formulated diets was not very high (25.8–52.6%). Considering high mortality of hatchery-raised PBT juvenile is the one of the biggest bottleneck of mass production of the PBT juveniles, further improvement of the feed performance of the formulated diet for PBT is desired.

5. Conclusions

In conclusion, it is important to understand whether the selection of a feed protein source and its heat treatment would improve the nutritional status and thus growth of PBT juveniles. The present study showed that heat treatment of the protein source in feed adversely affected growth performance and survival of PBT juveniles. Furthermore, the influence of heat treatment was more remarkable in squid meal than in krill meal. Taken together, the results of the present study suggest that protein source as well as heat treatment may influence the nutritional status of cultured juvenile PBT, and this should be taken into consideration for the management of seed production.

Author Contributions: J.-H.C.: conceptualization, methodology, data curation, writing—original draft, and visualization. T.K.: conceptualization, methodology, investigation, validation. Y.H.: Conceptualization, supervision, project administration, funding acquisition, writing, review, and editing. Y.K.: Investigation, validation, resources. A.I.: Conceptualization, funding acquisition, project administration. S.S.: Conceptualization, funding acquisition, project administration. All authors have read and agreed to the published version of the manuscript.

Funding: This research was funded by the Maruha Nichiro Co. and Fisheries Agency Japan.

Institutional Review Board Statement: All activities related to animal ethical considerations, such as anesthesia, dissection, and euthanasia, were conducted according to Handling Rules for Animal Experiments, etc., Tokyo University of Marine Science and Technology (13 March 2020, TUMSAT Regulations No. 8) based on Basic Guidelines for Conducting Animal Experiments at Research Institutes, etc. (Ministry of Education, Culture, Sports, Science and Technology), Act on Welfare and Management of Animals (Act No. 105), Guidelines for the Proper Implementation of Animal Experiments (Science Council of Japan), and Guidelines on How to Dispose of Animals (Prime Minister's Office).

Data Availability Statement: Not applicable.

Acknowledgments: The expense of this work was partly supported part by a project on the development of production techniques for healthy PBT juveniles from Maruha Nichiro Co. and a grant for Technological Development for Selection and Secure Stock of Brood stock for Culture of Bluefin Tuna from the Fisheries Agency.

Conflicts of Interest: The authors have no declaration of interest.

References

1. FAO. *Food and Agriculture Organization, Year Book of Fishery Statistics Summary Tables;* Sales and Marketing Group, Information Division, Food and Agriculture Organization: Rome, Italy, 2019; Available online: http://www.fao.org/3/ca5495t/CA5495T.pdf (accessed on 14 April 2021).
2. MAFF (Ministry of Agriculture, Forestry and Fisheries). *Statistics Division Annual Report of Fisheries and Aquaculture Industry Statistics;* Chiyoda: Tokyo, Japan, 2019. (In Japanese)
3. JFA (Japanese Fisheries Agency). Marino-Forum 21 General Incorporated Association. 2015. Available online: http://www.yousyokugyojyou.net/index4.htm (accessed on 14 April 2021).

4. Bransden, M.P.; Carter, C.G.; Nichols, P.D. Replacement of Fish Oil with Sunflower Oil in Feeds for Atlantic Salmon (*Salmo salar* L.): Effect on Growth Performance, Tissue Fatty Acid Composition and Disease Resistance. *Comp. Biochem. Physiol. B Biochem. Mol. Biol.* **2003**, *135*, 611–625. [CrossRef]
5. FAO. *Food and Agriculture Organization, The State of World Fisheries and Aquaculture*; Contributing to Food Security and Nutrition for All, Food and Agriculture Organization: Rome, Italy, 2016; Available online: http://www.fao.org/3/i5555e/i5555e.pdf (accessed on 14 April 2021).
6. Glencross, B.D.; Booth, M.; Allan, G.L. A Feed Is Only as Good as Its Ingredients—A Review of Ingredient Evaluation Strategies for Aquaculture Feeds. *Aquacult. Nutr.* **2007**, *13*, 17–34. [CrossRef]
7. Storebakken, T.; Refstie, S.; Ruyter, B. Soy products as fat and protein sources in fish feeds for intensive aquaculture. In *Soy in Animal Nutrition*; Drackly, J.K., Ed.; Federation of Animal Science Societies Savoy: Champaign, IL, USA, 2000; pp. 127–170.
8. Barrows, F.; Gaylord, T.; Stone, D.; Smith, C. Effect of Protein Source and Nutrient Density on Growth Efficiency, Histology and Plasma Amino Acid Concentration of Rainbow Trout (*Oncorhynchus mykiss* Walbaum). *Aquacult. Res.* **2007**, *38*, 1747–1758. [CrossRef]
9. Gatlin, D.; Barrows, F.; Brown, P.; Dabrowski, K.; Gaylord, T.; Hardy, R.; Herman, E.; Hu, G.; Krogdahl, A.; Nelson, R.; et al. Expanding the Utilization of Sustainable Plant Products in Aquafeeds: A Review. *Aquacult. Res.* **2007**, *38*, 551–579. [CrossRef]
10. Oliva-Teles, A.; Enes, P.; Peres, H. Replacing Fish Meal and Fish Oil in Industrial Aquafeed. In *Feed and Feeding Practices in Aquaculture*; Davis, A., Ed.; Woodhead Publishing Limited: Waltham, MA, USA, 2015; pp. 203–233.
11. Wilson, R.P. Amino Acids and Proteins. In *Fish Nutrition*, 3rd ed.; Halver, J.E., Hardy, R.W., Eds.; Academic Press Inc.: San Diego, CA, USA, 2002; pp. 143–179. [CrossRef]
12. Floreto, E.A.T.; Bayer, R.C.; Brown, P.B. The Effects of Soybean-Based Diets, With and Without Amino Acid Supplementation, on Growth and Biochemical Composition of Juvenile American Lobster, *Homarus americanus*. *Aquaculture* **2000**, *189*, 211–235. [CrossRef]
13. Francis, G.; Makkar, H.P.S.; Becker, K. Antinutritional Factors Present in Plant-Derived Alternate Fish Feed Ingredients and Their Effects in Fish. *Aquaculture* **2001**, *199*, 197–227. [CrossRef]
14. Ostaszewska, T.; Dabrowski, K.; Palacios, M.E.; Olejniczak, M.; Wieczorek, M. Growth and Morphological Changes in the Digestive Tract of Rainbow Trout (*Oncorhynchus mykiss*) and Pacu (*Piaractus mesopotamicus*) Due to Casein Replacement with Soybean Proteins. *Aquaculture* **2005**, *245*, 273–286. [CrossRef]
15. Krogdahl, Å.; Penn, M.; Thorsen, J.; Refstie, S.; Bakke, A.M. Important Antinutrients in Plant Feedstuffs for Aquaculture: An Update to Recent Findings Regarding Responses in Salmonids. *Aquacult. Res.* **2010**, *41*, 333–344. [CrossRef]
16. Komaki, Y.; Matsue, Y. Ecological Studies on the *Euphausia cea* Distributed in the Japan Sea. *Rep. Coop. Surv. Warm Tsushima Curr. Relat. Waters* **1958**, *2*, 146–159. (In Japanese)
17. Nicol, S.; Endo, Y. Krill Fisheries: Development, Management and Ecosystem Implications. *Aquat. Living Resour.* **1999**, *12*, 105–120. [CrossRef]
18. Rehbein, H. Amino Acid Composition and Pepsin Digestibility of Krill Meal. *J. Agric. Food Chem.* **1981**, *29*, 682–684. [CrossRef]
19. Van der Meeren, T.; Olsen, R.E.; Hamre, K.; Fyhn, H.J. Biochemical Composition of Copepods for Evaluation of Feed Quality in Production of Juvenile Marine Fish. *Aquaculture* **2008**, *274*, 375–397. [CrossRef]
20. Everson, I. *Krill: Biology, Ecology, and Fisheries*; Blackwell Science Publishing: Oxford, UK, 2008.
21. Hansen, J.Ø.; Shearer, K.D.; Øverland, M.; Penn, M.H.; Krogdahl, Å.; Mydland, L.T.; Storebakken, T. Replacement of LT Fish Meal with a Mixture of Partially Deshelled Krill Meal and Pea Protein Concentrates in Diets for Atlantic Salmon (*Salmo salar*). *Aquaculture* **2011**, *315*, 275–282. [CrossRef]
22. Tibbetts, S.M.; Olsen, R.E.; Lall, S.P. Effects of Partial or Total Replacement of Fish Meal with Freeze-Dried Krill (*Euphausia superba*) on Growth and Nutrient Utilization of Juvenile Atlantic Cod (*Gadus morhua*) and Atlantic Halibut (*Hippoglossus hippoglossus*) Fed the Same Practical Diets. *Aquac. Nutr.* **2011**, *17*, 287–303. [CrossRef]
23. Lu, H.F.S.; Bruheim, I.; Jacobsen, C. Oxidative Stability and Non-Enzymatic Browning Reactions in Antarctic Krill Oil (*Euphausia superba*). *Lipid Technol.* **2014**, *26*, 111–114. [CrossRef]
24. Watanabe, T.; Arakawa, T.; Kitajima, C.; Fujita, S. Effect of Nutritional Quality of Broodstock Diets on Reproduction of Red Seabream. *Nippon Suisan Gakkaishi* **1984**, *50*, 495–501. [CrossRef]
25. Watanabe, T.; Itoh, A.; Murakami, A.; Tsukashima, Y.; Kitajima, C.; Fujita, S. Effect of Nutritional Quality of Diets Given to Broodstock on the Verge of Spawning on Reproduction of Red Seabream. *Nippon Suisan Gakkaishi* **1984**, *50*, 1023–1028. [CrossRef]
26. Cho, S.H.; Lee, S.M.; Lee, S.M.; Park, B.H.; Park, I.S.; Choi, C.Y.; Min, B.H.; Hur, S.B.; Jo, J.Y. Effect of Partial Replacement of Fish Meal with Squid Liver meal™ in the Diet on Growth and Body Composition of Juvenile Olive Flounder (*Paralichthys olivaceus*) during Winter Season. *Fish. Aquat. Sci.* **2005**, *8*, 65–69. [CrossRef]
27. Seoka, M.; Bandou, T.; Suimon, M.; Sawada, Y. Toyama Squid *Watasenia scintillans* Meal as a Protein Source in Artificial Diet for Juvenile Pacific Bluefin Tuna *Thunnus orientalis*. *Aquac. Sci.* **2010**, *58*, 143–144. [CrossRef]
28. Cho, J.H.; Haga, Y.; Masuda, R.; Satoh, S. Periodic Changes in the Growth Performance and Biochemical Composition of Juvenile Red Sea Bream *Pagrus major* Fed Non-Heated and Heated Squid and Krill Meal-Based Diets. *Fish. Sci.* **2018**, *84*, 699–713. [CrossRef]
29. Tonheim, S.K.; Nordgreen, A.; Høgøy, I.; Hamre, K.; Rønnestad, I. In Vitro Digestibility of Water-Soluble and Water-Insoluble Protein Fractions of Some Common Fish Larvae Feeds and Feed Ingredients. *Aquaculture* **2007**, *262*, 426–435. [CrossRef]

30. Ishibashi, Y.; Izumi, T.; Kurata, M.; Okada, T. Effects of the Tank Wall Pattern on the Survival Rate, Bone Injury, and Stress Response of Pacific Bluefin Tuna, *Thunnus orientalis*. *Aquac. Eng.* **2013**, *56*, 13–17. [CrossRef]
31. AOAC. *Official Methods of Analysis of AOAC International*, 17th ed.; AOAC: Gaithersburg, MD, USA, 2000.
32. De Schrijver, R.; Ollevier, F. Protein Digestion in Juvenile Turbot (*Scophthalmus maximus*) and Effects of Dietary Administration of *Vibrio proteolyticus*. *Aquaculture* **2000**, *186*, 107–116. [CrossRef]
33. Folch, J.; Lees, M.; Sloan Stanley, G.H. A Simple Method for Isolation and Purification of Total Lipids from Animal Tissues. *J. Biol. Chem.* **1957**, *226*, 497–509. [CrossRef]
34. Morrison, W.R.; Smith, L.M. Preparation of Fatty Acid Methyl Esters and Dimethylacetals from Lipids with Boron Fluoride-Methanol. *J. Lipid Res.* **1964**, *5*, 600–608. [CrossRef]
35. Boonyoung, S.; Haga, Y.; Satoh, S. Preliminary Study on Effects of Methionine Hydroxyl Analog and Taurine Supplementation in a Soy Protein Concentrate-Based Diet on the Biological Performance and Amino Acid Composition of Rainbow Trout *Oncorhynchus mykiss* (Wallbaum). *Aquac. Res.* **2012**, *44*, 1339–1347. [CrossRef]
36. Kvåle, A.; Harboe, T.; Mangor-jensen, A.; Hamre, K. Effects of Protein Hydrolysate in Weaning Diets for Atlantic Cod (*Gadus morhua* L.) and Atlantic Halibut (*Hippoglossus hippoglossus* L.). *Aquac. Nutr.* **2009**, *15*, 218–227. [CrossRef]
37. Haga, Y.; Naiki, T.; Tazaki, Y.; Takebe, T.; Kumon, K.; Tanaka, Y.; Shiozawa, S.; Nakamura, T.; Ishida, S.; Ide, K.; et al. Effect of Feeding Microdiet and Yolk-Sac Larvae of Spangled Emperor *Lethrinus nebulosus* at Different Ages on Survival and Growth of Pacific Bluefin Tuna *Thunnus orientalis* Larvae. *Aquac. Sci.* **2010**, *58*, 491–499. [CrossRef]
38. Cho, J.H.; Haga, Y.; Kamimura, Y.; Akawaza, A.; Itoh, A.; Satoh, S. Production Performance of Pacific Bluefin Tuna *Thunnus orientalis* Larvae and Juveniles Fed Commercial Diets and Effects of Switching Diets. *Aquac. Sci.* **2016**, *64*, 359–370. [CrossRef]
39. Cho, J.H.; Haga, Y.; Kamimura, Y.; Itoh, A.; Satoh, S. Peruvian Fish Meal Has Comparative Potential to Enzyme-Treated Chilean Fish Meal as Protein Source of Diet for Larvae and Juvenile Pacific Bluefin Tuna *Thunnus orientalis*. *Fish. Sci.* **2022**, *186*, 107–116. [CrossRef]
40. Watanabe, Y. Intracellular Digestion of Horseradish Peroxidase by the Intestinal Cells of Teleost Larvae and Juveniles. *Bull. Jpn. Soc. Sci. Fish.* **1982**, *48*, 37–42. [CrossRef]
41. Fyhn, H.J.; Serigstad, B. Free Amino Acids as Energy Substrate in Developing Eggs and Larvae of the Cod *Gudas morhua*. *Mar. Biol.* **1987**, *96*, 335–341. [CrossRef]
42. Fyhn, H.J. First Feeding of Marine Fish Larvae: Are Free Amino Acids the Source of Energy? *Aquaculture* **1989**, *80*, 111–120. [CrossRef]
43. Satoh, K. Studies on Improvement of Composed Diet for Yellowtail Culture. *Bull. Oita Inst. Mar. Fish. Sci.* **2005**, *6*, 19–77, (In Japanese with English Abstract).
44. Kohbara, J.; Miyazaki, T.; Takii, K.; Hosokawa, H.; Ukawa, M.; Kumai, H. Gustatory Responses in Pacific Bluefin Tuna *Thunnus orientalis* (Temminck and Schlegel). *Aquac. Res.* **2006**, *37*, 847–854. [CrossRef]
45. Kristinsson, H.G.; Rasco, B.A. Fish Protein Hydrolysates: Production, Biochemical, and Functional Properties. *Crit. Rev. Food Sci. Nutr.* **2000**, *40*, 43–81. [CrossRef]
46. Dohmoto, N. Antarctic Krill and Its Function for Human Health. *Nippon Suisan Gakkaishi* **2002**, *68*, 714–718. [CrossRef]
47. Sánchez, C.A.O.; Zavaleta, E.B.; García, G.R.U.; Solano, G.L.; Díaz, M.P.R. Krill Oil Microencapsulation: Antioxidant Activity, Astaxanthin Retention, Encapsulation Efficiency, Fatty Acids Profile, In Vitro Bioaccessibility and Storage Stability. *LWT Food Sci. Technol.* **2021**, *147*, 111476. [CrossRef]
48. Huynh, M.D.; Kitts, D.D.; Hu, C.; Trites, A.W. Comparison of Fatty Acid Profiles of Spawning and Non-Spawning Pacific Herring, *Clupea harengus pallasi*. *Comp. Biochem. Physiol. B Biochem. Mol. Biol.* **2007**, *146*, 504–511. [CrossRef]
49. Zlatanos, S.; Laskaridis, K. Seasonal Variation in the Fatty Acid Composition of Three Mediterranean Fish—Sardine (*Sardina pilchardus*), Anchovy (*Engraulis incrasicholus*) and Picarel (*Spicara smaris*). *Food Chem.* **2007**, *103*, 725–728. [CrossRef]
50. Bell, J.G.; McEvoy, L.A.; Estevez, A.; Shields, R.J.; Sargent, J.R. Optimising Lipid Nutrition in First-Feeding Flatfish Larvae. *Aquaculture* **2003**, *227*, 211–220. [CrossRef]
51. Seoka, M.; Kurata, M.; Tamagawa, R.; Biswas, A.K.; Biswas, B.K.; Seok Kian Yong, A.S.K.; Kim, Y.-S.; Ji, S.-C.; Takii, K.; Kumai, H. Dietary Supplementation of Salmon Roe Phospholipid Enhances the Growth and Survival of Pacific Bluefin Tuna *Thunnus orientalis* Larvae and Juveniles. *Aquaculture* **2008**, *275*, 225–234. [CrossRef]
52. Tocher, D.R.; Sargent, J.R. Analysis of Lipids and Fatty Acids in Ripe Roes from Some Northwest European Marine Fish. *Lipids* **1984**, *19*, 492–499. [CrossRef]
53. Drillet, G.; Jørgensen, N.O.G.; Sørensen, T.F.; Ramløv, H.; Hansen, B.W. Biochemical and Technical Observations Supporting the Use of Copepods as Live Feed Organisms in Marine Larviculture. *Aquac. Res.* **2006**, *37*, 756–772. [CrossRef]

fishes

Review

Success of Aquaculture Industry with New Insights of Using Insects as Feed: A Review

Amna Hameed [1], Waqar Majeed [1,*], Muhammad Naveed [1], Uzma Ramzan [2,*], Matteo Bordiga [3], Maryam Hameed [1], Saud Ur Rehman [4] and Naureen Rana [1]

1 Department of Zoology, Wildlife and Fisheries, University of Agriculture, Faisalabad 38000, Pakistan
2 Institute of Zoology, University of the Punjab, Lahore 54000, Pakistan
3 Department of Pharmaceutical Sciences, Università degli Studi del Piemonte Orientale "A. Avogadro", Largo Donegani 2, 28100 Novara, Italy
4 Department of Animal Sciences, Purdue University, West Lafyette, IN 47906, USA
* Correspondence: waqar.majeed@uaf.edu.pk (W.M.); uzma.phd.zool@pu.edu.pk (U.R.)

Citation: Hameed, A.; Majeed, W.; Naveed, M.; Ramzan, U.; Bordiga, M.; Hameed, M.; Ur Rehman, S.; Rana, N. Success of Aquaculture Industry with New Insights of Using Insects as Feed: A Review. *Fishes* **2022**, *7*, 395. https://doi.org/10.3390/fishes7060395

Academic Editors: Marina Paolucci and Shunsuke Koshio

Received: 1 November 2022
Accepted: 8 December 2022
Published: 17 December 2022

Abstract: Most of world's fish and seafood are produced by aquaculture, which is one of the biggest contributors to the world's food security. The substantial increase in prices of conventional feed ingredients and the over-exploitation of natural resources are some of the biggest constraints to aquaculture production. To overcome this stress, different approaches and techniques are used, among which the use of non-conventional feed ingredients in the aquaculture sector is the most recent approach. Different non-conventional feed ingredients such as plant-based products, algae (both micro and macroalgae), single-cell protein (bacteria and yeast), and insect meal are currently used in aquaculture for sustainable food production. Amongst all these novel ingredients, insects have greater potential to replace fishmeal. The existence of about 1.3 billion tons of food and agriculture waste from the food chain supply poses a serious environmental threat. Insects are tiny creatures that can thrive on organic waste and thus can convert the waste to wealth by the bioconversion and nutritional upcycling of organic waste. Insects have the potential to recover nutrients from waste aquaculture products, and many fish species feed on insects naturally. Therefore, employing insects in the aquaculture sector to replace fishmeal is an eco-friendly approach. The present review briefly highlights emerging non-conventional feed ingredients, with special attention given to insects. The current review also focuses on the nutritional value of insects, factors affecting the nutritional value of insects, potential insects that can be employed in the aquaculture sector, the physiological response of fish when fed with insect meal, techno-functional properties of insect meal, and emerging approaches for addressing possible downsides of employing insect meal in fish diets. Finally, it suggests avenues for further research into these inventive fishmeal replacements.

Keywords: feed; insects; aquaculture; innovation; sustainability

1. Background

The largest sector of food production is aquaculture, which has the potential to contribute to sustainable food production in the future. In 2018, the aquaculture sector, which is expanding, produced 114.5 million metric tons of live weight [1]. By supplying the world's expanding population with high-quality food, aquaculture is essential to ensuring food security [2]. Making sure that everyone has access to enough food at all times is known as food security. While the global population is growing by 1.6 percent per year, the demand for seafood is rising at a rate of 3.1% annually. At a rate of 2.1 percent per year, the aquaculture industry is expanding more quickly than other areas of animal production, such as livestock [1]. This might be because the aquaculture and fisheries sectors generate more economy through production, selling, and marketing [3,4].

Despite the fact that aquaculture has many advantages for the availability of food, its sustainability depends on a variety of factors, including knowledge, experience, the regulatory environment, the ecological conditions for the growth of cultured species, and the presence of a sizable market for the sale of aquaculture products [5]. Currently, 50% of world fish production is generated from aquaculture, and it is one of the biggest suppliers of high-quality protein in terms of quantity and quality [6]. Organisms from terrestrial and aquatic environments provide about 43% of the world's protein and play a key role in preventing malnutrition, especially in developing nations [1]. Aquaculture production, like other types of terrestrial farming, is fully dependent on the availability of nutrients as feed [6].

Fishmeal has frequently been replaced in aquaculture with plant proteins. Previous research has shown that plant-based diets are unfavorable for fish growth and health, and this could be because of the presence of anti-nutritional substances (ANFs) [7–9]. Among plant-based diets, soybean is considered the best to replace fishmeal due to its cost-effectiveness and nutritional value [10]. However, the addition of high quantities of soybean meal in the fish diet has been shown to negatively impact growth, gut and liver integrity, intestinal microbiota composition, and immunological response in various carnivorous fish species [11–13]. When the salmonids were fed with SBM-based diets, severe gut health impairment with enteritis and increased inflammatory influx was observed in fish [13]. The researchers were therefore forced to create new aquafeeds with innovative feed ingredients that can replace fishmeal and minimize the detrimental effect caused by using vegetable protein because they were aware of the negative effects of utilizing a plant-based diet, even at a low inclusion level [14–16].

1.1. *Emergent Non-Conventional Feed Ingredients*

The aquaculture sector is one of the major contributors to food security worldwide [2], but the increased food demand has compelled scientists to look for novel and non-conventional feed ingredients such as algae, land animal by-products, single-cell proteins, and insects [17].

Bacteria and yeast have the highest potential to replace fishmeal due to their good nutritional value. Bacteria and yeast have good AA profiles comparable to fishmeal and the protein content ranges from 50–80% and 45–55%, respectively. They also have great potential to be utilized as raw materials and functional feed additives [18]. Microalgae have good nutritional profiles, which have emerged as good candidates for use in aquaculture [19,20]. Algae can promote growth, stimulate the immune system, and enhance fish color, in addition to utilizing waste-derived nutrients such as nitrate, nitrite, and TAN (total ammonia nitrogen) [21]. Yadav et al. [22] conducted a study to evaluate the effect of dried chlorella species on the survival and growth of common carp (*Cyprinus carpio*). The fish fed with the microalgal biomass showed a great improvement in weight gain, lower FCR, and protein efficiency ratio [22]. A cost/benefit analysis showed a significant reduction in the cost of formulated feed compared to commercial feed, which can enhance the profit margin of fish farmers. This research highlights the importance of algae as a novel non-conventional feed ingredient. Using this strategy, the waste carbon dioxide can be further utilized to formulate alternative feed supplements.

Biofloc production is a reasonable strategy to reduce environmental impacts by contributing low emissions. The Biofloc production and efficiency depend upon different factors, such as the cultured species [23,24]. Debbarma et al. [25] reported Biofloc production's importance for managing aquaculture nitrogen wastes. The Biofloc production system showed the best performance in terms of growth, condition factor, digestive enzyme activity, and survival. The production cost is one of the biggest causes of the limitation of the applicability of SCP and other newly emerged non-conventional feed ingredients in aquafeed [26,27].

Designing the new, highly effective aquafeed diets by combining plant- and animal-based diets (insect meal/poultry by-product meal) may be a revolutionary technique [12]. Poultry by-product meal (PBM) is a cost-effective and sustainable protein source that is rich in essential amino acids and has high nutrient digestibility [28]. Different studies have been conducted to replace the vegetable-protein-based diet with an animal-based diet (insect meal/poultry by-product meal). Intestinal histological changes and inflammatory responses were reduced in gilthead seabream due to this novel strategy when the amount of vegetable protein in the fish diet was decreased. Incorporating insect meal into the diet of seabream had a favorable impact on molecular inflammatory markers [12]. Poultry by-product meal positively improved lipid absorption in the seabream diet. Zarantoniello et al. [29] examined the effect of conventional and novel feed additives, which included Louisiana red claw crayfish (*Procambarus clarkii*) meal (RCM) and dried microbial biomass from *Tetraselmis suecica* (TS) and *Artrhospira platensis* (AP), on rainbow trout (*Oncorhynchus mykiss*). The fish's development was unaffected by the vegetable protein sources, including 0.25 percent of conventional feed additives, and AP-based diets, but their severely compromised gut health condition marginally improved. The fish's general welfare and intestinal integrity were preserved by RCM and TS-based diets, but the fish's growth performance was poor. This might be because these diets contain certain biogenic amines, which could negatively affect nutrient uptake. Gaudioso et al. [11] evaluated the effect of poultry by-product meal and *H. illucens* meal on gut microbiota inflammatory and immune markers of fish when substituted with vegetal protein. When poultry by-product meal was added to vegetal protein meal (VM), the growth performance and microbiota composition were maintained as in FM. Meals made from insect and poultry by-products demonstrated better fish health and were recommended as the ideal replacement for plant-based proteins. In the case of insect meal, the aquaculture industry faced some cost-effectiveness issues. Therefore, it was recommended to use a combination of insect meal and poultry by-product meal for better fish growth performance. Another study conducted by Pulido-Rodriguez et al. [30] showed the positive effect on growth performance of red swamp crayfish when replacing vegetal protein with insect meal by-product. However, *Tisochrisis lutea* and *Tetraselmis suecica* (marine algae) were added at a low inclusion level, which badly affected the fish growth performance and increased the FCR value.

A feed with balanced nutrients is vital for improving fish growth and development. By 2020, fishmeal production was expected to rise to 70,969 thousand tons globally, a nearly 10-fold increase from 1995 [31]. The fishmeal ingredients used to prepare feeds are under a lot of stress as a result of these expanding numbers. According to studies, insects may be a feasible alternative to fishmeal in the near future, as they are a good source of macromolecules and micromolecules [32–35]. The substitution of fishmeal with insect meal is a unique concept in aquaculture. However, aquaculture practices currently employ a diverse range of insect species. This category of insects includes maggots (*Musca domestica*), black soldier flies (*H. illucens*), silkworm pupae (*Bombyx mori*), grasshoppers, and mealworms (*Tenebrio molitor*).

A significant effect has been observed in the growth performance of fish when fish are fed an insect-based meal. Tran et al. [36] reported that when aquatic animals were fed with *Tenebrio molitor* and pupal full-fat silkworm *Bombyx mori* they showed a substantial increase in Hedges' g value for SGR with the reduction in FCR. The growth performance of aquatic animals fed with insects vary according to the stage and species of both the insects and aquatic animals. It is documented in many studies that when fishmeal is substituted with more than 30% insect meal, a negative effect on growth performance is observed [37,38].

The nutritional profile of insects has a major role in deciding the substitution level of fishmeal. The protein part of insects' diets ranges from 50–82% (dry matter basis), comparable to fishmeal ranging from 60–72% by weight. Insect meals are rich with essential amino acids, but different species have observed minor variations depending upon their orders. Hydroxyproline and taurine are unique ingredients in insects, absent in a plant-based diet but present in fishmeal [39]. The insect species are enriched with saturated fatty acids and less in the concentration of PUFAs. On average, the n-6/n-3 ratios in terrestrial insects are three times higher than in aquatic insects [40,41]. Fishmeal and fish oil are the main components of fish feed.

To preserve the sustainability of food, new protein substitutes must be used in replacement of fishmeal and fish oil due to their scarcity [42]. Although there is literature on the nutritional value of insects, there is a lack of information for a thorough assessment of the factors influencing that nutritional value. The challenges with employing insects as a feed for aquaculture to address fishmeal scarcity are discussed in this review.

1.2. Ecological Advantages of Using Insects in Fishmeal Diet

Environmental issues such as the overuse of natural resources, loss of biodiversity, and rising pollution levels directly affect global climatic conditions. This fact is evident from the scientific reports and data published in the last decades [43]. Different strategies and approaches have been applied to combating environmental problems, among which the concept of bioeconomy is the most prominent. The concept of bioeconomy focuses on utilizing food waste or algae [44].

As the world population increases rapidly, there is a need for a constant supply of protein with reduced environmental impacts [45]. Recent research suggested that insect-based meals could positively impact the environment by selecting a suitable diet. Incorporating yellow mealworms into the diet of rainbow trout decreased the net primary production use. Still, it did not lower the land use, global warming potential (GWP), eutrophication, and energy demand [46]. However, different results have been achieved in the case of *H. illucens*. When *H. illucens* is added to the arctic char diet, it reduces the environmental impacts by reducing abiotic depletion, GWP, acidification, and land use [47].

1.3. Waste to Wealth Concept

Agro-industrial waste's continuous production and proper disposal are major environmental issues. Still, the improved nutritional value of these agro-industrial wastes is paving the way for these wastes to be valorized in animal/fish feed [48]. Jayan et al. [49] proposed this concept by valorizing castor cake or meal by standard processes. The growth and nutrient utilization indices of *L. rohita* were significantly improved when fed with castor oil protein isolates. The biodegradation of plastic is also one of the biggest environmental problems. Using insects in the biodegradation of plastic is a novel approach. *Tenebrio molitor* has ability to biodegrade plastic waste. Both forms of *Tenebrio molitor* larvae and imago are plastic eaters. Bulak et al., 2021, also reported that *Tenebrio molitor* could actively degrade plastic waste.

Aquafeed has a finite shelf-life and is easily subjected to rancidity due to the peroxidation of lipids [50]. The oxidated lipids have a detrimental effect on the feed quality, reducing the palatability and nutritional value of feed [51]. Several insect species have shown the potential to recover the nutrients from expired fish feed. The black soldier fly has great potential to recover nutrients from expired fish feed [52].

Moreover, the black soldier fly is also considered an ideal candidate for converting waste into protein-rich biomass that can be used as animal feed [53,54]. This method relies on the black soldier fly larvae to feed on the wastes, which will reduce it to 50–80% (wet weight) residue and yield useable and high-quality biomass at a rate of 20–30% (solid basis) [55]. Black soldier fly larvae fed on waste are a good source of protein and are used as feed for fish [39], and the residue can be used as fertilizer for soil [56].

Insects are simple to grow and can be produced in large quantities with little need for water or land [39,57]. They grow quickly due to their fast reproduction rate [53,58] and have high feed conversion efficiency [57,58]. Insects can be grown on organic food waste, and they can convert this waste into protein and fat-rich biomass [59]. Additionally, this approach reduces environmental problems related to the reuse of food waste [53].

2. Nutritional Value of Insects

There are around one million insect species around the globe, and they are essential to the integrity of food chains and ecosystems. Although the majority of Europeans still view insects as harmful pests, several insect species are used as food in 100 different countries [33,60]. Insects are the largest class of arthropods and have been used as a food source for humans throughout history. Numerous insect species can be used as food, but only a few have been domesticated, such as silkworms and honeybees [61]. Insects, a high protein source, are also thought to be an excellent substitute for fishmeal for many freshwater and marine fish as aquaculture production rises day by day. Due to their adequate protein content, amino acid composition, and digestibility, animal by-product meals (AM) can be used in aquaculture instead of fishmeal [62].

Among the plant proteins, soybean meal is one of the most logical alternatives for high-quality fishmeal due to its high protein value, healthy amino acid profile, and low price [63]. Plant-based protein diets have been extensively studied in the past few years as a protein source for aquafeed formulations [64–66]. Plant-based diets are more economical as compared to animal proteins, but they have many limitations that make their substitution with fishmeal questionable, such as the absence of some vital amino acids, low palatability, the presence of anti-nutritional factors, and competitive prices with other food production sectors [66–68]. Additionally, essential components such as taurine and hydroxyproline which are rich in fishmeal are lacking in plant-based substances [69,70], whereas a variety of insect species contain high levels of taurine [71]. Taurine is the essential ingredient for the growth of brood stock and juvenile fish. A taurine-deficient diet results in increased demand for vitamin E and C, particularly in the larvae of marine water fish [72]. The insufficiency of taurine may cause serious psychological abnormalities, green liver disorder, and stunted growth [73]. Fish's principal physiological and behavioral responses are associated with taurine availability in the diet, including the survival rate [74], growth, immune response, and antioxidant activity [71].

As a result, interest in insect meal (IM) as a feasible feed alternative in fish aquaculture has developed dramatically. Insects are a rich source of energy due to macronutrients and micronutrients in sufficient amounts, making them the best suitable fish food [57,75]. It has been discovered that insects contain an adequate level of crude protein similar to soybean meal but slightly less than fishmeal by comparing the nutritional value of several insect orders, soybean meal, and fishmeal. The quality of protein depends upon the presence of essential amino acids, so fishmeal is used as an aquatic feed due to its healthy amino acid profile. Fishmeal is considered a source of high-quality protein because it contains a high amount of digestible essential amino acids, including lysine and methionine, which are thought to be absent in grains, the primary source of animal feed [76]. The amino acid composition of many insect species is currently unknown. With an adequate amount of methionine and lysine, Diptera's amino acid profile is comparable to fishmeal. Some insect species even have a higher amount of EAA than fishmeal, such as histidine and threonine in Diptera and leucine in Coleoptera and Orthopterans. Insects have a high level of polyunsaturated fatty acid n-6, lower than soybean meal but higher than fishmeal [77,78]. One critical point is that nutritional composition varies among the species. Their nutritional composition also varies according to order, diet, and developmental stage [79], as shown in Figure 1.

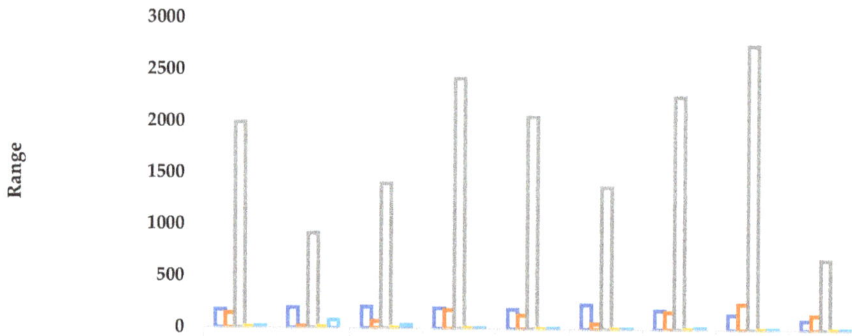

	Black soldier fly	House fly (A)	House cricket (A)	Super worm (A)	Mealworm (L)	Mealworm (A)	Giant mealworm (L)	Waxworm (L)	Silkworm (L)
Protein (g/kg)	175	197	205	197	187	237	184	141	93
Fat (g/kg)	140	19	68	177	134	54	168	249	144
Calories (kcal/kg)	1994	918	1402	2423	2056	1378	2252	2747	674
Thiamin (mg/kg)	7.7	11.3	0.4	0.6	2.4	1	1.2	2.3	3.3
Riboflavin (mg/kg)	16.2	77.2	34.1	7.5	8.1	8.5	16.1	7.3	9.4

Figure 1. Nutritional value of potential insects used as fish feed. Reproduce from sources [80,81] (L = larvae; A = adult).

2.1. Proteins

Insects are an excellent source of protein since they contain a good proportion of amino acids; their levels of protein range from 25 to 75 percent (DM) [82–85]. A protein factor (Kp) of 6.2 can be used to calculate crude protein content by multiplying nitrogen content by protein. In order to fulfill fish nutritional demands, we advise using Kp modification based on the quantity of amino acids when calculating insect-containing practical feed compositions [86]. Hung et al. [87] studied that increased protein content shown among the treatments does not seem to be important for the observed growth performance improvement. The erroneous measurement of amino acids and amino acid loss due to methodological difficulties such as hydrolysis might result in incorrect calculations of crude protein [88]. The nitrogen of chitin is the main factor in the over-estimation of protein [89]. When insect protein is compared with plant and meat proteins, it is found that insects contain a substantial amount of good-quality protein due to balanced amino acid proportions [90,91]. Protein content varies amongst different orders due to different factors, and Orthoptera has the maximum protein values, ranging from 60 to 77% [92–94]. The crude protein ranges from 40 to 50% in Diptera, such as in the case of larvae of *Musca domestica* where CP was 46.9% and in house fly where it was 37–57% [92,95]. *Holotrichia parallela* is an edible beetle with a protein content of about 66% and an amino acid score of 87 and 100, with threonine as a limiting amino acid [96]. The protein level of *T. molitor* larvae is 46% [97], *H. illucens* 39% [98], *A. domesticus* 64–71% [99], *L. migratoria* 48–52% [100], *B. mori* 48–55% [101], termites 20–43% [102,103], beetle species 26–50% [102,104], and grasshoppers 26–45% [102,103].

It has been found from recent studies that usually an animal-based diet contains 40% of essential amino acids, but the range of essential amino acids in edible insects is about 46–96% [105]. House cricket meal contains some essential amino acids even at a higher concentration than plant- and other animal-based meals. House cricket meal also meets all the amino acid requirements recommended by the WHO (World Health Organization) [106–108]. Leucine content in Orthoptera is similar to animal-based protein but greater than soybean [106]. Another important amino acid found in greater quantities in insect proteins than in plant proteins is lysine [105]. Beef and eggs are good sources of methionine, serine, and tryptophan, which cannot be found in adequate amounts in some edible insects such as house crickets [107]. The silkworm is an edible insect containing 17 amino acids, with all essential amino acids. The content of essential amino acids is 47% in silkworm pupal protein, according to the recommendation of the WHO. In both male and female silkworm pupae, lysine was the amino acid that was most prevalent [109]. Numerous research studies have indicated that protein and AAs act as mediators of the trade-off between growth and immune protection. They are considered growth-limiting nutrients, particularly in those insects that feed on plant material.

2.2. Carbohydrates

Chitin makes up between 5% and 20% of the dry weight of insects [110,111]. It is a polysaccharide that is present in the exoskeleton of many insects in varying amounts and makes up 6.71–15.98% of carbohydrates [112]. Chitin is impermeable in a liquid medium and is the most common carbohydrate in biomass after cellulose. Carbohydrates are a rich source of nitrogen and carbon [113,114]. Generally, carbohydrates computed as nitrogen-free extract are found in minor levels in insects [83,85]. Some of the carbohydrate levels of edible insect species are described in the following; yellow mealworm larvae have a carbohydrate content that ranges from 1–7% [115], *Rhynchophorus phoenicis* larvae 5.53% [116], *R phoenicis* 20.23%, *Heteroligus meles* 20.10% [104], *Gynanisa maja* larvae 10.70%, *Ruspolia differens* 8.40%, *Macrotermes falciger* 32.80% [103], *H. illucens* 10–20% [98], *Locusta migratoria* 4–5.5% [100], and *Bombyx mori* 23% [101]. However, these levels are probably due to food left in the gastrointestinal tract.

2.3. Lipids

Researchers have realized that certain insects contain a lot of lipids. Insects can therefore be used to increase lipid intake. When 20% ground yellow mealworm larvae were used with wheat flour to generate extruded cereals, the lipid content of the final product increased from 0.9 to 5.4% [117]. Insects contain a significant concentration of unsaturated fatty acids, which must be examined during the preparation and preservation of insect meals [118]. Triacylglycerols account for approximately 80% of lipids, which is a reservoir of energy during periods of heavy energy consumption. Orthoptera, Lepidoptera, Blattodea, Isoptera, and Coleoptera have average lipid contents of 13%, 28%, 29%, 32%, and 33% dry weight, respectively [119]. Insects at larval and pupal stages have more lipids than adults [111]. The lipid content of *T. molitor* larvae is 33% [97], *H. illucens* larvae 38% [98], *L. migratoria* 19–20% [100], crickets 12–25% [120], termites 22–43% [102,103], beetle species 21–32% [102,104], *A. domesticus* 18–22% [99], *B. mori* 19% [101], and grasshoppers 3–49% [102,103].

Phospholipids are commonly found in nature and are responsible for maintaining the structure of membranes [121]. The phospholipid concentration of lipid content is about 20%; however, it changes with life stage, insect order, and species [121,122]. In insects, lipid content comprises a comparatively high concentration of C_{18} fatty acids, including oleic, linoleic, and linolenic acid [121]. The palmitic acid level is slightly higher than other lipids, and the lipid profile is greatly influenced by the diet fed by insects [90].

Insects are rich in n-6 fatty acids but deficient in n-3 fatty acid contents. However, this n-6/n-3 ratio is unsuitable for human health [123]. Therefore, different strategies are being employed to improve the fatty acid profile of insects. The content of n-3 fatty acids can be increased by using a suitable substrate; thus, the nutritional value of insects can be improved. Different factors also affect the fatty acid profile of insects and substrates, such as the rearing environment. This fact was also supported by Ruschioni et al. [124], who reported that levels of linoleic (C18:2n6) and alpha-linolenic (C18:3n3) acids varied amongst the insects even when they were reared on the same substrate. PUFAs are abundant in edible insects, commonly known as linolenic and linoleic acids [125]. However, the insects do not contain eicosapentaenoic acid (EPA) or docosahexaenoic acid (DHA), which are necessary for the health of human beings [125]. Lauric acids are one of the most important medium-chain saturated fatty acids in most insects. HI contains a substantial amount of lauric acid, which has antimicrobial properties and thus can modulate the intestinal health of host organism [126].

2.4. Vitamins

Insects have a range of vitamins that are either water-soluble or lipophilic [127,128]. A variety of insects having thiamine were listed by [90]. Its concentration varies between 0.1–4 mg per 100 g (DM). Riboflavin, also known as vitamin B_2, is found in edible insects and plays an important role in metabolic function. The yellow mealworm and the house cricket *Acheta domesticus* have high contents of vitamin B_{12}. Many other species that have been studied possess only trace quantities of this vitamin [90]. Insects often have low quantities of retinol but high levels of biotin, pantothenic acid, and riboflavin. Another important vitamin, folic acid, is identified in significant concentrations in different insects [119]. This trade-off is thought to focus on dietary nutrients, a significant idea in studying nutritional immunology [129,130].

2.5. Minerals

Edible insects are a good source of minerals such as zinc, sodium, potassium, iron, magnesium, copper, calcium, manganese, and phosphorous [131,132]. A substantial amount of iron is found in the moth *G. belina* (31–77 mg per 100 g of dry matter) and *Locusta migratoria* (8–20 mg per 100 g of dry matter) [133]. Similarly, a good quantity of zinc is found in caterpillars of mopane (14 mg per 100 g of dry matter) and *R. phoenicis* (26.5 mg per 100 g of dry matter) [90]. Hyun et al. [134] reported that edible insects such as the grasshopper *Oxya chinensis* contain very little heavy metal that is safe for human consumption. The solubility of iron was found in the following order in different insects: cricket > mealworms > grasshopper = buffalo worms > sirloin beef. Grasshoppers, crickets, and mealworms are reported to have high solubilities for Cu, Mn, and Zn. Similarly, the high solubility of Ca and Mg were reported in mealworms [135].

Kosečková et al. [136] examined the mineral content of the house cricket, yellow mealworm, desert locust, and super worm. All these insect species were reported to be good sources of iron. They reported that a good quantity of Zn, Cu, and P was found in house crickets, yellow mealworms, and desert locusts, even greater than the DRV (dietary recommended values). These insect species are all very low in Cd and Pb content, indicating that eating them poses no risk. Eating insects can increase zinc, sodium, phosphorus, potassium, calcium, iron, copper, manganese, and magnesium availability in terms of mineral nutrition [137]. Mopani or mopane, a caterpillar of moths, contains an iron concentration of 31–77 mg per 100 g DM. The ranges of different minerals (mg/kg) in *T. molitor* are Ca (349.2), P (5600–5700), Mg (1290–1416), Zn (95–101), Fe (60–64), Cu (11.4), Mn (7.0), B (20.3–24.2), and Mo (0.6), and in *B. mori* Ca (987.2), P (8550–8860), Mg (2400–2600), Zn (130–151), Fe (49–50), Cu (9.3–9.5), Mn (16.2–17.6), B (10–18.8), and Mo (0.2) [101].

2.6. Functional Properties of Insect Meals

Insect proteins have many favorable characteristics that make them fit to be used as food and feed. However, it is very important to evaluate the functional properties of insect protein before their substitution in any food and feed [138]. Depending upon insect protein's techno-functional properties, which include solubility, water and oil holding capacity, gelling, and emulsification, different food processing methods can be applied to improve the quality of insect-based food [139]. Several studies have been conducted to investigate the improvement in the functional properties of insect protein by applying suitable processing strategies [138]. Drying, defatting, and extraction are the different processing methods commonly used to improve the functional properties of edible insect proteins [140]. By enhancing the protein content, fractionation processing enhances the functionality of insect proteins [138,141–143]. Kim et al. [142] reported that the functional properties of *T. molitor*, *P. brevitarsis*, and *Allomyrina dichotoma* increased with protein content. Another important method for increasing protein functionality is enzymatic hydrolysis [138,144]. Purschke et al. [145] examined the effect of enzymatic hydrolysis on the functional properties of *L. migratoria*. They observed the different changes in functional properties of *L. migratoria* by applying hydrolysis conditions. The functional properties of *L. migratoria* (solubility, gelling, foaming, and water and oil holding capacity) were increased with the decrease in the molecular weight of protein.

There was no difference in the protein functionality, but a reduction in emulsifying activity was observed with the application of enzymatic hydrolysis [146]. Mintah et al. [143] examined the enzymatic hydrolysis of soldier flies that improved the dispersibility, turbidity, and particle size by changing the protein secondary structure. Lipids and chitin removal also improve foaming and emulsifying properties, respectively.

3. Factors Influencing the Nutritional Value of Insects

3.1. Insect's Species

An insect's species and order affect its nutritional value. The majority of the insects studied have a significant amount of protein [123], almost similar to soybean meal but inferior to fishmeal [147]. Usually, Orthoptera has a higher CP, between 50 to 55 g/100 g [148], Diptera amino acid content is analogous to fishmeal [92], and Coleoptera has a better amino acid profile than soy meal [149], although not comparable to fishmeal. The ratio of amino acids varies widely amongst the various insect orders, plant meals, and fishmeals. Histidine and threonine are more deficient in insect meals than in fishmeals out of all the essential amino acids. However, insect meals contain a sufficient level of tyrosine, lysine, and methionine compared to soy meals. Diptera has a similar amount of histidine, lysine, and threonine as fishmeal. Diptera contains a higher level of phenylalanine than soybean meal and fishmeal, but the methionine level is comparable to fishmeal. Leucine is abundant in the orders Orthoptera and Coleoptera, but deficient in the order Diptera. The percentages of tyrosine and valine are higher in Orthoptera, Coleoptera, and Diptera than in fishmeal [150,151]. Some of the values of the amino acids in different edible insects are presented in Figure 2 and Table S1.

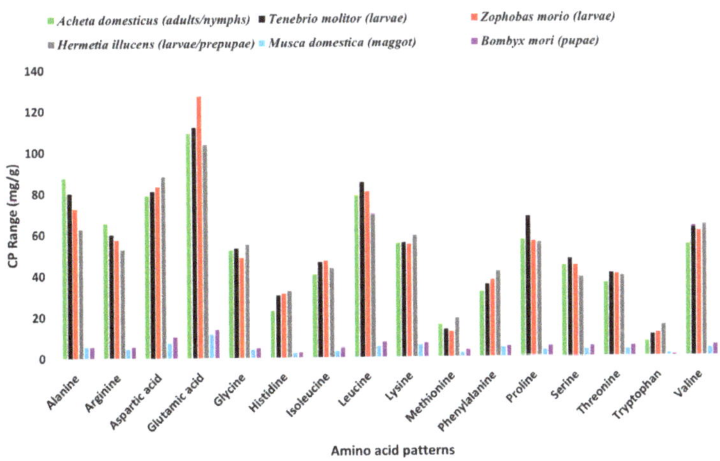

Figure 2. Patterns of amino acids (mg/g CP) of six commonly used edible insect species. Sources [80,82,152–155].

3.2. The Developmental Stage of Insects

An insect's developmental stage may also have an impact on how nutritious it is. With advanced developmental stages, it has been demonstrated that the protein content is generally increased. On the other hand, lipid content is the reverse. The orthopteran species is a high-protein species in its adult form. In Diptera, CP levels range from 40 to 50% in larvae and 50–62% in pupae in several species [150]. In general, larval stages have higher lipid contents than adults. At various phases of development, the lipid percentage in *A. domestica* varies between 14% and 22% [152]. Five-day-old larvae of the black soldier fly contain a maximum percentage of crude protein which is about 61%, and as the insect becomes older its crude protein content gradually declines [156].

The pupae of most insects contain a sufficient amount of saturated fatty acid but a lower amount of unsaturated fatty acid, however, adults behave differently [157]. Ademolu et al. [158] investigated differences in the nutritional composition of distinct post-embryonic developmental stages of the grasshopper *Zonoceros variegatus* nymphs and adults. There is an increase in protein content but a decrease in lipids from nymphs to adults. Similar to this, several Blattodea species show a correlation between developmental stages and higher protein and lower lipid levels [159]. One of the most significant features of insect development is the distinction between larvae and adult forms. Larvae and adults show diverse amino acid profiles; for example, yellow mealworm (*Tenebrio molitor*) adults are rich in protein and chitin, whereas larvae are rich in lipids [160,161].

3.3. Diet

Diet appears to affect the lipid composition of insects. Larvae and adults may consume a diverse range of diets, nonetheless, pupae usually do not, explaining the differences in amino acid, lipid, and mineral contents [162]. For example, in male bees amino acid, protein, and mineral contents vary with the development stage [157]. The orthopteran breeds in the wild have an 8.2% lipid content, while orthopterans bred in captivity have greater contents.

According to Barroso et al. [163], saturated fatty acids are abundant in terrestrial insects but they are deficient in polyunsaturated fatty acids, limiting their use as feed. In this study, the diet composition of BSFL was altered to improve lipid composition. This study suggests that diet-fed larvae contained three times more n-3 fatty acids and a reduction in the n-6: n-3 ratio compared to the control diet group. Dietary manipulation could quickly affect the lipid composition of the experimental larvae group. Van Huis et al. [164] prepared experimental diets from different organic products. Three types of edible mealworms were fed these diets to examine the effects of the experimental diet on the nutritional composition of the insects. Larval growth and survival rates as well as the protein and carbohydrate contents of experimental meals differed. The manipulated diet improved the FCE, crude protein, lipid composition, weight gain, and survival. Zarantoniello et al. [165] studied the rearing of BSF on coffee and *Schizochytrium* sp., and they found significant results in improving the FA profile of zebrafish larvae. The nutritional addition of BSF meal significantly impacted the fatty acid classes of the zebrafish larvae fed on the various diets. For example, the proportion of SFAs increased when insect meal was added to the diets, but the percentages of MUFAs and PUFAs frequently reduced as BSF inclusion levels increased. The higher inclusion of the BSFL in the diet reduces the n-3 and increases the n-6 FA. Chemello et al. [166] explained the effect of BSF mixed with *Schizochytrium* and stated that oocyte maturation stages, fish stress responses, and spawning and hatching success were all negatively impacted by higher replacement levels (BSF 75% and BSF 100%).

3.4. Processing of Insects

The ability of insect meal to replace fishmeal depends not only on the nutrition of insects but also on the bioavailability of nutrients. Many insect species have very high lipid levels. Different processing techniques are applied, including mechanical pressure for the defatting of insects by using different organic solvents such as petroleum ether. These processing methods can improve the protein content of IM further by generating various combinations of free amino acids (AAs) and decreasing the amount of undesirable lipids [167].

Insect meals have low palatability, which can be attributed to various factors, including chemical or biological contamination, anti-nutritional factors such as flavonoids and terpenoids in insect feedstuffs, and high monounsaturated fatty acid [92]. All these factors make insect meals susceptible to oxidation and subsequent rancidity problems. Drying, hydrolyzing, ensiling, and defatting are the different processing methods that can improve fish nutrition by improving the nutrient availability, digestibility, palatability, and composition of insect meal [168]. The processing technique of edible insects can affect their nutritional content. Toasting and solar drying of grasshoppers decreased their protein digestibility and niacin content, but winged termites retained their riboflavin and retinol content [169]. The processing step directly affects the protein quality and content of insect extracts. The nutritional composition of the mopane caterpillar was improved in terms of digestibility and crude protein after degutting, but the roasting resulted in decreased crude protein level and digestibility of the insect [170]. Thus, exploring and developing efficient processing procedures to enhance insect meal commercialization and consumer acceptability is necessary [131].

Fermentation, drying, extraction, enzymatic modifications, and thermal processing are the most common methods that improve insect meal quality, flavor, and texture. It is interesting to note that different insects react to processing methods differently, demonstrating various outcomes for in vitro digestibility [171]. Pre-processing technologies are the first steps in the processing of edible insects. These initial steps are normally performed before food production and are stated as pre-processing phases. Insect degutting, drying, and defatting are pre-processing steps [172,173]. Protein extractability is frequently improved with defatting. This pre-processing stage is completed in some studies following measurements of other protein-related parameters impacted by defatting [142,174]. Additional innovative processes, including pulsed electric field, ultrasound, cold plasma, and high-pressure processing, are extensively applied to replace standard operations and can be used as a pre-processing step [175]. These techniques improve the product's quality, digestibility, and preservation capacity [138,176]. At the industrial level, protein extraction is performed by utilizing different enzymes and organic and inorganic solvents. Newly emerging processing techniques are currently utilized to extract chitin, protein, and lipids in edible insects. The extraction rate varies among the insect species and can affect the extraction yield and insects' physical, chemical, functional, and bioactive properties [174]. A layout of the processing techniques is illustrated in Figure S1.

3.4.1. Enzymatic Hydrolysis

Enzymatic hydrolysis is principally performed to understand the bioactive properties of edible insect hydrolysate after stimulation of digestion in the gastrointestinal tract [177], specifically angiotensin-converting enzyme inhibition [178] and antioxidant capacities [179]. Techno-functional properties of enzyme hydrolysates also change during the processing, and this area needs to be investigated further. In the case of the locust *L. migratoria* L. [145], techno-functionality was increased by utilizing different food-grade proteases alone or in combination with varied enzyme–substrate ratios.

3.4.2. Drying and Thermal Processing

Drying is one of the best processing techniques that increases the shelf-life of insect meals. This processing technique ranges from conventional methods such as sun drying to contemporary methods, including microwave-assisted drying and freeze-drying. Drying stops food spoilage by reducing the water content, inhibiting the enzymatic and microbial reactions. Microbial growth depends upon the availability of water content; when the water content is low, microorganisms stop growing [172]. As the drying process reduces the water content, there is an increase in the concentration of dry matter, which does not damage the physical and chemical properties of insect meal and thus increases the shelf-life [180].

Thermal processing is also a widely used technique in the food production industry that inhibits microbial reactions and extends the shelf-life of insect meals [181]. Thermal processing induces physical and chemical changes without damaging the quality of food [182]. Different types of thermal processing such as hot air drying and oven-assisted boiling show the highest result in maintaining the quality and standard of mealworm larvae [183]. Thermal processing can process the silkworm larvae into a yellowish powder by using warm water, drying, and grinding. Thermal processing can also damage some important bioactive compounds [184] and the nutritional content of insect meals [185]. Thermal processing increases the bioavailability of biomolecules by increasing the process of protein hydrolysis [186]. When *A. domesticus* and *T. molitor* are heated at 200 °C for 10 min, improvements in the nutritional value and AA profile were observed [187].

3.4.3. Fermentation and Antibiotic Resistance

Recently, fermentation technology has been used to produce insect meal, increasing the nutritional value and quality of the meal [188]. Considerable literature is currently available on fermented insect meal, including other insect strategies such as feed additives and biofuel [189]. A particular culture of microorganisms is utilized to ferment the biomolecules in the substrate during the fermentation of the insect meal, improving its nutritional value and digestibility. Antibiotic-resistant bacteria can undoubtedly come from insects [190,191]. However, there are currently not many studies that have been published that demonstrate how frequent antibiotic resistance is in edible insects [192–195] and the antibiotic-resistant (AR) genes in different edible insects. The existence of transferable AR genes in edible insects has not yet been precisely linked by studies to rearing conditions. Additionally, no research has been conducted to ascertain the loads and dynamics of antibiotic resistance in certain microbial communities associated with edible insects. This aspect of edible insect processing needs much attention for the healthy production of food and feed.

4. Potential Insects Used as Fishmeal

Over the last two decades, numerous studies have been conducted to lower the demand for fishmeal, fish oil, and their by-products in the aquaculture industry [196], which resulted in increased incorporation of plant-derived ingredients in fish feed [197]. However, compared to fishmeal-based diets, adding these substances to aquafeeds imposes greater pressure on water and land resources [197] and generates more waste [198]. Several protein alternatives have been examined for fish feed, insect meal, and fishery by-products, showing the greatest potential to fulfill aquafeed's protein requirements in the next decades [18]. For several aquatic species, the successful incorporation of insect meal preference to fishmeal in the diet of many fresh and marine water fish has been widely reviewed [39,199,200].

4.1. *Black Soldier Fly*

The black soldier fly (BSF) (*Hermetia illucens*) is one of the best options currently used as an alternative source of protein in the aquaculture diet. It is one of the extensively investigated insect species due to its healthy nutritional value [199]. The insect's balanced nutritional composition makes it an ideal and possibly significant alternative to fishmeal, and the larval form is what is used to prepare BSF meals.

Bioactive compounds with nutraceutical properties, such as lauric acid, chitin, and antimicrobial peptides, are present in the BSF larvae meal. These compounds are very important for improving the growth of the fish [201]. Previous research suggests that these chemical compounds have either prebiotic or probiotic properties; they are fermented by beneficial commensal bacteria in the intestine and produce metabolites that improve the host's health [202,203]. Black soldier fly (BSF) larvae growth as fish feed is a promising approach, since it uses organic wastes while also being safe for humans and animals [201,204]. BSFLM also has an amino acid profile comparable to fishmeal [39], making it a good candidate for long-term mass production. Many experimental studies have indicated that substituting FM with BSFLM in aquaculture has no adverse impact on fish development [205–208]. One of the best features of BSFLM is that it can utilize organic wastes as the substrate, such as animal dung [209,210] and plant waste, which includes vegetables and fruit wastes [211], algae [212], and fish [163,213]. The inclusion of fish offal in the BSF diet causes black soldier fly prepupae to absorb different lipids. In the modified prepupae, omega-3 fatty acids are found in reasonable amounts, which could be used as feed for carnivorous fish and other animals. Furthermore, this insect species may offer new approaches for reducing and recycling fish offal using different processing operations.

BSFL meal was found to have a greater growth response and feed conversion ratio than FM in Nile tilapia [214]. Previous studies have shown that when BSFL meal is included in the diet of blue tilapia, *Oreochromis aureus*, it shows positive results. The replacement of FM with BSFLM for different inclusion levels greatly increases *O. niloticus* growth without causing detrimental effects. It has also been studied that BSFL meals can completely replace FM [215]. It is obvious from previous research that in yellow catfish, FM can be substituted for up to 20% of the diet [216]. There were no remarkable changes in sea bass growth, feed consumption, survival, and hematological parameters when the fishmeal was partially replaced with black soldier fly meal. Additionally, it was discovered that switching from fishmeal to BSFM at 50% for eight weeks reduced feed costs by 15.5% compared to fishmeal pricing [217]. The HI has good protein solubility, water binding capacity, and lipid binding capacity [174].

4.2. Common Housefly

Housefly (*Musca domestica*) maggot meal has a high nutritional value as an insect protein source. The housefly contains a substantial amount of proteins, lipids, and carbohydrates similar to fishmeal, improving fish growth [53]. It also comprises several biologically active compounds such as antimicrobial proteins, lectin, and chitin [218]. Maggot meal shows good potential to replace fishmeal, as the housefly has a fast reproduction rate, balanced nutrient proportion, and simple processing method [219]. Energy, protein, and micronutrients EAAs and FAs are abundant in housefly larvae. Housefly larvae are less expensive, have healthy nutritional contents, and are easier to produce than other alternative animal proteins. HFL meals, high in lysine, threonine, and methionine, can supplement protein-deficient cereals and legume-based feed for aquatic animals [220].

Houseflies can quickly digest food waste and cattle dung waste, which is organic, using nutrients from waste to reduce the volume of waste in its entirety. Maggot meal enriched with protein and oil from dried fly larvae could be used as important cattle and aquaculture fodder [150]. The fish efficiently absorb maggot meal, and its inclusion in tilapia diets appears to have no oxidative-stress-inducing effect on fish metabolism. It can be used effectively as a rich source of protein for the growth of tilapia fingerlings [151]. Wang et al. [221] observed the effects of four experimental diets supplemented with 25%, 50%, 75%, and 100% *M. domestica* (MD) larva meal on Nile tilapia. About 75% of MD meals can be included in the diet of fish without causing substantial adverse effects on the growth and development of fish. Adding maggot meal (MM) slightly enhanced fillet quality by making them tougher. It is also worth noting that several experimental diets significantly improved water quality compared to the control diet. Several investigations, including MM-based feeding experiments in different fish species, have been conducted over the last decade. Maggot meal can replace 100% fishmeal in the diet of Nile tilapia fingerlings [222] and African catfish [95] without damaging growth or nutrient utilization ability or causing oxidative stress [151,222]. Partially substituting FM diets with blowfly (*Chrysomya megacephala*) MM increased juvenile tilapia growth, feed efficiency, and survival [223].

Feeding tests were conducted on juvenile Asian bass to evaluate physiological responses to growth and fillet composition to dietary FM partly supplemented with housefly-maggot-based meal. According to the research, replacing dietary FM with up to 300 g/kg of housefly maggot meal might be achieved without harming development [224]. An eight-week rearing experiment on swamp eels (*Monopterus albus*) indicated that supplementing the food with housefly larvae had positive influences on the swamp eels' development and immunity [225]. In conclusion, several feeding studies on various aquaculture species have shown that adding housefly maggots to fish diets may boost growth and FCR while limiting physiological stress. Feeding fish diets with housefly maggot meal is also less expensive. It is a different protein source that may be used to replace FM in aquafeeds, depending on its nutritional content, availability, growth potential, and feed efficiency. This is especially helpful in underdeveloped nations when FM imports are expensive.

4.3. Mealworm

The yellow mealworm (*Tenebrio molitor*) is a fast-growing and rapidly reproducing insect that feeds on bread and cereals. In *T. molitor* (TM), the protein content varies between 47% and 63% and lipid content between 31% and 41%. The amino acid profile is in accordance with the nutritional demand of aquatic animals [120]. Mealworms are simple to produce, have a low environmental impact, have easily manipulable nutritional content, and are highly efficient.

Recent research has shown that TM meal is an innovative protein source that can partially substitute FM in chickens [226] and aquatic species [53]. A percentage ranging from approximately 33 to 74 of FM can be replaced by TM larvae without harming the growth of gilthead seabream [227]. They also reported to have tremendous potential for replacing FM in cattle and fish [39,53,228]. It is evident from previous studies that mealworms can substitute the fishmeal of rainbow trout [229], European sea bass [230], gilthead seabream [227], and blackspot seabream [231].

An approximately 35 percent substitution of yellow mealworm may replace fishmeal in European sea bass without slowing fish growth, whereas a 70 percent of fishmeal substitution reduces fish growth [230]. According to previous studies, higher inclusion levels of mealworms in rainbow trout meal had no negative effect on weight gain but enhanced the protein content and lowered the lipid content compared to the control group [229]. A 100% replacement of fishmeal with yellow mealworm meal enhances the Pacific white shrimp lipid content but does not affect the growth rate and feed conversion ratio [232]. When a considerable amount of the fish feed is replaced with a mealworm in common catfish fingerlings and African catfish, its growth retards [233]. The HI has good protein solubility, water binding capacity, and lipid binding capacity [174]. Insects have substantially greater phosphorus levels than calcium levels. They have higher phosphorus content than mammals. Mealworms lack calcium, so primarily feeding fish with mealworms might result in calcium deficiencies and body deformities. Despite this, the mealworm insect is an excellent choice to replace fishmeal [53].

4.4. Cricket

Crickets (*Acheta domesticus*) belong to the Orthoptera order, having a good amount of crude protein ranging from 55–73% and a sufficient amount of indispensable AA except lysine and methionine, which can be supplemented in the feed [80,92]. Cricket food is high in proteins and lipids and contains vitamins and minerals [234]. Cricket meal contains a significant amount of crude protein (64.9%) and lipids (17.4%) with a good proportion of amino acids, including lysine and methionine, which are deficient in a plant-based diet [235]. Recent studies reported that 8.7% of chitin is present in cricket meals, and its supplementation in the fish diet improved the interaction of chitosan glucosamine with bacterial cell walls; in addition, the consequent alterations in the permeability of cell walls reduced the bacterial population [236,237].

Due to their excellent nutritional value, live crickets are currently sold at the commercial level in big pet stores and markets as fish bait or supplemental feed for differential ornamental fish species. Cricket meal can substitute up to 100% of fishmeal in African catfish, showing better results than the control diet [238–240]. It has the potential to partially or completely replace fishmeal in fish feed [235,241]. A recent study investigated whether substituting insects (house cricket and super worm in equal proportions) for FM in perch feed could affect fish survival, growth parameters, and fatty acid composition [242]. Cricket meal might contain any xenobiotic ingredient, thereby altering enzyme activity.

Hanan et al. [243] described meat quality and the fatty acid profile of the insects *Zophbas morio* and *A. domestica*. These species were specifically selected for their simple breeding, standardization, and higher AA, FA, and vitamin contents. Only super worms and crickets possessed a significant concentration of vitamin A for the animals that feed on insects [82]. Fishmeal can be substituted with up to 50% cricket meal in red Nile tilapia without significantly affecting growth performance or feed consumption.

4.5. Locust

Locusts (*Locusta migratoria*) are generally produced to feed domestic and zoo animals and have also been studied for cattle feeding. They vary greatly in protein composition (29 to 70% on a dry matter basis) and lipid content (4 to 22% on a dry matter basis) [53,244]. Locusts are among the most commonly consumed insect species globally, but they have also been utilized as a supplementary protein source for chickens and fish [131]. Aquaculture development in Africa and Asia and the search for alternate protein sources led to feeding experiments with locusts and grasshoppers for catfish and tilapia [245].

A 50% inclusion level of locust meal yielded the optimum results in Nile tilapia in terms of growth. As a result, it is recommended that locust meals at a 50% inclusion rate can be added to the diet of tilapia fish without impairing fish growth. When about 20% of adult Orthoptera (grasshoppers or locusts) was added in the diet of catfish and tilapia, it did not affect catfish or tilapia digestibility or growth [246,247]. However, the use of locust meal as an alternative source of protein in aquaculture diets is still understudied, and no data on its use as feed for marine fish species are known [246,248].

4.6. Silkworm

The silkworm (*Bombyx mori*) (SW) is believed to have originated in China more than 4000 years ago [249]. Its pupae are the primary by-product of the silk industry, with about 8 kg of pupae produced for each kilogram of silk. Silkworm meal is prepared by properly grinding and drying boiled cocoons of silkworm larvae. SW meal is obtained by drying and grinding the larvae's uncoiled boiled cocoons, containing 56% protein and a substantial amount of essential amino acids [250]. Kurbanov et al. [251] investigated the effect of substitution of fishmeal with silkworm pupa protein on the growth and feed utilization ability of African catfish (*Clarias gariepinus*) fingerlings. Five isonitrogenous diets containing approximately 40% crude protein and different levels of fishmeal replacement (0–100%) were fed for 40 days at 5% live fish body weight. The growth rate and feed utilization efficiency of fingerlings fed diets with a 50:50 mixture of fishmeal and SPP were significantly higher than those fed diets containing SPP or fishmeal alone. The fishmeals of different fish species have been substituted with silkworm pupae meal and promising results were obtained, thus showing the good potential of silkworms to replace fishmeal. About 10% SWM can be successfully substituted in the diet of chum salmon and olive flounder [252].

The snakeskin gourami diet can be substituted with up to 15% SWP (silkworm pupae) without adverse effects on growth, but a 22% inclusion level of SWPM decreases growth rate and protein digestibility [253]. A 30–50% successful inclusion of defatted or non-defatted SWP can be achieved in rohu, common carp, mahseer, and rainbow trout [254,255]. A total FM replacement without adverse effects on growth has been observed in common carp and Japanese sea bass [256]. Some fish species show negative responses even when a low level of SWM is incorporated into their diet. In tilapia, a 5% inclusion level drastically decreased growth and development [257]. Jian carp also showed reduced digestive enzyme activity, heat shock protein activities, and increased oxidative stress when the diet was incorporated with about 8% of SWPM [258]. In walking catfish, the protease activity result is comparable to FM when 58% FM is substituted with SWP meal. SWP appears to be a potential source of protein for fish diets in general, except for Nile tilapia and Jian carp gourami. SWP meal was found to be an advantageous sustainable feed element in carp diets, with benefits for boosting growth performance and specific physiological markers, according to an eleven-week feeding research on *Cyprinus carpio* [259]. In the rainbow shark (*Epalzeorhynchos frenatum*), feeding experiments revealed that SPM could substitute up to 30% of FM in the diet [260].

5. Insects as Feed for Crustaceans

Fishmeal is a commercial diet for rearing prawns and shrimp because of its high protein content, favorable amino acid and fatty acid profiles, high digestibility, and palatability [205]. Farm animals' diets have recently begun to include insect larvae meals, and previous studies showed that insect meals are a promising addition to the diets of farmed crustaceans [205,232].

5.1. Shrimp

Studies have found different results comparing the growth rate performances of commercially produced shrimp and fish fed with FM substituted with insect meal. This could result from changes in meal preparation methods or diet formulation. The nutritional value of animal protein meals is influenced by the processing technique and the quality of the raw materials [261]. For example, research on rainbow trout showed that the BSF larval diet could replace 50% of fishmeal without impairing growth performance. However, a decrease in protein utilization efficiency was reported in shrimp [262]. Without impairing shrimp growth, a full-fat BSF (*H. illucens*) meal can substitute up to 25% of fishmeal [205]. Like this study, subsequent studies on *T. molitor* full-fat meal-based diets found that when methionine was added in adequate amounts, Pacific white shrimp's nutritional value and growth performance were on par with or even superior to those of FM-based diets [167,232].

Motte et al. [263] reported that utilizing TM meal in shrimp diets significantly enhances growth when paired with FM without affecting shrimp survival or feed consumption, even when TM replaces 100% of FM. Moreover, this insect meal's very important key feature is that it improves the immune system and resistance to infection, leading to increased shrimp survival. Improving immunity is critical when shrimp production intensity is very high, causing stress and increasing sensitivity to illnesses. TM meal has good potential as an efficient feed source for commercial shrimp farming. The 50% substitution of fishmeal with mealworms for the shrimp diet is ideal for boosting shrimp growth performance, while having no negative effects [167]. TM is a suitable replacement for FM that has no negative effects on the expression of important digestive enzymes, gut microbiota, or the Pacific white shrimp's immune system [264].

5.2. Prawns

Efforts have been made to supplement the diet of prawns with insect meal. A recent study examined the effect of replacing the diet of prawns with insect meal [265]. The insect meals were prepared from the larvae of three insects: house fly, BSF, and mealworm. Incorporating the black soldier fly into the fishmeal diets of prawns resulted in considerably higher prawn growth rate and survival. The shrimp survival rate was significantly reduced when fed a diet consisting of house fly larvae, but growth performance was incomparably higher. Insect inclusion in plant-based foods had little effect on growth performance, while survival was greater in the TM and HI inclusion diets. When TM and HI were added to fishmeal and plant meal diets, they increased the protein content of the prawns' muscles. Langer et al. [266] studied the consequences of replacing fishmeal with silkworm pupae, soybean meal, and earthworm meal in the freshwater prawn *Macrobrachium dayanum* for 90 days. The diet including silkworm pupae showed the second best result, closely followed by fishmeal.

Two studies investigated the utilization of seven diets for prawns farmed at suboptimal temperatures [267,268]. All seven diets supplied to the juvenile prawns were quickly devoured. Shrimp- and silkworm-based meals provided the best growth and survivorship, while fishmeal- and black soldier fly larvae meal (BSFLM)-based diets provided the poorest growth. Survival rates were comparable for diets based on soybeans, crickets, mealworms, and fishmeal. BSFLM inclusion in the diet caused the mortality in prawns after 1 day of feeding, which might be due to some antimicrobial or toxic compounds produced by BSFL. Similarly, as the termite meal inclusion level increased in the diet, decreasing order in growth was observed in prawns. Partial replacement of FM with 35% TM in prawns showed positive results but was lower than FM. Termite meal cannot replace fishmeal as an alternate protein source for *M. rosenbergii* juveniles. However, it can be added to the diet of prawns for amino acid supplementation since its addition can improve the growth rate [268].

6. Physiological Responses of Fish Using Insects as Fishmeal

Diet formulation is the key factor for sustainable fish production, as the composition and nutrition of the diet will directly influence the fish growth performance and health status [269]. An accurate and detailed examination of the organs involved in digestion and absorption of feed, immunological response, and metabolic processes should be conducted while testing new dietary formulations [270]. It is generally known that insects contain bioactive substances such as chitin, which, at specific concentrations, can strengthen fish immune systems and promote the diversification of the gut flora of fish [271]. Recently, scientists have been working on new aquafeed formulations such as insect meal and examining their effect on the physiological response of different fish species. A review of some of these studies is presented in the subsequent sections.

Recent research has shown that rainbow trout's inflammatory response can be lowered by including HI-based meal (*H. illucens* meal) in low-FM diets [272,273]. Additionally, it has been noted that HI-based meal positively impacts the intestinal physiology of various farmed fish species [274]. According to studies by Osimani et al. [275], Zarantoniello et al. [276], and Zarantoniello et al. [165], HI-based meals contain some bioactive substances, such as chitin and medium-short FAs. These useful substances have immune-stimulating, antimicrobial, and/or anti-inflammatory properties, which have positive and beneficial effects on fish gut health [277]. It is evident from recent research that the quality of the nutrients can affect the reproductive system (with an emphasis on oocyte quality). A good aquafeed composition is essential for brood stock's proper growth and development [270]. Chemello et al. [166] observed the physiological effect of different levels of BSF prepupae meal (0, 25, 50, 75, and 100%) on fish growth performance, lipid metabolism, stress response, and reproduction. For determination of the physiological response of fish to experimental diets, a multidisciplinary approach (biometric, gas-chromatographic, histological, and molecular analyses) was used. There was not any detrimental effect on physiological responses of fish when the fishmeal was replaced with 50% BSF meal in zebrafish. However, the inclusion of a higher level of BSF meal (75% and 100%) showed a negative effect on stress response, oocyte maturation, and spawning and hatching of zebrafish.

Sudha et al. [278] revealed the inclusion of different levels of BSFM in the catfish diet and its effect on the physiological response of catfish. When the fishmeal was replaced with 20%, 40%, and 60% diets, no significant difference in growth performance was observed compared to the control group. However, the inclusion of 100% BSFM in the diet of catfish showed a reduction in growth and feed efficiency. No significant differences in catfish's amino acid profile and hematological responses were observed when fed with different BSFM. The non-significant difference in the liver and intestinal amylase and lipase activity was observed in the inclusion of BSFM. Higher levels of BSFM (80% and 100%) significantly lowered the proximate body composition of fish and increased lipase activity. Fish fed with 80% and 100% BSFM showed increased congestion in the hepatocyte. Zarantoniello et al. [276] observed a significant growth and survival reduction when fish were fed with a diet containing 50% *H. illucens*.

Additionally, a 50% HI diet substantially decreased the hepatic lipid and glycogen content but increased hepatic *hsp* 70.1 gene expression. The inclusion of 50% HI in sturgeon's diet also badly affected gut histological morphometric parameters. Zarantoniello et al. [165] reported on the inclusion of insect meal and its effect on the physiological responses of fish. The different inclusion levels of BSF meal (0, 25, 50, 75, and 100%) in the diet of zebrafish showed a significant difference in the physiological response of fish. When the fish were fed with a high inclusion level of BSF meal (75 and 100%), severe hepatic steatosis was observed in the fish's liver. Inclusion of a high level of BSF meal resulted in a reduction in gut microbiota biodiversity, high lipid content, and significant upregulation of genes involved in immune response.

7. Challenges of Using Insects as Feed in Aquaculture

Chitin, a non-protein nitrogen molecule found in most insect cuticles, causes reduced meal digestibility and growth performance [279]. Increasing the chitin concentration in IM products to ensure a favorable response in fed species requires more investigation. Alkaline extraction may easily remove chitin [280], but it is costly and leaves chemical residues and contaminants [281].

In addition to the presence of chitin, a negative effect in the growth of aquatic animals is observed when insect meals are included in aquafeeds. This fact can be attributed to lower levels of fatty acids in the insect-based diets compared to the fishmeal control diet [13]. Many studies have reported that insects have lower levels of n-3 PUFAs [165], which results in lower n-3 PUFA levels in aquafeeds if substrates enriched with n-3 PUFAs are not used during the insect rearing. Zarantoniello et al. [276] found that a 50% substitution of fishmeal with BSFM caused a significant reduction in n-3 fatty acids, the result of which was a marked reduction in sturgeon weight and SGR observed when fed with BSF-substituted diets. In addition to fatty acids, insect meals are also deficient in some essential amino acids as compared to fishmeal, and manipulation of these components through the rearing substrate is still a big challenge [282].

Prices of insect meals are predicted to become competitive by 2023, but at present, IM prices are very high [283,284]. The main limitation of using insects is the existence of toxic compounds which can negatively affect fish physiologically, such as lowering growth and altering hematological parameters [285]. On the other hand, the substitution of FM and FO with *H. illucens* meal can reduce the content of some potentially toxic elements such as Ni, As, and Pb in fish feed, causing levels of harmful chemicals in animal feed to be below the permitted limit [286]. It is now widely acknowledged that dietary metal(loid) intake can cause chronic toxicity in aquatic species [287]. These pollutants can bioaccumulate or bio-magnify in food chains [288]. Consumption of PTE contaminates fish, thus is a major risk to human health. These PTEs should be examined in the environment and feed or food because of their hazard to feed/food safety and, eventually, human health. Non-defatted insect meals and oils show a dramatic change in FA profile, which shows huge variation in the quality of insect meal and its composition, limiting the usage of insects as feed ingredients.

Another important aspect of incorporating insect meal in aquaculture depends upon the aquaculture producer and consumer acceptance. Without consumer acceptance, adopting insects in the aquaculture sector is difficult. An accurate evaluation of factors involved in insects' production is also a big challenge, as it involves transitioning from the wild catching of potential edible insects from their habitats to their large-scale production. Intensive insect rearing requires intensive labor, as most commonly used insect-rearing methods are labor intensive. Scaling up the insect industry requires the availability of labor/staff, as only a few stages of insect rearing are automated. The safety risks (allergens, chemicals, and microbial hazards) of insect-fed animals are also increasing with the increasing demand for the inclusion of insect meals in the diet of aquatic animals [289]. The regulation of insect feed is also a critical issue for the insect-rearing industry. The regulations and guidelines regarding insect feed are different in different countries. For example, in several states of the United States, insect-based feed is allowed, while other states are waiting for FDA permission. There is a lack of proper guidelines, regulations, and legislation regarding insect rearing and consumption [290] (Figure 3).

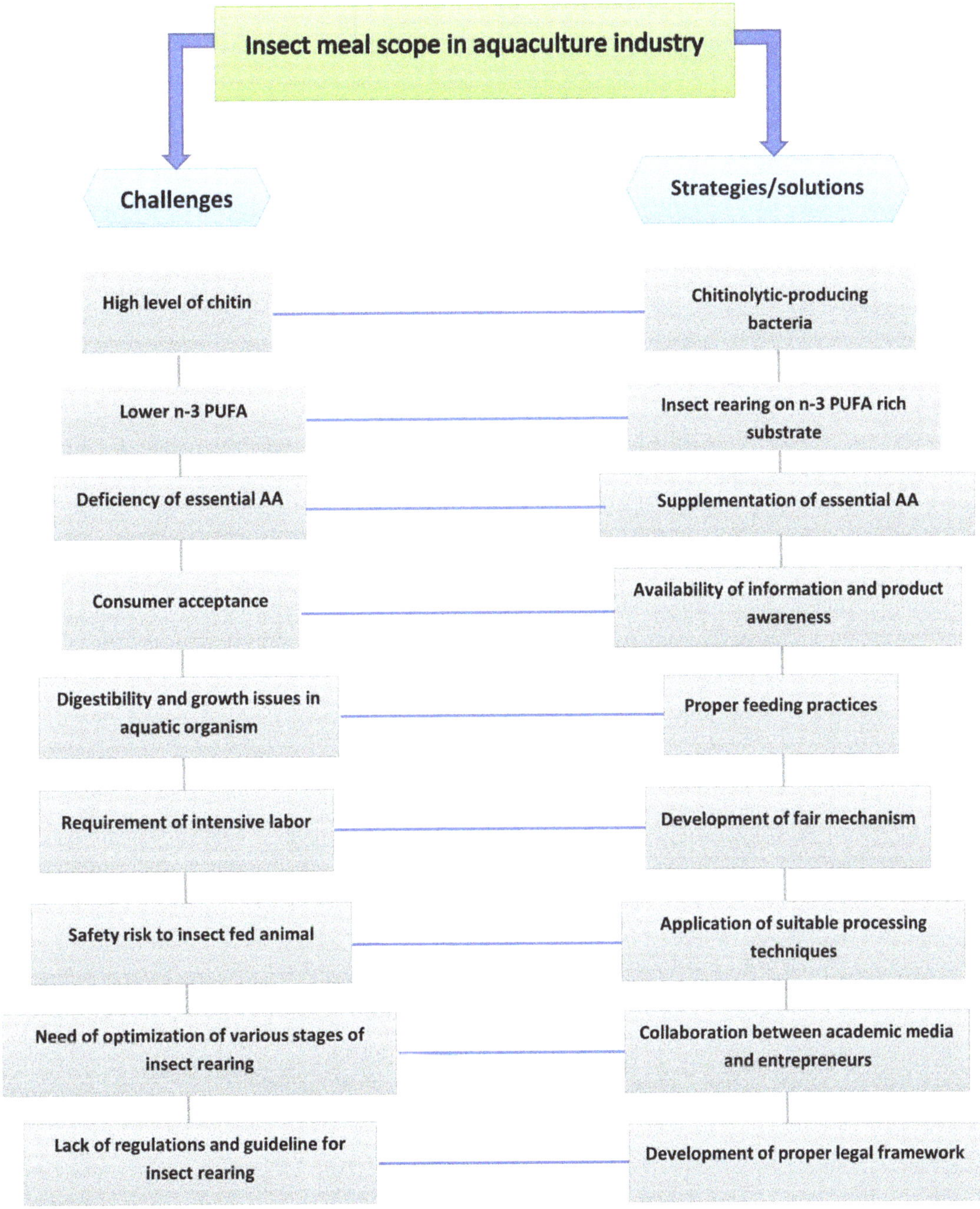

Figure 3. Challenges and solutions of insect-based meal in aquaculture industry.

8. Possible Solutions to Challenges in Introducing the Insects in Aquafeed

Chitin, which has a detrimental effect on the growth of fish, can be removed by supplementation with chitinase/chitinolytic-producing bacteria. This approach is very beneficial in inducing an immune response to fish and mitigating the environmental impact of fish waste [281]. Two major factors that can accelerate consumer acceptance and positive perception regarding the aquatic products produced through insect-based feed are the availability of information and product awareness [291,292]. To promote wider consumer acceptance, it is very important to reduce information asymmetry [292].

The environmental impact can be reduced by proper upscaling of insect production, and as a result, the insects can compete with conventional ingredients [293]. Environmental impact categories of *T. molitor* [46], *M. domestica* [294], and *H. illucens* [295] largely contribute to feeding produced from the insect-rearing industry. Therefore, providing a suitable substrate for insect rearing and increasing the efficiency of facilities will be major contributors to attaining insect meal's environmental benefits [293,296].

Another method for improving the amino acid content is defatting, but it requires intensive energy usage, increasing the environmental impact and feed cost [297]. It is better to add complementary raw material to insect meals or supplement the IM with deficient amino acids. Defatted mealworm and aquatic insect meals have higher arginine content than full-fat BSF diets [298,299]. As many AA compositions are found in insects, it is critical to analyze them before producing any insect-based feed [233,300]. More trials are needed to find a suitable feed to boost the omega-3 lipid content in insect meals, notably in HI, which tends to acclimate more easily to different diets [212,301].

Among the manufacturing techniques, extrusion can play a vital role in nutrition utilization [196]. The efficiency of extruded insect-based feed is also documented in recent studies [302,303]. Environmental impacts can be reduced by properly addressing feeding practices to minimize feed waste and FCR. Novel processing techniques are needed to address the safety concerns regarding insect feed consumption. It has been observed that various chemicals can be degraded if the insects are grown properly [289]. Preparing insect protein meal or dry pellets would save money and extend their shelf-life (relative to live food).

To optimize various stages of insect rearing, academic researchers and institutes collaborate with different EU companies to foster innovations. Studies have indicated that collaboration between academic media and entrepreneurs resulted in the development new innovative techniques. For example, a Dutch company has developed a centralized system for controlling oviposition in HI adult colonies with the help of olfactory triggers (web source). Larvae from the adult cages are transferred to a separate system by a newly developed automated device that allows counting larvae, dozing, and proper analysis. Breeding cages with pipes have also been developed to efficiently clean insect-rearing cages (web source). For promoting the insect business, there is a dire need to develop the proper legal framework to help the feed industry flourish fast (Figure 3).

9. Future Perspective

First, the insect industry must greatly increase its production capacity. Insect feed prices are currently too high. Fishmeal, high-quality SBM extracts, and soybean meal production volumes are thousands of times higher than insect meal protein and its by-products. By expanding the production scale, insect producers can compete on price and product stability with other protein sources. Industrialization and controlled manufacturing technologies will help shareholders to scale up the industry by reducing labor-intensive insect production [304].

Further cost–benefit research will need to be conducted regularly to determine the economic effects of adding insects to animal feed to see how these alternative protein ingredients affect total production costs. Additional research is needed on insect meal to modify and improve the digestive tract of the fed animal. Multiple approaches, including histomorphology, molecular biology, and histochemistry, will be advised to assess the gastrointestinal tract health of insect-fed organisms.

It is crucial to clarify and comprehend the factors behind consumer concerns about farm animal welfare worldwide. First, one rationale is that customers give farm animal welfare a higher priority than other traits when evaluating the quality of food [305–307]. One key reason consumers buy animal-welfare-friendly items is a correlation between better human health benefits and farm animal wellbeing [308–310].

The bioconversion technique for insect production is one of the sustainable solutions to food security. In this sense, waste is a valuable resource for producing high-quality protein (insect meal) for the food system. As a result, the technology produces zero waste and lowers the need for costly protein sources such as soy meal and fishmeal in aquafeed. Utilization of organic wastes for rearing edible insects is an attractive approach that would facilitate the SDGs promoting female entrepreneurs. This approach will help build sustainable and smart areas with reduced greenhouse gas emissions, indirectly reducing the carbon footprints. The potential of the insect meal industry to meet the increasing demand of the fish feed industry is unclear and data deficient. Research on this topic is limited to tiny plots or cages with immature or juvenile fish. More research is needed to verify that insects can be produced effectively and efficiently to feed young and adult fish.

Moreover, the insect's requirement level for different fish species varies according to stages and culture systems, and more studies are needed to fill this gap. This gap needs to be filled to commercialize aquaculture insects. The long-term sustainability impact of insect rearing is unknown. The data are not available for the proper understanding of production aspects of potential insects and the requirements and risks of their accidental release, which opens a new avenue for further research. Some insects have the potential to replace fishmeal completely, such as the BSF. However, the species-specific threshold limits the complete substitution of fishmeal with insect meal. Therefore, it is very important to explore new fractionation, separation, and biorefining schemes to extract useful products from insects.

Recent studies have demonstrated that different parts of insect's meal, such as meal, oil, pulp, and paste, can be used in aquaculture. The most literature is available on insect meal and, to some extent, oil. However, the data regarding the utilization of pulp and paste are scarce, and this area of research needs to be addressed. More collaboration is needed between the feed industry, government, academia, and local farmers to explore new insect protein sources and build efficient, cost-effective, sustainable rearing systems.

10. Conclusions

The present review analyzed that insects have good potential to replace fishmeal due to their nutritional value. Among the non-conventional feed ingredients, insects have the biggest potential to replace conventional feed ingredients. Proper processing technologies can further improve the nutritional value of insects. Insect-based diets showed a positive physiological response in many fish species. Insects have great potential to use agro-industrial and plastic wastes, thus contributing to combating pollution-related environmental problems. Our review also highlighted the challenges or hurdles in using insects in aquafeed, and possible solutions to these challenges were also addressed. However, more studies need to be conducted to determine the required level for aquatic animals, which varies from species to species and with developmental stages. Most of the studies on replacing fishmeal with insect meal focus on juvenile stages, not adult ones. For this reason, this area still needs to be investigated. It is important to scale up insect farming at the industrial level with standard, cost-effective, and eco-friendly facilities and to develop suitable substrates for insects to deal with nutritional and environmental issues.

Supplementary Materials: The following are available online at https://www.mdpi.com/article/10.3390/fishes7060395/s1, Table S1: Patterns of amino acids (mg/g CP) of six commonly used edible insect species. Sources [80,82,152–155]; Figure S1: The processing techniques of the insect's-based feed.

Author Contributions: Conceptualization, W.M. and A.H; Writing-Original Draft Preparation, W.M., A.H., M.N., M.B., U.R., M.H., S.U.R. and N.R.; Writing-Review & Editing, W.M., A.H., M.N., M.B., U.R., M.H., S.U.R. and N.R.; Funding Acquisition, N.R. and W.M. All authors have read and agreed to the published version of the manuscript.

Funding: This research received no external funding.

Institutional Review Board Statement: Not applicable.

Data Availability Statement: Not applicable.

Conflicts of Interest: All authors declare that they have no conflict of interest.

References

1. FAO. *The State of World Fisheries and Aquaculture, Sustainability in Action; 2020*; FAO: Rome, Italy, 2020; ISBN 9789251326923.
2. Pradeepkiran, J.A. Aquaculture Role in Global Food Security with Nutritional Value: A Review. *Transl. Anim. Sci.* **2019**, *3*, 903–910. [CrossRef] [PubMed]
3. Azra, M.N.; Okomoda, V.T.; Tabatabaei, M.; Hassan, M.; Ikhwanuddin, M. The Contributions of Shellfish Aquaculture to Global Food Security: Assessing Its Characteristics from a Future Food Perspective. *Front. Mar. Sci.* **2021**, *8*, 365. [CrossRef]
4. Cai, J.; Leung, P.S. A Note on Linkage between Gross Value Added and Final Use at the Industry Level. *Econ. Syst. Res.* **2020**, *32*, 428–437. [CrossRef]
5. Broitman, B.R.; Halpern, B.S.; Gelcich, S.; Lardies, M.A.; Vargas, C.A.; Vásquez-Lavín, F.; Widdicombe, S.; Birchenough, S.N.R. Dynamic Interactions among Boundaries and the Expansion of Sustainable Aquaculture. *Front. Mar. Sci.* **2017**, *4*, 15. [CrossRef]
6. Tacon, A.G.J. Trends in Global Aquaculture and Aquafeed Production: 2000–2017. *Rev. Fish. Sci. Aquac.* **2020**, *28*, 43–56. [CrossRef]
7. Serra, C.R.; Oliva-Teles, A.; Enes, P.; Tavares, F. Gut Microbiota Dynamics in Carnivorous European Seabass (*Dicentrarchus labrax*) Fed Plant-Based Diets. *Sci. Rep.* **2021**, *11*, 447. [CrossRef]
8. Molinari, G.S.; McCracken, V.J.; Wojno, M.; Rimoldi, S.; Terova, G.; Kwasek, K. Can Intestinal Absorption of Dietary Protein Be Improved through Early Exposure to Plantbased Diet? *PLoS ONE* **2020**, *15*, e0228758. [CrossRef]
9. Magbanua, T.O.; Ragaza, J.A. Selected Dietary Plant-Based Proteins for Growth and Health Response of Nile Tilapia Oreochromis Niloticus. *Aquac. Fish.* **2022**, in press. [CrossRef]
10. Rawles, S.D.; Green, B.W.; McEntire, M.E.; Gaylord, T.G.; Barrows, F.T. Reducing Dietary Protein in Pond Production of Hybrid Striped Bass (*Morone chrysops* × *M. saxatilis*): Effects on Fish Performance and Water Quality Dynamics. *Aquaculture* **2018**, *490*, 217–227. [CrossRef]
11. Gaudioso, G.; Marzorati, G.; Faccenda, F.; Weil, T.; Lunelli, F.; Cardinaletti, G.; Marino, G.; Olivotto, I.; Parisi, G.; Tibaldi, E.; et al. Processed Animal Proteins from Insect and Poultry By-products in a Fish Meal-free Diet for Rainbow Trout: Impact on Intestinal Microbiota and Inflammatory Markers. *Int. J. Mol. Sci.* **2021**, *22*, 5454. [CrossRef]
12. Randazzo, B.; Zarantoniello, M.; Cardinaletti, G.; Cerri, R.; Giorgini, E.; Belloni, A.; Contò, M.; Tibaldi, E.; Olivotto, I. *Hermetia illucens* and Poultry By-Product Meals as Alternatives to Plant Protein Sources in Gilthead Seabream (*Sparus aurata*) Diet: A Multidisciplinary Study on Fish Gut Status. *Animals* **2021**, *11*, 677. [CrossRef]
13. Randazzo, B.; Zarantoniello, M.; Gioacchini, G.; Cardinaletti, G.; Belloni, A.; Giorgini, E.; Faccenda, F.; Cerri, R.; Tibaldi, E.; Olivotto, I. Physiological Response of Rainbow Trout (*Oncorhynchus mykiss*) to Graded Levels of *Hermetia illucens* or Poultry by-Product Meals as Single or Combined Substitute Ingredients to Dietary Plant Proteins. *Aquaculture* **2021**, *538*, 736550. [CrossRef]
14. Gasco, L.; Gai, F.; Maricchiolo, G.; Genovese, L.; Ragonese, S.; Bottari, T.; Caruso, G. Feeds for the Aquaculture Sector Current Situation and Alternative Sources. In *Feeds for the Aquaculture Sector*; Springer International Publishing: New York, NY, USA, 2018; pp. 1–103. ISBN 9783319779409.
15. López-Pedrouso, M.; Lorenzo, J.M.; Cantalapiedra, J.; Zapata, C.; Franco, J.M.; Franco, D. Aquaculture and By-Products: Challenges and Opportunities in the Use of Alternative Protein Sources and Bioactive Compounds. *Adv. Food Nutr. Res.* **2020**, *92*, 127–185. [CrossRef] [PubMed]
16. Parisi, G.; Tulli, F.; Fortina, R.; Marino, R.; Bani, P.; Dalle Zotte, A.; De Angeli, A.; Piccolo, G.; Pinotti, L.; Schiavone, A.; et al. Protein Hunger of the Feed Sector: The Alternatives Offered by the Plant World. *Ital. J. Anim. Sci.* **2020**, *19*, 1205–1227. [CrossRef]
17. Ameixa, O.M.C.C.; Duarte, P.M.; Rodrigues, D.P. Insects, Food Security, and Sustainable Aquaculture. In *Zero Hunger*; Springer: Berlin/Heidelberg, Germany, 2020; pp. 425–435.
18. Hua, K.; Cobcroft, J.M.; Cole, A.; Condon, K.; Jerry, D.R.; Mangott, A.; Praeger, C.; Vucko, M.J.; Zeng, C.; Zenger, K.; et al. The Future of Aquatic Protein: Implications for Protein Sources in Aquaculture Diets. *One Earth* **2019**, *1*, 316–329. [CrossRef]
19. Younis, E.S.M.; Al-Quffail, A.S.; Al-Asgah, N.A.; Abdel-Warith, A.W.A.; Al-Hafedh, Y.S. Effect of Dietary Fish Meal Replacement by Red Algae, Gracilaria Arcuata, on Growth Performance and Body Composition of Nile Tilapia Oreochromis Niloticus. *Saudi J. Biol. Sci.* **2018**, *25*, 198–203. [CrossRef] [PubMed]

20. Oswald, A.T.O.; Ishikawa, M.; Koshio, S.; Yokoyama, S.; Moss, A.S.; Serge, D. Nutritional Evaluation of Nannochloropsis Powder and Lipid as Alternative to Fish Oil for Kuruma Shrimp, Marsupenaeus Japonicus. *Aquaculture* **2019**, *504*, 427–436. [CrossRef]
21. Meitei, M.M.; Singh, S.K.; Mangang, Y.A.; Meena, D.K.; Debbarma, R.; Biswas, P.; Waikhom, G.; Patel, A.B.; Ngasotter, S.; Newmei, T.; et al. Effective Valorization of Precision Output of Algaquaculture towards Eco-Sustainability and Bioeconomy Concomitant with Biotechnological Advances: An Innovative Concept. *Clean. Waste Syst.* **2022**, *3*, 100026. [CrossRef]
22. Yadav, G.; Meena, D.K.; Sahoo, A.K.; Das, B.K.; Sen, R. Effective Valorization of Microalgal Biomass for the Production of Nutritional Fish-Feed Supplements. *J. Clean. Prod.* **2020**, *243*, 118697. [CrossRef]
23. Mandal, A.; Das, S.K. Comparative Efficacy of Neem (Azadirachta Indica) and Non-Neem Supplemented Biofloc Media in Controlling the Harmful Luminescent Bacteria in Natural Pond Culture of Litopenaeus Vannaemei. *Aquaculture* **2018**, *492*, 157–163. [CrossRef]
24. Hosain, M.E.; Amin, S.M.N.; Arshad, A.; Kamarudin, M.S.; Karim, M. Effects of Carbon Sources on the Culture of Giant River Prawn in Biofloc System during Nursery Phase. *Aquac. Rep.* **2021**, *19*, 100607. [CrossRef]
25. Debbarma, R.; Meena, D.K.; Biswas, P.; Meitei, M.M.; Singh, S.K. Portioning of Microbial Waste into Fish Nutrition via Frugal Biofloc Production: A Sustainable Paradigm for Greening of Environment. *J. Clean. Prod.* **2022**, *334*, 130246. [CrossRef]
26. Hardy, R.W.; Patro, B.; Pujol-Baxley, C.; Marx, C.J.; Feinberg, L. Partial Replacement of Soybean Meal with Methylobacterium Extorquens Single-Cell Protein in Feeds for Rainbow Trout (*Oncorhynchus mykiss* Walbaum). *Aquac. Res.* **2018**, *49*, 2218–2224. [CrossRef]
27. Jones, S.W.; Karpol, A.; Friedman, S.; Maru, B.T.; Tracy, B.P. Recent Advances in Single Cell Protein Use as a Feed Ingredient in Aquaculture. *Curr. Opin. Biotechnol.* **2020**, *61*, 189–197. [CrossRef] [PubMed]
28. Irm, M.; Taj, S.; Jin, M.; Luo, J.; Andriamialinirina, H.J.T.; Zhou, Q. Effects of Replacement of Fish Meal by Poultry By-Product Meal on Growth Performance and Gene Expression Involved in Protein Metabolism for Juvenile Black Sea Bream (*Acanthoparus schlegelii*). *Aquaculture* **2020**, *528*, 735544. [CrossRef]
29. Zarantoniello, M.; Pulido Rodriguez, L.F.; Randazzo, B.; Cardinaletti, G.; Giorgini, E.; Belloni, A.; Secci, G.; Faccenda, F.; Pulcini, D.; Parisi, G.; et al. Conventional Feed Additives or Red Claw Crayfish Meal and Dried Microbial Biomass as Feed Supplement in Fish Meal-Free Diets for Rainbow Trout (*Oncorhynchus mykiss*): Possible Ameliorative Effects on Growth and Gut Health Status. *Aquaculture* **2022**, *554*, 738137. [CrossRef]
30. Pulido-Rodriguez, L.F.; Cardinaletti, G.; Secci, G.; Randazzo, B.; Bruni, L.; Cerri, R.; Olivotto, I.; Tibaldi, E.; Parisi, G. Appetite Regulation, Growth Performances and Fish Quality Are Modulated by Alternative Dietary Protein Ingredients in Gilthead Sea Bream (*Sparus aurata*) Culture. *Animals* **2021**, *11*, 1919. [CrossRef] [PubMed]
31. Tacon, A.G.J.; Hasan, M.R.; Metian, M. Demand and Supply of Feed Ingredients for Farmed Fish and Crustaceans: Trends and Prospects. *FAO Fish. Aquac. Tech. Pap.* **2011**, *564*, 87.
32. Ayensu, J.; Annan, R.A.; Edusei, A.; Lutterodt, H. Beyond Nutrients, Health Effects of Entomophagy: A Systematic Review. *Nutr. Food Sci.* **2019**, *49*, 2–17. [CrossRef]
33. De Castro, R.J.S.; Ohara, A.; dos SantosAguilara, J.G.; Domingues, M.A.F. Nutritional, Functional and Biological Properties of Insect Proteins: Processes for Obtaining, Consumption and Future Challenges. *Trends Food Sci. Technol.* **2018**, *76*, 82–89. [CrossRef]
34. Giron, R.C.; Hidalgo, G.G.; Garcia, J.P.B.; Pino-Hernández, E.; Villa, P. Exploring the Food and Nutritional Potential of Three Edibles Amazonian Arthropods. *Rev. Etnobiol.* **2017**, *15*, 26–31.
35. Patel, S.; Suleria, H.A.R.; Rauf, A. Edible Insects as Innovative Foods: Nutritional and Functional Assessments. *Trends Food Sci. Technol.* **2019**, *86*, 352–359. [CrossRef]
36. Tran, H.Q.; Nguyen, T.T.; Prokešová, M.; Gebauer, T.; Van Doan, H.; Stejskal, V. Systematic Review and Meta-Analysis of Production Performance of Aquaculture Species Fed Dietary Insect Meals. *Rev. Aquac.* **2022**, *14*, 1637–1655. [CrossRef]
37. Hua, K. A Meta-Analysis of the Effects of Replacing Fish Meals with Insect Meals on Growth Performance of Fish. *Aquaculture* **2021**, *530*, 735732. [CrossRef]
38. Liland, N.S.; Araujo, P.; Xu, X.X.; Lock, E.J.; Radhakrishnan, G.; Prabhu, A.J.P.; Belghit, I. A Meta-Analysis on the Nutritional Value of Insects in Aquafeeds. *J. Insects Food Feed* **2021**, *7*, 743–759. [CrossRef]
39. Henry, M.; Gasco, L.; Piccolo, G.; Fountoulaki, E. Review on the Use of Insects in the Diet of Farmed Fish: Past and Future. *Anim. Feed Sci. Technol.* **2015**, *203*, 1–22. [CrossRef]
40. Fontaneto, D.; Tommaseo-Ponzetta, M.; Galli, C.; Risé, P.; Glew, R.H.; Paoletti, M.G. Differences in Fatty Acid Composition between Aquatic and Terrestrial Insects Used as Food in Human Nutrition. *Ecol. Food Nutr.* **2011**, *50*, 351–367. [CrossRef]
41. Oonincx, D.G.A.B.; Laurent, S.; Veenenbos, M.E.; van Loon, J.J.A. Dietary Enrichment of Edible Insects with Omega 3 Fatty Acids. *Insect Sci.* **2020**, *27*, 500–509. [CrossRef]
42. Tacon, A.G.J.; Metian, M. Global Overview on the Use of Fish Meal and Fish Oil in Industrially Compounded Aquafeeds: Trends and Future Prospects. *Aquaculture* **2008**, *285*, 146–158. [CrossRef]
43. Almond, R.; Grooten, M.; Petersen, T. *Living Planet Report 2020: Bending the Curve of Biodiversity Loss*; World Wildlife Fund: Gland, Switzerland, 2020; ISBN 294052999X.
44. Fraga-Corral, M.; Ronza, P.; Garcia-Oliveira, P.; Pereira, A.G.; Losada, A.P.; Prieto, M.A.; Quiroga, M.I.; Simal-Gandara, J. Aquaculture as a Circular Bio-Economy Model with Galicia as a Study Case: How to Transform Waste into Revalorized by-Products. *Trends Food Sci. Technol.* **2022**, *119*, 23–35. [CrossRef]
45. Tilman, D.; Clark, M. Global Diets Link Environmental Sustainability and Human Health. *Nature* **2014**, *515*, 518–522. [CrossRef]

46. Le Féon, S.; Thévenot, A.; Maillard, F.; Macombe, C.; Forteau, L.; Aubin, J. Life Cycle Assessment of Fish Fed with Insect Meal: Case Study of Mealworm Inclusion in Trout Feed, in France. *Aquaculture* **2019**, *500*, 82–91. [CrossRef]
47. Smárason, B.Ö.; Ögmundarson, Ó.; Árnason, J.; Björnsdóttir, R.; Davíðsdóttir, B. Life Cycle Assessment of Icelandic Arctic Char Fed Three Different Feed Types. *Turkish J. Fish. Aquat. Sci.* **2017**, *17*, 79–90. [CrossRef]
48. Jayant, M.; Hassan, M.A.; Srivastava, P.P.; Meena, D.K.; Kumar, P.; Kumar, A.; Wagde, M.S. Brewer's Spent Grains (BSGs) as Feedstuff for Striped Catfish, Pangasianodon Hypophthalmus Fingerlings: An Approach to Transform Waste into Wealth. *J. Clean. Prod.* **2018**, *199*, 716–722. [CrossRef]
49. Jayant, M.; Sahu, N.P.; Deo, A.D.; Gupta, S.; Rajendran, K.V.; Garg, C.K.; Meena, D.K.; Wagde, M.S. Effective Valorization of Bio-Processed Castor Kernel Meal Based Fish Feed Supplements Concomitant with Oil Extraction Processing Industry: A Prolific Way towards Greening of Landscaping/Environment. *Environ. Technol. Innov.* **2021**, *21*, 101320. [CrossRef]
50. Miyashita, K. Prevention of Fish Oil Oxidation. *J. Oleo Sci.* **2019**, *68*, 1–11. [CrossRef]
51. Phung, N.; Félix, R.; Meggiolaro, D.; Al-Ashouri, A.; Sousa, E.; Silva, G.; Hartmann, C.; Hidalgo, J.; Köbler, H.; Mosconi, E.; et al. The Doping Mechanism of Halide Perovskite Unveiled by Alkaline Earth Metals. *J. Am. Chem. Soc.* **2020**, *142*, 2364–2374. [CrossRef]
52. Rodrigues, D.P.; Calado, R.; Pinho, M.; Rosário Domingues, M.; Antonio Vázquez, J.; Ameixa, O.M.C.C. Bioconversion and Performance of Black Soldier Fly (*Hermetia illucens*) in the Recovery of Nutrients from Expired Fish Feeds. *Waste Manag.* **2022**, *141*, 183–193. [CrossRef]
53. Makkar, H.P.S.; Tran, G.; Heuzé, V.; Ankers, P. State-of-the-Art on Use of Insects as Animal Feed. *Anim. Feed Sci. Technol.* **2014**, *197*, 1–33. [CrossRef]
54. Smetana, S.; Palanisamy, M.; Mathys, A.; Heinz, V. Sustainability of Insect Use for Feed and Food: Life Cycle Assessment Perspective. *J. Clean. Prod.* **2016**, *137*, 741–751. [CrossRef]
55. Lohri, C.R.; Diener, S.; Zabaleta, I.; Mertenat, A.; Zurbrügg, C. Treatment Technologies for Urban Solid Biowaste to Create Value Products: A Review with Focus on Low- and Middle-Income Settings. *Rev. Environ. Sci. Biotechnol.* **2017**, *16*, 81–130. [CrossRef]
56. Diener, S.; Zurbrügg, C.; Tockner, K. Conversion of Organic Material by Black Soldier Fly Larvae: Establishing Optimal Feeding Rates. *Waste Manag. Res.* **2009**, *27*, 603–610. [CrossRef] [PubMed]
57. FAO. *Edible Insects. Future Prospects for Food and Feed Security*; FAO: Rome, Italy, 2013; Volume 171, ISBN 9789251075951.
58. Bruni, L.; Pastorelli, R.; Viti, C.; Gasco, L.; Parisi, G. Characterisation of the Intestinal Microbial Communities of Rainbow Trout (*Oncorhynchus mykiss*) Fed with *Hermetia illucens* (Black Soldier Fly) Partially Defatted Larva Meal as Partial Dietary Protein Source. *Aquaculture* **2018**, *487*, 56–63. [CrossRef]
59. FAO. *Food Wastage Footprint*; FAO: Rome, Italy, 2013; 62p.
60. Durst, P.B.; Shono, K. Edible Forest Insects: Exploring New Horizons and Traditional Practices. In *Forest Insects as Food Humans Bite Back, Proceedings of the a Workshop on Asia-Pacific Resources and Their Potential for Development, Chiang Mai, Thailand, 19–21 February 2008*; Food and Agriculture Organization of the United Nations, Regional Office for Asia and the Pacific: Rome, Italy, 2017; Volume 1, pp. 1–4.
61. Lecocq, T. Insects: The Disregarded Domestication Histories. In *Animal Domestication*; IntechOpen: London, UK, 2019.
62. EL-Haroun, E.R.; Azevedo, P.A.; Bureau, D.P. High Dietary Incorporation Levels of Rendered Animal Protein Ingredients on Performance of Rainbow Trout *Oncorhynchus mykiss* (Walbaum, 1972). *Aquaculture* **2009**, *290*, 269–274. [CrossRef]
63. Jang, E.H.; Ko, J.H.; Ahn, C.W.; Lee, H.H.; Shin, J.K.; Chang, S.J.; Park, C.S.; Kang, J.H. In Vivo and in Vitro Application of Black Soybean Peptides in the Amelioration of Endoplasmic Reticulum Stress and Improvement of Insulin Resistance. *Life Sci.* **2010**, *86*, 267–274. [CrossRef]
64. Oliva-Teles, A.; Enes, P.; Peres, H. Replacing Fishmeal and Fish Oil in Industrial Aquafeeds for Carnivorous Fish. In *Feed and Feeding Practices in Aquaculture*; Woodhead Publishing: Oxford, UK, 2015; pp. 203–233. [CrossRef]
65. Barrows, F.T.; Gaylord, T.G.; Stone, D.A.J.; Smith, C.E. Effect of Protein Source and Nutrient Density on Growth Efficiency, Histology and Plasma Amino Acid Concentration of Rainbow Trout (*Oncorhynchus mykiss* Walbaum). *Aquac. Res.* **2007**, *38*, 1747–1758. [CrossRef]
66. Gatlin, D.M.; Barrows, F.T.; Brown, P.; Dabrowski, K.; Gaylord, T.G.; Hardy, R.W.; Herman, E.; Hu, G.; Krogdahl, Å.; Nelson, R.; et al. Expanding the Utilization of Sustainable Plant Products in Aquafeeds: A Review. *Aquac. Res.* **2007**, *38*, 551–579. [CrossRef]
67. Krogdahl, Å.; Penn, M.; Thorsen, J.; Refstie, S.; Bakke, A.M. Important Antinutrients in Plant Feedstuffs for Aquaculture: An Update on Recent Findings Regarding Responses in Salmonids. *Aquac. Res.* **2010**, *41*, 333–344. [CrossRef]
68. Li, X.; Rahimnejad, S.; Wang, L.; Lu, K.; Song, K.; Zhang, C. Substituting Fish Meal with Housefly (*Musca domestica*) Maggot Meal in Diets for Bullfrog *Rana* (*Lithobates*) *Catesbeiana*: Effects on Growth, Digestive Enzymes Activity, Antioxidant Capacity and Gut Health. *Aquaculture* **2019**, *499*, 295–305. [CrossRef]
69. Kousoulaki, K.; Albrektsen, S.; Langmyhr, E.; Olsen, H.J.; Campbell, P.; Aksnes, A. The Water Soluble Fraction in Fish Meal (Stickwater) Stimulates Growth in Atlantic Salmon (*Salmo salar* L.) given High Plant Protein Diets. *Aquaculture* **2009**, *289*, 74–83. [CrossRef]
70. Pinto, W.; Rønnestad, I.; Dinis, M.T.; Aragão, C. Taurine and Fish Development: Insights for the Aquaculture Industry. In *Advances in Experimental Medicine and Biology*; Springer: Berlin/Heidelberg, Germany, 2013; Volume 776, pp. 329–334. ISBN 9781461460923.

71. Zhang, J.; Hu, Y.; Ai, Q.; Mao, P.; Tian, Q.; Zhong, L.; Xiao, T.; Chu, W. Effect of Dietary Taurine Supplementation on Growth Performance, Digestive Enzyme Activities and Antioxidant Status of Juvenile Black Carp (*Mylopharyngodon piceus*) Fed with Low Fish Meal Diet. *Aquac. Res.* **2018**, *49*, 3187–3195. [CrossRef]
72. Izquierdo, M.; Domínguez, D.; Jiménez, J.I.; Saleh, R.; Hernández-Cruz, C.M.; Zamorano, M.J.; Hamre, K. Interaction between Taurine, Vitamin E and Vitamin C in Microdiets for Gilthead Seabream (*Sparus aurata*) Larvae. *Aquaculture* **2019**, *498*, 246–253. [CrossRef]
73. Takagi, S.; Murata, H.; Goto, T.; Endo, M.; Yamashita, H.; Ukawa, M. Taurine Is an Essential Nutrient for Yellowtail Seriola Quinqueradiata Fed Non-Fish Meal Diets Based on Soy Protein Concentrate. *Aquaculture* **2008**, *280*, 198–205. [CrossRef]
74. Rotman, F.; Stuart, K.; Drawbridge, M. Effects of Taurine Supplementation in Live Feeds on Larval Rearing Performance of California Yellowtail Seriola Lalandi and White Seabass Atractoscion Nobilis. *Aquac. Res.* **2017**, *48*, 1232–1239. [CrossRef]
75. Chen, X.; Feng, Y.; Chen, Z. Common Edible Insects and Their Utilization in China: Invited Review. *Entomol. Res.* **2009**, *39*, 299–303. [CrossRef]
76. Nguyen, T.N.; Davis, D.A.; Saoud, I.P. Evaluation of Alternative Protein Sources to Replace Fish Meal in Practical Diets for Juvenile Tilapia, *Oreochromis* Spp. *J. World Aquac. Soc.* **2009**, *40*, 113–121. [CrossRef]
77. Ramos-Elorduy, J. Energy Supplied by Edible Insects from Mexico and Their Nutritional and Ecological Importance. *Ecol. Food Nutr.* **2008**, *47*, 280–297. [CrossRef]
78. Shah, A.A.; Totakul, P.; Matra, M.; Cherdthong, A.; Hanboonsong, Y.; Wanapat, M. Nutritional Composition of Various Insects and Potential Uses as Alternative Protein Sources in Animal Diets. *Anim. Biosci.* **2022**, *35*, 317–331. [CrossRef]
79. Fischer, A.R.H.; Steenbekkers, L.P.A. All Insects Are Equal, but Some Insects Are More Equal than Others. *Br. Food J.* **2018**, *120*, 852–863. [CrossRef]
80. Finke, M.D. Complete Nutrient Composition of Commercially Raised Invertebrates Used as Food for Insectivores. *Zoo Biol.* **2002**, *21*, 269–285. [CrossRef]
81. Shockley, M.; Dossey, A.T. Insects for Human Consumption. In *Mass Production of Beneficial Organisms*; Elsevier: Amsterdam, The Netherlands, 2014; pp. 617–652. ISBN 9780123914538.
82. Finke, M.D. Complete Nutrient Content of Four Species of Commercially Available Feeder Insects Fed Enhanced Diets during Growth. *Zoo Biol.* **2015**, *34*, 554–564. [CrossRef]
83. Finke, M.D.; Oonincx, D. Insects as Food for Insectivores. In *Mass Production of Beneficial Organisms: Invertebrates and Entomopathogens*; Elsevier: Amsterdam, The Netherlands, 2013; pp. 583–616. ISBN 9780123914538.
84. Oonincx, D.G.A.B.; de Boer, I.J.M. Environmental Impact of the Production of Mealworms as a Protein Source for Humans—A Life Cycle Assessment. *PLoS ONE* **2012**, *7*, e51145. [CrossRef] [PubMed]
85. Oonincx, D.G.A.B.; Van Der Poel, A.F.B. Effects of Diet on the Chemical Composition of Migratory Locusts (*Locusta migratoria*). *Zoo Biol.* **2011**, *30*, 9–16. [CrossRef] [PubMed]
86. Rawski, M.; Mazurkiewicz, J.; Kierończyk, B.; Józefiak, D. Black Soldier Fly Full-Fat Larvae Meal as an Alternative to Fish Meal and Fish Oil in Siberian Sturgeon Nutrition: The Effects on Physical Properties of the Feed, Animal Growth Performance, and Feed Acceptance and Utilization. *Animals* **2020**, *10*, 2119. [CrossRef] [PubMed]
87. Hung, S.S.O. Recent Advances in Sturgeon Nutrition. *Anim. Nutr.* **2017**, *3*, 191–204. [CrossRef]
88. Oonincx, D.G.A.B.; van Broekhoven, S.; van Huis, A.; van Loon, J.J.A. Correction: Feed Conversion, Survival and Development, and Composition of Four Insect Species on Diets Composed of Food Byproducts. *PLoS ONE* **2015**, *10*, e0144601. [CrossRef]
89. Boulos, S.; Tännler, A.; Nyström, L. Nitrogen-to-Protein Conversion Factors for Edible Insects on the Swiss Market: *T. molitor*, *A. domesticus*, and *L. migratoria*. *Front. Nutr.* **2020**, *7*, 89. [CrossRef]
90. Bukkens, S.G.F. Insects in the Human Diet: Nutritional Aspects. In *Ecological Implications of Minilivestock*; CRC Press: Boca Raton, FL, USA, 2005; pp. 559–592. [CrossRef]
91. Rumpold, B.A.; Schlüter, O.K. Potential and Challenges of Insects as an Innovative Source for Food and Feed Production. *Innov. Food Sci. Emerg. Technol.* **2013**, *17*, 1–11. [CrossRef]
92. Barroso, F.G.; de Haro, C.; Sánchez-Muros, M.J.; Venegas, E.; Martínez-Sánchez, A.; Pérez-Bañón, C. The Potential of Various Insect Species for Use as Food for Fish. *Aquaculture* **2014**, *422–423*, 193–201. [CrossRef]
93. Blásquez, J.R.-E.; Moreno, J.M.P.; Camacho, V.H.M. Could Grasshoppers Be a Nutritive Meal? *Food Nutr. Sci.* **2012**, *3*, 164–175. [CrossRef]
94. Ramos Elorduy, J.; Costa Neto, E.M.; Cuevas Correa, M.S.; García-Figueroa, J.; Zentina, D.H. Conocimiento de La Entomofauna Útil En El Poblado La Purísima Palmar de Bravo, Estado de Puebla, México. *Biotemas* **2007**, *20*, 121–134.
95. Okey Aniebo, A.; Erondu, E.S.; Owen, O.J. Replacement of Fish Meal with Maggot Meal in African Catfish (*Clarias gariepinus*) Diets. *Rev. UDO Agrícola* **2009**, *9*, 666–671.
96. Yang, Q.; Liu, S.; Sun, J.; Yu, L.; Zhang, C.; Bi, J.; Yang, Z. Nutritional Composition and Protein Quality of the Edible Beetle Holotrichia Parallela. *J. Insect Sci.* **2014**, *14*, 139. [CrossRef] [PubMed]
97. Ravzanaadii, N.; Kim, S.-H.; Choi, W.-H.; Hong, S.-J.; Kim, N.-J. Nutritional Value of Mealworm, *Tenebrio molitor* as Food Source. *Int. J. Ind. Entomol.* **2012**, *25*, 93–98. [CrossRef]
98. Sadykova, E.O.; Shumakova, A.A.; Shestakova, S.I.; Tyshko, N.V. Nutritional and Biological Value of *Hermetia illucens* Larvae Biomass. *Vopr. Pitan.* **2021**, *90*, 73–82. [CrossRef] [PubMed]

99. Dossey, A.T.; Tatum, J.T.; McGill, W.L. Modern Insect-Based Food Industry: Current Status, Insect Processing Technology, and Recommendations Moving Forward. In *Insects as Sustainable Food Ingredients*; Springer: Amsterdam, The Netherlands, 2016; pp. 113–152.
100. Mohamed, E. Determination of Nutritive Value of the Edible Migratory Locust *Locusta migratoria*, Linnaeus, 1758 (Orthoptera: Acrididae) Impact of Environmental Changes on Fish Availability and Fishery Landings in Khartoum State View Project Impact of Environmental Cha. *Int. J. Adv. Pharm. Biol. Chem.* **2015**, *4*, 144–148.
101. Kim, S.-K.; Weaver, C.M.; Choi, M.-K. Proximate Composition and Mineral Content of Five Edible Insects Consumed in Korea. *CyTA J. Food* **2016**, *15*, 143–146. [CrossRef]
102. Banjo, A.D.; Lawal, O.A.; Songonuga, E.A. The Nutritional Value of Fourteen Species of Edible Insects in Southwestern Nigeria. *African J. Biotechnol.* **2006**, *5*, 298–301. [CrossRef]
103. Siulapwa, N.; Mwambungu, A.; Lungu, E.; Sichilima, W. Nutritional Value of Four Common Edible Insects in Zambia. *Int. J. Sci. Res.* **2012**, *3*, 2319–7064.
104. Adeolu, A. Proximate and Anti-Nutritional Composition of Two Common Edible Insects: Yam Beetle (*Heteroligus meles*) and Palm Weevil (*Rhynchophorus phoenicis*). *Elixir Food Sci.* **2012**, *49*, 9782–9786.
105. Kinyuru, J.N.; Mogendi, J.B.; Riwa, C.A.; Ndung'u, N.W. Edible Insects-A Novel Source of Essential Nutrients for Human Diet: Learning from Traditional Knowledge. *Anim. Front.* **2015**, *5*, 14–19. [CrossRef]
106. Churchward-Venne, T.A.; Pinckaers, P.J.M.; van Loon, J.J.A.; van Loon, L.J.C. Consideration of Insects as a Source of Dietary Protein for Human Consumption. *Nutr. Rev.* **2017**, *75*, 1035–1045. [CrossRef] [PubMed]
107. Hackewitz, L. von The House Cricket Acheta Domesticus, a Potential Source of Protein for Human Consumption. *Mol. Sci.* **2018**, *17*, 33.
108. Rumpold, B.A.; Schlüter, O. Insect-Based Protein Sources and Their Potential for Human Consumption: Nutritional Composition and Processing. *Anim. Front.* **2015**, *5*, 20–24. [CrossRef]
109. Gangopadhyay, D.; Ray, M.; Sinha, S. Comparison of Amino Acid Profiles and Vitamin Contents of Male and Female Prepupae and Pupae of Eri Silkworm, Samia Ricini. *J. Food Compos. Anal.* **2022**, *113*, 104723. [CrossRef]
110. Kouřimská, L.; Adámková, A. Nutritional and Sensory Quality of Edible Insects. *NFS J.* **2016**, *4*, 22–26. [CrossRef]
111. Mlcek, J.; Rop, O.; Borkovcova, M.; Bednarova, M. A Comprehensive Look at the Possibilities of Edible Insects as Food in Europe—A Review. *Polish J. Food Nutr. Sci.* **2014**, *64*, 147–157. [CrossRef]
112. Raksakantong, P.; Meeso, N.; Kubola, J.; Siriamornpun, S. Fatty Acids and Proximate Composition of Eight Thai Edible Terricolous Insects. *Food Res. Int.* **2010**, *43*, 350–355. [CrossRef]
113. Hajji, S.; Younes, I.; Ghorbel-Bellaaj, O.; Hajji, R.; Rinaudo, M.; Nasri, M.; Jellouli, K. Structural Differences between Chitin and Chitosan Extracted from Three Different Marine Sources. *Int. J. Biol. Macromol.* **2014**, *65*, 298–306. [CrossRef]
114. Khoushab, F.; Yamabhai, M. Chitin Research Revisited. *Mar. Drugs* **2010**, *8*, 1988–2012. [CrossRef]
115. Ramos-Elorduy, J.; González, E.A.; Hernández, A.R.; Pino, J.M. Use of *Tenebrio molitor* (Coleoptera: Tenebrionidae) to Recycle Organic Wastes and as Feed for Broiler Chickens. *J. Econ. Entomol.* **2002**, *95*, 214–220. [CrossRef]
116. Ekpo, K.E.; Onigbinde, A.O. Nutritional Potentials of the Larva of *Rhynchophorus phoenicis* (F). *Pak. J. Nutr.* **2005**, *4*, 287–290. [CrossRef]
117. Azzollini, D.; Derossi, A.; Fogliano, V.; Lakemond, C.M.M.; Severini, C. Effects of Formulation and Process Conditions on Microstructure, Texture and Digestibility of Extruded Insect-Riched Snacks. *Innov. Food Sci. Emerg. Technol.* **2018**, *45*, 344–353. [CrossRef]
118. Zhao, X.; Vázquez-Gutiérrez, J.L.; Johansson, D.P.; Landberg, R.; Langton, M. Yellow Mealworm Protein for Food Purposes—Extraction and Functional Properties. *PLoS ONE* **2016**, *11*, e0147791. [CrossRef] [PubMed]
119. Rumpold, B.A.; Schlüter, O.K. Nutritional Composition and Safety Aspects of Edible Insects. *Mol. Nutr. Food Res.* **2013**, *57*, 802–823. [CrossRef]
120. Ghosh, S.; Lee, S.M.; Jung, C.; Meyer-Rochow, V.B. Nutritional Composition of Five Commercial Edible Insects in South Korea. *J. Asia. Pac. Entomol.* **2017**, *20*, 686–694. [CrossRef]
121. Tzompa-Sosa, D.A.; Yi, L.; van Valenberg, H.J.F.; van Boekel, M.A.J.S.; Lakemond, C.M.M. Insect Lipid Profile: Aqueous versus Organic Solvent-Based Extraction Methods. *Food Res. Int.* **2014**, *62*, 1087–1094. [CrossRef]
122. Ekpo, K.E.; Onigbinde, A.O.; Asia, I.O. Pharmaceutical Potentials of the Oils of Some Popular Insects Consumed in Southern Nigeria. *Afr. J. Pharm. Pharmacol.* **2009**, *3*, 051–057.
123. Hawkey, K.J.; Lopez-Viso, C.; Brameld, J.M.; Parr, T.; Salter, A.M. Insects: A Potential Source of Protein and Other Nutrients for Feed and Food. *Annu. Rev. Anim. Biosci.* **2021**, *9*, 333–354. [CrossRef]
124. Ruschioni, S.; Loreto, N.; Foligni, R.; Mannozzi, C.; Raffaelli, N.; Zamporlini, F.; Pasquini, M.; Roncolini, A.; Cardinali, F.; Osimani, A. Addition of Olive Pomace to Feeding Substrate Affects Growth Performance and Nutritional Value of Mealworm (*Tenebrio molitor* L.) Larvae. *Foods* **2020**, *9*, 317. [CrossRef]
125. Dobermann, D.; Swift, J.A.; Field, L.M. Opportunities and Hurdles of Edible Insects for Food and Feed. *Nutr. Bull.* **2017**, *42*, 293–308. [CrossRef]
126. Borrelli, L.; Varriale, L.; Dipineto, L.; Pace, A.; Menna, L.F.; Fioretti, A. Insect Derived Lauric Acid as Promising Alternative Strategy to Antibiotics in the Antimicrobial Resistance Scenario. *Front. Microbiol.* **2021**, *12*, 620798. [CrossRef] [PubMed]

127. Chen, B.Y.; Lin, K.W.; Wang, Y.M.; Yen, C.Y. Revealing Interactive Toxicity of Aromatic Amines to Azo Dye Decolorizer Aeromonas Hydrophila. *J. Hazard. Mater.* **2009**, *166*, 187–194. [CrossRef] [PubMed]
128. Oonincx, D.G.; Dierenfeld, E.S. An Investigation into the Chemical Composition of Alternative Invertebrate Prey. *Zoo Biol.* **2012**, *31*, 40–54. [CrossRef] [PubMed]
129. Ponton, F.; Wilson, K.; Cotter, S.C.; Raubenheimer, D.; Simpson, S.J. Nutritional Immunology: A Multi-Dimensional Approach. *PLoS Pathog.* **2011**, *7*, e1002223. [CrossRef] [PubMed]
130. Ponton, F.; Wilson, K.; Holmes, A.J.; Cotter, S.C.; Raubenheimer, D.; Simpson, S.J. Integrating Nutrition and Immunology: A New Frontier. *J. Insect Physiol.* **2013**, *59*, 130–137. [CrossRef]
131. Van Huis, A.; Van Itterbeeck, J.; Klunder, H.; Mertens, E.; Halloran, A.; Muir, G.; Vantomme, P. *Edible Insects: Future Prospects for Food and Feed Security*; Food and Agriculture Organization of the United Nations: Rome, Italy, 2013; ISBN 9251075964.
132. Zielińska, E.; Baraniak, B.; Karaś, M.; Rybczyńska, K.; Jakubczyk, A. Selected Species of Edible Insects as a Source of Nutrient Composition. *Food Res. Int.* **2015**, *77*, 460–466. [CrossRef]
133. Oonincx, D.G.A.B.; van Itterbeeck, J.; Heetkamp, M.J.W.; van den Brand, H.; van Loon, J.J.A.; van Huis, A. An Exploration on Greenhouse Gas and Ammonia Production by Insect Species Suitable for Animal or Human Consumption. *PLoS ONE* **2010**, *5*, e14445. [CrossRef]
134. Hyun, S.H.; Kwon, K.H.; Park, K.H.; Jeong, H.C.; Kwon, O.; Tindwa, H.; Han, Y.S. Evaluation of Nutritional Status of an Edible Grasshopper, Oxya Chinensis Formosana. *Entomol. Res.* **2012**, *42*, 284–290. [CrossRef]
135. Latunde-Dada, G.O.; Yang, W.; Vera Aviles, M. In Vitro Iron Availability from Insects and Sirloin Beef. *J. Agric. Food Chem.* **2016**, *64*, 8420–8424. [CrossRef]
136. Kosečková, P.; Zvěřina, O.; Pěchová, M.; Krulíková, M.; Duborská, E.; Borkovcová, M. Mineral Profile of Cricket Powders, Some Edible Insect Species and Their Implication for Gastronomy. *J. Food Compos. Anal.* **2022**, *107*, 104340. [CrossRef]
137. Van Huis, A. Potential of Insects as Food and Feed in Assuring Food Security. *Annu. Rev. Entomol.* **2013**, *58*, 563–583. [CrossRef] [PubMed]
138. Gravel, A.; Doyen, A. The Use of Edible Insect Proteins in Food: Challenges and Issues Related to Their Functional Properties. *Innov. Food Sci. Emerg. Technol.* **2020**, *59*, 102272. [CrossRef]
139. Villaseñor, V.M.; Enriquez-Vara, J.N.; Urías-Silva, J.E.; Mojica, L. Edible Insects: Techno-Functional Properties Food and Feed Applications and Biological Potential. *Food Rev. Int.* **2021**, 1–27. [CrossRef]
140. Kim, T.-K.; Cha, J.Y.; Yong, H.I.; Jang, H.W.; Jung, S.; Choi, Y.-S. Application of Edible Insects as Novel Protein Sources and Strategies for Improving Their Processing. *Food Sci. Anim. Resour.* **2022**, *42*, 372–388. [CrossRef]
141. Kim, T.-K.; Yong, H.I.; Kim, Y.-B.; Kim, H.-W.; Choi, Y.-S. Edible Insects as a Protein Source: A Review of Public Perception, Processing Technology, and Research Trends. *Food Sci. Anim. Resour.* **2019**, *39*, 521–540. [CrossRef]
142. Kim, T.-K.; Yong, H.I.; Chun, H.H.; Lee, M.-A.; Kim, Y.-B.; Choi, Y.-S. Changes of Amino Acid Composition and Protein Technical Functionality of Edible Insects by Extracting Steps. *J. Asia. Pac. Entomol.* **2020**, *23*, 298–305. [CrossRef]
143. Mintah, B.K.; He, R.; Agyekum, A.A.; Dabbour, M.; Golly, M.K.; Ma, H. Edible Insect Protein for Food Applications: Extraction, Composition, and Functional Properties. *J. Food Process Eng.* **2020**, *43*, e13362. [CrossRef]
144. Purschke, B.; Meinlschmidt, P.; Horn, C.; Rieder, O.; Jäger, H. Improvement of Techno-Functional Properties of Edible Insect Protein from Migratory Locust by Enzymatic Hydrolysis. *Eur. Food Res. Technol.* **2018**, *244*, 999–1013. [CrossRef]
145. Purschke, B.; Tanzmeister, H.; Meinlschmidt, P.; Baumgartner, S.; Lauter, K.; Jäger, H. Recovery of Soluble Proteins from Migratory Locust (*Locusta migratoria*) and Characterisation of Their Compositional and Techno-Functional Properties. *Food Res. Int.* **2018**, *106*, 271–279. [CrossRef]
146. Dion-Poulin, A.; Laroche, M.; Doyen, A.; Turgeon, S.L. Functionality of Cricket and Mealworm Hydrolysates Generated after Pretreatment of Meals with High Hydrostatic Pressures. *Molecules* **2020**, *25*, 5366. [CrossRef]
147. Li, X.; Zheng, S.; Cheng, K.; Ma, X.; Wu, G. Use of Alternative Protein Sources for Fishmeal Replacement in the Diet of Largemouth Bass (*Micropterus salmoides*). Part II: Effects of Supplementation with Methionine or Taurine on Growth, Feed Utilization, and Health. *Amino Acids* **2021**, *53*, 49–62. [CrossRef] [PubMed]
148. Lokeshkumar, V.; Daniel, B.A.; Jayanthi, J.; Ragunathan, M.G. Nutritive Evaluations of Laboratory-Reared Edible Field Cricket *Coiblemmus compactus* Chopard, 1928 (Orthoptera: Gryllidae), for Utilising Them as an Alternate Protein Source. *J. Basic Appl. Zool.* **2022**, *83*, 26. [CrossRef]
149. Kurečka, M.; Kulma, M.; Petříčková, D.; Plachý, V.; Kouřimská, L. Larvae and Pupae of Alphitobius Diaperinus as Promising Protein Alternatives. *Eur. Food Res. Technol.* **2021**, *247*, 2527–2532. [CrossRef]
150. Ogunji, J.O.; Kloas, W.; Wirth, M.; Neumann, N.; Pietsch, C. Effect of Housefly Maggot Meal (Magmeal) Diets on the Performance, Concentration of Plasma Glucose, Cortisol and Blood Characteristics of Oreochromis Niloticus Fingerlings. *J. Anim. Physiol. Anim. Nutr.* **2008**, *92*, 511–518. [CrossRef] [PubMed]
151. Ogunji, J.O.; Nimptsch, J.; Wiegand, C.; Schulz, C. Evaluation of the Influence of Housefly Maggot Meal (Magmeal) Diets on Catalase, Glutathione S-Transferase and Glycogen Concentration in the Liver of *Oreochromis niloticus* Fingerling. *Comp. Biochem. Physiol. A Mol. Integr. Physiol.* **2007**, *147*, 942–947. [CrossRef]
152. Finke, M.D. Estimate of Chitin in Raw Whole Insects. *Zoo Biol.* **2007**, *26*, 105–115. [CrossRef]
153. Spranghers, T.; Michiels, J.; Vrancx, J.; Ovyn, A.; Eeckhout, M.; De Clercq, P.; De Smet, S. Gut Antimicrobial Effects and Nutritional Value of Black Soldier Fly (*Hermetia illucens* L.) Prepupae for Weaned Piglets. *Anim. Feed Sci. Technol.* **2018**, *235*, 33–42. [CrossRef]

154. Tran, G.; Heuzé, V.; Makkar, H.P.S. Insects in Fish Diets. *Anim. Front.* **2015**, *5*, 37–44. [CrossRef]
155. Yi, L.; Lakemond, C.M.M.; Sagis, L.M.C.; Eisner-Schadler, V.; Van Huis, A.; Boekel, M.A.J.S.V. Extraction and Characterisation of Protein Fractions from Five Insect Species. *Food Chem.* **2013**, *141*, 3341–3348. [CrossRef]
156. Rachmawati, R.; Buchori, D.; Hidayat, P.; Hem, S.; Fahmi, M.R. Perkembangan Dan Kandungan Nutrisi Larva *Hermetia illucens* (Linnaeus) (Diptera: Stratiomyidae) Pada Bungkil Kelapa Sawit. *J. Entomol. Indones.* **2015**, *7*, 28. [CrossRef]
157. Ghosh, S.; Sohn, H.-Y.; Pyo, S.-J.; Jensen, A.B.; Meyer-Rochow, V.B.; Jung, C. Nutritional Composition of Apis Mellifera Drones from Korea and Denmark as a Potential Sustainable Alternative Food Source: Comparison Between Developmental Stages. *Foods* **2020**, *9*, 389. [CrossRef] [PubMed]
158. Ademolu, K.O.; Idowu, A.B.; Olatunde, G.O. Nutritional Value Assessment of Variegated Grasshopper, *Zonocerus variegatus* (L.) (Acridoidea: Pygomorphidae), during Post-Embryonic Development. *Afr. Entomol.* **2010**, *18*, 360–364. [CrossRef]
159. Kulma, M.; Plachý, V.; Kouřimská, L.; Vrabec, V.; Bubová, T.; Adámková, A.; Hučko, B. Nutritional Value of Three Blattodea Species Used as Feed for Animals. *J. Anim. Feed Sci.* **2016**, *25*, 354–360. [CrossRef]
160. Nowak, V.; Persijn, D.; Rittenschober, D.; Charrondiere, U.R. Review of Food Composition Data for Edible Insects. *Food Chem.* **2016**, *193*, 39–46. [CrossRef] [PubMed]
161. Finke, M.D. Nutrient Content of Insects. In *Encyclopedia of Entomology*; Wageningen Academic Publishers: Wageningen, The Netherlands, 2006; pp. 1563–1575.
162. Ghosh, S.; Jung, C.; Meyer-Rochow, V.B. Nutritional Value and Chemical Composition of Larvae, Pupae, and Adults of Worker Honey Bee, Apis Mellifera Ligustica as a Sustainable Food Source. *J. Asia. Pac. Entomol.* **2016**, *19*, 487–495. [CrossRef]
163. Barroso, F.G.; Sánchez-Muros, M.J.; Segura, M.; Morote, E.; Torres, A.; Ramos, R.; Guil, J.L. Insects as Food: Enrichment of Larvae of *Hermetia illucens* with Omega 3 Fatty Acids by Means of Dietary Modifications. *J. Food Compos. Anal.* **2017**, *62*, 8–13. [CrossRef]
164. Van Huis, A.; Dicke, M.; van Loon, J.J.A. Insects to Feed the World. *J. Insects Food Feed* **2015**, *1*, 3–5. [CrossRef]
165. Zarantoniello, M.; Zimbelli, A.; Randazzo, B.; Compagni, M.D.; Truzzi, C.; Antonucci, M.; Riolo, P.; Loreto, N.; Osimani, A.; Milanović, V.; et al. Black Soldier Fly (*Hermetia illucens*) Reared on Roasted Coffee by-Product and *Schizochytrium* Sp. as a Sustainable Terrestrial Ingredient for Aquafeeds Production. *Aquaculture* **2020**, *518*, 734659. [CrossRef]
166. Chemello, G.; Zarantoniello, M.; Randazzo, B.; Gioacchini, G.; Truzzi, C.; Cardinaletti, G.; Riolo, P.; Olivotto, I. Effects of Black Soldier Fly (*Hermetia illucens*) Enriched with Schizochytrium Sp. on Zebrafish (Danio Rerio) Reproductive Performances. *Aquaculture* **2022**, *550*, 737853. [CrossRef]
167. Choi, I.H.; Kim, J.M.; Kim, N.J.; Kim, J.D.; Park, C.; Park, J.H.; Chung, T.H. Replacing Fish Meal by Mealworm (*Tenebrio molitor*) on the Growth Performance and Immunologic Responses of White Shrimp (*Litopenaeus vannamei*). *Acta Sci. Anim. Sci.* **2018**, *40*, e35015. [CrossRef]
168. Newton, G.L.; Sheppard, D.C.; Watson, D.W.; Burtle, G.J.; Dove, C.R.; Tomberlin, J.K.; Thelen, E.E. The Black Soldier Fly, *Hermetia illucens*, as a Manure Management/Resource Recovery Tool. In Proceedings of the Symposium on the state of the science of Animal Manure and Waste Management, San Antonio, TX, USA, 16 September 2005; Volume 57, pp. 1–5.
169. Kinyuru, J.N.; Kenji, G.M.; Njoroge, S.M.; Ayieko, M. Effect of Processing Methods on the in Vitro Protein Digestibility and Vitamin Content of Edible Winged Termite (*Macrotermes subhylanus*) and Grasshopper (*Ruspolia differens*). *Food Bioprocess Technol.* **2010**, *3*, 778–782. [CrossRef]
170. Madibela, O.R.; Seitiso, T.K.; Thema, T.F.; Letso, M. Effect of Traditional Processing Methods on Chemical Composition and in Vitro True Dry Matter Digestibility of the Mophane Worm (*Imbrasia belina*). *J. Arid Environ.* **2007**, *68*, 492–500. [CrossRef]
171. Meshulam-Pascoviche, D.; David-Birman, T.; Refael, G.; Lesmes, U. Big Opportunities for Tiny Bugs: Processing Effects on the Techno-Functionality and Digestibility of Edible Insects. *Trends Food Sci. Technol.* **2022**, *122*, 265–274. [CrossRef]
172. Grabowski, N.T.; Klein, G. Microbiology of Cooked and Dried Edible Mediterranean Field Crickets (*Gryllus bimaculatus*) and Superworms (*Zophobas atratus*) Submitted to Four Different Heating Treatments. *Food Sci. Technol. Int.* **2017**, *23*, 17–23. [CrossRef]
173. Nantanga, K.K.M.; Amakali, T. Diversification of Mopane Caterpillars (*Gonimbrasia belina*) Edible Forms for Improved Livelihoods and Food Security. *J. Arid Environ.* **2020**, *177*, 104148. [CrossRef]
174. Bußler, S.; Rumpold, B.A.; Jander, E.; Rawel, H.M.; Schlüter, O.K. Recovery and Techno-Functionality of Flours and Proteins from Two Edible Insect Species: Meal Worm (*Tenebrio molitor*) and Black Soldier Fly (*Hermetia illucens*) Larvae. *Heliyon* **2016**, *2*, e00218. [CrossRef]
175. Ojha, S.; Bußler, S.; Psarianos, M.; Rossi, G.; Schlüter, O.K. Edible Insect Processing Pathways and Implementation of Emerging Technologies. *J. Insects Food Feed* **2021**, *7*, 877–900. [CrossRef]
176. Mishyna, M.; Chen, J.; Benjamin, O. Sensory Attributes of Edible Insects and Insect-Based Foods–Future Outlooks for Enhancing Consumer Appeal. *Trends Food Sci. Technol.* **2020**, *95*, 141–148. [CrossRef]
177. Nongonierma, A.B.; FitzGerald, R.J. Unlocking the Biological Potential of Proteins from Edible Insects through Enzymatic Hydrolysis: A Review. *Innov. Food Sci. Emerg. Technol.* **2017**, *43*, 239–252. [CrossRef]
178. Lu, Y.; Yan, R.; Ma, X.; Wang, Y. Synthesis and Characterization of Raffinose Fatty Acid Monoesters under Ultrasonic Irradiation. *Eur. Food Res. Technol.* **2013**, *237*, 237–244. [CrossRef]
179. Zielińska, E.; Karaś, M.; Baraniak, B. Comparison of Functional Properties of Edible Insects and Protein Preparations Thereof. *LWT Food Sci. Technol.* **2018**, *91*, 168–174. [CrossRef]
180. Lamidi, R.O.; Jiang, L.; Pathare, P.B.; Wang, Y.D.; Roskilly, A.P. Recent Advances in Sustainable Drying of Agricultural Produce: A Review. *Appl. Energy* **2019**, *233–234*, 367–385. [CrossRef]

181. Hashim, M.A.; Yahya, F.; Mustapha, W.A.W. Effect of Different Drying Methods on the Morphological Structure, Colour Profile and Citral Concentration of Lemongrass (*Cymbopogon citratus*) Powder. *Asian J. Agric. Biol* **2019**, *7*, 93–102.
182. Tantiyani, N.; Othman, A.; Elmi, M.; Mohd, F. Drying of Instant Coffee in a Spray Dryer. *J. Kejuruter.* **2019**, *31*, 295–301.
183. Baek, M.; Yoon, Y.-I.; Kim, M.A.; Hwang, J.-S.; Goo, T.-W.; Yun, E.-Y. Physical and Sensory Evaluation of *Tenebrio molitor* Larvae Cooked by Various Cooking Methods. *Korean J. Food Cook. Sci.* **2015**, *31*, 534–543. [CrossRef]
184. Zahwal, E.N.; Fang, C.P.; Ibrahim, N.N.A.; Razali, N.S.M.; Rahman, H.A. Effect of Different Drying Methods and Solvents on Phenolic Content, Antioxidant and Anti-Hyperglycemic Activities of Melastoma Malabathricum Leaves Extract. *Malays. Appl. Biol.* **2018**, *47*, 47–53.
185. Norafida, A.; Aminah, A. Effect of Blanching Treatments on Antioxidant Activity of Frozen Green Capsicum (*Capsicum annuum* L. var bell pepper). *Int. Food Res. J.* **2018**, *25*, 1427–1434.
186. Van Boekel, M.; Fogliano, V.; Pellegrini, N.; Stanton, C.; Scholz, G.; Lalljie, S.; Somoza, V.; Knorr, D.; Jasti, P.R.; Eisenbrand, G. A Review on the Beneficial Aspects of Food Processing. *Mol. Nutr. Food Res.* **2010**, *54*, 1215–1247. [CrossRef]
187. Poelaert, C.; Francis, F.; Alabi, T.; Caparros Megido, R.; Crahay, B.; Bindelle, J.; Beckers, Y. Protein Value of Two Insects, Subjected to Various Heat Treatments, Using Growing Rats and the Protein Digestibility-Corrected Amino Acid Score. *J. Insects Food Feed* **2018**, *4*, 77–87. [CrossRef]
188. Kewuyemi, Y.O.; Kesa, H.; Chinma, C.E.; Adebo, O.A. Fermented Edible Insects for Promoting Food Security in Africa. *Insects* **2020**, *11*, 283. [CrossRef]
189. Castro-López, C.; Santiago-López, L.; Vallejo-Cordoba, B.; González-Córdova, A.F.; Liceaga, A.M.; García, H.S.; Hernández-Mendoza, A. An Insight to Fermented Edible Insects: A Global Perspective and Prospective. *Food Res. Int.* **2020**, *137*, 109750. [CrossRef]
190. Osimani, A.; Milanović, V.; Cardinali, F.; Garofalo, C.; Clementi, F.; Ruschioni, S.; Riolo, P.; Isidoro, N.; Loreto, N.; Galarini, R.; et al. Distribution of Transferable Antibiotic Resistance Genes in Laboratory-Reared Edible Mealworms (*Tenebrio molitor* L.). *Front. Microbiol.* **2018**, *9*, 2702. [CrossRef]
191. Zurek, L.; Ghosh, A. Los Insectos Representan Un Vínculo Entre Las Granjas de Animales de Consumo y El Entorno Urbano Para Los Rasgos de Resistencia a Los Antibióticos. Müller V., Ed. *Appl. Environ. Microbiol.* **2014**, *80*, 3562–3567. [CrossRef] [PubMed]
192. Vandeweyer, D.; Milanović, V.; Garofalo, C.; Osimani, A.; Clementi, F.; Van Campenhout, L.; Aquilanti, L. Real-Time PCR Detection and Quantification of Selected Transferable Antibiotic Resistance Genes in Fresh Edible Insects from Belgium and the Netherlands. *Int. J. Food Microbiol.* **2019**, *290*, 288–295. [CrossRef] [PubMed]
193. Milanović, V.; Osimani, A.; Pasquini, M.; Aquilanti, L.; Garofalo, C.; Taccari, M.; Cardinali, F.; Riolo, P.; Clementi, F. Getting Insight into the Prevalence of Antibiotic Resistance Genes in Specimens of Marketed Edible Insects. *Int. J. Food Microbiol.* **2016**, *227*, 22–28. [CrossRef]
194. Osimani, A.; Cardinali, F.; Aquilanti, L.; Garofalo, C.; Roncolini, A.; Milanović, V.; Pasquini, M.; Tavoletti, S.; Clementi, F. Occurrence of Transferable Antibiotic Resistances in Commercialized Ready-to-Eat Mealworms (*Tenebrio molitor* L.). *Int. J. Food Microbiol.* **2017**, *263*, 38–46. [CrossRef] [PubMed]
195. Osimani, A.; Garofalo, C.; Aquilanti, L.; Milanović, V.; Cardinali, F.; Taccari, M.; Pasquini, M.; Tavoletti, S.; Clementi, F. Transferable Antibiotic Resistances in Marketed Edible Grasshoppers (*Locusta migratoria migratorioides*). *J. Food Sci.* **2017**, *82*, 1184–1192. [CrossRef] [PubMed]
196. Turchini, G.M.; Trushenski, J.T.; Glencross, B.D. Thoughts for the Future of Aquaculture Nutrition: Realigning Perspectives to Reflect Contemporary Issues Related to Judicious Use of Marine Resources in Aquafeeds. *N. Am. J. Aquac.* **2019**, *81*, 13–39. [CrossRef]
197. Fry, J.P.; Love, D.C.; MacDonald, G.K.; West, P.C.; Engstrom, P.M.; Nachman, K.E.; Lawrence, R.S. Environmental Health Impacts of Feeding Crops to Farmed Fish. *Environ. Int.* **2016**, *91*, 201–214. [CrossRef] [PubMed]
198. Kokou, F.; Fountoulaki, E. Aquaculture Waste Production Associated with Antinutrient Presence in Common Fish Feed Plant Ingredients. *Aquaculture* **2018**, *495*, 295–310. [CrossRef]
199. Chen, H.Y.; Li, H.L.; Pang, H.; Zhu, C.D.; Zhang, Y.Z. Investigating the Parasitoid Community Associated with the Invasive Mealybug Phenacoccus Solenopsis in Southern China. *Insects* **2021**, *12*, 290. [CrossRef] [PubMed]
200. Sánchez-Muros, M.J.; Barroso, F.G.; Manzano-Agugliaro, F. Insect Meal as Renewable Source of Food for Animal Feeding: A Review. *J. Clean. Prod.* **2014**, *65*, 16–27. [CrossRef]
201. Surendra, K.C.; Tomberlin, J.K.; van Huis, A.; Cammack, J.A.; Heckmann, L.H.L.; Khanal, S.K. Rethinking Organic Wastes Bioconversion: Evaluating the Potential of the Black Soldier Fly (*Hermetia illucens* (L.)) (Diptera: Stratiomyidae) (BSF). *Waste Manag.* **2020**, *117*, 58–80. [CrossRef]
202. Kumari, M.; Singh, P.; Nataraj, B.H.; Kokkiligadda, A.; Naithani, H.; Azmal Ali, S.; Behare, P.V.; Nagpal, R. Fostering Next-Generation Probiotics in Human Gut by Targeted Dietary Modulation: An Emerging Perspective. *Food Res. Int.* **2021**, *150*, 110716. [CrossRef] [PubMed]
203. Peng, M.; Tabashsum, Z.; Anderson, M.; Truong, A.; Houser, A.K.; Padilla, J.; Akmel, A.; Bhatti, J.; Rahaman, S.O.; Biswas, D. Effectiveness of Probiotics, Prebiotics, and Prebiotic-like Components in Common Functional Foods. *Compr. Rev. Food Sci. Food Saf.* **2020**, *19*, 1908–1933. [CrossRef]
204. Singh, A.; Kumari, K. An Inclusive Approach for Organic Waste Treatment and Valorisation Using Black Soldier Fly Larvae: A Review. *J. Environ. Manag.* **2019**, *251*, 109569. [CrossRef]

205. Cummins, V.C.; Rawles, S.D.; Thompson, K.R.; Velasquez, A.; Kobayashi, Y.; Hager, J.; Webster, C.D. Evaluation of Black Soldier Fly (*Hermetia illucens*) Larvae Meal as Partial or Total Replacement of Marine Fish Meal in Practical Diets for Pacific White Shrimp (*Litopenaeus vannamei*). *Aquaculture* **2017**, *473*, 337–344. [CrossRef]
206. Dumas, A.; Raggi, T.; Barkhouse, J.; Lewis, E.; Weltzien, E. The Oil Fraction and Partially Defatted Meal of Black Soldier Fly Larvae (*Hermetia illucens*) Affect Differently Growth Performance, Feed Efficiency, Nutrient Deposition, Blood Glucose and Lipid Digestibility of Rainbow Trout (*Oncorhynchus mykiss*). *Aquaculture* **2018**, *492*, 24–34. [CrossRef]
207. Lock, E.R.; Arsiwalla, T.; Waagbø, R. Insect Larvae Meal as an Alternative Source of Nutrients in the Diet of a Tlantic Salmon (*Salmo salar*) Postsmolt. *Aquac. Nutr.* **2016**, *22*, 1202–1213. [CrossRef]
208. Renna, M.; Schiavone, A.; Gai, F.; Dabbou, S.; Lussiana, C.; Malfatto, V.; Prearo, M.; Capucchio, M.T.; Biasato, I.; Biasibetti, E.; et al. Evaluation of the Suitability of a Partially Defatted Black Soldier Fly (*Hermetia illucens* L.) Larvae Meal as Ingredient for Rainbow Trout (*Oncorhynchus mykiss* Walbaum) Diets. *J. Anim. Sci. Biotechnol.* **2017**, *8*, 57. [CrossRef]
209. Moula, N.; Scippo, M.L.; Douny, C.; Degand, G.; Dawans, E.; Cabaraux, J.F.; Hornick, J.L.; Medigo, R.C.; Leroy, P.; Francis, F.; et al. Performances of Local Poultry Breed Fed Black Soldier Fly Larvae Reared on Horse Manure. *Anim. Nutr.* **2018**, *4*, 73–78. [CrossRef]
210. Parodi, A.; Gerrits, W.J.J.; Van Loon, J.J.A.; De Boer, I.J.M.; Aarnink, A.J.A.; Van Zanten, H.H.E. Black Soldier Fly Reared on Pig Manure: Bioconversion Efficiencies, Nutrients in the Residual Material, Greenhouse Gas and Ammonia Emissions. *Waste Manag.* **2021**, *126*, 674–683. [CrossRef]
211. Meneguz, M.; Gasco, L.; Tomberlin, J.K. Impact of PH and Feeding System on Black Soldier Fly (*Hermetia illucens*, L.; Diptera: Stratiomyidae) Larval Development. *PLoS ONE* **2018**, *13*, e0202591.
212. Liland, N.S.; Biancarosa, I.; Araujo, P.; Biemans, D.; Bruckner, C.G.; Waagbø, R.; Torstensen, B.E.; Lock, E.J. Modulation of Nutrient Composition of Black Soldier Fly (*Hermetia illucens*) Larvae by Feeding Seaweed-Enriched Media. *PLoS ONE* **2017**, *12*, e0183188. [CrossRef] [PubMed]
213. Danieli, P.P.; Lussiana, C.; Gasco, L.; Amici, A.; Ronchi, B. The Effects of Diet Formulation on the Yield, Proximate Composition, and Fatty Acid Profile of the Black Soldier Fly (*Hermetia illucens* L.) Prepupae Intended for Animal Feed. *Animals* **2019**, *9*, 178. [CrossRef]
214. Aini, N.; Nugroho, R.A.; Hariani, N. Growth and Survival Evaluation of *Oreochromis* Sp Fed *Hermetia Illucens* Larva and Manihot Esculenta Leaves Meal. *Biosaintifika J. Biol. Biol. Educ.* **2018**, *10*, 565–573. [CrossRef]
215. Tippayadara, N.; Dawood, M.A.O.; Krutmuang, P.; Hoseinifar, S.H.; Van Doan, H.; Paolucci, M. Replacement of Fish Meal by Black Soldier Fly (*Hermetia illucens*) Larvae Meal: Effects on Growth, Haematology, and Skin Mucus Immunity of Nile Tilapia, Oreochromis Niloticus. *Animals* **2021**, *11*, 193. [CrossRef]
216. Hu, J.; Wang, G.; Huang, Y.; Sun, Y.; He, F.; Zhao, H.; Li, N. Effects of Substitution of Fish Meal with Black Soldier Fly (*Hermetia illucens*) Larvae Meal, in Yellow Catfish (*Pelteobagrus fulvidraco*) Diets. *Isr. J. Aquac. Bamidgeh* **2017**, *69*, 1–9.
217. Abdel-Tawwab, M.; Khalil, R.H.; Metwally, A.A.; Shakweer, M.S.; Khallaf, M.A.; Abdel-Latif, H.M.R. Effects of Black Soldier Fly (*Hermetia illucens* L.) Larvae Meal on Growth Performance, Organs-Somatic Indices, Body Composition, and Hemato-Biochemical Variables of European Sea Bass, *Dicentrarchus labrax*. *Aquaculture* **2020**, *522*, 735136. [CrossRef]
218. Zhu, F.X.; Yao, Y.L.; Wang, S.J.; Du, R.G.; Wang, W.P.; Chen, X.Y.; Hong, C.L.; Qi, B.; Xue, Z.Y.; Yang, H.Q. Housefly Maggot-Treated Composting as Sustainable Option for Pig Manure Management. *Waste Manag.* **2015**, *35*, 62–67. [CrossRef] [PubMed]
219. Campbell, M.; Ortuño, J.; Stratakos, A.C.; Linton, M.; Corcionivoschi, N.; Elliot, T.; Koidis, A.; Theodoridou, K. Impact of Thermal and High-Pressure Treatments on the Microbiological Quality and in Vitro Digestibility of Black Soldier Fly (*Hermetia illucens*) Larvae. *Animals* **2020**, *10*, 682. [CrossRef] [PubMed]
220. Moreki, J.C.; Tiroesele, B.; Chiripasi, S.C. Prospects of Utilizing Insects as Alternative Sources of Protein in Poultry Diets in Botswana: A Review. *J. Anim. Sci. Adv.* **2012**, *2*, 649–658.
221. Wang, L.; Li, J.; Jin, J.N.; Zhu, F.; Roffeis, M.; Zhang, X.Z. A Comprehensive Evaluation of Replacing Fishmeal with Housefly (*Musca domestica*) Maggot Meal in the Diet of Nile Tilapia (*Oreochromis niloticus*): Growth Performance, Flesh Quality, Innate Immunity and Water Environment. *Aquac. Nutr.* **2017**, *23*, 983–993. [CrossRef]
222. Ogunji, J.; Schulz, C.; Kloas, W. Growth Performance, Nutrient Utilization of Nile Tilapia Oreochromis Niloticus Fed Housefly Maggot Meal (Magmeal) Diets. *Turkish J. Fish. Aquat. Sci.* **2008**, *8*, 141–147.
223. Sing, K.W.; Kamarudin, M.S.; Wilson, J.J.; Sofian-Azirun, M. Evaluation of Blowfly (*Chrysomya megacephala*) Maggot Meal as an Effective, Sustainable Replacement for Fishmeal in the Diet of Farmed Juvenile Red Tilapia (*Oreochromis* Sp.). *Pak. Vet. J.* **2014**, *34*, 288–292.
224. Lin, Y.H.; Mui, J.J. Evaluation of Dietary Inclusion of Housefly Maggot (*Musca domestica*) Meal on Growth, Fillet Composition and Physiological Responses for Barramundi, Lates Calcarifer. *Aquac. Res.* **2017**, *48*, 2478–2485. [CrossRef]
225. Xiang, J.; Qin, L.; Zhao, D.; Xiong, F.; Wang, G.; Zou, H.; Li, W.; Li, M.; Song, K.; Wu, S. Growth Performance, Immunity and Intestinal Microbiota of Swamp Eel (*Monopterus albus*) Fed a Diet Supplemented with House Fly Larvae (*Musca domestica*). *Aquac. Nutr.* **2020**, *26*, 693–704. [CrossRef]
226. Dabbou, S.; Gasco, L.; Lussiana, C.; Brugiapaglia, A.; Biasato, I.; Renna, M.; Cavallarin, L.; Gai, F.; Schiavone, A. Yellow Mealworm (*Tenebrio molitor* L.) Larvae Inclusion in Diets for Free-Range Chickens: Effects on Meat Quality and Fatty Acid Profile. *Renew. Agric. Food Syst.* **2020**, *35*, 571–578. [CrossRef]

227. Piccolo, G.; Iaconisi, V.; Marono, S.; Gasco, L.; Loponte, R.; Nizza, S.; Bovera, F.; Parisi, G. Effect of *Tenebrio molitor* Larvae Meal on Growth Performance, in Vivo Nutrients Digestibility, Somatic and Marketable Indexes of Gilthead Sea Bream (*Sparus aurata*). *Anim. Feed Sci. Technol.* **2017**, *226*, 12–20. [CrossRef]
228. Nogales-Mérida, S.; Gobbi, P.; Józefiak, D.; Mazurkiewicz, J.; Dudek, K.; Rawski, M.; Kierończyk, B.; Józefiak, A. Insect Meals in Fish Nutrition. *Rev. Aquac.* **2019**, *11*, 1080–1103. [CrossRef]
229. Belforti, M.; Gai, F.; Lussiana, C.; Renna, M.; Malfatto, V.; Rotolo, L.; De Marco, M.; Dabbou, S.; Schiavone, A.; Zoccarato, I.; et al. *Tenebrio molitor* Meal in Rainbow Trout (*Oncorhynchus mykiss*) Diets: Effects on Animal Performance, Nutrient Digestibility and Chemical Composition of Fillets. *Ital. J. Anim. Sci.* **2015**, *14*, 670–676. [CrossRef]
230. Gasco, L.; Henry, M.; Piccolo, G.; Marono, S.; Gai, F.; Renna, M.; Lussiana, C.; Antonopoulou, E.; Mola, P.; Chatzifotis, S. *Tenebrio molitor* Meal in Diets for European Sea Bass (*Dicentrarchus labrax* L.) Juveniles: Growth Performance, Whole Body Composition and in Vivo Apparent Digestibility. *Anim. Feed Sci. Technol.* **2016**, *220*, 34–45. [CrossRef]
231. Jeong, S.M.; Khosravi, S.; Yoon, K.Y.; Kim, K.W.; Lee, B.J.; Hur, S.W.; Lee, S.M. Mealworm, *Tenebrio molitor*, as a Feed Ingredient for Juvenile Olive Flounder, Paralichthys Olivaceus. *Aquac. Rep.* **2021**, *20*, 100747. [CrossRef]
232. Panini, R.L.; Freitas, L.E.L.; Guimarães, A.M.; Rios, C.; da Silva, M.F.O.; Vieira, F.N.; Fracalossi, D.M.; Samuels, R.I.; Prudêncio, E.S.; Silva, C.P.; et al. Potential Use of Mealworms as an Alternative Protein Source for Pacific White Shrimp: Digestibility and Performance. *Aquaculture* **2017**, *473*, 115–120. [CrossRef]
233. Roncarati, A.; Gasco, L.; Parisi, G.; Terova, G. Growth Performance of Common Catfish (*Ameiurus melas* Raf.) Fingerlings Fed Mealworm (*Tenebrio molitor*) Diet. *J. Insects Food Feed* **2015**, *1*, 233–240. [CrossRef]
234. Stull, V.J.; Finer, E.; Bergmans, R.S.; Febvre, H.P.; Longhurst, C.; Manter, D.K.; Patz, J.A.; Weir, T.L. Impact of Edible Cricket Consumption on Gut Microbiota in Healthy Adults, a Double-Blind, Randomized Crossover Trial. *Sci. Rep.* **2018**, *8*, 10762. [CrossRef]
235. Jeong, S.M.; Khosravi, S.; Mauliasari, I.R.; Lee, B.J.; You, S.G.; Lee, S.M. Nutritional Evaluation of Cricket, *Gryllus bimaculatus*, Meal as Fish Meal Substitute for Olive Flounder, Paralichthys Olivaceus, Juveniles. *J. World Aquac. Soc.* **2021**, *52*, 859–880. [CrossRef]
236. Phesatcha, B.; Phesatcha, K.; Viennaxay, B.; Matra, M.; Totakul, P.; Wanapat, M. Cricket Meal (*Gryllus bimaculatus*) as a Protein Supplement on In Vitro Fermentation Characteristics and Methane Mitigation. *Insects* **2022**, *13*, 129. [CrossRef]
237. Wang, D.; Bai, Y.Y.; Li, J.H.; Zhang, C.X. Nutritional Value of the Field Cricket (Gryllus Testaceus Walker). *Insect Sci.* **2004**, *11*, 275–283. [CrossRef]
238. Taufek, N.M.; Aspani, F.; Muin, H.; Raji, A.A.; Razak, S.A.; Alias, Z. The Effect of Dietary Cricket Meal (*Gryllus bimaculatus*) on Growth Performance, Antioxidant Enzyme Activities, and Haematological Response of African Catfish (*Clarias gariepinus*). *Fish Physiol. Biochem.* **2016**, *42*, 1143–1155. [CrossRef] [PubMed]
239. Taufek, N.M.; Muin, H.; Raji, A.A.; Md Yusof, H.; Alias, Z.; Razak, S.A. Potential of Field Crickets Meal (*Gryllus bimaculatus*) in the Diet of African Catfish (*Clarias gariepinus*). *J. Appl. Anim. Res.* **2018**, *46*, 541–546. [CrossRef]
240. Taufek, N.M.; Razak, S.A.; Alias, Z. Potential Value of Black Crickets M Eal as Protein Replacement for Fish M Eal in African Catfish, (*Clarias gariepinus*) Fingerlings Nutrition. In Proceedings of the Advancement in Marine and Freshwater Sciences, Terengganu, Malaysia, 8–10 October 2013; pp. 520–525.
241. Permatahati, D.; Mutia, R.; Astuti, D.A. Effect of Cricket Meal (*Gryllus bimaculatus*) on Production and Physical Quality of Japanese Quail Egg. *Trop. Anim. Sci. J.* **2019**, *42*, 53–58. [CrossRef]
242. Tilami, S.K.; Turek, J.; Červený, D.; Lepič, P.; Kozák, P.; Burkina, V.; Sakalli, S.; Tomčala, A.; Sampels, S.; Mráz, J. Insect Meal as a Partial Replacement for Fish Meal in a Formulated Diet for Perch Perca Fluviatilis. *Turkish J. Fish. Aquat. Sci.* **2020**, *20*, 867–878. [CrossRef]
243. Hanan, M.Y.; Amatul-Samahah, M.A.; Jaapar, M.Z.; Mohamad, S.N. The Effects of Field Cricket (*Gryllus bimaculatus*) Meal Substitution on Growth Performance and Feed Utilization of Hybrid Red Tilapia (*Oreochromis* Spp.). *Appl. Food Res.* **2022**, *2*, 100070.
244. Paul, A.; Frederich, M.; Uyttenbroeck, R.; Malik, P.; Filocco, S.; Richel, A.; Heuskin, S.; Alabi, T.; Megido, R.C.; Franck, T.; et al. Nutritional Composition and Rearing Potential of the Meadow Grasshopper (Chorthippus Parallelus Zetterstedt). *J. Asia. Pac. Entomol.* **2016**, *19*, 1111–1116. [CrossRef]
245. Heuzé, V.; Tran, G. *Locust Meal, Locusts, Grasshoppers and Crickets*; Feedipedia, a Programme by INRA, CIRAD, AFZ and FAO; FAO: Rome, Italy, 2016.
246. Alegbeleye, W.O.; Obasa, S.O.; Olude, O.O.; Otubu, K.; Jimoh, W. Preliminary Evaluation of the Nutritive Value of the Variegated Grasshopper (*Zonocerus variegatus* L.) for African Catfish *Clarias gariepinus* (Burchell, 1822) Fingerlings. *Aquac. Res.* **2012**, *43*, 412–420. [CrossRef]
247. Emehinaiye, P.A. Growth Performance of Oreochromis Niloticus Fingerlings Fed with Varying Levels of Migratory Locust (*Locusta migratoria*) Meal. Bachelor's Thesis, Federal University of Agriculture Abeokuta, Abeokuta, Nigeria, 2012.
248. Grace, O.I. Effects of Grasshopper Meal in the Diet of Clarias Gariepinus Fingerlings. *J. Aquac. Res. Dev.* **2015**, *6*, 321. [CrossRef]
249. Sun, W.; Yu, H.S.; Shen, Y.H.; Banno, Y.; Xiang, Z.H.; Zhang, Z. Phylogeny and Evolutionary History of the Silkworm. *Sci. China Life Sci.* **2012**, *55*, 483–496. [CrossRef]
250. Tomotake, H.; Katagiri, M.; Yamato, M. Silkworm Pupae (*Bombyx mori*) Are New Sources of High Quality Protein and Lipid. *J. Nutr. Sci. Vitaminol.* **2010**, *56*, 446–448. [CrossRef]

251. Kurbanov, A.R.; Milusheva, R.; Rashidova, S.; Kamilov, B. Effect of Replacement of Fish Meal with Silkworm (*Bombyx mori*) Pupa Protein on the Growth of Clarias Gariepinus Fingerling. *Int. J. Fish. Aquat. Stud.* **2015**, *2*, 25–27.
252. Lee, J.; Choi, I.C.; Kim, K.T.; Cho, S.H.; Yoo, J.Y. Response of Dietary Substitution of Fishmeal with Various Protein Sources on Growth, Body Composition and Blood Chemistry of Olive Flounder (*Paralichthys olivaceus*, Temminck & Schlegel, 1846). *Fish Physiol. Biochem.* **2012**, *38*, 735–744. [CrossRef]
253. Jintasataporn, O.; Chumkam, S.; Jintasataporn, O. Substitution of Silkworm Pupae (*Bombyx mori*) for Fish Meal in Broodstock Diets for Snakeskin Gourami (*Trichogaster pectoralis*). *J. Agric. Sci. Technol.* **2011**, *1*, 1341–1344.
254. Dheke, S.; Gubhaju, S.R. Growth Response of Rainbow Trout (*Oncorhynchus mykiss*) on Substitution of Shrimp Meal by Different Protein Sources. *Nepal. J. Zool. Online* **2013**, *1*, 24–29.
255. Sawhney, S. Effect of Partial Substitution of Expensive Ingredient Ie Fishmeal on the Growth of *Tor putitora* Fed Practical Diets. *J. Int. Acad. Res. Multidiscipl.* **2014**, *2*, 482–489.
256. Xu, X.; Xu, B.; Wang, J.; Su, Y.; Zhang, Z.; Chen, X. Studies on Blood Chemistry Indices and Histopathology of Pseudosciaena Crocea Artificially Challenged with Vibrio Harveyi. *J. Fish. China* **2010**, *34*, 618–625. [CrossRef]
257. Boscolo, W.R.; Hayashi, C.; Meurer, F. Digestibilidade Aparente Da Energia e Nutrientes de Alimentos Convencionais e Alternativos Para a Tilápia Do Nilo (*Oreochromis niloticus*, L.). *Rev. Bras. Zootec.* **2002**, *31*, 539–545. [CrossRef]
258. Ji, H.; Zhang, J.L.; Huang, J.Q.; Cheng, X.F.; Liu, C. Effect of Replacement of Dietary Fish Meal with Silkworm Pupae Meal on Growth Performance, Body Composition, Intestinal Protease Activity and Health Status in Juvenile Jian Carp (*Cyprinus carpio* var. *jian*). *Aquac. Res.* **2015**, *46*, 1209–1221. [CrossRef]
259. Wan, A.H.L.; Snellgrove, D.L.; Davies, S.J. A Comparison between Marine and Terrestrial Invertebrate Meals for Mirror Carp (Cyprinus Carpio) Diets: Impact on Growth, Haematology and Health. *Aquac. Res.* **2017**, *48*, 5004–5016. [CrossRef]
260. Raja, P.K.; Aanand, S.; Sampathkumar, J.S.; Padmavathy, P. Effect of Silkworm (*Bombyx mori*) Pupae on the Growth and Maturation of Rainbow Shark Epalzeorhynchos Frenatum (Fowler, 1934) under Captive Rearing. *Indian J. Fish.* **2020**, *67*, 89–96. [CrossRef]
261. Lewis, M.J.; Francis, D.S.; Blyth, D.; Moyano, F.J.; Smullen, R.P.; Turchini, G.M.; Booth, M.A. A Comparison of In-Vivo and in-Vitro Methods for Assessing the Digestibility of Poultry by-Product Meals Using Barramundi (*Lates calcarifer*); Impacts of Cooking Temperature and Raw Material Freshness. *Aquaculture* **2019**, *498*, 187–200. [CrossRef]
262. Stadtlander, T.; Stamer, A.; Buser, A.; Wohlfahrt, J.; Leiber, F.; Sandrock, C. *Hermetia illucens* Meal as Fish Meal Replacement for Rainbow Trout on Farm. *J. Insects Food Feed* **2017**, *3*, 165–175. [CrossRef]
263. Motte, C.; Rios, A.; Lefebvre, T.; Do, H.; Henry, M.; Jintasataporn, O. Replacing Fish Meal with Defatted Insect Meal (Yellow Mealworm *Tenebrio molitor*) Improves the Growth and Immunity of Pacific White Shrimp (*Litopenaeus vannamei*). *Animals* **2019**, *9*, 258. [CrossRef]
264. Ríos, J.M.; Attademo, A.M.; Mammana, S.B.; Altamirano, J.C.; Lajmanovich, R.C. Impact of Dietary Lipid Level on Esterase Enzyme Activities in the Non-Target Freshwater Shrimp Macrobrachium Borellii Exposed to Chlorpyrifos. *Environ. Sci. Pollut. Res.* **2021**, *28*, 19497–19504. [CrossRef]
265. Mastoraki, M.; Mollá Ferrándiz, P.; Vardali, S.C.; Kontodimas, D.C.; Kotzamanis, Y.P.; Gasco, L.; Chatzifotis, S.; Antonopoulou, E. A Comparative Study on the Effect of Fish Meal Substitution with Three Different Insect Meals on Growth, Body Composition and Metabolism of European Sea Bass (*Dicentrarchus labrax* L.). *Aquaculture* **2020**, *528*, 735511. [CrossRef]
266. Langer, S.; Bakhtiyar, Y.; Lakhnotra, R. Replacement of Fishmeal with Locally Available Ingredients in Diet Composition of *Macrobrachium dayanum*. *Afr. J. Agric. Res.* **2011**, *6*, 1080–1084.
267. Mccallum, M.L.; Weston, S.D.; Tilahun, Y. Performance of Early Juvenile Giant River Prawns (*Macrobrachium rosenbergii*) Fed Fish, Soybean, Shrimp and Four Insect Based Diets While Under Low Temperature Stress. *Proc. Okla. Acad. Sci.* **2020**, *100*, 5–15.
268. Serrano, A.; Poku, G. Nutritive Value of Termite as Fish Meal Supplement in the Diet of Freshwater Prawns (*Macrobrachium rosenbergii* de Man) Juveniles. *Extrem. Life Biospeology Astrobiol.* **2014**, *6*, 63–71.
269. Long, S.; You, Y.; Dong, X.; Tan, B.; Zhang, S.; Chi, S.; Yang, Q.; Liu, H.; Xie, S.; Yang, Y.; et al. Effect of Dietary Oxidized Fish Oil on Growth Performance, Physiological Homeostasis and Intestinal Microbiome in Hybrid Grouper (♀*Epi-Nephelus fuscoguttatus* × ♂*Epinephelus lanceolatus*). *Aquac. Rep.* **2022**, *24*, 101130. [CrossRef]
270. Randazzo, B.; Zarantoniello, M.; Gioacchini, G.; Giorgini, E.; Truzzi, C.; Notarstefano, V.; Cardinaletti, G.; Huyen, K.T.; Carnevali, O.; Olivotto, I. Can Insect-Based Diets Affect Zebrafish (Danio Rerio) Reproduction? A Multidisciplinary Study. *Zebrafish* **2020**, *17*, 287–304. [CrossRef] [PubMed]
271. Terova, G.; Rimoldi, S.; Ascione, C.; Gini, E.; Ceccotti, C.; Gasco, L. Rainbow Trout (*Oncorhynchus mykiss*) Gut Microbiota Is Modulated by Insect Meal from *Hermetia illucens* Prepupae in the Diet. *Rev. Fish Biol. Fish.* **2019**, *29*, 465–486. [CrossRef]
272. Bruni, L.; Randazzo, B.; Cardinaletti, G.; Zarantoniello, M.; Mina, F.; Secci, G.; Tulli, F.; Olivotto, I.; Parisi, G. Dietary Inclusion of Full-Fat *Hermetia illucens* Prepupae Meal in Practical Diets for Rainbow Trout (*Oncorhynchus mykiss*): Lipid Metabolism and Fillet Quality Investigations. *Aquaculture* **2020**, *529*, 735678. [CrossRef]
273. Cardinaletti, G.; Randazzo, B.; Messina, M.; Zarantoniello, M.; Giorgini, E.; Zimbelli, A.; Bruni, L.; Parisi, G.; Olivotto, I.; Tulli, F. Effects of Graded Dietary Inclusion Level of Full-Fat *Hermetia illucens* Prepupae Meal in Practical Diets for Rainbow Trout (*Oncorhynchus mykiss*). *Animals* **2019**, *9*, 251. [CrossRef]
274. Vogel, H.; Müller, A.; Heckel, D.G.; Gutzeit, H.; Vilcinskas, A. Nutritional Immunology: Diversification and Diet-Dependent Expression of Antimicrobial Peptides in the Black Soldier Fly *Hermetia illucens*. *Dev. Comp. Immunol.* **2018**, *78*, 141–148. [CrossRef]

275. Osimani, A.; Milanović, V.; Roncolini, A.; Riolo, P.; Ruschioni, S.; Isidoro, N.; Loreto, N.; Franciosi, E.; Tuohy, K.; Olivotto, I.; et al. *Hermetia illucens* in Diets for Zebrafish (Danio Rerio): A Study of Bacterial Diversity by Using PCR-DGGE and Metagenomic Sequencing. *PLoS ONE* **2019**, *14*, e0225956. [CrossRef]
276. Zarantoniello, M.; Randazzo, B.; Secci, G.; Notarstefano, V.; Giorgini, E.; Lock, E.J.; Parisi, G.; Olivotto, I. Application of Laboratory Methods for Understanding Fish Responses to Black Soldier Fly (*Hermetia illucens*) Based Diets. *J. Insects Food Feed* **2021**, *8*, 1173–1195. [CrossRef]
277. Abdel-Ghany, H.M.; Salem, M.E.S. Effects of Dietary Chitosan Supplementation on Farmed Fish: A Review. *Rev. Aquac.* **2020**, *12*, 438–452. [CrossRef]
278. Sudha, C.; Ahilan, B.; Felix, N.; Uma, A.; Prabu, E. Effects of Dietary Protein Substitution of Fishmeal with Black Soldier Fly Larval Meal on Growth and Physiological Responses of Juvenile Striped Catfish, *Pangasianodon hypophthalmus*. *Aquac. Res.* **2022**, *53*, 2204–2217. [CrossRef]
279. Kroeckel, S.; Harjes, A.G.E.; Roth, I.; Katz, H.; Wuertz, S.; Susenbeth, A.; Schulz, C. When a Turbot Catches a Fly: Evaluation of a Pre-Pupae Meal of the Black Soldier Fly (*Hermetia illucens*) as Fish Meal Substitute—Growth Performance and Chitin Degradation in Juvenile Turbot (*Psetta maxima*). *Aquaculture* **2012**, *364–365*, 345–352. [CrossRef]
280. Caligiani, A.; Marseglia, A.; Leni, G.; Baldassarre, S.; Maistrello, L.; Dossena, A.; Sforza, S. Composition of Black Soldier Fly Prepupae and Systematic Approaches for Extraction and Fractionation of Proteins, Lipids and Chitin. *Food Res. Int.* **2018**, *105*, 812–820. [CrossRef] [PubMed]
281. Zhang, Y.; Zhou, Z.; Liu, Y.; Cao, Y.; He, S.; Huo, F.; Qin, C.; Yao, B.; Ringø, E. High-Yield Production of a Chitinase from Aeromonas Veronii B565 as a Potential Feed Supplement for Warm-Water Aquaculture. *Appl. Microbiol. Biotechnol.* **2014**, *98*, 1651–1662. [CrossRef]
282. Ewald, N.; Vidakovic, A.; Langeland, M.; Kiessling, A.; Sampels, S.; Lalander, C. Fatty Acid Composition of Black Soldier Fly Larvae (*Hermetia illucens*)—Possibilities and Limitations for Modification through Diet. *Waste Manag.* **2020**, *102*, 40–47. [CrossRef]
283. Arru, B.; Furesi, R.; Gasco, L.; Madau, F.A.; Pulina, P. The Introduction of Insect Meal into Fish Diet: The First Economic Analysis on European Sea Bass Farming. *Sustainability* **2019**, *11*, 1697. [CrossRef]
284. Ferrer Llagostera, P.; Kallas, Z.; Reig, L.; Amores de Gea, D. The Use of Insect Meal as a Sustainable Feeding Alternative in Aquaculture: Current Situation, Spanish Consumers' Perceptions and Willingness to Pay. *J. Clean. Prod.* **2019**, *229*, 10–21. [CrossRef]
285. Joshi, P.K.; Bose, M.; Harish, D. Haematological Changes in the Blood of Clarias Batrachus Exposed to Mercuric Chloride. *J. Ecotoxicol. Environ. Monit.* **2002**, *12*, 119–122.
286. Truzzi, C.; Girolametti, F.; Giovannini, L.; Olivotto, I.; Zarantoniello, M.; Scarponi, G.; Annibaldi, A.; Illuminati, S. New Eco-Sustainable Feed in Aquaculture: Influence of Insect-Based Diets on the Content of Potentially Toxic Elements in the Experimental Model Zebrafish (Danio Rerio). *Molecules* **2022**, *27*, 818. [CrossRef]
287. Boyle, D.; Hogstrand, C.; Bury, N.R. Physiological Response to a Metal-Contaminated Invertebrate Diet in Zebrafish: Importance of Metal Speciation and Regulation of Metal Transport Pathways. *Aquat. Toxicol.* **2011**, *105*, 21–28. [CrossRef] [PubMed]
288. Ali, H.; Khan, E.; Ilahi, I. Environmental Chemistry and Ecotoxicology of Hazardous Heavy Metals: Environmental Persistence, Toxicity, and Bioaccumulation. *J. Chem.* **2019**, *2019*, 6730305. [CrossRef]
289. Van der Fels-Klerx, H.J.; Camenzuli, L.; Belluco, S.; Meijer, N.; Ricci, A. Food Safety Issues Related to Uses of Insects for Feeds and Foods. *Compr. Rev. Food Sci. Food Saf.* **2018**, *17*, 1172–1183. [CrossRef] [PubMed]
290. Usman, H.S.; Yusuf, A.A. Legislation and Legal Frame Work for Sustainable Edible Insects Use in Nigeria. *Int. J. Trop. Insect Sci.* **2021**, *41*, 2201–2209. [CrossRef]
291. Rumbos, C.I.; Mente, E.; Karapanagiotidis, I.T.; Vlontzos, G.; Athanassiou, C.G. Insect-Based Feed Ingredients for Aquaculture: A Case Study for Their Acceptance in Greece. *Insects* **2021**, *12*, 586. [CrossRef]
292. Baldi, L.; Mancuso, T.; Peri, M.; Gasco, L.; Trentinaglia, M.T. Consumer Attitude and Acceptance toward Fish Fed with Insects: A Focus on the New Generations. *J. Insects Food Feed* **2022**, *8*, 1249–1263. [CrossRef]
293. Smetana, S.; Schmitt, E.; Mathys, A. Sustainable Use of *Hermetia illucens* Insect Biomass for Feed and Food: Attributional and Consequential Life Cycle Assessment. *Resour. Conserv. Recycl.* **2019**, *144*, 285–296. [CrossRef]
294. Roffeis, M.; Fitches, E.C.; Wakefield, M.E.; Almeida, J.; Alves Valada, T.R.; Devic, E.; Koné, N.G.; Kenis, M.; Nacambo, S.; Koko, G.K.D.; et al. Ex-Ante Life Cycle Impact Assessment of Insect Based Feed Production in West Africa. *Agric. Syst.* **2020**, *178*, 102710. [CrossRef]
295. Maiolo, S.; Parisi, G.; Biondi, N.; Lunelli, F.; Tibaldi, E.; Pastres, R. Fishmeal Partial Substitution within Aquafeed Formulations: Life Cycle Assessment of Four Alternative Protein Sources. *Int. J. Life Cycle Assess.* **2020**, *25*, 1455–1471. [CrossRef]
296. Thévenot, A.; Rivera, J.L.; Wilfart, A.; Maillard, F.; Hassouna, M.; Senga-Kiesse, T.; Le Féon, S.; Aubin, J. Mealworm Meal for Animal Feed: Environmental Assessment and Sensitivity Analysis to Guide Future Prospects. *J. Clean. Prod.* **2018**, *170*, 1260–1267. [CrossRef]
297. Feng, W.; Xiong, H.; Wang, W.; Duan, X.; Yang, T.; Wu, C.; Yang, F.; Xiong, J.; Wang, T.; Wang, C. Energy Consumption Analysis of Lipid Extraction from Black Soldier Fly Biomass. *Energy* **2019**, *185*, 1076–1085. [CrossRef]
298. Józefiak, A.; Nogales-Mérida, S.; Mikołajczak, Z.; Rawski, M.; Kierończyk, B.; Mazurkiewicz, J. The Utilization of Full-Fat Insect Meal in Rainbow Trout (*Oncorhynchus mykiss*) Nutrition: The Effects on Growth Performance, Intestinal Microbiota and Gastrointestinal Tract Histomorphology. *Ann. Anim. Sci.* **2019**, *19*, 747–765. [CrossRef]

299. Fabrikov, D.; Sánchez-Muros, M.J.; Barroso, F.G.; Tomás-Almenar, C.; Melenchón, F.; Hidalgo, M.C.; Morales, A.E.; Rodriguez-Rodriguez, M.; Montes-Lopez, J. Comparative Study of Growth Performance and Amino Acid Catabolism in *Oncorhynchus mykiss*, *Tinca tinca* and *Sparus aurata* and the Catabolic Changes in Response to Insect Meal Inclusion in the Diet. *Aquaculture* **2020**, *529*, 735731. [CrossRef]
300. Basto, A.; Matos, E.; Valente, L.M.P. Nutritional Value of Different Insect Larvae Meals as Protein Sources for European Sea Bass (*Dicentrarchus labrax*) Juveniles. *Aquaculture* **2020**, *521*, 735085. [CrossRef]
301. Sealey, W.M.; Gaylord, T.G.; Barrows, F.T.; Tomberlin, J.K.; McGuire, M.A.; Ross, C.; St-Hilaire, S. Sensory Analysis of Rainbow Trout, *Oncorhynchus mykiss*, Fed Enriched Black Soldier Fly Prepupae, *Hermetia illucens*. *J. World Aquac. Soc.* **2011**, *42*, 34–45. [CrossRef]
302. Weththasinghe, P.; Hansen, J.; Nøkland, D.; Lagos, L.; Rawski, M.; Øverland, M. Full-Fat Black Soldier Fly Larvae (*Hermetia illucens*) Meal and Paste in Extruded Diets for Atlantic Salmon (*Salmo salar*): Effect on Physical Pellet Quality, Nutrient Digestibility, Nutrient Utilization and Growth Performances. *Aquaculture* **2021**, *530*, 735785. [CrossRef]
303. Irungu, F.G.; Mutungi, C.M.; Faraj, A.K.; Affognon, H.; Kibet, N.; Tanga, C.; Ekesi, S.; Nakimbugwe, D.; Fiaboe, K.K.M. Physico-Chemical Properties of Extruded Aquafeed Pellets Containing Black Soldier Fly (*Hermetia illucens*) Larvae and Adult Cricket (*Acheta domesticus*) Meals. *J. Insects Food Feed* **2018**, *4*, 19–30. [CrossRef]
304. IPIFF. *The European Insect Sector Today: Challenges, Opportunities and Regulatory Landscape*; IPIFF Vision Paper on the Future of the Insect Sector towards 2030; IPIFF: Brussels, Belgium, 2019; p. 16.
305. Alonso, M.E.; González-Montaña, J.R.; Lomillos, J.M. Consumers' Concerns and Perceptions of Farm Animal Welfare. *Animals* **2020**, *10*, 385. [CrossRef]
306. Buller, H.; Blokhuis, H.; Jensen, P.; Keeling, L. Towards Farm Animal Welfare and Sustainability. *Animals* **2018**, *8*, 81. [CrossRef]
307. Weary, D.M.; Robbins, J.A. Understanding the Multiple Conceptions of Animal Welfare. *Anim. Welf.* **2019**, *28*, 33–40. [CrossRef]
308. Grunert, K.G.; Sonntag, W.I.; Glanz-Chanos, V.; Forum, S. Consumer Interest in Environmental Impact, Safety, Health and Animal Welfare Aspects of Modern Pig Production: Results of a Cross-National Choice Experiment. *Meat Sci.* **2018**, *137*, 123–129. [CrossRef] [PubMed]
309. Mellor, D.J.; Beausoleil, N.J.; Littlewood, K.E.; McLean, A.N.; McGreevy, P.D.; Jones, B.; Wilkins, C. The 2020 Five Domains Model: Including Human–Animal Interactions in Assessments of Animal Welfare. *Animals* **2020**, *10*, 1870. [CrossRef] [PubMed]
310. Miranda-de la Lama, G.C.; Estévez-Moreno, L.X.; Villarroel, M.; Rayas-Amor, A.A.; María, G.A.; Sepúlveda, W.S. Consumer Attitudes Toward Animal Welfare-Friendly Products and Willingness to Pay: Exploration of Mexican Market Segments. *J. Appl. Anim. Welf. Sci.* **2019**, *22*, 13–25. [CrossRef] [PubMed]

fishes

MDPI

Review

Lamiaceae as Feed Additives in Fish Aquaculture

Graziella Orso, Roberta Imperatore, Elena Coccia, Ghasem Ashouri and Marina Paolucci *

Department of Science and Technologies (DST), University of Sannio, 82100 Benevento, Italy
* Correspondence: paolucci@unisannio.it

Abstract: The growing demand for high-quality food has induced a rapid expansion of the aquaculture sector. On the other hand, this sector has to overcome numerous challenges and problems triggered by the adoption of intensive farming systems, such as stress and high susceptibility to diseases. The improper use of chemicals and antibiotics has led to the development of antibiotic resistance in fish, with consequent health risks for consumers. Natural additives are increasingly used in aquaculture and, among these, medicinal plants are constantly under investigation as safe and environmentally friendly alternatives to chemicals. Great attention has been paid to *Lamiaceae* plants as feed additives capable of enhancing the growth performance, immune system, and antioxidant status of farmed fish. The aim of this review is to provide an updated picture of the employment of the *Lamiaceae* species (oregano, rosemary, sage, thyme, and mint) to enhance farmed fish health. The benefits of oregano, rosemary, sage, thyme, and mint feed supplementation on growth performance, immune system, antioxidant status, hemato-biochemical parameters, and resistance to stress, parasites, and bacteria have been described, highlighting weaknesses and drawbacks and proposing possible implementations.

Keywords: *Lamiaceae*; fish; health; growth performance; antioxidant; immunity; nutrition

Citation: Orso, G.; Imperatore, R.; Coccia, E.; Ashouri, G.; Paolucci, M. *Lamiaceae* as Feed Additives in Fish Aquaculture. *Fishes* **2022**, *7*, 349. https://doi.org/10.3390/fishes7060349

Academic Editor: Francisco J. Moyano

Received: 27 October 2022
Accepted: 24 November 2022
Published: 26 November 2022

1. Introduction

In recent decades, the aquaculture sector has shown rapid expansion in order to meet the food needs of the growing human population [1]. Aquaculture products represent an important source of high-quality animal proteins, as well as essential macro- and micronutrients. The growing demand for fish, both salty and freshwater, has prompted the aquaculture industry to adopt intensive and even ultra-intensive farming systems to increase productivity. However, intensive practices are responsible for numerous problems, such as poor water quality, overcrowding, high temperature, and poor nutrition, that contribute to lowering the growth performances of fish health and immune competence, with consequently increased stress and high susceptibility to diseases. Although aquaculture plays an important role today, it is a sector that must overcome numerous challenges that hinder its expansion, such as the spreading of infectious diseases, fish health problems, and consequent economic damage [2]. In recent years, disinfectants, chemotherapeutics and synthetic antibiotics have been used in order to prevent or mitigate the economic losses caused by diseases in farmed fish [3,4]. Unfortunately, as there are no antibiotics developed specifically for fish [4], veterinary or human antibiotics have been used, contributing to the onset of antibiotic-resistance [5]. Furthermore, the recurrent and uncontrolled use of antibiotics in farmed fish leads to the accumulation of residues of these substances in fish products, with consequent health risks for consumers [6].

Recently, researchers have paid great attention to identifying safe and environmentally friendly alternatives to antibiotics [7–9]. The use of natural additives capable of replacing pharmaceutical substances in intensive farming appears to have many potential benefits, including immunostimulation, the inhibition of pathogens in the intestinal tract, and the improvement of the absorption and utilization of nutrients [10,11]. Numerous studies have evaluated the effects of several natural feed additives, including probiotics [12], prebiotics [13], synbiotics [14], functional amino acids [15], minerals [16], and additives of origin vegetable or phytochemicals [8,17–20].

Medicinal plants, which include herbs, spices, and their extracts, have been increasingly used in aquaculture due to their low cost and simple use. In farmed fish, medicinal plants promote a vast array of effects, including the improvement of growth [21], immunity [22], antimicrobial and anti-stress activities [23], as well as resistance against pathogens [24]. In general, the efficacy of medicinal plants and their derivatives (extracts, and essential oils) is closely related to the abundance of bioactive substances such as alkaloids, quinones, lectins, steroids, phenolic compounds, tannins, terpenoids, saponins and flavonoids [25]. Among the medicinal plants used as a feed supplement for fish, growing scientific interest is directed to aromatic plants, both as extracts and essential oils [26]. Many of these aromatic plants belong to the *Lauraceae*, *Umbelliferae*, *Myrtaceae* and *Lamiaceae* families. In particular, the plant species of the *Lamiaceae* family are among the most studied and are frequently used as feed additives in aquaculture [27]. The aim of this review is to furnish a general outlook of the main *Lamiaceae* species (oregano, rosemary, sage, thyme, and mint) on the health and pathogen resistance in farmed fish, with the aim of providing a picture as complete as possible of *Lamiaceae* employment in aquaculture, highlighting the weaknesses and drawbacks of their implementation and proposing possible alternatives.

2. *Lamiaceae* Family

Lamiaceae are distributed all over the world, although the best environmental conditions for their growth were found in the Mediterranean basin [28]. The *Lamiaceae* family includes 245 genera and approximately 8000 species [29]. Since ancient times, the dried herb, leaves, and essential oils of *Lamiaceae* plants have been used in humans to treat various respiratory diseases, rheumatoid arthritis, gastrointestinal disorders, and urinary tract infections [29]. Plants of the *Lamiaceae* family represent a natural, economical, sustainable, and safe source of feed integrators capable of enhancing the growth performance, immune system, and antioxidant status of farmed fish [30–32]. Such beneficial effects are attributable to the bioactive molecules present in *Lamiaceae* plants, such as terpenes, terpenoids, alkaloids, and flavonoids [2]. For example, the immunomodulatory properties of the *Lamiaceae* plants are mediated by the predominant terpenes, carvacrol, and thymol, which are capable of modulating inflammatory processes through the activation of ion channels, such as TRP (Transient Receptor Potential) channels, and consequently activate the NFkB pathway [26]. Moreover, carvacrol and thymol show strong antioxidant activity due to their ability to neutralize the oxygen free radicals (ROS) in tissues and cells [26].

Lamiaceae, as feed additives, can be administered in different forms, as a whole plant or parts (leaves or seeds), as active compounds extracted from the plant, and individually or as a combination of extracted compounds [2]. It should be emphasized that the efficacy of *Lamiaceae* plants as feed additives depends on several crucial factors such as dose, duration, time schedule of administration, and fish species [2]. In particular, the most important factor is represented by the dose which, if suitable, can induce beneficial effects, while if too low or too high, may induce either no response or even be toxic [11]. As reported in a meta-analysis study on fish diets enriched with plants, the dosages used in aquaculture vary according to the plant species used. The higher dosages are used with powdered plants (0.1–420 mg/100 g of fish × day), followed by ethanolic and aqueous extracts (0.2–160 mg/100 g of fish × day; 0.03–200 mg/100 g of fish × day, respectively), while the lower doses are used with essential oils (0.005–30 mg/100 g of fish × day) [21]. Thus, to improve the growth performances and health of a specific fish species, the challenge for researchers is to identify the optimal conditions in terms of the part of the plant to be used, doses, duration, and time schedule.

3. Oregano

Among the *Lamiaceae* family, the oregano (*Origanum vulgare* L.) is the most worldwide spread species, distributed throughout Eurasia and North Africa, and particularly abundant in the Mediterranean area [33]. The richness of the chemical composition and aromatic compounds of oregano have led to its use, since ancient times, in the pharmaceutical and cosmetic fields, as well as in the food industry as a flavoring substance [34]. The efficacy of oregano in the treatment of a wide range of human diseases has been reported in both in vitro and in vivo studies [35,36]. Furthermore, several studies have reported the growth and health-promoting role of oregano in farmed animals, primarily in terrestrial monogastric animals (poultry and pigs) [37,38] and fish (Table 1).

The biological activities attributed to oregano are related to its bioactive components, which include a wide variety of secondary metabolites, most of which are monoterpenes (carvacrol and thymol) and polyphenols (rosmarinic acid, luteolin and derivatives, chlorogenic acid, quercetin and derivatives, caffeic acid, hyperoside, rutin, *p*-coumaric, ferulic, carnosic, ursoli acids) [29,39].

Table 1. Studies of oregano products added to the feed of aquaculture species.

Feeding			Fish			Effects	Ref.
Form	Inclusion Doses (% in Feed)	Period (Days)	Species	Vital Stage	Stress/Toxicant/ Pathogen Challenge		
O. vulgare essential oil	0.05, 0.1, 0.15, 0.2, 0.25	90	*Astyanax altiparanae*			↑ WG, SGR, PER	[40]
Origanum spp. essential oil	0.5, 1, 1.5, 2	56	*Cyprinus carpio*	Juvenile	*Aeromonas Hydrophila*	↑ LYS activity, phagocytic activity, and phagocytic index ↑ SOD and CAT activities ↑ IL-1β and IL-10	[41]
Origanum spp. essential oil	0.5, 1, 1.5, 2	60	*Cyprinus carpio*	Fingerling		↑ WG and SGR	[42]
O. vulgare essential oil	0.05, 0.15, 0.45	56	*Cyprinus carpio*	Juvenile	*Aeromonas Hydrophila*	↑ LYS and complement activities ↑ SOD, CAT and GPx activities ↓ TNFα and TGFβ	[43]
O. majorana hydroalcoholic extract	0.01, 0.02, 0.04	56	*Cyprinus carpio*	Juvenile	*Aeromonas Hydrophila*	↑ FW, WG and SGR ↑ RBC, WBC, Hct and Hb ↑ LYS and complement activities ↑ Total Ig levels ↑ SOD and CAT activities	[44]
O. vulgare ethanolic extract	0.5, 1, 2	56	*Danio rerio*	Adult	*Aeromonas Hydrophila*	↑ WG and SGR ↑ LYS and complement activities ↑ Total Ig levels ↑ SOD and CAT activities	[45]
O. vulgare essential oil	0.01, 0.02	60	*Dicentrarchus labrax*	Juvenile		↑ WG, SGR and PER ↓ Cholesterol and tryglicerides serum levels	[46]
O. vulgare essential oil	0.01, 0.02	150	*Dicentrarchus labrax*	Juvenile	Temperature change	↑ WG and SGR ↑ SOD and CAT activities	[47]
O. vulgare hydroalcoholic extract	1	56	*Oncorynchus mykiss*	Juvenile		↑ Serum total protein, respiratory burst activity, phagocytic activity and LYS activity	[48]
O. vulgare hydroalcoholic extract	1	56	*Oncorynchus mykiss*	Juvenile		↑ Serum total protein, respiratory burst activity, phagocytic activity and LYS activity	[49]
O. vulgare hydroalcoholic extract	0.2, 0.6, 1, 1.4	60	*Oncorynchus mykiss*	Juvenile	Diazinon (25% of the LC50 or 0.287 mg/L or 0.942 μM)	↑ BWI and SGR ↑ SOD, CAT and GPx activities	[50,51]

Table 1. *Cont.*

Feeding			Fish			Effects	Ref.
Form	Inclusion Doses (% in Feed)	Period (Days)	Species	Vital Stage	Stress/Toxicant/ Pathogen Challenge		
O. vulgare hydroalcoholic extract	0.5, 1, 1.5	70	*Oreochromis niloticus*	Fingerling	*Pseudomonas aeruginosa; Pseudomonas flourscence*	↑ FCR, PER and energy utilization	[52]
O. vulgare essential oil	0.1, 0.2	70	*Oreochromis niloticus*	Juvenile		↑ FW, SGR and FCR ↑ SOD activity	[53]
O. vulgare leaves powder	0.025, 0.5, 0.075, 0.1, 0.125, 0.15	30	*Oreochromis niloticus*	Juvenile	*Streptococcus agalactiae*	NS on growth ↑ LYS activity	[54]
O. vulgare hydroalcoholic extract	0.2, 0.5	60	*Oreochromis niloticus*	Juvenile	*Aeromonas Hydrophila*	↑ FW, WG and SGR ↑ RBC, WBC, Hct and Hb ↑ LYS and complement activities ↑ SOD, CAT and GPx activities	[55]
O. vulgare leaves powder	0.5, 1	30	*Sparus aurata*	Juvenile		NS on growth NS on LYS activity ↑ Ig levels	[56]

WG: Weight gain, FW: Final weight; FCR: Feed conversion ratio; SGR: Specific growth rate; PER: Protein efficiency ratio; BWI: Body weight index; RBC: Red blood cell count; WBC: White blood cell count; Hct: Hematocrit; Hb: Hemoglobin; LYS: Lysozyme; Ig: Immunoglobulin; IL-1β: Interleukin-1β; IL-10: Interleukin-10; TNFα: Tumor necrosis factor-α; TGFβ: Transforming growth factor-β; SOD: Superoxide dismutase; CAT: Catalase; GPx: Glutathione peroxidase; NS, non-significant effects.

3.1. Oregano's Effects on Growth Performance

Great attention has been paid by researchers to the use of oregano essential oil (OEO) in farmed fish (Table 1). OEO feed inclusion stimulates the growth performance of fish, primarily by improving the feed utilization rate and by acting on metabolic processes. Zhang et al. [43] reported that 0.15 and 0.45% of OEO supplementation, for 56 days, stimulated digestive enzymes in koi carp juveniles (*Cyprinus carpio*), increasing the activation of proteases, amylases and lipases. The same beneficial effects on intestinal enzymes have been reported for the hydroalcoholic extract of oregano (at a dose of 3% in 85 days feeding trial) in rainbow trout (*Oncorynchus mykiss*) [57]. In addition, OEO dietary supplementation may promote growth due to its beneficial effects on intestinal health. The inclusion of 1.5% of OEO in the diet for 60 days significantly improved growth performance and intestinal histomorphometry (villous height and width) in common carp fry [42]. Similarly, the addition of 0.05% of OEO to the diet of yellow-tailed (*Astyanax altiparanae)* for 90 days increased the absorption area of the intestine [40]. The study by Huley et al. [58] showed that the inclusion of different OEO concentrations (0.075, 0.15, 0.225, and 0.3%) in Nile tilapia (*Oreochromis niloticus*) juveniles for 64 days acted as a developmental stimulant of intestinal villi and, consequently, as a growth promoter.

The beneficial effects of OEO supplementation on growth performance are also, most likely, linked to the improvement of the gut microbial community [43]. Fish gut microbiota serves crucial functions in host health, growth, and development, aiding digestive functions and protecting against intestinal infections [59]. Dietary supplementation with the major monoterpenes of oregano (thymol and carvacrol) positively altered the gut microbiota of Nile tilapia [60], and resulted in improved nutrient digestibility and absorption, as well as protein conversion [50,61]. The OEO inhibited some pathogenic bacterial groups and increased commensal beneficial communities of *Corynebacterium*, *Brevinema*, and *Propionibacterium* in koi carp juveniles [43].

In contrast to the beneficial effects of OEO, Santo et al. [54] reported no significant improvement in growth performances and no significant alterations in intestinal villous height in Nile Tilapia juveniles fed with different percentages (0.025, 0.05, 0.075, 0.1, 0.125, and 0.15%) of dried oregano leaves for 30 days. Similarly, weight gain (WG) and specific growth rate (SGR) did not significantly differ in seabream juveniles (*Sparus aurata*) fed with 0.5 and 1% oregano leaves powder for 15 or 30 days [56].

The hydroalcoholic oregano leaf extract also appeared to counteract the toxic effects of Diazinon, an organophosphate pesticide, on growth and liver metabolic enzymes (aspartate aminotransferase (AST), alanine aminotransferase (ALT) and lactate dehydrogenase (LDH) in rainbow trout juveniles; in fact, doses between 0.2 and 1%, but not higher, of oregano hydroalcoholic extract dietary inclusion significantly increased the body weight index (BWI) and the SGR compared to the standard diet in a 60 day feeding trial [51].

Based on these results, the best forms of oregano feed supplement for fish to stimulate growth rate and feed conversion parameters are essential oils and hydroalcoholic extracts, while powdered oregano leaves have no beneficial effects. A possible explanation may reside in the similar percentage of bioactive constituents (carvacrol 63%; thymol 4.7%; ρ-cimene 12.8%; γ-terpinene 8.4%) in essential oils and hydroalcoholic extracts [40,54,62].

3.2. Oregano's Effects on Oxidative Stress

Oregano essential oil or hydroalcoholic extracts administered in the diet reduced the oxidative stress in different fish species, including common carp [41,43,44], rainbow trout [31,49,50,61], Nile tilapia [53,55], and catfish [63]. Oregano acted as an antioxidant activity enhancer, promoting the activities of serum and hepatic superoxide dismutase (SOD), catalase (CAT), and glutathione peroxidase (GPX) enzymes accompanied by a reduction in malonaldehyde (MDA) levels [41,43].

The choice of the administered dose plays an important role in the antioxidant effect of oregano when used as a feed additive. For example, in rainbow trout juveniles, after a 60 day feeding trial with 0.6 and 1% doses of hydroalcoholic oregano leaf extract, the activity of antioxidant enzymes SOD, CAT, and glutathione peroxidase (GPX) increased, while high doses (1.4%) caused a decrease in their activities [50]. Similarly, the particular part of the plant being used appears to play an important role in determining its antioxidant effect. The use of 0.5 and 1% of oregano leaf powder for 30 days in sea bream juveniles, for example, did not cause any significant effects on liver antioxidant enzymes activities [56]; this is likely a result of the lower number of bioactive components with respect to the essential oil and the hydroalcholic extract. Further, the presence of the bioactive molecules within the vegetable matrix, as it occurs in the leaves, makes their extraction and absorption during the digestive processes difficult, with a consequent limited action. Indeed, the level of bioactive molecule uptake in the intestine represents only a limited percentage of the total quantity.

3.3. Oregano's Effects on Immune Response

The effects of dietary oregano supplementation on the immune status of farmed fish have been widely reported. The results of numerous studies carried out on rainbow trout [31,48,49], Nile tilapia [55], and koi carp [43] reported that both oregano essential oil and hydroalcoholic extract increased the non-specific immune response, mainly via improving lysozyme, protease and complement system activities. In comparison to mammals, the innate immune system represents a fundamental defense weapon in fish [64]. For example, lysozyme is capable of destroying the bacterial cell, by splitting the β-1,4 glycosidic bonds of the peptidoglycan, providing protection against fish pathogens [65]. The use of 0.02% of hydroalcoholic oregano (*O. majorana*) leaf extract for 56 days enhanced the activity of lysozyme in common carp juveniles [44]. In another study, dietary integration of 0.1% of OEO in red-bellied tilapia (*Tilapia zillii*) improved lysozyme activity levels, accompanied by the increase in proteases, antiproteases, and bactericidal activities [66].

The supplementation with immunostimulants in fish diets also beneficially improved the expression of specific immune elements, such as IgM [56] and pro-inflammatory cytokyne Interleukin-1β (IL-1β) [66]. In particular, in sea bream juveniles, supplementation with oregano leaf powder at 0.5 and 1% for 30 days improved both the innate (complement system and antibacterial activity) and adaptive (IgM) responses of skin mucus immunity compared to the control group, while the oregano leaf powder integration did not alter the humoral immune response in the serum [56]. From this difference in the results, authors have suggested that the immune defense against pathogens resides in the antibody response of the skin mucus, which increases proportionally with the concentration of oregano leaf powder in the diet (resulting highest at the dose of 1%).

3.4. Oregano's Effects on Hemato-Biochemical Parameters

Hemato-biochemical parameters are reliable biomarkers of the health and immunity conditions of farmed fish species [67]. The incorporation of hydroalcoholic extract of the oregano aerial part into fish feed had no effect on red and white blood cell counts (RBC and WBC, respectively), leukocytes count (monocyte, lymphocyte and neutrophile), hematocrit (Hct), and hemoglobin (Hb) in rainbow trout juveniles treated with a dose of 1% [48,49] and in Nile tilapia treated with 0.2 and 0.5% [55]. On the contrary, the hematological parameters were significantly enhanced in red-bellied tilapia fed with 0.1% of OEO for 15 days [66]. Similar augmentation of RBC, WBC, thrombocytes, and hemoglobin was recorded in common carp juveniles fed a diet containing 0.02% of oregano (*O. majorana*) leaf hydroalcoholic extract for 56 days [44]. It has been suggested that the increase in the hematological parameters RBC, Hct, and Hb, may favor the tissue oxygenation and the elimination of carbon dioxide, contributing to growth [68]. Moreover, homeostasis, or the increase in such hematological parameters, indicates that the oregano supplementation had no negative effects on erythrocytes production and the destruction of mature RBC, therefore indicating that it is non-toxic [69]. Serum biochemical parameters, such as total protein, albumin, and globulin values, were enhanced by hydroalcoholic oregano leaf extract, added to the diet at the dose of 1% in rainbow trout juveniles [48,49]. Similarly, in sea bass (*Dicentrarchus labrax*) juveniles fed with 0.01% of OEO for 150 days, the improvement of total protein, glucose, triglycerides, and cholesterol occurred [47].

3.5. Oregano's Effects against Pathogen Infections

Several studies have revealed that the use of OEO, in addition to increasing growth and feed utilization, improves resistance to pathogens in common carp [41,43,44], channel catfish (*Ictalurus punctatus*) [63], zebrafish (*Danio rerio*) [45], Nile Tilapia [55], rainbow trout [31], and red-bellied tilapia [66]. Carvacrol and thymol, the most abundant phenolic components, are likely responsible for the antimicrobial activities of oregano, being able to alter the bacterial outer membrane and consequently its permeability [39]. Carvacrol, in particular, is involved in the disintegration of bacterial cells by altering the synthesis and mobility of the flagella, the fatty acid composition of the membranes, membrane proteins, and periplasmic enzymes [55,70]. The flavonoids and terpenoids contained in the oregano also contribute to the antimicrobial power, as demonstrated by the terpenoids, ρ-cymene [70].

3.6. Conclusions

Based on the literature, it appears that the best dietary supplement is represented by oregano's essential oil and hydroalcoholic leaf extract. It should be noted that fish fed the diet supplemented with oregano showed improved growth performance, immunological parameters, and antioxidant status in a dose-dependent manner and that an excessive amount of oregano could cause immunosuppression and toxicity. In this regard, on the basis of the results examined, it can be stated that it is convenient to use oregano as a feed additive for fish in the form of essential oil and hydroalcoholic extract in a concentration ranging between 0.5 and 1%, for a minimum duration of 8 weeks.

4. Rosemary

Rosemary (*Rosmarinus officinalis* L.) is a small evergreen medicinal herb, widespread in the Mediterranean region. It is widely used for farmed animals' nutrition. Both in vitro and in vivo studies have shown that rosemary-based food supplements improve oxidative stress and immune responses [71]. In particular, rosemary extract possesses anti-inflammatory, anticancer, antidiabetic, hepato- and blood-protective activities [72,73]. Among *Lamiaceae* medicinal plants, rosemary presents the least chemical composition variability. Rosemary is primarily composed of terpenes β- and α-pinene, camphene, camphor, and limonene [74]. Many other compounds are also extracted from rosemary, such as polyphenols and steroids [71,75]. Among polyphenols, the most abundant are rosmarinic acid, 7-methylrosmanol naringin, and also, at lower concentrations, rutin, and ferulic acid [76]. Although there is numerous evidence to show the beneficial effects of rosemary as a feed additive in terrestrial animals, its application in aquaculture is still scarce. However, significant interest in the use of rosemary in aquatic animals has recently increased due to its efficacy as a stimulant of growth, the immune system, and health status [72,77].

4.1. Rosemary's Effects on Growth Performance

Several studies have confirmed that the oral administration of rosemary could enhance growth performances in farmed fish, such as common carp [72,77], Nile tilapia [78–80], and sea bream [81] (Table 2). Among the rosemary-based products, rosemary leaf powder is the most commonly investigated as a fish feed additive. In common carp fingerlings, different doses (1, 2, and 3%) of rosemary leaf powder positively increased, in a dose-dependent manner, the growth performances and feed conversion parameters (WG, SGR, final weight (FG), feed conversion ratio (FCR) levels) after a trial of 65 days [72].

Table 2. Studies of rosemary products added to the feed of aquaculture species.

Feeding			Fish			Effects	Ref.
Form	Inclusion Doses (% in Feed)	Period (Days)	Species	Vital Stage	Stress/Toxicant/ Pathogen Challenge		
R. officinalis leaf powder	1, 2, 3	65	*Cyprinus carpio*	Juvenile		↑ WG, SGR and PER ↑ RBC, WBC, Hct and Hb ↑ LYS and complement activities ↑ Total Ig and protein levels ↑ SOD and CAT activities	[72]
R. officinalis hydroalcoholic extract	0.25, 0.5, 1	65	*Cyprinus carpio*	Juvenile		↑ WG and SGR ↑ LYS and complement activites, phagocytic activity, and phagocytic index ↑ RBC, WBC, Hct and Hb ↓ Cholesterol and tryglicerides serum levels	[77]
R. officinalis leaf powder	1	140	*Oreochromis niloticus*	Fingerling		↑ WG, SGR and PER ↑ Hct and leukocrit levels	[78]
R. officinalis commercial leaf extract	0.1, 0.25, 0.5	90	*Oreochromis niloticus*	Juvenile		NS on growth	[79]
R. officinalis leaf powder	0.25, 0.5, 1	60	*Oreochromis niloticus*	Fingerling	*Aeromonas Hydrophila*	↑ WG, SGR and FCR ↑ Total serum protein ↑ LYS and complement activities ↑ Total Ig levels ↑ CAT activity	[80]
R. officinalis commercial extract	0.06, 0.12, 0.18, 0.24	84	*Sparus aurata*			NS on growth ↓ Total serum protein	[81]

WG: Weight gain; FCR: Feed conversion ratio; SGR: Specific growth rate; PER: Protein efficiency ratio; RBC: Red blood cell count; WBC: White blood cell count; Hct: Hematocrit; Hb: Hemoglobin; LYS: Lysozyme; Ig: Immunoglobulin; SOD: Superoxide dismutase; CAT: Catalase; NS, non-significant effects.

The same findings with rosemary leaf powder supplementation were also obtained in Nile tilapia fingerlings [78,80]; in particular, Naiel et al. [80] recorded better growth performance in fish fed on 0.5 and 1% of rosemary leaf powder for 60 days. Similarly, in a

65-day feeding trial, common carp juveniles fed on hydroalcoholic rosemary leaf extract (0.01, 0.25, 0.5, and 1%) showed an increase in growth performances [77]. Various studies have shown that herbal plants not only improved fish growth and nutrition, but also enhanced appetite and modified the gut microbiota composition, increasing the diversity and activity of the beneficial bacteria, while inhibiting pathogenic bacteria [2,75,82]. In agreement with these findings, rosemary leaf powder also showed a positive role in controlling nutrient uptake and enhancing the intestinal mucosal condition in rats ([83]. On the contrary, in Nile tilapia juveniles fed 90-day diets with different amounts (0.1, 0.25, and 0.5%) of commercial rosemary extract, Yilmaz et al. [79] did not report significant changes in growth performances. In addition, in gilthead seabream, growth performances and feed intake were not modified by the inclusion of different doses (0.06, 0.12, 0.18, 0.24%) of commercial rosemary extract for 84 days [81]. Such differences could be attributed to different fish species, feeding trial length, source and rosemary doses. In this regard, it is necessary to emphasize that, in the experiment conducted by Hernández et al. [81], a commercial rosemary extract powder made of a blend at the ratio 1:1 of two diterpenes (carnosic acid and carnosol) was used. Similarly, Yilmaz et al. [79] used a commercial rosemary extract composed of rosmarinic acid at 5.32%. Therefore, the lack of results may be associated with the small amount of the chemical active principles in the feed additive used. In contrast to the inclusion of powder or fresh leaf extract, it is also interesting to underline that rosemary oil did not result in an increase in growth performance, as well as growth rate (GR) and FCR in sturgeon juveniles (*Huso huso*) [84] and seabass [85].

4.2. Rosemary's Effects on Oxidative Stress

The beneficial effects of rosemary dietary-inclusion also resulted in the improvement of the antioxidant status in common carp [72] and in Nile tilapia [80]. Rosemary leaf powder supplementation at the doses of 0.5 and 1% in the diet of Nile tilapia fingerlings for 60 days significantly improved the antioxidant status via an increase in CAT activity [80]. Similarly, in a 65-day feeding trial in common carp juveniles, different doses (1, 2, and 3%) of rosemary leaf powder induced an increase, in a dose-dependent manner, of blood CAT activity, but the higher dose (3%) led to a decrease in blood SOD activity [86].

The effect of powdered rosemary leaves as antioxidant defense enhancers could be linked to its several beneficial compounds, such as rosmarinic and carnosic acids [76].

4.3. Rosemary's Effects on Immune Response

Dietary supplement with rosemary products showed an enhancement of the immune system in fish. The elevation of total immunoglobulin (Ig) levels, lysozyme and alternative complement activities of common carp juveniles fed on diets containing rosemary leaf powder in various doses (1, 2, and 3%), for 65 days, was reported [72]. The findings of Dezfoulnejad and Molayemraftar [77] confirmed the potential of oral administration of hydroalcoholic rosemary leaf extract as a stimulatory agent of the non-specific immune system in common carp juveniles. Similarly, in tilapia (*O. mossambicus*) fingerlings, the inclusion of 0.25 and 0.5% hydroalcoholic rosemary extracts for 60 days led to an improvement in the principal non-specific immunity elements (lysozyme, immunoglobulin and alternative complement) [87]. In addition, in Nile tilapia fingerlings, the oral administration of 1% of rosemary leaf powder for 60 days induced a significant increase in the expression of the immune genes related to innate and adaptive immune response, such as lysozyme, complement and immunoglobulin M (IgM) [80].

4.4. Rosemary's Effects on Hemato-Biochemical Parameters

It has been reported that rosemary bioactive compounds, such as rosmarinic acid, could positively affect thymus and spleen activities, leading to a significant increase in the WBC counts (lymphocytes T and B, monocytes and neutrophils) [88]. In fact, after 65 days of oral administration of 2 and 3% of rosemary leaf powder, WBC markedly increased in common carp juveniles [72]. Similarly, tilapia fingerlings treated with 1%

of rosemary leaf powder showed a significant increase in both haematological (WBC, haematocrit and leukocrit levels) and serum biochemical (total protein, albumin and globulin levels) parameters [78]. Serum biochemical parameters are good fish health indicators [68]. Several studies have suggested the possible correlation between enhanced fish growth performance and the simultaneous increase in total protein, albumin and globulin levels due to dietary herbal inclusion [68,89]. Findings on the oral supplementation of rosemary in common carp [72,77] and in Nile tilapia [78,80] confirmed the hypothesis of the combination effects of health and growth performance in fish treated with herbal supplementation. Moreover, several vitamins (A, B, and C) and minerals (K, Ca, and Fe) present in significant quantities in rosemary could positively modulate other blood biochemical parameters due to their hypocholesterolaemic effects [90]. For example, the levels of triglycerides and LDL (low-density lipoprotein cholesterol) diminished, while HDL (high-density lipoprotein cholesterol) augmented in common carp juveniles fed on hydroalcoholic extract of rosemary at 1% in a 65-day feeding trial [77].

4.5. Rosemary's Effects against Pathogen's Infections

Some studies have evaluated the effects of rosemary as an alternative antimicrobial agent in aquaculture. The dietary application of dried rosemary leaves for 20 days improved the resistance against *Streptococcus iniae* at the 8% dose and against *Streptococcus agalactiae* at 16% dose in tilapia fingerlings [91]. Similarly, the 60-day feeding supplementation of 1% of rosemary leaf powder provided adequate protection to Nile tilapia fingerlings against the infection of *Aeromonas hydrophila* [80]. Numerous in vitro studies demonstrated that rosemary possessed antibacterial properties against Gram-positive and Gram-negative bacteria, mainly linked to its composition in phenolic compounds [92,93].

4.6. Conclusions

Based on the reported literature, both leaf powder and extract of rosemary positively affect growth performance, antioxidant status, and the general health of farmed fish. In order to improve the haemato-biochemical and non-specific immune parameters and increase the resistance against bacterial diseases, a dosage of 1% of rosemary extract or leaf powder and 60 days of administration can be recommended as useful fish feed additives.

5. Sage

Sage is the largest genus of *Lamiaceae* and includes approximately 900 species, among which *Salvia officinalis* is globally widespread and highly considered for its medical relevance [29]. In fact, *Salvia officinalis* is cultivated in numerous countries and its dried leaves are used as raw material in medicine, the food industry, and perfumery [94]. It has been found that the essential oil and leaf extract of *Salvia officinalis* have strong antimicrobial and antioxidant effects, and also exhibit immunomodulatory and anti-inflammatory activities [29]. These beneficial effects may be due to the particular chemical constituents of sage, such as tannic acid, oleic acid, carnosol and carnosic acid and some polyphenols, such as caffeic acid, p-coumaric acid, rutin, rosmarinic acid, quercetin, luteolin, and apigenin [29]. Other compounds, such as monoterpenes and terpenoids, including 1.8-cineole, α-thujone, β-thujone, β-pinene, and camphor, are present in sage [74]; α-Thujone is a neurotoxic monoterpene ketone whose amount may vary according to the harvesting time, being high after flowering and low before flowering [95]. Therefore, an important parameter that must be considered is the variation of the chemical composition of medicinal plants. This variation, in fact, could influence the biological properties of the herb.

5.1. Sage's Effects on Growth Performance

As reported for *O. vulgare* L. and *R. officinalis* L., *Salvia officinalis* has also been studied in several experiments in farmed fish [30,96,97] (Table 3). One study by Sönmez et al. [30] reported the positive effects of a 60-day dietary supplementation of sage oil (0.05, 0.1, and 0.15%) on growth performance and parameters such as SGR and FCR in rainbow trout juveniles. The same results were shown in beluga after 42 days of dietary inclusion of sage ethanolic extract (3, 6, and 12%) [97]. This growth-promoting action could be partially attributed to the sage polyphenolic compounds, such as ursolic acid, a pentacyclic triterpenoid carboxylic acid, which induces muscular hypertrophy in rainbow trout [98]. An increase in growth performance was also reported in gilthead seabream juveniles fed for 92 days with 0.01% of a combined extract of sage and lemon verbena (*Lippia citriodora*) leaf [96].

Table 3. Studies of sage products added to the feed of aquaculture species.

Feeding			Fish			Effects	Ref.
Form	**Inclusion Doses (% in Feed)**	**Period (Days)**	**Species**	**Vital Stage**	**Stress/Toxicant/ Pathogen Challenge**		
S. officinalis ethanolic extract	3, 6, 12	42	*Huso huso*	Juvenile		↑ FW, BWI and FCR ↑ RBC, WBC, Hct and Hb ↑ LYS and complement activities ↑ Total Ig levels ↓ Serum ALT and AST levels	[97]
S. officinalis and *Arthrospira platensis* leaf	0.00075, 0.001	28	*Oreochromis niloticus*	Juvenile	*Pseudomonas aeruginosa*	↑ LYS and nitric oxide activites, IgM levels ↑ TNFα and IL-1β	[99]
S. officinalis essential oil	0.05, 0.1, 0.15	60	*Oncorynchus mykiss*	Juvenile		↑ WG, SGR and FCR ↑ SOD, G6PD and GPx activities	[30]
S. officinalis ethanolic extract	0.5, 1, 1.5	30	*Oncorynchus mykiss*	Juvenile	*Streptococcus. iniae*	↑ Total Ig levels	[5]
S. officinalis and *Lippia citriodora* ethanolic extract	0.1	92	*Sparus aurata*	Fingerling	LPS	↑ SGR and FCR ↑ Total serum protein NS on bacteriolytic and complement activities ↑ LYS, IgM, TNFα, IL-1β, TGFβ and IL-10 ↑ SOD and CAT activities	[96]

WG: Weight gain; FW: Final weight; FCR: Feed conversion ratio; SGR: Specific growth rate; BWI: Body weight index; RBC: Red blood cell count; WBC: White blood cell count; Hct: Hematocrit; Hb: Hemoglobin; LYS: Lysozyme; Ig: Immunoglobulin; ALT: Alanine aminotransferase; AST: Aspartate aminotransferase; IL-1β: Interleukin-1β; IL-10: Interleukin-10; TNFα: Tumor necrosis factor-α; TGFβ: Transforming growth factor-β; SOD: Superoxide dismutase; CAT: Catalase; GPx: Glutathione peroxidase; G6PD: glucose-6-phosphate dehydrogenase; NS, non-significant effects.

5.2. Sage's Effects on Oxidative Stress

The dietary inclusion of sage protects against reactive oxygen species (ROS) by stimulating the antioxidant defenses in farmed fish [30,96]. In rainbow trout juveniles, different concentrations (0.05, 0.1, and 0.15%) of sage oil added to the diet for 30 days significantly increased liver enzyme SOD, glucose-6-phosphate dehydrogenase (G6PD) and glutathione peroxidase (GPx) activities, while an extension of the feeding trial to 60 days induced a reduction in the antioxidant enzymes activities [30]. A positive modulation on the antioxidant defense system was also reported in gilthead seabream [96]. The findings of Salomón et al. [96] have shown that a 92-day administration of 0.1% dietary additives made of sage and lemon verbena hydroethanolic leaf extract stimulated SOD and CAT gene expression in gilthead seabream fingerlings. According to the authors, the up-regulation of SOD and CAT genes could be linked to the triterpenic and polyphenolic compounds, mainly ursolic acid, present in the sage [96].

5.3. Sage's Effects on Immune Response

Great attention has been given to the utilization of the dietary inclusion of sage to fortify innate immunity in farmed fish. In beluga juveniles, the immunomodulation through the oral administration of sage ethanolic extract for 42 days (3, 6, and 12%) enhanced lysozyme and alternative complement activities, and serum immunoglobulin levels [97].

In addition, in rainbow trout juveniles, 30 days of dietary supplementation of 0.5, 1, and 1.5% of hydroethanolic extracts of sage positively affected the immune system indices (lysozyme and complement activities and total immunoglobulin levels) in a dose-dependent manner [5].

In fish, the immunomodulatory properties of the dietary supplementation of sage combined with other medicinal herbs have also been demonstrated. After a feeding trial of 28 days, the combination of sage and Spirulina platensis (*Arthrospira platensis*) increased the non-specific (lysozyme, IgM and complement) and specific (IL-1β and TNFα cytokines) immune response in Nile tilapia juveniles [99]. In sea bream fingerlings, the dietary administration of 0.1% sage and verbena hydroethanolic leaf extract stimulated the expression of lysozyme, IgM, Il-1β and TNFα, and also increased the anti-inflammatory cytokines TGF-1β and IL-10 levels [96].

5.4. Sage's Effects on Hemato-Biochemical Parameters

The dietary inclusion of sage leads to the improvement of the hemato-biochemical parameters in beluga [97] and seabream [96]. Sage ethanolic extract (3, 6, and 12%), administered for 42 days, stimulated RBC, Hct, Hb, total protein, albumin, and globulin levels in beluga juveniles [97]. Moreover, Dadras et al. [97] reported that the dietary inclusion of sage ethanolic extract decreased the serum ALT and AST levels, supporting the beneficial effect of sage on the physiological status of fish. In fact, AST and ALT enzyme activities are used as stress indicators and the increase in their blood levels indicates liver impairment and hepatocellular damage [69].

5.5. Sage's Effects against Pathogen's Infections

The positive impact of the dietary inclusion of 0.5, 1, and 1.5% hydroethanolic extracts of sage for 30 days on the non-specific and specific immune responses led to an increase in rainbow trout juveniles' resistance against infection with *S. iniae* [5]. The 28-day dietary treatment with sage leaf inclusion protected Nile tilapia juveniles against infection with *Pseudomonas aeruginosa*, causing a significant elevation of the expression of lysozyme, IgM, and pro-inflammatory cytokines (IL-1β and TNFα) [99].

5.6. Conclusions

Based on the reported literature, it could be concluded that the dietary inclusion of sage can improve immune response, antioxidant system activity and stimulate feed intake, leading to enhanced growth performance. The feed incorporation of sage extract at a dosage of between 6–12%, for 42 days, shows important immunomodulatory properties. Regarding the use of sage essential oil, the optimal dose seems to be at 0.05%, with a duration of feed supplementation between 30 and 60 days. In addition, the combination of leaf extracts from sage and other medicinal plants added at low concentrations (0.1%) in the fish diet for long periods (>90 days) could be useful for its beneficial effects in aquafeeds. However, further studies are needed to understand doses and timing of administration in order to optimize the beneficial effects of using sage as a fish feed additive.

6. Thyme

Among the *Lamiaceae* family, the use of the aromatic plant thyme (*Thymus vulgaris*) is common in traditional medicine, food, as well as the pharmaceutical and cosmetic industries [29]. Fresh or dry thyme leaves can be used, and the essential oil can be extracted from flowers. Thyme possesses antiseptic, antinflammatory, antimicrobial and antioxidative properties [100]. Thyme is characterized by well recognized and documented in vitro antibacterial potential [101,102], showing that both thyme extract and essential oil have strong activity against *Escherichia coli, Staphylococcus aureus, Citrobacter freundii, P. aeruginosa, Proteus mirabilis, Proteus vulgaris* and *Salmonella typhimurium*. Thyme is rich in monoterpenes such as ρ-cymene, γ-terpinene, carvacrol, and thymol. The concentrations of these four main compounds remain very stable in plants harvested in different seasons, suggesting that they are the compounds that functionally and biologically support the plant [103]. Thyme is also rich in polyphenols, including p-hydroxybenzoic acid, caffeic acid, rosmarinic acid, catechin, luteolin, apigenin, and quercetin [29]. The observed activities of thyme can be ascribed, in particular, to the presence of the caffeic and rosmarinic acids, quercetin and luteolin [100]. Thyme is an immunostimulator and growth promoter in poultry and swine farming; however, knowledge concerning its efficacy in aquatic species is limited to a few studies reporting that the dietary inclusion of thyme was effective in growth stimulation, immune responses, disease resistance and antioxidant enzyme profile in different farmed fish, including Nile tilapia, rainbow trout, and common carp [78,104,105] (Table 4).

Table 4. Studies of thyme products added to the feed of aquaculture species.

Feeding			Fish			Effects	Ref.
Form	Inclusion Doses (% in Feed)	Period (Days)	Species	Vital Stage	Stress/Toxicant/ Pathogen Challenge		
T. vulgaris leaf	0.5, 1, 1.5, 2	56	*Cyprinus carpio*	Fingerling	*Saprolegnia* spp.	↑ WG, SGR and PER ↑ RBC, WBC, Hct and Hb ↑ Total serum protein levels	[104]
T. vulgaris essential oil	0.05, 0.1, 0.15	60	*Oncorynchus mykiss*	Juvenile		↑ WG, SGR and FCR ↑ SOD, CAT, G6PD and GPx activities	[30]
T. vulgaris essential oil	0.05, 0.1, 0.2	60	*Oncorynchus mykiss*	Juvenile	*Aeromonas Hydrophila*	↑ WG and SGR ↑ LYS and complement activities	[106]
T. vulgaris essential oil	1	30	*Oncorynchus mykiss*	Juvenile	Aflatoxin B1	↑ FW, WG, FCR and SGR ↑ LYS and complement activities ↑ Total serum protein levels ↓ TNFα, TGFβ and IL-8 ↑ IL-1β	[107]
T. vulgaris essential oil	1, 2	60	*Oncorynchus mykiss*	Juvenile		↑ FW, WG and SGR ↑ WBC and Hb ↓ ALT, ALP and AST ↑ LYS and complement and total Ig levels ↑ SOD, CAT, GR, and GPx activities	[105]
T. vulgaris leaf	1	140	*Oreochromis niloticus*	Fingerling		↑ WG, SGR and PER	[78]
T. vulgaris essential oil	0.1, 0.5, 1	15	*Oreochromis niloticus*	Juvenile		↑ WBC NS on ALT and AST levels	[108]

WG: Weight gain; FW: Final weight; FCR: Feed conversion ratio; SGR: Specific growth rate; PER: Protein efficiency ratio; RBC: Red blood cell count; WBC: White blood cell count; Hct: Hematocrit; Hb: Hemoglobin; LYS: Lysozyme; Ig: Immunoglobulin; ALT: Alanine aminotransferase; AST: Aspartate aminotransferase; ALP: Alkaline phosphatase activity; IL-1β: Interleukin-1β; IL-8: Interleukin-8; TNFα: Tumor necrosis factor-α; TGFβ: Transforming growth factor-β; SOD: Superoxide dismutase; CAT: Catalase; GPx: Glutathione peroxidase; GR: Glutathione reductase; NS, non-significant effects.

6.1. Thyme's Effects on Growth Performance

Several studies have investigated the effect of dietary thyme inclusion on fish growth parameters [78,104–107]. These scientific findings have shown that thyme does not possess adverse or toxic effects and is able to maintain the physiological conditions of the alimentary tract in fish [32,108]. As for the other herbal products, the optimal concentration of thyme is a critically important factor. Rainbow trout juveniles fed on 0.05, 0.1, and 0.2% of thyme essential oil for 60 days showed the best growth performance and parameters (weight gain, SGR, and feed intake) with the dose of 0.05% [30,106]. In common carp fingerlings, the dietary administration of 1.5% of thyme leaf led to the improvement of growth performances after a 56-day feeding trial when compared to the other experimental dietary concentrations tested (0.5, 1, and 2%) [104]. In sturgeon (*Acipenser stellatus*) juveniles, 58 days of feed thyme application improved fish growth at the concentration of 2% [109] compared to the 1% inclusion dose [110].

6.2. Thyme's Effects on Oxidative Stress

The positive role of thyme in enhancing antioxidant capacity has been demonstrated in rainbow trout juveniles [30,105]. For example, 0.05 and 0.1% of thyme essential oil supplementation provided enhanced antioxidant protection, improving liver CAT, SOD, GPx, and glutathione reductase (GR) activities and decreasing MDA production after 30 days of the feeding trial [30]. Thyme essential oil or water extract could successfully mitigate oxidative stress, likely due to their high concentrations of thymol and carvacrol [32]. The antioxidant effects of thymol and its isomer carvacrol have been well documented in several in vitro and in vivo studies, including cell lines [111] and animal models, such as weaning piglets [112].

6.3. Thyme's Effects on Immune Response

Several studies have been carried out to understand the immunomodulatory effects of thyme in fish. Thyme dietary inclusion is capable of stimulating the non-specific immune response in rainbow trout, including lysozyme, alternative complement and total immunoglobulin levels [105,106]. Furthermore, dietary 1% of thyme essential oil counteracted the negative effects on immunity and intestinal inflammation induced by aflatoxin B1 in rainbow trout juveniles, significantly lowering the expression levels of TNFα, IL-8 and TGF-β [107]. The immunomodulatory effects of thyme are linked to its major bioactive components, such as carvacrol, thymol, eugenol, and cymene [106]. Thymol feed supplementation, for example, improved the immunoglobulin levels in broiler chickens [113] and in pigs' guts [114].

On the contrary, the feeding inclusion of 0.1, 0.5, and 1% of thyme essential oil for a short period (15 days) did not alter respiratory burst activity, lysozyme concentration, or alternative complement activity in Nile tilapia juveniles [108]. These results confirm the importance of the optimal choice of the duration of immunostimulant administration.

6.4. Thyme's Effects on Hemato-Biochemical Parameters

In farmed fish, the increase in blood parameters (Hb, RBC, and WBC counts) and the improvement of biochemical profile (total protein, albumin and globulin levels) suggest that the dietary inclusion of thyme products are safe feed additives able to enhance fish health and welfare. In Nile tilapia juveniles, 0.1, 0.5, and 1% of thyme essential oil for 15 days led to a significant increase in total leukocytes (monocytes, neutrophils, basophils and lymphocytes), especially at the highest dose (1%) [108]. The safety of thyme as a fish feed additive is also confirmed by the absence of negative or toxic effects on ALT and AST levels [106,108]. For example, the inclusion of up to 0.2% of thyme oils over 2 months did not alter the activity of these enzymes in rainbow trout juveniles, suggesting that thyme oils at up to 0.2% in feed can be considered as a safe additive for trout [106].

6.5. Thyme's Effects against Pathogen's Infections

Thyme also improves fish disease resistance against several bacteria and fungi, such as *Saprolegnia* spp. [104], *A. hydrophila* [106], *Yersinia ruckeri* [115], and *S. iniae* [116]. The efficacy of thyme essential oil or leaf powder could be a consequence of the increasing levels of the main immunity factors (lysozyme, alternative complement, immunoglobulin and cytokynes) and hemato-biochemical parameters. Feed supplementation of 0.05% thyme essential oil improved the resistance of rainbow trout juveniles against motile *Aeromonas* septicemia caused by *A. hydrophila* via the upregulation of the C3 and CD4 immune genes and the increase in IL-1β cytokine gene expression [106]. In fish, CD4 T helper cells provide a protective response against bacteria, fungi, and protozoa and C3 protein is crucial for the activation of both classical and alternative complement pathways [117].

6.6. Conclusions

Based on the reported literature, it can be observed that periods of 60 days of feeding supplementation with 0.5-1% of thyme essential oil can be considered a proper length of time and percentage to stimulate the cellular components of the non-specific immune response, enhance the growth performance and disease resistance against pathogens. On the contrary, short-term supplementation (such as 15 days) of 1% of thyme essential oil does not show beneficial effects. Regarding the use of thyme leaf powder, the feeding supplementation dose of 1% shows positive effects even in very long administration periods (140 days).

7. Mint

Another aromatic plant belonging to the family *Laminaceae* that captured the attention of researchers for its use in aquaculture is mint, also known as mentha or peppermint (*Mentha piperita*). Mint is a perennial herbaceous plant and is widely cultivated [118]. Peppermint is a crucial medicinal and aromatic plant, used in food since ancient times, and more recently in sanitary and cosmetic industries [119]. Several studies have confirmed its antimicrobial, antioxidant, and immunomodulatory effects [118]. The beneficial activities of mint, especially its antimicrobial effect, are due to its major compounds, such as menthol (33.8%), menthone (15.8%) and pulegone (8.3%) [119,120]. Used in perfumery and aromatherapy, pulegone and menthol are potentially toxic compounds when administered in large amount, causing liver damage in rats [121]. On the contrary, menthone has a digestive favoring effect and is non-toxic [120]. Mint also presents a high polyphenolic content (19–23%), primarily characterized by rosmarinic acid, luteolin, hesperidin and apigenin [122].

7.1. Mint's Effects on Growth Performance

The incorporation of mint into the diets of fish showed positive growth- stimulating effects, improving GR, WGR, and FCR in several fish species, such as Asian sea bass (*L. calcarifer*) [123], Nile tilapia [124], Caspian brown trout (*Salmo trutta caspius*) [125], and Caspian white fish (*Rutilus frisii kutum*) [126]. In Caspian white fish juveniles fed with 1, 2, and 3% of peppermint hydroalcoholic extracts for eight weeks, the growth parameters increased in a dose-dependent manner [126]. Mint could be considered as an appetite activator that significantly increases the daily feed intake [123]. Furthermore, the beneficial effect of mint feed inclusion on growth parameters could be attributed to its influence on intestinal enzymes (amylase and protease) and microbiota, leading to an improvement of the digestibility and availability of nutrients [126]. Interestingly, after a 60-day feeding trial, the dietary inclusion of 0.01 and 0.025% of mint essential oil enhanced intestinal health and increased the length of the intestinal villi [127]. On the contrary, the dietary supplementation, for 50 days, of mint essential oil at 0.075, 0.125, and 0.25% did not cause significant differences in growth compared to the control group in Nile tilapia fingerlings [120]. The possible explanation for this lack of beneficial effects on growth performance may be caused by small quantities employed in the study as the effect of mint is dose-dependent [126,128].

7.2. Mint's Effects on Oxidative Stress

Mint as a feed additive is effective in improving oxidative stress induced by the main environmental stressors, such as the water pollutants ammonia [129] and pesticides [124]. Nile tilapia juveniles fed on 0.25% of mint essential oil for 30 days displayed enhanced CAT, SOD, and GPx gene expression levels, allowing a reduction in the oxidative stress induced by pesticides toxicity exposure [124]. Similarly, oral administration of menthol at 0.25% improved the antioxidative status in common carp juveniles, mitigating the ammonia-induced alterations on antioxidant enzymes activities [129]. The antioxidant effects of mint are mainly attributable to the monoterpenic ketones mentone and isomentone [118].

7.3. Mint's Effects on Immune Response

Recently, great attention has been given to the immunostimulating effect of mint on different fish species, including rainbow trout [130,131], tilapia [120,127], common carp [132], sea bass [133], Caspian brown trout [125], and Caspian kutum [126] (Table 5).

Table 5. Studies of mint products added to the feed of aquaculture species.

Feeding			Fish			Effects	Ref.
Form	Inclusion Doses (% in Feed)	Period (Days)	Species	Vital Stage	Stress/Toxicant/ Pathogen Challenge		
M. piperita essential oil	0.01, 0.025	60	*Lates calcarifer*	Fingerling	*Vibrio harveyi*	↑ WG and FCR ↑ RBC, WBC, Hct and Hb ↑ Phagocytic activity, respiratory burst, LYS, anti-protease and bactericidal activities ↑ Total serum protein and globulin levels	[123]
M. piperita essential oil	0.075, 0.125, 0.25	50	*Oreochromis niloticus*	Fingerling		↑ Complement activity ↑ SOD, CAT, G6PD and GPx activities NS on ALT and AST levels NS on RBC, WBC, Hct and Hb NS on total protein levels	[127]
M. piperita essential oil	0.25	30	*Oreochromis niloticus*	Fingerling	*Streptococcus agalactiae*	↑ Total serum protein levels NS on growth parameters NS on LYS activity	[120]
M. piperita hydroalcoholic extract	1, 2, 3	56	*Oreochromis niloticus*	Juvenile	Chlorpyrifos	↑ FW, WG and SGR ↑ RBC, WBC, Hct and Hb ↓ ALT, ALP and AST	[124]
M. piperita hydroalcoholic extract	1, 2, 3	56	*Oncorynchus mykiss*	Juvenile	*Yersinia ruckeri*	↑ LYS activiys and total Ig levels ↑ Total serum protein and albumin levels NS on ALT, ALP and AST levels ↑ SOD, CAT, GR, and GPx activities	[130]
M. piperita hydroalcoholic extract	1, 2, 3	56	*Rutilus frisii kutum*	Juvenile		↑ WG and SGR ↑ LYS and respiratory burst activities ↑ RBC, WBC, Hct and Hb	[126]
M. piperita hydroalcoholic extract	1, 2, 3	56	*Salmo trutta caspius*	Juvenile		↑ WG, SGR and FCR ↑ LYS and alkaline phosphatase activities ↑ RBC, WBC, Hct and Hb ↑ Total serum protein levels NS on ALT, ALP and AST levels	[125]

WG: Weight gain; FW: Final weight; FCR: Feed conversion ratio; SGR: Specific growth rate; RBC: Red blood cell count; WBC: White blood cell count; Hct: Hematocrit; Hb: Hemoglobin; LYS: Lysozyme; Ig: Immunoglobulin; ALT: Alanine aminotransferase; AST: Aspartate aminotransferase; ALP: Alkaline phosphatase activity; SOD: Superoxide dismutase; CAT: Catalase; GPx: Glutathione peroxidase; GR: Glutathione reductase; G6PD: glucose-6-phosphate dehydrogenase; NS, non-significant effects.

A four-week period of dietary supplementation of horsemint (*Mentha longifolia*) hydroalcoholic extract at 0.1, 0.2, and 0.3% improved the non-specific immunity response (lysozyme and complement) and immune-related genes (TNFα) in rainbow trout juveniles, especially at the dose of 0.3% [131]. Similarly, in the same fish species, the improvement of lysozyme activity in a dose-dependent manner was reported after a 56-day feeding trial with 1, 2, and 3% of mint hydroalcoholic extract [130]. The enhancement of the immune system was also observed in juveniles of Caspian brown trout [125] and Caspian kutum [126]. Results of both studies showed that 56 days of dietary inclusion of 1, 2, and 3% of dried mint powder improved the immunological parameters in a dose-dependent manner. On the contrary, in Nile tilapia fingerlings, 60 days of feeding with 0.01 and 0.025% of mint essential oil did not alter the lysozyme levels, while the activation of the complement system was significantly increased, especially at the concentration of 0.025% [127].

7.4. Mint's Effects on Hemato-Biochemical Parameters

The improvement of the haematological and biochemical parameters due to the dietary inclusion of mint dried leaf, essential oil and hydroalcoholic extract were also observed [123–126,130]. Dietary administration, for eight weeks, of mint hydroalcoholic extract at 3% improved the RBC, Hb, and WBC levels in the rainbow trout juveniles [130], Caspian brown trout [125], and Caspian white fish [126]. Mint is rich in vitamins, such as vitamins A, C, and E, and in mineral salts, such as iron, potassium, and calcium [119]. Some studies have suggested that mint used as food additive favors the intestinal absorption of iron and vitamins, increases hematopoiesis and, consequently, the hematological indices [134,135]. Moreover, mint-integrated diets enhanced serum biochemical parameters, resulting in the reduction in serum glucose, lipids, triglycerides and cholesterol levels, and in an increase in total protein, albumin and globulin levels [123,127,130].

7.5. Mint's Effects against Pathogen's Infections

Several studies have revealed the efficacy of mint dried leaf, essential oil and hydroalcoholic extract on the protection against *S. agalactiae* [120], *Vibrio harveyi* [123], and *Y. ruckeri* [130]. In rainbow trout juveniles, the oral supplementation of 1, 2, and 3% of mint hydroalcoholic extract for 56 days significantly enhanced serum bactericidal and antiprotease activity, protecting from infections and giving resistance to the pathogens [130]. Moreover, the oral supplementation of mint essential oil or hydroalcoholic extract increased fish survival in experimental challenge tests, modulated haematological RBC and WBC counts, parameters of non-specific and specific immunity (lysozyme and complement activities and Ig levels) and increased cytokines expression (TNFα, IL-1β, IL-8) [120,130,131].

7.6. Conclusions

Based on the reported literature, the fish dietary incorporation of mint is able to improve the haematological and immune response parameters and provide resistance against pathogenic infections in a dose-dependent manner. Mint hydroalcoholic extracts should be used in fish feed at a concentration range of between 2–3% for a duration of 56 days. Regarding the use of mint essential oils as a fish feed additive, low doses (0.075–0.125%) of feed supplementation shows benefits to intestinal health and immune response, while higher doses (0.25%) are necessary to stimulate growth and improve haematological parameters.

8. General Conclusions and Future Perspectives

This review summarizes the findings regarding the role of the species of *Lamiaceae* family as feed additives in aquaculture. According to studies conducted with medical herbs, oregano, rosemary, sage, thyme, and mint (whole plant, extract or essential oil) have the potential for use as safe additives in fish feed, showing benefits on growth performance, immune system, antioxidant status, hemato-biochemical parameters and resistance to stress, parasites and bacteria.

Considering the scientific literature reported in this review, it is possible to indicate that great importance must be given to the choice of suitable dosage and administration time to obtain positive effects on fish health. A specific dose may induce beneficial impacts such as immunostimulation, whereas an unfavorable dose may not cause any responses, or may even be cytotoxic. Consequently, the optimization of the dosage according to the plant and the type of material chosen is strongly recommended. Moreover, as several *Lamiaceae* plants have been shown to have a dose-dependent effect, further studies are required to understand the toxicological safety of these feed additives.

The employment of *Lamiaceae* plants is an interesting field in aquaculture; however, there are still numerous research gaps. Foremost, comparative studies concerning the part of the *Lamiaceae* plants and the type of extraction (leaves, extract, mixed, or essential oil), the optimal administration method (immersion, injection or oral administration), and the duration of administration are necessary to gather information about the best beneficial effects on fish health and the parameters of interest in aquaculture, primarily growth performance and immune response. Additionally, as the impact of a feed supplement is species-specific, further research is required on the use of the *Lamiaceae* family in order to identify the plant species and products with the best beneficial potential for each fish species of interest in aquaculture.

Moreover, in a considerable number of the reviewed studies, the chemical characterization of the fish feed supplements is absent; thus, the chemical analysis of *Lamiaceae* products used as feed additives should be encouraged, with the aim of identifying and quantifying the active molecules and establishing their proper dosages and the duration of administration. Detailed information about the chemical compositions of *Lamiaceae* species could help critically analyze their effectiveness as growth promoters, immunostimulants, and antioxidant agents. In addition, the knowledge of the chemical composition could open the way to a possible correlation between the bioactive compounds present in the fish feed supplement used and the results obtained. As has previously been reported, the chemical composition of plants is influenced by numerous factors, such as the form and type of extraction (Table 6).

Table 6. The major bioactive compounds identified in essential oil, hydrolcoholic extract and leaves of *Lamiaceae* plants (oregano, rosemary, sage, thyme and mint).

Lamiaceae	Form	Main Bioactive Components	Method of Analysis	Ref.
Oregano	Essential oil	Carvacrol (63%), ρ-Cymene (12.8%), γ-Terpinene (8.4%), Thymol (4.7%)	H-R GC	[40]
	Hydroalcoholic extract	Carvacrol (59.4%), Thymol (25%), ρ-Cymene (6.9%), 1-Octacosanol (4%)	GC/MS	[62]
	Dry leaves	Carvacrol (63%), ρ-Cymene (12.8%), γ-Terpinene (8.4%), Thymol (4.7%)	H-R GC	[54]
Rosemary	Hydroalcoholic extract	Camphor (4.8%), Phytol (3.28%), Borneol (3.27%), Caryophillene (3.20%)	GC-MS	[136]
	Dry leaves	α-Pinene (21.65%), β-Pinene (12.58%), Camphene (12.54%), Limonene (7.22%), Camphor (5.29%)	HS-SPME-GC-MS	[74]
Sage	Essential oil	α-Thujone (10–60%), β-Thujone (4–36%), Camphor (5–20%), 1.8 -Cineole (2–15%)	H-R GC	[29]
	Hydroalcoholic extract	Manool (7%), β-Thujone (6.2%), Carnosol (2.4%), Camphor (4.8%)	GC/MS	[137]
	Dry leaves	β-Thujone (27%), 1.8 -Cineole (19.55%), β-Pinene (11.36%), Camphor (8.62%)	HS-SPME-GC-MS	[74]
Thyme	Essential oil	Thymol (37–55%), ρ-Cymene (14–28%), γ-Terpinene (4–12%), Carvacrol (0.5–5.5%)	H-R GC	[105]
	Hydroalcoholic extract	Thymol (42.6%)	GC/MS	[138]
	Dry leaves	ρ-Cymene (30.35%), γ-Terpinene (11.85%)	HS-SPME-GC/MS	[74]
Mint	Essential oil	Menthol (33.8%), Menthone (15.2%), Methyl acetate (13%), Pulegone (8.3%)	H-R GC	[120]
	Hydroalcoholic extract	Menthone (25.4%), 1,8-cineole (17.7%), Menthol (12.1%)	GC/MS	[139]
	Dry leaves	Menthol (35–60%), Menthone (2–44%), Methyl acetate (0.7–23%), 1.8—Cineole (1-13%), Menthofuran (0.3–14%)	GC/MS	[140]

H-R GC: High-Resolution Gas Chromatography; GC/MS: Gas chromatography/mass spectrometry; HS-SPME-GC/MS: Headspace solid-phase microextraction- Gas chromatography/mass spectrometry.

This variation of bioactive compounds could be reflected in the biological properties of the herb used as feed additives. In our opinion, the knowledge of the chemical composition of *Lamiaceae* plants or products represents an important parameter that must be considered in order to standardize the use of medicinal herbs in fish nutrition.

In addition, the knowledge of the mechanism of action of the bioactive molecules present in medicinal plants is still scarce. The understanding of the mechanisms of action of the bioactive compounds contained in *Lamiaceae* plants used as feed additives could elucidate the cellular and molecular processes underlying their capabilities of enhancing growth performance, immune system, and antioxidant status.

In conclusion, plants of the *Lamiaceae* family represent an exciting research field in aquaculture and a natural, economical, sustainable, and safe source of feed integrators capable of enhancing the health of farmed fish. However, although these are natural products, it is necessary to take into account the criteria for the safe use of plant ingredients in diets for farmed fish according to legislation, which differs among countries. Within the European Union, the safe use of oregano, rosemary, sage, thyme, and mint as feed additives for animal nutrition is approved and governed by regulation (EC) 1831/2003 of the European Parliament and of the Council of 22 September 2003 (https://www.efsa.europa.eu/it/applications/feedadditives/regulationsandguidance accessed on 1 November 2022).

Author Contributions: Conceptualization, M.P.; writing—original draft preparation, G.O. and M.P.; writing—review and editing, G.O., R.I., E.C., G.A. and M.P. All authors have read and agreed to the published version of the manuscript.

Funding: This research received no external funding.

Conflicts of Interest: The authors declare no conflict of interest.

References

1. FAO. *The State of World Fisheries and Aquaculture*; Food and Agriculture Organization of the United Nations: Rome, Italy, 2020.
2. Awad, E.; Awaad, A. Role of Medicinal Plants on Growth Performance and Immune Status in Fish. *Fish Shellfish Immunol.* **2017**, *67*, 40–54. [CrossRef] [PubMed]
3. Lieke, T.; Meinelt, T.; Hoseinifar, S.H.; Pan, B.; Straus, D.L.; Steinberg, C.E.W. Sustainable Aquaculture Requires Environmental-Friendly Treatment Strategies for Fish Diseases. *Rev. Aquac.* **2020**, *12*, 943–965. [CrossRef]
4. Santos, L.; Ramos, F. Antimicrobial Resistance in Aquaculture: Current Knowledge and Alternatives to Tackle the Problem. *Int. J. Antimicrob. Agents* **2018**, *52*, 135–143. [CrossRef] [PubMed]
5. Tafi, A.A.; Meshkini, S.; Tukmechi, A.; Alishahi, M.; Noori, F. Immunological and Antistreptococcal Effects of *Salvia Officinalis* and *Aloe Vera* Extracts Supplemented Feed in Rainbow Trout (*Oncorhynchus mykiss*). *Kafkas Univ. Vet. Fak. Derg.* **2018**, *24*, 365–370.
6. Stratev, D.; Zhelyazkov, G.; Noundou, X.S.; Krause, R.W.M. Beneficial Effects of Medicinal Plants in Fish Diseases. *Aquacult. Int.* **2018**, *26*, 289–308. [CrossRef]
7. Hoseinifar, S.H.; Jahazi, M.A.; Nikdehghan, N.; van Doan, H.; Volpe, M.G.; Paolucci, M. Effects of Dietary Polyphenols from Agricultural By-Products on Mucosal and Humoral Immune and Antioxidant Responses of Convict Cichlid (*Amatitlania Nigrofasciata*). *Aquaculture* **2020**, *517*, 734790. [CrossRef]
8. Taheri Mirghaed, A.; Hoseini, S.M.; Hoseinifar, S.H.; van Doan, H. Effects of Dietary Thyme (*Zataria multiflora*) Extract on Antioxidant and Immunological Responses and Immune-Related Gene Expression of Rainbow Trout (*Oncorhynchus mykiss*) Juveniles. *Fish Shellfish Immunol.* **2020**, *106*, 502–509. [CrossRef]
9. Hoseinifar, S.H.; Shakouri, M.; Yousefi, S.; Van Doan, H.; Shafiei, S.; Yousefi, M.; Mazandarani, M.; Mozanzadeh, M.T.; Tulino, M.G.; Faggio, C. Humoral and skin mucosal immune parameters, intestinal immune related genes and antioxidant defense in rainbow trout (*Oncorhynchus mykiss*) fed dietary olive (*Olea europea* L.) waste. *Fish Shellfish Immunol.* **2020**, *100*, 171–178. [CrossRef]
10. Dawood, M.A.O. Nutritional Immunity of Fish Intestines: Important Insights for Sustainable Aquaculture. *Rev. Aquac.* **2021**, *13*, 642–663. [CrossRef]
11. Dawood, M.A.O.; Koshio, S.; Esteban, M.Á. Beneficial Roles of Feed Additives as Immunostimulants in Aquaculture: A Review. *Rev. Aquac.* **2018**, *10*, 950–974. [CrossRef]
12. Al-Hisnawi, A.; Rodiles, A.; Rawling, M.D.; Castex, M.; Waines, P.; Gioacchini, G.; Carnevali, O.; Merrifield, D.L. Dietary Probiotic Pediococcus Acidilactici MA18/5M Modulates the Intestinal Microbiota and Stimulates Intestinal Immunity in Rainbow Trout (*Oncorhynchus mykiss*). *J. World Aquac. Soc.* **2019**, *50*, 1133–1151. [CrossRef]
13. van Doan, H.; Hoseinifar, S.H.; Naraballobh, W.; Paolucci, M.; Wongmaneeprateep, S.; Charoenwattanasak, S.; Dawood, M.A.O.; Abdel-Tawwab, M. Dietary Inclusion of Watermelon Rind Powder and *Lactobacillus plantarum*: Effects on Nile Tilapia's Growth, Skin Mucus and Serum Immunities, and Disease Resistance. *Fish Shellfish Immunol.* **2021**, *116*, 107–114. [CrossRef] [PubMed]
14. Ghafarifarsani, H.; Hoseinifar, S.H.; Talebi, M.; Yousefi, M.; van Doan, H.; Rufchaei, R.; Paolucci, M. Combined and Singular Effects of Ethanolic Extract of Persian Shallot (*Allium Hirtifolium Boiss*) and Synbiotic Biomin®IMBO on Growth Performance, Serum-and Mucus-Immune Parameters and Antioxidant Defense in Zebrafish (*Danio rerio*). *Animals* **2021**, *11*, 2995. [CrossRef] [PubMed]

15. Hoseini, S.M.; Moghaddam, A.A.; Ghelichpour, M.; Pagheh, E.; Haghpanah, A.; Gharavi, B.; Mansouri, B.; Arghideh, M. Dietary Glycine Supplementation Modulates Antioxidant and Immune Responses of Beluga, *Huso Huso*, Juveniles. *Aquac. Rep.* **2022**, *23*, 101026. [CrossRef]
16. Evliyaoğlu, E.; Kilercioğlu, S.; Yılmaz, H.A.; Turchini, G.M.; Paolucci, M.; Clark, T.D.; Demirkale, İ.; Eroldoğan, O.T. Iron Supplementation in Plant-Based Aquafeed: Effects on Growth Performance, Tissue Composition, Iron-Related Serum Parameters and Gene Expression in Rainbow Trout (*Oncorhynchus mykiss*). *Aquaculture* **2022**, *550*, 737884. [CrossRef]
17. Hoseinifar, S.H.; Sohrabi, A.; Paknejad, H.; Jafari, V.; Paolucci, M.; van Doan, H. Enrichment of Common Carp (*Cyprinus carpio*) Fingerlings Diet with Psidium Guajava: The Effects on Cutaneous Mucosal and Serum Immune Parameters and Immune Related Genes Expression. *Fish Shellfish Immunol.* **2019**, *86*, 688–694. [CrossRef]
18. Motlagh, H.A.; Safari, O.; Paolucci, M. Effect of Different Levels of Milkweed (*Calotropis persica*) Seed Powder on the Growth Parameters, Immunity and Gut Microbiota of *Oncorhynchus mykiss*. *Iran J. Vet. Sci. Technol.* **2019**, *11*, 43–50.
19. Orso, G.; Solovyev, M.M.; Facchiano, S.; Tyrikova, E.; Sateriale, D.; Kashinskaya, E.; Pagliarulo, C.; Hoseinifar, H.S.; Simonov, E.; Varricchio, E.; et al. Chestnut Shell Tannins: Effects on Intestinal Inflammation and Dysbiosis in Zebrafish. *Animals* **2021**, *11*, 1538. [CrossRef]
20. Imperatore, R.; Orso, G.; Facchiano, S.; Scarano, P.P.; Hoseinifar, S.H.; Ashouri, G.; Guarino, C.; Paolucci, M. Anti-inflammatory and immunostimulant effect of different timing-related administration of dietary polyphenols on intestinal inflammation in zebrafish, *Danio rerio*. *Aquaculture* **2023**, *563*, 738878. [CrossRef]
21. Reverter, M.; Tapissier-Bontemps, N.; Sarter, S.; Sasal, P.; Caruso, D. Moving towards More Sustainable Aquaculture Practices: A Meta-Analysis on the Potential of Plant-Enriched Diets to Improve Fish Growth, Immunity and Disease Resistance. *Rev. Aquac.* **2021**, *13*, 537–555. [CrossRef]
22. van Hai, N. The Use of Medicinal Plants as Immunostimulants in Aquaculture: A Review. *Aquaculture* **2015**, *446*, 88–96. [CrossRef]
23. Abdel-Tawwab, M.; Adeshina, I.; Jenyo-Oni, A.; Ajani, E.K.; Emikpe, B.O. Growth, Physiological, Antioxidants, and Immune Response of African Catfish, *Clarias Gariepinus* (B.), to Dietary Clove Basil, *Ocimum gratissimum*, Leaf Extract and Its Susceptibility to *Listeria Monocytogenes* Infection. *Fish Shellfish Immunol.* **2018**, *78*, 346–354. [CrossRef] [PubMed]
24. Sarhadi, I.; Alizadeh, E.; Ahmadifar, E.; Adineh, H.; Dawood, M.A.O. Skin Mucosal, Serum Immunity and Antioxidant Capacity of Common Carp (*Cyprinus carpio*) Fed Artemisia (*Artemisia annua*). *Ann. Anim. Sci.* **2020**, *20*, 1011–1027. [CrossRef]
25. Ahmadifar, E.; Yousefi, M.; Karimi, M.; Fadaei Raieni, R.; Dadar, M.; Yilmaz, S.; Dawood, M.A.O.; Abdel-Latif, H.M.R. Benefits of Dietary Polyphenols and Polyphenol-Rich Additives to Aquatic Animal Health: An Overview. *Rev. Fish. Sci. Aquac.* **2021**, *29*, 478–511. [CrossRef]
26. Sutili, F.J.; Gatlin, D.M.; Heinzmann, B.M.; Baldisserotto, B. Plant Essential Oils as Fish Diet Additives: Benefits on Fish Health and Stability in Feed. *Rev. Aquac.* **2018**, *10*, 716–726. [CrossRef]
27. Firmino, J.P.; Galindo-Villegas, J.; Reyes-López, F.E.; Gisbert, E. Phytogenic Bioactive Compounds Shape Fish Mucosal Immunity. *Front. Immunol.* **2021**, *12*, 695973. [CrossRef]
28. Barbero, F.; Maffei, M. Biodiversity and chemotaxonomic significance of specialized metabolites. In *Plant Specialized Metabolism: Genomics, Biochemistry, and Biological Functions*, 1st ed.; Arimura, G., Maffei, M., Eds.; CRC Press: Boca Raton, FL, USA, 2016; pp. 24–63.
29. Napoli, E.; Siracusa, L.; Ruberto, G. New Tricks for Old Guys: Recent Developments in the Chemistry, Biochemistry, Applications and Exploitation of Selected Species from the *Lamiaceae* Family. *Chem. Biodivers.* **2020**, *17*, e1900677. [CrossRef]
30. Sönmez, A.Y.; Bilen, S.; Alak, G.; Hisar, O.; Yanik, T.; Biswas, G. Growth performance and antioxidant enzyme activities in rainbow trout (*Oncorhynchus mykiss*) juveniles fed diets supplemented with sage, mint and thyme oils. *Fish Physiol. Biochem.* **2015**, *41*, 165–175. [CrossRef]
31. Diler, O.; Gormez, O.; Diler, I.; Metin, S. Effect of oregano (*Origanum onites* L.) essential oil on growth, lysozyme and antioxidant activity and resistance against *Lactococcus garvieae* in rainbow trout, *Oncorhynchus mykiss* (Walbaum). *Aquac. Nutr.* **2017**, *23*, 844–851. [CrossRef]
32. Hoseini, S.M.; Yousefi, M. Beneficial effects of thyme (*Thymus vulgaris*) extract on oxytetracycline-induced stress response, immunosuppression, oxidative stress and enzymatic changes in rainbow trout (*Oncorhynchus mykiss*). *Aquac. Nutr.* **2019**, *25*, 298–309. [CrossRef]
33. Chishti, S.; Kaloo, Z.A.; Sultan, P. Medicinal importance of genus *Origanum*: A review. *J. Pharmacogn. Phytother.* **2013**, *5*, 170–177.
34. García-Beltrán, J.M.; Esteban, M.A. Properties and applications of plants of *Origanum* sp. Genus. *SM J. Biol.* **2016**, *2*, 1006.
35. Dundar, E.; Olgun, E.G.; Isiksoy, S.; Kurkcuoglu, M.; Baser, K.; Bal, C. The Effects of intra-rectal and intra-peritoneal application of *Origanum onites* L. essential oil on 2, 4, 6-trinitrobenzenesulfonic acid-induced colitis in the rat. *Exp. Toxicol. Pathol.* **2008**, *59*, 399–408. [CrossRef]
36. Pezzani, R.; Vitalini, S.; Iriti, M. Bioactivities of *Origanum vulgare* L.: An update. *Phytochem. Rev.* **2017**, *16*, 1253–1268. [CrossRef]
37. Scocco, P.; Forte, C.; Franciosini, M.P.; Mercati, F.; Casagrande-Proietti, P.; Dall'Aglio, C.; Acuti, G.; Tardella, F.M.; Trabalza-Marinucci, M. Gut complex carbohydrates and intestinal microflora in broiler chickens fed with oregano (*Origanum vulgare* L.) aqueous extract and Vitamin E. *J. Anim. Physiol. Anim. Nutr.* **2017**, *101*, 676–684. [CrossRef]
38. Mercati, F.; Dall'Aglio, C.; Acuti, G.; Faeti, V.; Tardella, F.M.; Pirino, C.; De Felice, E.; Scocco, P. Oregano Feed Supplementation Affects Glycoconjugates Production in Swine Gut. *Animals* **2020**, *10*, 149. [CrossRef] [PubMed]

39. Oniga, I.; Pușcaș, C.; Silaghi-Dumitrescu, R.; Olah, N.K.; Sevastre, B.; Marica, R.; Marcus, I.; Sevastre-Berghian, A.; Benedec, D.; Pop, C.; et al. *Origanum vulgare* ssp. vulgare: Chemical Composition and Biological Studies. *Molecules* **2018**, *23*, 2077. [CrossRef]
40. Ferreira, P.M.; da Silva Nascimento, L.; Coelho Dias, D.; da Veiga Moreira, D.M.; Lúcia Salaro, A.; Duca de Freitas, M.B.; Souza Carneiro, A.P.; Sampaio Zuanon, J.A. Essential Oregano Oil as a Growth Promoter for the *Yellowtail tetra, Astyanax altiparanae*. *J. World Aquac. Soc.* **2014**, *45*, 28–34. [CrossRef]
41. Abdel-Latif, H.M.R.; Abdel-Tawwab, M.; Khafaga, A.F.; Dawood, M.A.O. Dietary oregano essential oil improved antioxidative status, immune-related genes, and resistance of common carp (*Cyprinus carpio* L.) to *Aeromonas hydrophila* infection. *Fish Shellfish Immunol.* **2020**, *104*, 1–7. [CrossRef] [PubMed]
42. Abdel-Latif, H.M.R.; Abdel-Tawwab, M.; Khafaga, A.F.; Dawood, M.A.O. Dietary oregano essential oil improved the growth performance via enhancing the intestinal morphometry and hepato-renal functions of common carp (*Cyprinus carpio* L.) fingerlings. *Aquaculture* **2020**, *526*, 735432. [CrossRef]
43. Zhang, R.; Wang, X.W.; Liu, L.L.; Cao, Y.C.; Zhu, H. Dietary oregano essential oil improved the immune response, activity of digestive enzymes, and intestinal microbiota of the koi carp, *Cyprinus carpio*. *Aquaculture* **2020**, *518*, 734781. [CrossRef]
44. Yousefi, M.; Ghafarifarsani, H.; Hoseinifar, S.H.; Rashidian, G.; Van Doan, H. Effects of dietary marjoram, *Origanum majorana* extract on growth performance, hematological, antioxidant, humoral and mucosal immune responses, and resistance of common carp, *Cyprinus carpio* against *Aeromonas hydrophila*. *Fish Shellfish Immunol.* **2020**, *108*, 127–133. [CrossRef]
45. Rashidian, G.; Boldaji, J.T.; Rainis, S.; Proki'c, M.D.; Faggio, C. Oregano (*Origanum vulgare*) Extract Enhances Zebrafish (*Danio rerio*) Growth Performance, Serum and Mucus Innate Immune Responses and Resistance against *Aeromonas hydrophila* Challenge. *Animals* **2021**, *11*, 299. [CrossRef]
46. Dinardo, F.R.; Deflorio, M.; Casalino, E.; Crescenzo, G.; Centoducati, G. Effect of feed supplementation with *Origanum vulgare* L. essential oil on sea bass (*Dicentrarchus labrax*): A preliminary framework on metabolic status and growth performances. *Aquac. Rep.* **2020**, *18*, 100511. [CrossRef]
47. Dinardo, F.R.; Maggiolino, A.; Casalino, E.; Deflorio, M.; Centoducati, G. A Multi-Biomarker Approach in European Sea Bass Exposed to Dynamic Temperature Changes under Dietary Supplementation with *Origanum vulgare* Essential Oil. *Animals* **2021**, *11*, 982. [CrossRef]
48. Pourmoghim, H.; Haghighi, M.; Rohani, M.S. Effect of dietary inclusion of *Origanum vulgare* extract on nonspecific immune responses and hematological parameters of rainbow trout (*Oncorhynchus mykiss*). *Bull. Pharm. Sci.* **2015**, *4*, 33–39.
49. Haghighi, M.; Pourmoghim, H.; Sharif Rohani, M. Effect of *Origanum vulgare* extract on immune responses and hematological parameters of rainbow trout (*Oncorhynchus mykiss*). *Oceanogr. Fish. Open Access J.* **2018**, *6*, 555687.
50. Rafieepour, A.; Hajirezaee, S.; Rahimi, R. Dietary oregano extract (*Origanum vulgare* L.) enhances the antioxidant defence in rainbow trout, *Oncorhynchus mykiss* against toxicity induced by organophosphorus pesticide, diazinon. *Toxin Rev.* **2019**, *39*, 397–407. [CrossRef]
51. Rafieepour, A.; Hajirezaee, S.; Rahimi, R. Moderating Effects of Dietary Oregano Extract (*Origanum vulgare*) on the Toxicity Induced by Organophosphate Pesticide, Diazinon in Rainbow Trout, *Oncorhynchus mykiss*: Metabolic Hormones, Histology and Growth Parameters. *Turk. J. Fish. Aquat. Sci.* **2020**, *20*, 207–219. [CrossRef] [PubMed]
52. El Araby, D.A.; El Arabey, A.A. New Approach to use *Origanum vulgare* Extract as Immunostimulant to Increase Resistance to *Pseudomonas aeruginosa* and *Pseudomonas flourscence*. *J. Mar. Sci. Res. Dev.* **2015**, *6*, 1000182. [CrossRef]
53. El-Hawarry, W.N.; Mohamed, R.A.; Ibrahim, S.A. Collaborating effects of rearing density and oregano oil supplementation on growth, behavioral and stress response of Nile tilapia (*Oreochromis niloticus*). *Egypt. J. Aquat. Res.* **2018**, *44*, 173–178. [CrossRef]
54. Santo, E.; Amanda, H.; Brito, T.S.; Brandão, L.L.; Tavares, G.C.; Leibowitz, M.P.; Prado, S.A.; Ferraz, V.P.; Hoyos, D.C.M.; Turra, E.M.; et al. Dietary supplementation of dry oregano leaves increases the innate immunity and resistance of Nile tilapia against *Streptococcus agalactiae* infection. *J. World Aquac. Soc.* **2019**, *51*, 418–436. [CrossRef]
55. Mohammadi, G.; Rafiee, G.; El Basuini, M.F.; Van Doan, H.; Ahmed, H.A.; Dawood, M.A.O.; Abdel-Latif, H.M.R. Oregano (*Origanum vulgare*), St John's-wort (*Hypericum perforatum*), and lemon balm (*Melissa officinalis*) extracts improved the growth rate, antioxidative, and immunological responses in Nile tilapia (*Oreochromis niloticus*) infected with *Aeromonas hydrophila*. *Aquac. Rep.* **2020**, *18*, 100445. [CrossRef]
56. García Beltrán, J.M.; Silvera, D.G.; Ruiz, C.E.; Campo, V.; Chupani, L.; Faggio, C.; Esteban, M.Á. Effects of dietary *Origanum vulgare* on gilthead seabream (*Sparus aurata* L.) immune and antioxidant status. *Fish Shellfish Immunol.* **2020**, *99*, 452–461. [CrossRef]
57. AziziBasir, M.; Imani, A.; Moghanloo, K.S.; Meshkini, S. The effect of Marjoram (*Origanum vulgare* L.) and St John's Wort (*Hypericum perforatum*) hydro-alcoholic extracts on growth, immunity, digestive and antioxidative enzymes activity of rainbow trout (*Oncorhynchus mykiss*) under the commercial farming condition. *J. Fish.* **2020**, *72*, 339–351.
58. Heluy, G.M.; Ramos, L.R.V.; Pedrosa, V.F.; Sarturi, C.; Figueiredo, P.G.P.; Vidal, L.G.P.; Franca, I.d.F.; Pereira, M.M. Oregano (*Origanum vulgare*) essential oil as an additive in diets for Nile tilapia (*Oreochromis niloticus*) fingerlings reared in salinized water. *Aquac. Res.* **2020**, *51*, 3237–3243. [CrossRef]
59. Romero, J.; Ringø, E. The gut microbiota of fish. In *Aquaculture Nutrition (Gut Health, Probiotics and Prebiotics)*, 1st ed.; Merrifield, D., Ringø, E., Eds.; John Wiley & Sons Ltd.: Hoboken, NJ, USA, 2014; pp. 75–100.
60. Ran, C.; Hu, J.; Liu, W.; Liu, Z.; He, S.; Dan, B.C.T.; Diem, N.N.; Ooi, E.L.; Zhou, Z. Thymol and Carvacrol Affect Hybrid Tilapia through the Combination of Direct Stimulation and an Intestinal Microbiota-Mediated Effect: Insights from a Germ-Free Zebrafish Model. *Nutr. J.* **2016**, *115*, 229377. [CrossRef]

61. Giannenas, I.; Triantafillou, E.; Stavrakakis, S.; Margaroni, M.; Mavridis, S.; Steiner, T.; Karagouni, E. Assessment of dietary supplementation with carvacrol or thymol containing feed additives on performance, intestinal microbiota and antioxidant status of rainbow trout (*Oncorhynchus mykiss*). *Aquaculture* **2012**, *350–353*, 26–32. [CrossRef]
62. Coccimiglio, J.; Alipour, M.; Jiang, Z.H.; Gottardo, C.; Suntres, Z. Antioxidant, Antibacterial, and Cytotoxic Activities of the Ethanolic *Origanum vulgare* Extract and Its Major Constituents. *Oxidative Med. Cell. Longev.* **2016**, *2016*, 1404505. [CrossRef] [PubMed]
63. Zheng, Z.L.; Justin, Y.W.T.; Liu, H.Y.; Zhou, X.H.; Xiang, X.; Wang, K.Y. Evaluation of oregano essential oil (*Origanum heracleoticum* L.) on growth, antioxidant effect and resistance against *Aeromonas hydrophila* in channel catfish (*Ictalurus punctatus*). *Aquaculture* **2009**, *292*, 214–218. [CrossRef]
64. Swain, P.; Dash, S.; Sahoo, P.K.; Routray, P.; Sahoo, S.K.; Gupta, S.D.; Meher, P.K.; Sarangi, N. Nonspecific immune parameters of brood Indian major carp *Labeo rohita* and their seasonal variations. *Fish Shellfish Immunol.* **2007**, *22*, 38–43. [CrossRef]
65. Song, Q.; Xiao, Y.; Xiao, Z.; Liu, T.; Li, J.; Li, P.; Han, F. Lysozymes in Fish. *J. Agric. Food Chem.* **2021**, *69*, 15039–15051. [CrossRef] [PubMed]
66. Mabrok, M.A.E.; Wahdan, A. The immune modulatory effect of oregano (*Origanum vulgare* L.) essential oil on *Tilapia zillii* following intraperitoneal infection with *Vibrio anguillarum*. *Aquacult. Int.* **2018**, *26*, 1147–1160. [CrossRef]
67. Pastorino, P.; Bergagna, S.; Vercelli, C.; Pagliasso, G.; Dellepiane, L.; Renzi, M.; Barbero, R.; Re, G.; Elia, A.C.; Dondo, A.; et al. Changes in Serum Blood Parameters in Farmed Rainbow Trout (*Oncorhynchus mykiss*) Fed with Diets Supplemented with Waste Derived from Supercritical Fluid Extraction of Sweet Basil (*Ocimum basilicum*). *Fishes* **2022**, *7*, 89. [CrossRef]
68. Hoseini, S.M.; Taheri Mirghaed, A.; Iri, Y.; Hoseinifar, S.H.; van Doan, H.; Miriam Reverter, M. Effects of dietary Russian olive, *Elaeagnus angustifolia*, leaf extract on growth, hematological, immunological, and antioxidant parameters in common carp, *Cyprinus carpio*. *Aquaculture* **2021**, *536*, 736461. [CrossRef]
69. Al-Asgah, N.A.; Abdel-Warith, A.W.; Younis, E.-S.M.; Allam, H.Y. Haematological and biochemical parameters and tissue accumulations of cadmium in *Oreochromis niloticus* exposed to various concentrations of cadmium chloride. *Saudi J. Biol. Sci.* **2015**, *22*, 543–550. [CrossRef]
70. Alagawany, M.; Farag, M.R.; Salah, A.S.; Mahmoud, M.A. The role of oregano herb and its derivatives as immunomodulators in fish. *Rev. Aquac.* **2020**, *12*, 2481–2492. [CrossRef]
71. Sánchez-Camargo, A.D.P.; Herrero, M. Rosemary (*Rosmarinus officinalis*) as a functional ingredient: Recent scientific evidence. *Curr. Opin. Food Sci.* **2017**, *14*, 13–19. [CrossRef]
72. Yousefi, M.; Hoseini, S.M.; Vatnikov, Y.A.; Kulikov, E.V.; Drukovsky, S.G. Rosemary leaf powder improved growth performance, immune and antioxidant parameters, and crowding stress responses in common carp (*Cyprinus carpio*) fingerlings. *Aquaculture* **2019**, *505*, 473–480. [CrossRef]
73. Takayama, C.; de-Faria, F.M.; de Almeida, A.C.A.; Dunder, R.J.; Manzo, L.P.; Socca, E.A.R.; Batista, L.M.; Salvador, M.J.; Souza-Brito, A.R.M.; Luiz-Ferreira, A. Chemical composition of *Rosmarinus officinalis* essential oil and antioxidant action against gastric damage induced by absolute ethanol in the rat. *Asian Pac. J. Trop. Biomed.* **2016**, *6*, 677–681. [CrossRef]
74. D'Auria, M.; Racioppi, R. The Effect of Drying of the Composition of Volatile Organic Compounds in *Rosmarinus officinalis*, *Laurus nobilis*, *Salvia officinalis* and *Thymus serpyllum*. A HS-SPME-GC-MS Study. *J. Essent. Oil-Bear* **2015**, *18*, 1209–1223. [CrossRef]
75. Ahmadifar, E.; Pourmohammadi Fallah, H.; Yousefi, M.; Dawood, M.A.O.; Hoseinifar, S.H.; Adineh, H.; Yilmaz, S.; Paolucci, M.; Doan, H.V. The Gene Regulatory Roles of Herbal Extracts on the Growth, Immune System, and Reproduction of Fish. *Animals* **2021**, *11*, 2167. [CrossRef] [PubMed]
76. Barbieri, J.B.; Goltz, C.; Cavalheiro, F.B.; Toci, A.T.; Igarashi-Mafra, L.M.; Mafra, M.R. Deep eutectic solvents applied in the extraction and stabilization of rosemary (*Rosmarinus officinalis* L.) phenolic compounds. *Ind. Crop. Prod.* **2020**, *144*, 112049. [CrossRef]
77. Dezfoulnejad, M.C.; Molayemraftar, T. Use of dietary rosemary (*Rosmarinus officinalis* L.) extract as a growth promotor and immunostimulant in common carp. *Aquac. Res.* **2022**, *53*, 1553–1562. [CrossRef]
78. Hassan, A.A.M.; Yacout, M.H.; Khalel, M.S.; Hafsa, S.H.A.; Ibrahim, M.A.R.; Mocuta, D.N.; Turek Rahoveanu, A.; Dediu, L. Effects of some herbal plant supplements on growth performance and the immune response in Nile tilapia (*Oreochromis niloticus*). *Life Agric. Conf. Proc.* **2018**, *1*, 134–141. [CrossRef]
79. Yilmaz, E.; Çoban, D.; Kirim, B.; Güler, M. Effects of Extracts of Feed Additives Including Rosemary (*Rosmarinus officinalis*) and Aloe Vera (*Aloe barbadensis*) on the Growth Performance and Feed Utility of Nile Tilapia (*Oreochromis niloticus*). *Turk. J. Agric.-Food Sci. Technol.* **2019**, *7*, 866–870. [CrossRef]
80. Naiel, M.A.E.; Ismael, N.E.M.; Negm, S.S.; Ayyat, M.S.; Al-Sagheer, A.A. Rosemary leaf powder supplemented diet enhances performance, antioxidant properties, immune status, and resistance against bacterial diseases in Nile Tilapia (*Oreochromis niloticus*). *Aquaculture* **2020**, *526*, 735370. [CrossRef]
81. Hernández, A.; García García, B.; Caballero, M.J.; Hernández, M.D. Preliminary insights into the incorporation of rosemary extract (*Rosmarinus officinalis* L.) in fish feed: Influence on performance and physiology of gilthead seabream (*Sparus aurata*). *Fish Physiol. Biochem.* **2015**, *41*, 1065–1074. [CrossRef]
82. Hosseini, S.M.; Hoseinifar, S.H.; Mazandarani, M.; Paknejad, H.; Van Doan, H.; El-Haroun, E. The potential benefits of orange peels derived pectin on serum and skin mucus immune parameters, antioxidant defence and growth performance in common carp (*Cyprinus carpio*). *Fish Shellfish Immunol.* **2020**, *103*, 17–22. [CrossRef]

83. Koga, K.; Shibata, H.; Yoshino, K.; Nomoto, K. Effects of 50% ethanol extract from rosemary (*Rosmarinus officinalis*) on α-glucosidase inhibitory activity and the elevation of plasma glucose level in rats, and its active compound. *J. Food Sci.* **2006**, *71*, 507–512. [CrossRef]
84. Ebrahimi, E.; Haghjou, M.; Nematollahi, A.; Goudarzian, F. Effects of rosemary essential oil on growth performance and hematological parameters of young great sturgeon (*Huso huso*). *Aquaculture* **2020**, *521*, 734909. [CrossRef]
85. Di Turi, L.; Ragni, M.; Caputi Jambrenghi, A.; Lastilla, M.; Vicenti, A.; Colonna, M.A.; Giannico, F.; Vonghia, G. Effect of dietary rosemary oil on growth performance and flesh quality of farmed seabass (*Dicentrarchus labrax*), *Ital. J. Anim. Sci.* **2009**, *8*, 857–859.
86. Naiel, M.A.; Ismael, N.E.; Shehata, S.A. Ameliorative effect of diets supplemented with rosemary (*Rosmarinus officinalis*) on aflatoxin B1 toxicity in terms of the performance, liver histopathology, immunity and antioxidant activity of Nile Tilapia (*Oreochromis niloticus*). *Aquaculture* **2019**, *511*, 734264. [CrossRef]
87. Turan, F.; Yiğitarslan, D. The Effects of Rosemary Extract (*Rosemaria officinalis*) as a Feed Additive on Growth and Whole-body Composition of the African Catfish (*Clarias gariepinus* (Burchell, 1822)). *Nat. Eng. Sci.* **2016**, *1*, 49–55. [CrossRef]
88. Eftekhar, N.; Moghimi, A.; Boskabady, M.H. Prophylactic effect of rosmarinic acid on tracheal responsiveness, white blood cell count and oxidative stress markers in lung lavage of sensitized rats. *Pharmacol. Rep.* **2018**, *70*, 119–125. [CrossRef]
89. Gabriel, N.N.; Qiang, J.; Ma, X.Y.; He, J.; Xu, P.; Liu, K. Dietary Aloe vera improves plasma lipid profile, antioxidant, and hepatoprotective enzyme activities in GIFT-tilapia (*Oreochromis niloticus*) after *Streptococcus iniae* challenge. *Fish Physiol. Biochem.* **2015**, *41*, 1321–1332. [CrossRef]
90. Farkhondeh, T.; Samarghandian, S.; Pourbagher-Shahri, A.M. Hypolipidemic effects of *Rosmarinus officinalis* L. *J. Cell. Physiol.* **2019**, *234*, 14680–14688. [CrossRef]
91. Zilberg, D.; Tal, A.; Froyman, N.; Abutbul, S.; Dudai, N.; Golan-Goldhirsh, A. Dried leaves of Rosmarinus officinalis as a treatment for streptococcosis in tilapia. *J. Fish Dis.* **2010**, *33*, 361–369. [CrossRef]
92. Jiang, Y.; Wu, N.; Fu, Y.-J.; Wang, W.; Luo, M.; Zhao, C.-J.; Zu, Y.-G.; Liu, X.-L. Chemical composition and antimicrobial activity of the essential oil of rosemary. *Environ. Toxicol. Pharmacol.* **2011**, *32*, 63–68. [CrossRef]
93. Rozman, T.; Jersek, B. Antimicrobial activity of rosemary extracts (*Rosmarinus officinalis* L.) against different species of listeria. *Acta Agric. Slov.* **2009**, *93*, 51. [CrossRef]
94. Bruneton, J. *Pharmacognosy, Phytochemistry, Medicinal Plants*, 2nd ed.; Lavoisier Publishing: Paris, France, 1999; pp. 540–544.
95. Cuvelier, M.E.; Berset, C.; Richard, H. Antioxidant constituents in sage (*Salvia officinalis*). *J. Agric. Food Chem.* **1994**, *42*, 665–669. [CrossRef]
96. Salomón, R.; Firmino, J.P.; Reyes-López, F.E.; Andre, K.B.; González-Silvera, D.; Esteband, M.A.; Tort, L.; Quintela, J.C.; Pinilla-Rosas, J.M.; Vallejos-Vidal, E.; et al. The growth promoting and immunomodulatory effects of a medicinal plant leaf extract obtained from *Salvia officinalis* and *Lippia citriodora* in gilthead seabream (*Sparus aurata*). *Aquaculture* **2020**, *524*, 735291. [CrossRef]
97. Dadras, H.; Hayatbakhsh, M.R.; Golpour, A. Dietary Administration of Common Sage (*Salvia officinalis*) and Coneflower (*Echinacea angustifolia)* Extracts Affects Growth, Blood Parameters and Immune Responses of Beluga, *Huso huso*. *Turk. J. Fish. Aquat. Sci.* **2020**, *20*, 367–374. [CrossRef] [PubMed]
98. Fernández-Navarro, M.; Peragón, J.; Esteban, F.J.; de la Higuera, M.; Lupiáñez, J.A. Maslinic acid as a feed additive to stimulate growth and hepatic proteinturnover rates in rainbow trout (*Onchorhynchus mykiss*). *Comp. Biochem. Physiol. Toxicol. Pharmacol.* **2006**, *144*, 130–140. [CrossRef]
99. Abdellatief, A.N.; Rahman, A.; Abdallah, F.D.M. Evaluation of Immunostimulant Activity of Spirulina platensis (*Arthrospira platensis*) and Sage (*Salvia officinalis*) in Nile Tilapia (*Oreochromis niloticus*). *Zagazig Vet. J.* **2018**, *46*, 25–36. [CrossRef]
100. Boruga, O.; Jianu, C.; Mişcă, C.; Goleţ, I.; Gruia, A.; Horhat, F. *Thymus vulgaris* essential oil: Chemical composition and antimicrobial activity. *J. Med. Life* **2014**, *7*, 56–60.
101. Antih, J.; Houdkova, M.; Urbanova, K.; Kokoska, L. Antibacterial Activity of *Thymus vulgaris* L. Essential Oil Vapours and Their GC/MS Analysis Using Solid-Phase Microextraction and Syringe Headspace Sampling Techniques. *Molecules* **2021**, *26*, 6553. [CrossRef]
102. Abdollahzadeh, E.; Rezaei, M.; Hosseini, H. Antibacterial activity of plant essential oils and extracts: The role of thyme essential oil, nisin, and their combination to control *Listeria monocytogenes* inoculated in minced fish meat. *Food Control.* **2014**, *35*, 177–183. [CrossRef]
103. Hossain, M.A.; AL-Raqmi, K.A.S.; AL-Mijizy, Z.H.; Weli, A.M.; Al-Riyami, Q. Study of total phenol, flavonoids contents and phytochemical screening of various leaves crude extracts of locally grown *Thymus vulgaris*. *Asian Pac. J. Trop. Biomed.* **2013**, *3*, 705–710. [CrossRef]
104. ALsafah, A.H.; AL-Faragi, J.K. Influence of thyme (*Thymus vulgaris*) as feed additives on growth performance and antifungal activity on *Saprolegnia* spp. in *Cyprinus carpio* L. *J. Entomol. Zool. Stud.* **2017**, *5*, 1598–1602.
105. Yousefi, M.; Ghafarifarsani, H.; Hoseini, S.M.; Hoseinifar, S.H.; Abtahi, B.; Vatnikov, Y.A.; Kulikov, E.V.; Van Doan, H. Effects of dietary thyme essential oil and prebiotic administration on rainbow trout (*Oncorhynchus mykiss*) welfare and performance. *Fish Shellfish Immunol.* **2022**, *120*, 737–744. [CrossRef]
106. Zargar, A.; Rahimi-Afzal, Z.; Soltani, E.; Taheri Mirghaed, A.; Ebrahimzadeh-Mousavi, H.A.; Soltani, M.; Yuosefi, P. Growth performance, immune response and disease resistance of rainbow trout (*Oncorhynchus mykiss*) fed *Thymus vulgaris* essential oils. *Aquac. Res.* **2019**, *50*, 3097–3106. [CrossRef]

107. Ghafarifarsani, H.; Kachuei, R.; Imani, A. Dietary supplementation of garden thyme essential oil ameliorated the deteriorative effects of aflatoxin B1 on growth performance and intestinal inflammatory status of rainbow trout (*Oncorhynchus mykiss*). *Aquaculture* **2021**, *531*, 735928. [CrossRef]
108. Valladão, G.M.R.; Gallani, S.U.; Kotzent, S.; Assane, I.M.; Pilarski, F. Effects of dietary thyme essential oil on hemato-immunological indices, intestinal morphology, and microbiota of Nile tilapia. *Aquac. Int.* **2019**, *27*, 399–411. [CrossRef]
109. Dorojan, O.V.; Cristea, V.; Petrea, Ş.M. The influence of some phytobiotics (Thyme, Seabuckthorn) on growth performance of stellate sturgeon. *Int. J. Recirc. Aquac.* **2014**, *47*, 205–210.
110. Dorojan, O.G.; Cristea, V.; Creţu, M.; Coadă, M.T.; Dediu, L.; Grecu, I. Effect of thyme (*Thymus vulgaris*) and vitamin E on growth performance and body composition of *Acipenser stellatus* juveniles. *Aquac. Aquar. Conserv. Legis.* **2015**, *8*, 195–202.
111. Perez-Roses, R.; Risco, E.; Vila, R.; Penalver, P.; Canigueral, S. Biological and non- biological antioxidant activity of some essential oils. *J. Agric. Food Chem.* **2016**, *64*, 4716–4724. [CrossRef]
112. Wei, H.K.; Xue, H.X.; Zhou, Z.X.; Peng, J. A carvacrol-thymol blend decreased intestinal oxidative stress and influenced selected microbes without changing the messenger RNA levels of tight junction proteins in jejunal mucosa of weaning piglets. *Animal* **2017**, *11*, 193–201. [CrossRef]
113. Hashemipour, H.; Kermanshahi, H.; Golian, A.; Veldkamp, T. Effect of thymol and carvacrol feed supplementation on performance, antioxidant enzyme activities, fatty acid composition, digestive enzyme activities, and immune response in broiler chickens. *Poult. Sci.* **2013**, *92*, 2059–2069. [CrossRef]
114. Li, S.; Ru, Y.; Liu, M.; Xu, B.; Peron, A.; Shi, X. The effect of essential oils on performance, immunity and gut microbial population in weaner pigs. *Livest. Sci.* **2012**, *145*, 119–123. [CrossRef]
115. Gulec, A.; Azime, K.; Durali, D.; Mesut, U.; Engin, S.; Ali, A. Effect of mixed use of thyme and fennel oils on biochemical properties and electrolytes in rainbow trout as a response to *Yersinia ruckeri* infection. *Acta Vet. Brno* **2013**, *82*, 297–302. [CrossRef]
116. Gültepe, N.; Bilen, S.; Yılmaz, S.; Güroy, D.; Aydın, S. Effects of herbs and spice on health status of tilapia (*Oreochromis mossambicus*) challenged with *Streptococcus iniae*. *Acta Vet. Brno* **2014**, *83*, 125–131. [CrossRef]
117. Ashfaq, H.; Soliman, H.; Saleh, M.; El-Matbouli, M. CD4: A vital player in the teleost fish immune system. *Vet. Res.* **2019**, *7*, 1. [CrossRef]
118. Singh, R.R.; Shushni, M.A.M.; Belkheir, A. Antibacterial and antioxidant activities of *Mentha piperita* L. *Arab. J. Chem.* **2015**, *8*, 322–328. [CrossRef]
119. Brahmi, F.; Khodir, M.; Mohamed, C.; Pierre, D. Chemical composition and biological activities of Mentha species. In *Aromatic and Medicinal Plants—Back to Nature*, 1st ed.; El-Shemy, H.A., Ed.; InTechOpen: London, UK, 2017; pp. 47–79.
120. de Souza Silva, L.T.; de Pádua Pereira, U.; de Oliveira, H.M.; Brasil, E.M.; Pereira, S.A.; Chagas, E.C.; Jesus, G.F.A.; Cardoso, L.; Mouriño, J.L.P.; Martins, M.L. Hemato- immunological and zootechnical parameters of Nile tilapia fed essential oil of *Mentha piperita* after challenge with *Streptococcus agalactiae*. *Aquaculture* **2019**, *506*, 205–211. [CrossRef]
121. Balakrishnan, A. Therapeutic uses of peppermint—A review. *J. Pharm. Sci.* **2015**, *7*, 474–476.
122. McKay, D.L.; Blumberg, J.B. A review of the bioactivity and potential health benefits of peppermint tea (*Mentha piperita* L.). *Phytother. Res.* **2006**, *20*, 619–633. [CrossRef]
123. Talpur, A.D. *Mentha piperita* (Peppermint) as feed additive enhanced growth performance, survival, immune response and disease resistance of Asian seabass, *Lates calcarifer* (Bloch) against *Vibrio harveyi* infection. *Aquaculture* **2014**, *420–421*, 71–78. [CrossRef]
124. Dawood, M.A.; Metwally, A.E.-S.; Elkomy, A.H.; Gewaily, M.S.; Abdo, S.E.; Abdel-Razek, M.A.; Soliman, A.A.; Amer, A.A.; Abdel-Razik, N.I.; Abdel-Latif, H.M. The impact of menthol essential oil against inflammation, immunosuppression, and histopathological alterations induced by chlorpyrifos in Nile tilapia. *Fish Shellfish Immunol.* **2020**, *102*, 316–325. [CrossRef]
125. Adel, M.; Safari, R.; Pourgholam, R.; Zorriehzahra, J.; Esteban, M.Á. Dietary peppermint (*Mentha piperita*) extracts promote growth performance and increase the main humoral immune parameters (both at mucosal and systemic level) of Caspian brown trout (*Salmo trutta caspius* Kessler, 1877). *Fish Shellfish Immunol.* **2015**, *47*, 623–629. [CrossRef]
126. Adel, M.; Abedian Amiri, A.; Zorriehzahra, J.; Nematolahi, A.; Esteban, M.Á. Effects of dietary peppermint (*Mentha piperita*) on growth performance, chemical body composition and hematological and immune parameters of fry Caspian white fish (*Rutilus frisii kutum*). *Fish Shellfish Immunol.* **2015**, *45*, 841–847. [CrossRef] [PubMed]
127. Valladão, G.M.R.; Gallani, S.U.; Pala, G.; Jesus, R.B.; Kotzent, S.; Costa, J.C.; Silva, T.F.A.; Pilarski, F. Practical diets with essential oils of plants activate the complement system and alter the intestinal morphology of Nile tilapia. *Aquac. Res.* **2017**, *48*, 5640–5649. [CrossRef]
128. Harikrishnan, R.; Balasundaram, C.; Heo, M.-S. Impact of plant products on innate and adaptive immune system of cultured finfish and shellfish. *Aquaculture* **2011**, *317*, 1–15. [CrossRef]
129. Hoseini, S.M.; Yousefi, M.; Hoseinifar, S.H.; Van Doan, H. Antioxidant, enzymatic and hematological responses of common carp (Cyprinus carpio) fed with myrcene- or menthol-supplemented diets and exposed to ambient ammonia. *Aquaculture* **2019**, *506*, 246–255. [CrossRef]
130. Adel, M.; Pourgholam, R.; Zorriehzahra, J.; Ghiasi, M. Hemato—Immunological and biochemical parameters, skin antibacterial activity, and survival in rainbow trout (*Oncorhynchus mykiss*) following the diet supplemented with *Mentha piperita* against *Yersinia ruckeri*. *Fish Shellfish Immunol.* **2016**, *55*, 267–273. [CrossRef]
131. Heydari, M.; Firouzbakhsh, F.; Paknejad, H. Effects of Mentha longifolia extract on some blood and immune parameters, and disease resistance against yersiniosis in rainbow trout. *Aquaculture* **2020**, *515*, 734586. [CrossRef]

132. Mohamad, S.; Abasali, H. Effect of plant extracts supplemented diets on immunity and resistance to *aeromonas hydrophila* in common carp (*Cyprinus carpio*). *Res. J. Anim. Sci.* **2010**, *4*, 26–34.
133. Serradell, A.; Torrecillas, S.; Makol, A.; Valdenegro, V.; Fernández-Montero, A.; Acosta, F.; Izquierdo, M.; Montero, D. Prebiotics and phytogenics functional additives in low fish meal and fish oil based diets for European sea bass (*Dicentrarchus labrax*): Effects on stress and immune responses. *Fish Shellfish Immunol.* **2020**, *100*, 219–229. [CrossRef]
134. Hrubec, T.C.; Smith, S.A. Hematology of fishes. In *Schalm's Veterinary Hematology*, 6th ed.; Weiss, D.J., Wardrop, J., Schalm, O.W., Eds.; Wiley-Blackwell: Ames, Iowa, 2010; pp. 994–1003.
135. Sharifzadeh, S.A.; Khara, H.; Ghobadi, S. Effects of vitamins E and riboflavin (B2) and combinations of them on the hematological parameters of common carp, *Cyprinus carpio* L., fingerlings. *Arch. Pol. Fish* **2015**, *23*, 107–111. [CrossRef]
136. Mohamed, W.A.M.; Abd-Elhakim, Y.M.; Farouk, S.M. Protective effects of ethanolic extract of rosemary against lead-induced hepato-renal damage in rabbits. *Exp. Toxicol. Pathol.* **2016**, *68*, 451–461. [CrossRef]
137. Glisic, S.B.; Ristic, M.; Skala, D.U. The combined extraction of sage (*Salvia officinalis* L.): Ultrasound followed by supercritical CO2 extraction. *Ultrason. Sonochem.* **2011**, *18*, 318–326. [CrossRef] [PubMed]
138. Al Hashmi, L.S.; Hossain, M.A.; Weli, A.M.; Al-Riyami, Q.; AlSabahi, J.N. Gas chromatography–mass spectrometry analysis of different organic crude extracts from the local medicinal plant of Thymus vulgaris L. *Asian Pac. J. Trop. Biomed.* **2013**, *3*, 69–73. [CrossRef] [PubMed]
139. Buleandra, M.; Oprea, E.; Popa, D.E.; David, I.G.; Moldovan, Z.; Mihai, I.; Badea, I.A. Comparative Chemical Analysis of *Mentha piperita* and *M. spicata* and a Fast Assessment of Commercial Peppermint Teas. *Nat. Prod. Commun.* **2016**, *11*, 551–555. [PubMed]
140. Riachi, L.G.; De Maria, C.A.B. Peppermint antioxidants revisited. *Food Chem.* **2015**, *176*, 72–81. [CrossRef] [PubMed]

fishes

MDPI

Review

In Pursuit of Fish-Free Feeds: A Multi-Species Evaluation

Kelly B. Campbell [1], Ewen McLean [2,*] and Frederic T. Barrows [3]

1 Anthropocene Institute, Palo Alto, CA 94303, USA
2 Aqua Cognoscenti LLC, West Columbia, SC 29170, USA
3 Aquatic Feed Technologies LLC, Bozeman, MT 59718, USA
* Correspondence: ewen.mclean@gmail.com; Tel.: +1-803-463-9362

Abstract: The future growth and sustainability of fed aquaculture, and especially that for carnivorous species, will be highly dependent upon the industry stepping away from its reliance upon forage fishes as major feed ingredients. With this goal in mind, the F3 Feed Innovation Network—a consortium of researchers; businesses, including feed manufacturers and ingredient providers; NGOs; and others—energizes industry to adopt novel and promising aquafeed ingredients and formulations. All evaluated formulae are open-source and freely available on the F3 website. Moreover, the F3 diets can be readily retailored to suit user demands and/or local conditions (i.e., ingredient availability/restrictions). This presentation summarizes completed F3 trials undertaken with five species of cultured and candidate fishes. With reference to eight studies, findings are compared against conventional fishmeal (FM)/fish oil (FO)-based feeds. The described research documents the response of test animals to aquafeeds containing traditional FM/FO alternatives (e.g., soybean meal and poultry by-product meal) as well as innovative ingredients (e.g., microalgae and single-cell proteins). Depending on the species examined, account is given to the overall growth performance, health aspects, and product quality. The F3 trials demonstrate the feasibility of the complete removal of FM/FO from the diets of the tested animals.

Keywords: largemouth bass; pompano; amberjack; red drum; algal oil

Citation: Campbell, K.B.; McLean, E.; Barrows, F.T. In Pursuit of Fish-Free Feeds: A Multi-Species Evaluation. *Fishes* **2022**, *7*, 336. https://doi.org/10.3390/fishes7060336

Academic Editors: Marina Paolucci and Shunsuke Koshio

Received: 6 October 2022
Accepted: 13 November 2022
Published: 17 November 2022

1. Introduction

The desire to optimize aquafeeds has a long history. Like today, although not necessarily directly articulated, there was an aspiration to develop a more sustainable aquaculture from as early as the 1920s. Concerns included reducing water pollution caused by raw meats (fish, horse, seal, and sheep) and offal (liver, spleen, heart, and lungs), which were commonly used as hatchery feeds. Around the same time, and especially during war years, fish and meats employed as feeds were also rationed and or becoming more expensive [1–4]. Concurrently, culturists sought to confront problems related to the effective storage and dissemination of feed [5–8] and disease transmission from trash to cultured fish [9]. Thus, cheaper feeds based on alternative ingredients were sought. Investigations with animal-protein-free diets, however, resulted in inferior growth and feed conversion, changes in animal physiology, and increased mortalities (e.g., [10,11]). These adverse reactions were generally attributed to plant-derived toxins and nutritional inadequacies, such as vitamin deficiencies [12,13], and were so commonly described that some suggested the use of plant meals, especially in fingerling feeds, was inadvisable [14].

In the intervening years, various dietary formulations were evaluated [15,16], with pelleted feeds such as the Oregon Moist Pellet [17] and dry preparations being used in US state hatcheries and at commercial farms in the 1950s [18–21]. In the late 1950s, Edward Grassl [22–24] evaluated the use of dry diets both as feeds and medicated diets. He compared the growth of trout fed either wet chopped meats or dry pelleted animal/vegetable feeds and reported identical growth even when the pellet was fed at 50% of the amount recommended for raw feeds, provided that chopped liver was fed once a month or so. Implementation of the pelleted feeds by state hatcheries resulted in 60% improved production and a 40% reduction in food costs. Use of a vitamin mixture in the dry pellet, as recorded by Phillips et al. [25], eliminated the need for chopped liver supplements and formed the basis for trout dry pellet formulations and the development of mechanized feeders [26,27]. Across the Atlantic, similar trials were being undertaken with salmon [28]. These pioneering studies, together with the elucidation of the nutritional requirements of some species, represent foundational moments in the elaboration of the global aquafeed industry.

Between 2001 and 2010, global aquaculture production increased at an annual average of 5.8%, and between 2011 and 2018, by 4.5%. This astonishing rate of expansion, while having moderated to around 3% during 2020, continues to grow [29]. Significantly, much of the growth experienced this century has occurred in the fed aquaculture sector, which accounted for 86% and 73% of global FM and fish oil (FO) supplies, respectively, in 2020 [29]. Future projections suggest that if the use of FM/FO is to remain as-is in aquafeeds, demand will outstrip supply by 2030 [30]. This scenario has led to major reductions in the use of FM/FO over the last two decades through wiser use allocations and their replacement with alternatives, including fish processing by-products among many others [31]. Together with enhanced feed conversion efficiencies, genetic selection programs [32], and even commodity price risk hedging [33], among other strategies, feed costs have been reduced, with some of the alternative products illustrating adequate environmental performance [34–36], thereby having potential to improve industry sustainability.

The drive to replace FM/FO from aquafeeds is based not only on projected availability, which also influences raw material prices, but also on the growing concerns of active environmental and consumer lobbies. These groups point to the fact that forage fisheries influence the health and sustainability of marine and coastal ecosystems, while their prey are vital to the sustenance of marine predators including other fishes, birds, marine mammals, and humans. Seafood buyers are also becoming more knowledgeable of the range of potential contaminants that may impact food safety, including those of raw materials used during aquafeed production (review: [37]). Well-informed consumers are learned of human rights infractions that occur in some industrial fisheries and across aquaculture supply chains [38] and aware of the negative consequences of at-sea discarding [39], ghost nets [40], harmful fisheries subsidies [41], carbon emissions from fleets and feed manufacturers [42,43], and animal welfare issues [44]. These worries have resulted in the creation of a sustainability imperative driven by consumers who demand safe and ethically and environmentally responsive food production systems and base purchase decisions on these principles. The aquafeed industry is in an influential position to ensure that consumer and environmental desires are achieved. In the interim, the search for and the evaluation of suitable alternatives to FM and FO must be unwavering.

Since 2014, the F3 Feed Innovation Network (f3fin.org (accessed on 11 November 2022)) has encouraged sustainable initiatives to reduce the dependence of the aquafeed industry on forage fishes and embolden the sector to adopt novel and promising ingredients and formulations. One way in which the F3 consortium accomplishes this is through openly sharing recipes and experimental findings through its website and publications. Aquafeeds do not have a requirement for any specific ingredient but must satisfy the nutritional prerequisites of the target animal. Feeds, therefore, must provide a combination of nutrients in the correct proportions to fulfill the metabolic needs of the species in question [45]. Bearing this in mind, the F3 consortium has completed several trials in efforts to eliminate FM/FO from the feeds of a variety of widely cultivated and candidate species. Here, we recount the findings of eight of these trials. The formulations considered herein for each species studied are all open-source and freely available on the F3 website. All F3 diets may be retailored to suit the user's demands; other tested formulations are similarly deposited on the F3 website.

2. Species Evaluated

Five well-established and candidate cultured species of teleost were examined for their sensitivity to fish-free feeds (F3). They included the largemouth bass *Micropterus salmoides*, which is the preeminent farmed Perciformes, representing over 50% or around 432,000 tons of total production [46]; the Florida pompano *Trachinotus carolinus*, a strong candidate species for US aquaculture; and other members of the family Carangidae, including two species of amberjack, namely the California yellowtail *Seriola dorsalis* and kampachi *S. rivoliana*. More than 170,000 tons of pompano is cultured annually, with most production being in Asia [47], while 150,000 tons of amberjack is farmed globally, with most production being dependent on raw fish, although the availability of formulated diets has recently increased. As a major portion of compounded feeds used in Chinese mariculture is taken by farmers of pompano and red drum *Sciaenops ocellatus* [48], the latter species was also evaluated. Global production of Sciaenids exceeds 340,000 tons, with the red drum representing around 25% of the total [49]. As adults, the species evaluated herein are considered as obligate carnivores. For many marine carnivores, only a few studies have investigated the potential for concurrently replacing FM/FO in diets.

3. Selection of Dietary Ingredients

The ingredients employed in the various diets are presented in Table 1. All experimental feeds were evaluated against the response of the test species to an FM/FO-based diet. The least expensive and best performing experimental feeds from each reference were chosen for comparison for the sake of this paper; however, additional diets and corresponding performance data are available in the references listed. Dietary protein for the investigational diets was derived from both animal and plant sources and based on availability, demonstrated utility, and/or promise as a dietary component. For example, poultry by-product meal (PBM) represents a resource of considerable potential as an FM alternative. Global production of chickens is estimated to be 33 billion individuals, equivalent to 101 million tons for 2022 [50], with the largest producers being the USA, Brazil, China, and the EU Raw materials leftover from slaughterhouses and processing facilities represent about 30% of liveweight [51] or around 30 million tons. The clean unused parts of butchered poultry, including the voided intestines and culled laying hens, are ground and then rendered into meal. Corn gluten meal (CGM), a by-product of corn processing containing about 65% crude protein, and corn protein concentrate (CPC), comprising around 67% protein, have both enjoyed success as components of a variety of commercial and investigational aquafeeds, including those for amberjack [52], Florida pompano [53], red drum [54], and largemouth bass [55]. No negative effects of CGM or CPC have been reported, even when used at relatively high levels of dietary incorporation. Soybean meal (SBM; ~50% crude protein) is an excellent substitute for animal proteins in aquafeeds, even though the presence of anti-nutritional factors (ANFs) may disrupt gut function in some species [56]. Nevertheless, SBM and soy products, such as soy protein concentrate (SPC; ~63% crude protein), which has a reduced concentration of ANFs, have garnered wide use and are well represented in commercial and experimental aquafeeds across the board [57,58]. MrFeed Pro 50 (~51% crude protein), a bacterial hydrolysate made from soybean-derived cellulosic sugars, and similar products have received increased attention due to their availability, high digestibility, lower costs, safety, and sustainability [59,60]. *Spirulina* expresses 55%+ crude protein and an elevated PUFA content and, when used at high levels of supplementation, has been observed to provide a beneficial effect on animal growth, body composition, pigmentation, immunity, and reproductive performance [61]. A wide variety of FO alternatives have been assessed with a broad range of species, and flax, canola, and algal oils are not exceptional, being widely available and competitively priced. Each has been successfully used as an FO substitute with pompano, yellowtail, largemouth bass, and others [62–64].

Table 1. Formulations of experimental diets in which fishmeal and fish oil were replaced with a combination of different protein sources and oils. GMO = Genetically Modified Organism; ARS = Agricultural Research Service. For formulation information on FM/FO-based diets, the reader is directed to f3fin.org/resources/open-feed-formulas/ (accessed on 11 November 2022).

Ingredient	Kampachi [65]	Yellowtail [66]	Largemouth Bass 1 [67]	Largemouth Bass 2 [68]	Largemouth Bass 3 [46]	Red Drum [49]	Pompano 1 [47]	Pompano 2 [69]
Poultry by-product meal	36.12	23.12	25.62	25.62	28.80	28.8	36.12	23.12
Wheat, whole ground	20.53	16.75	20.43	22.7	18.41	18.41	17.77	16.75
Corn gluten meal	-	-	8.16	8.16	-	-	-	-
Corn protein concentrate	13.56	7.14	-	-	-	-	8.22	7.14
Non-GMO soybean meal	-	-	11	11	-	-	-	-
Soy protein concentrate	7.86	-	17.93	-	24.32	24.32	5.96	-
MrFeed Pro50	-	-	-	15	12.5	12.5	12.5	-
Algae meal	-	-	-	6	-	-	-	-
Spirulina	-	30	-	-	-	-	-	30
Algal oil, Veramaris	5.32	10.80	-	-	2.28	2.28	2.13	10.80
Flax oil	2.71	-	-	-	-	4.52	1.86	-
Non-GMO soy oil	-	-	4.73	2.7	-	-	-	-
Canola oil	2.38	-	-	-	5.42	0.9	1.32	-
Fish oil—Menhaden	-	-	3	-	-	-	-	-
Dicalcium phosphate	3.1	4.16	-	-	-	-	3.10	4.16
Monocalcium phosphate	-	-	1.97	1.35	1.8	1.8	-	-
Lysine-HCL	2.67	2.68	1.66	1.97	1.62	1.62	2.27	2.68
Taurine	2	2	1	1	-	-	2	2
DL-Methionine	0.69	0.64	0.64	0.64	0.74	0.74	0.77	0.64
Threonine	0.46	0.31	0.31	0.31	0.21	0.21	0.38	0.31
Choline CL	0.6	0.6	0.6	0.6	0.6	0.6	0.60	0.60
Lecithin	-	-	2	2	2	2	-	-
Stay-C	0.2	0.2	0.2	0.2	0.2	0.2	0.20	0.20
Vitamin Premix ARS 702	1.5	1.5	0.5	0.5	1	1	1.50	1.50
Trace min premix ARS 1520	0.10	0.10	0.25	0.25	0.10	0.10	0.10	0.10
Trace min premix F3	0.20	-	-	-	-	-	0.20	-
TOTAL	100	100	100	100	100	100	100	100

4. Fish Holding and Husbandry

Other than for a study with kampachi, all feeding trials were undertaken in tanks configured as recirculating systems. The study lengths, which varied from 56 to 126 days; water quality parameters, including temperature, salinity, and dissolved oxygen levels; and start weights of experimental animals are summarized in Table 2. Water quality parameters were collected using standard methods. The feeding schedule for the experimental and control diets for each species is likewise presented in Table 2. All studies were executed with appropriate regard to Institutional Animal Care and Use Committee regulations and complied with all relevant international animal welfare laws, guidelines, and policies.

Table 2. Experimental systems employed, stocking densities, starting weights, study lengths, water quality parameters, and feeding schedules in various studies undertaken to evaluate the impact of dietary fishmeal and fish oil replacement on the performance of established and candidate species of teleost for aquaculture.

	System	# of Fish/Tank	Start Weight (g)	Study Length (d)	Temperature (°C)	DO_2 (mg L^{-1})	Salinity (g L^{-1})	Daily Feed Schedule
Kampachi	Tanks	30 → 15	282	84	-	-	-	2× → 1× to satiety
Yellowtail	RAS	15	20	64	22	10-12	34.5	5–10% body wt
Largemouth bass 1	RAS	20	25	84	28	7.7	3.1	3× to satiety
Largemouth bass 2	RAS	60–64	48	126	28	8.2	3.7	3× using feed tables
Largemouth bass 3	RAS	20	15.2	70	28	6.0	1.2	2× to satiety
Red drum	RAS	15	3	56	27.6	6.9	5.4	2× to satiety
Pompano 1	RAS	10	15	84	26.6	7.4	3	2× for 5 min
Pompano 2	RAS	20	4.1	84	28	8.0	34	4× up to 5% body wt

5. Data Collection

Details relating to the precise procedures employed in data acquisition for each species may be found in the relevant publications (see Table 1 for references). Depending on the trial under consideration, the following information was compiled to assess the performance of experimental animals with each dietary treatment:

$$\text{Weight gain (\%)} = [(\text{Final body weight} - \text{initial body weight})/(\text{initial body weight})] \times 100;$$

$$\text{Survival (\%)} = [\text{final population}/\text{initial population}] \times 100;$$

$$\text{Feed efficiency (FE)} = \text{weight gain (g)}/\text{dry feed consumed (g)};$$

$$\text{Feed conversion ratio (FCR)} = \text{weight of feed consumed (g)}/\text{weight gained by the animal (g)};$$

$$\text{Protein efficiency ratio (PER, \%)} = [\text{weight gain (g, wet weight)}/\text{protein intake (g, dry weight)}] \times 100;$$

$$\text{Fillet yield (\%)} = [\text{fillet weight (g)}/\text{gutted weight (g)}] \times 100;$$

$$\text{Hepatosomatic index (HSI, \%)} = [\text{liver weight (g)}/\text{body weight (g)}] \times 100;$$

$$\text{Interperitoneal fat ratio (IPF, \%)} = [\text{IPF weight (g)}/\text{body weight (g)}] \times 100;$$

$$\text{Fulton condition factor (K)} = [\text{fish weight (g)}/(\text{fish length, cm})^3] \times 100;$$

$$\text{Viscerosomatic index (VSI, \%)} = [\text{weight of viscera (g)}/\text{body weight (g)}] \times 100.$$

After gauging the above-mentioned indices, all the remains of fish samples ($n \geq 5$) were homogenized as a composite sample and analyzed for proximate composition, when measured, using established methods: the Dumas protocol for crude protein (6.25 × N) [70], and chloroform–methanol (4:1) extraction for crude lipid [71]. A lipid droplet subsample was isolated from these ingredients and conserved in N_2 at −80 °C for identification of their fatty acid profile by flame ionization gas chromatography. Fatty acid methyl esters (FAMEs) were prepared as described previously [72] and modified to include an additional saponification step [73]. Ash was determined after heating samples at 650 °C in a muffle furnace for 3 h [70].

Histological analyses were undertaken on the guts and livers of California yellowtail and largemouth bass. Samples were collected immediately following gross necropsies for performance characteristics ($n \leq 6$ per treatment). Sections of liver and distal intestine (2 cm × 2 cm) were preserved in Bouin's fixative for 24 h and subsequently transferred to 70% ethanol for final fixation. Tissues were then dehydrated, embedded in paraffin, and sectioned at 5 μm before staining with H&E using standard procedures. Rankings were then performed to differentiate histopathologic changes in the liver and intestine between diets.

Criteria assessed included intestinal goblet cell density and inflammation, hepatic glycogen content, and cellular changes [66]. For largemouth bass, spleen samples were stained using Gomori's modified iron procedure for hemosiderin [74] to evaluate the staining intensity of melano-macrophage centers (MMCs), which were graded from 0 to 2 for low, medium, and high, respectively.

Taste tests of largemouth bass were informal and used 25 active consumers who were provided with blind samples and asked to prepare fish using plain methods. Each was then requested to determine whether there were differences in taste, texture, or aroma between samples. Similar studies were undertaken with kampachi. The collected data were subjected to various statistical analyses with significance set at the $p < 0.05$ level. Readers are directed to the papers noted in Table 1 for complete details.

6. Observations and Discussion

Weight gain in all marine species fed F3 diets, except pompano 2, was less than that achieved by animals fed conventional FM/FO feeds (Table 3). However, there was no impact discerned on FCR, survival, fillet yield, HSI, or *K*. In kampachi, a significantly higher VSI in the control group accounted for the increased weight gain such that once this was taken into account, no differences in weight were apparent. In largemouth bass, the only freshwater species examined, the weight gain in fish fed F3 and F2 (FO included) was equivalent to that in animals fed the conventional diet (Table 3). The FCR in F3/F2-fed largemouth bass was equivalent to that in the conventional group in two of the three trials and was elevated in one of the trials, while survival was lower in one study. Accordingly, the trials described here illustrate the potential to severely reduce, and perhaps eliminate, FM/FO from aquafeeds of facultative carnivores. Importantly, the evidence presented to support this statement originated from investigations that employed a constrained list of possible FM/FO alternatives. Additionally, the F3 recipes used were derived from a formulator's experience rather than from experiments designed to determine the optimal inclusion rates for specific ingredients. Undoubtedly, with dietary refinement, perhaps involving the inclusion of other proteins and oils or modification to their concentrations/combinations, even greater benefits than those achieved will accrue. This supposition is supported by the findings of other researchers who have successfully replaced FM/FO in diets for an ever-increasing number of species (e.g., [75–81]).

The results considered here with the F3 feeds, together with the experience of others, imply that marine species will be more demanding than freshwater fishes regarding the complete removal of dietary FM/FO. It is probable that the largemouth bass were indifferent to lipid exchange due to their essential fatty acid (EFA) requirements being met by dietary 18:3n-3 and/or 18:2n-6 PUFA [82]. Similar observations have been made with other species of freshwater fish, where a wide variety of alternative dietary lipids have been shown to facilitate growth [83–85]. These results thus provide support for the idea that FO can already be totally removed from largemouth bass diets. However, a precautionary approach should be taken since some substitute oils have been demonstrated to cause physiological disturbance [86,87]. Marine species lack the enzymatic machinery necessary to elongate or desaturate PUFAs, such that EFA requirements are met by long-chain PUFAs, *viz.* 20:5n-3 and/or 22:6n-3 [88], which, in some diets, may have been limiting. Nonetheless, the substitution of fish oil with vegetable and/or algal oil in all species examined had no significant impact on survival, suggesting that the dietary fatty acid composition, even though varying, achieved the n-3 HUFA requirements of the species examined, at least over the study length. Importantly, lipid exchange had either no impact or only a marginal impact on feed palatability, thereby underscoring the flexibility that exists for the substitution of dietary lipids. An additional advantage of using *Schizochytrium* sp.-derived algal oil, produced by controlled heterotrophic fermentation, is its contaminant-free status, which contrasts to that of some FOs [89].

Table 3. Response of various species to experimental diets in which fishmeal and fish oil were replaced with alternatives. FCR = feed conversion efficiency; FE = feed efficiency; HSI = hepatosomatic index; IPF = intraperitoneal fat ratio; *K* = condition factor; PER = protein efficiency ratio; VSI = viscerosomatic index. Up- and downward-pointing arrows indicate significant differences ($p < 0.05$) from fish fed a control diet.

Ingredient	Kampachi [65]	Yellowtail [66]	Largemouth Bass 1 [67]	Largemouth Bass 2 [68]	Largemouth Bass 3 [46]	Red Drum [49]	Pompano 1 [47]	Pompano 2 [69]
Wt gain (%)	419↓	633.6↓	201.4	149.3	398	666↓	243.1↓	1149
FE	-	-	-	-	-	1.09	0.52	-
FCR	1.31	1.33	1.28	1.95↑	0.89	-	-	1.6
PER	-	-	-	-	2.34	-	-	1.2↓
Survival (%)	-	100	99	84↓	100	90	-	100
Fillet yield (%)	60.9	-	-	-	32.3	31.5	31.2	-
HSI (%)	-	-	1.49	1.66	3.0	1.99	2.60	-
IPF ratio (%)	-	-	-	-	3.0	0.39↓	0.01	-
K factor	-	-	1.16	1.19	1.29		1.34	1.59
VSI (%)	5.7↓	-	4.55	1.93	-	-	-	-
				Proximate composition				
Moisture	-	70.9	-	-	68.8	74.9	68.6	-
Protein	-	20.98	45.4	41.5	17.9	17.4	18.1	-
Lipid	-	7.37↓	14.7	15	8.8	3.87↓	9.72	-
Ash	-	2.46↑	6.96	7.39	4.0↓	3.90	3.29	-

As recorded previously for a wide variety of species [90,91], the fatty acid profiles of fillets of the assessed fish correlated well with those of their feeds (Table 4, Figure 1). One negative aspect of this trait, however, was that while n6:n3 ratios remained stable, the EPA/DHA fractions were inferior to those of control fillets. Fish oil substitution, therefore, may negatively affect the nutritional value of fillets [92,93]. Were it to be considered necessary, fillet lipids (types and levels) might be tailored to a specific use with finishing diets [94,95]. Such an eventuality might occur where significant changes in flesh quality, including firmness, juiciness, and fresh oily taste, deviate following large fluctuations in proximate composition, or, for example, when higher fillet lipid levels are required for reasons of processing, such as smoking [96]. Even given differences between the control and treatment group fillet fatty acid profiles, and subtle modifications to proximate the composition of largemouth bass, organoleptic evaluation by 25 habitual consumers resulted in 48% preferring the fishmeal–fish-oil-fed fish based on the taste, texture, and aroma, while 40% favored the F3-fed animals and 12% indicated no preference [68]. Thus, for largemouth bass, the deletion of FO from their diet had no apparent impact on consumer acceptance. Similarly, a blind taste test of kampachi resulted in 62% of participants preferring the F3-fed fish, 19% having a preference for *S. rivoliana* fed on a traditional diet, and 19% being unable to discriminate between the two dietary groups [65].

Table 4. Identified feed and fillet fatty acids of largemouth bass, Florida pompano, and red drum following 56–84 days of feeding with fishmeal- and fish-oil-free diets. Values are expressed as a percentage of total fatty acids. Up- and downward-pointing arrows indicate significantly ($p < 0.05$) higher and lower values, respectively, than control fillet levels.

	Largemouth Bass 3		Pompano 1		Red Drum	
	F3 Feed	F3 Fillet	F3 Feed	F3 Fillet	F3 Feed	F3 Fillet
C14:0	1.10	1.72↓	1.32	2.07↓	1.12	0.87↓
C14:1	0.05	0.31	-	0.58	-	-
C16:0	16.74	18.13	15.85	21.42	17.3	17.6↓
C16:1	2.23	4.02↓	2.96	3.40↓	2.13	2.65↓
C18:0	4.42	2.81	4.51	5.65	5.90	5.72
C18:1n9	25.55	34.35↑	24.28	41.32	20.80	24.60↑
C18:2n6	19.47	17.80↑	17.44	19.74	20.7	17.3↑
C18:3n3	13.89	2.97↑	19.29	5.46	14.7	9.81↑
C20:0	0.42	0.31	0.51	0.58	0.48	0.33
C20:1n9	0.55	1.62	0.36	1.13	-	-
C20:2n6	0.07	0.57	0.04	0.71	0.12	0.28
C20:3n3	-	0.36	-	0.81	-	-
C20:4n6	0.86	0.89↑	0.83	0.87	1.06	0.95
C20:5n3	4.13	1.52↓	3.57	0.87↓	4.10	2.66↓
C22:0	0.31	0.31	0.32	0.58	0.37	0.32
C22:1	-	-	-	0.58	-	-
C22:6n3	7.51	8.09↓	6.36	6.08↓	7.14	5.39
C24:0	0.56	0.31	0.39	0.58	-	-
C24:1n9	0.44	0.31	0.43	0.58	0.41	1.85
Total ω-3 Isomers	25.53	22.53	29.22	19.58	26.0	17.9
Total ω-6 Isomers	20.40	18.62	18.31	16.64	21.9	18.6
EPA/DHA	0.55	0.19	0.56	0.14	0.57	0.49
n3:n6	0.78	0.83	0.63	0.85	1.19	0.96

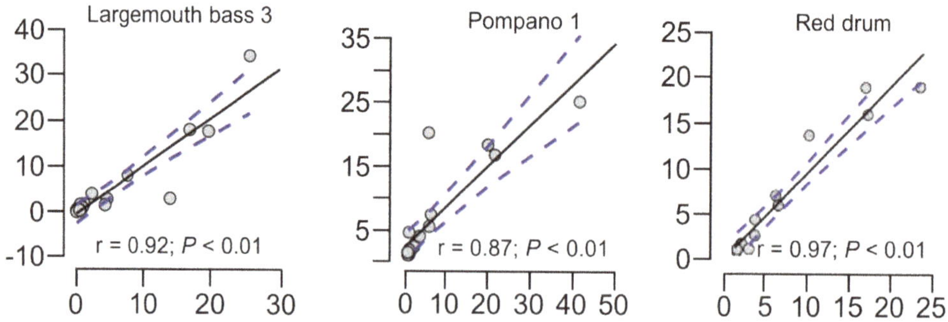

Figure 1. Scatter plots depicting the relationship between measured feed (y-axes) and fillet (x-axes) fatty acid content and 95% confidence intervals (dashed lines) for largemouth bass, Florida pompano, and red drum (see Table 4 for data).

Although the main goal of the F3 initiative is to eliminate the use of forage fishes in aquafeed production, an aspiration that is close to attainment for the species evaluated, some still question the practice of using animal by-products as alternative proteins. While this may be achievable with lower trophic species, a consistent observation with carnivores has been poorer overall performance when diets comprise vegetable proteins only. This is undoubtedly related to the presence of poorly digested carbohydrates and imbalances in essential amino acids (EAAs), the presence of a wide variety of anti-nutritional factors, and structural differences between plant and animal proteins [97]. These have negative impacts on growth, feed efficiency, metabolism, and health [31], and it is feasible that these effects may partly account for the reduced growth observed in the described trials herein. However, even given the presence of PBM in all F3 feeds, the marine test species failed to attain the growth recorded by control groups. Due to the variety of generally unsegregated material that is employed in PBM production, together with differences in processing and equipment, meals vary widely in their protein content and nutritional quality, lacking certain EAAs, being high in ash, and expressing variable digestibility [98,99]. Nonetheless, PBM has been successfully employed to replace relatively high levels of FM [100], although growth penalties coupled with higher FCRs and changes in body composition are known to occur in various species (e.g., [101–103]), and this may have been witnessed here.

The new and emerging technologies that modify raw materials, together with advances in process engineering, are starting to overcome many of the constraints encountered with alternative vegetable proteins, which bodes well for the future. For example, the production of plant protein concentrates and isolates removes carbohydrates, fiber, and anti-nutritional factors, resulting in products that, while more expensive, generally express an augmented EAA balance and have enhanced digestibility. However, the use of plant proteins for aquafeeds is disapproved by some who raise concerns relating to forest transitions, displacement of land use, increased use of fertilizers, eutrophication, environmental degradation, carbon footprinting, and others [104]. Given the current production strategies of established and emergent alternative proteins and their projected growth potential, it has been suggested that no single substitute protein will be able to source future demands of the animal feed industry, just as reliance on a few sources of ingredients, namely FM/FO, has created the bottlenecks we see today. Accordingly, the availability of a broad range of replacement proteins represents the soundest approach to overcome future supply constraints. Indeed, today, feed formulation scientists have a wide assortment of FM/FO alternatives [31,51,105]. Nevertheless, the aquafeed sector retains a significant dependency upon marine products [106], and it is likely that this addiction will remain for some time. Although their use will probably continue to decline in grower feeds, FM/FO will remain significant ingredients in specialty feeds, as exemplified by broodstock diets, and, perhaps, finishing feeds that may overcome fillet quality issues.

To date, most successful FM/FO replacement trials with carnivores have used diets containing blends of proteins and/or lipids that have been formulated to meet the nutrient requirements of the target species [77–79,107–110]. The broad range of potential aquafeed ingredients currently available, however, while providing strategic opportunities for formulating FM/FO-free feeds, also brings headaches for predicting optimal nutritional and economic blends, especially when mixtures might include a range of functional ingredients. Methods for overcoming some of these complexities are considered elsewhere [111–113]. One aspect of feed blending that has received limited attention is the potential to impact gut flora and fauna colonization and how this may influence nutrient absorption, etc., leading to potential for gut dysfunction. Clearly, there must be no consequences to the health of the target species when using alternative dietary ingredients. In one study with largemouth bass (Table 3), however, survival was apparently compromised by F3 feeds, although in a further two studies, no such effect was observed. Nonetheless, the detected anomaly prompted more detailed analyses of fish health. One indicator of immune function in teleosts is the status of splenic MMC [114], but evaluations thereof failed to detect differences between control and F3 treatments [67]. Moreover, the splenic index and hematocrit levels in examined fish were similar, and histological observations of the liver and distal intestine did not reveal any microscopic changes for the F3-fed group. In California yellowtail, slight hepatic inflammation and microscopic structural changes were encountered, with F3-fed animals also expressing higher glycogen accumulation. In contrast, control fish exhibited increased hepatocellular vacuolization and eccentric nuclei, together with a higher number of goblet cells in the distal intestine [66]. The decreased presence of goblet cells in F3 fish was not associated with inflammation, which the authors suggested might have indicated a protective effect of the *Spirulina* and/or algal oils incorporated into the diets. Notable is that the inclusion of soybean meal and concentrate in great amberjack *S. lalandi* diets was also associated with increased goblet cell numbers [115].

Since it is likely that animals cultured using sustainable marine-resource-free diets, such as organically certified and other premium foods, will represent quality products [116], methods for verifying their authenticity and traceability will become an imperative [117,118]. Animal tissue $\delta^{15}N$ is commonly employed to designate trophic position in food [119–121], and the technique has been applied to examine the relative contributions of plant and animal proteins in feeds for crustaceans [122–124] and fishes [117,125,126]. Thus, when the contribution of dietary FM declines, a corresponding decline in $\delta^{15}N$ is encountered. This response thereby potentially provides a method for verifying the integrity of animals reared using F3 diets. To substantiate this possibility, a study was undertaken with largemouth bass [68] (Figure 2). The trial examined fish fed a commercial feed, an FM/FO-based control diet, an FM-free feed containing FO, and an F3 diet. The FM control and commercial feeds both expressed final $\delta^{15}N$ values that were significantly higher than those for the FM-free feeds (Figure 2), no doubt reflecting the relative proportion and isotopic values of their ingredients. Substitution of the PBM from the F3 feed with another plant protein would likely shift the $\delta^{15}N$ values lower still. The use of stable isotope ratios to discriminate between aquacultured animals fed on more sustainable feeds, therefore, is apparently operational but should probably be restricted to animals reared in contained environments.

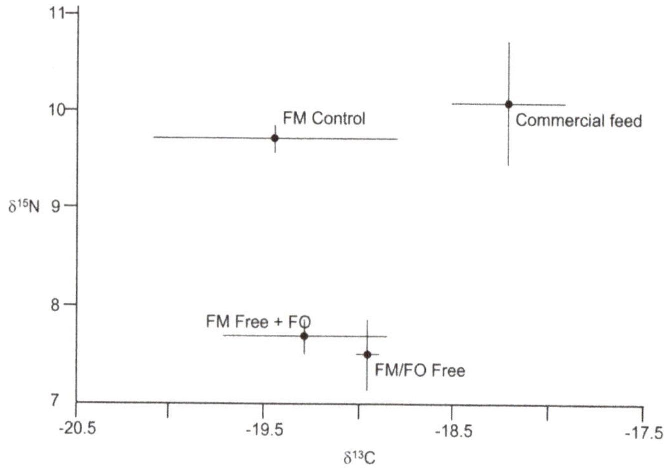

Figure 2. Isotope values for largemouth bass fed one of four diets. Values are means ± 95% confidence intervals (redrawn from [68]).

Based on the findings presented using essentially carnivorous species of cultured fish, total replacement of FM/FO appears more than just a convincing and economically viable proposition. Even so, further production-length research, perhaps with adjusted dietary formulae, is warranted to ensure that such diets have no negative consequences to the overall health and welfare of farmed animals. The potential adverse outcomes that dietary changes may have on various quality attributes, which may influence wholesale, retail, and consumer purchasing choices, also demand greater attention. Lucid though, from the considered trials, is that replacement protein/oil combinations provide products that are more secure in terms of food safety and more acceptable to discriminating consumers. The use of such nutrients will bridge gaps between the future supply and demand for FM/FO while serving global sustainability initiatives. While this might appear an over-enthusiastic conclusion, we have already demonstrated the potential for aquafeed mindset change with Pacific whiteleg shrimp *Litopenaeus vanammei* production [127,128], where F3 feeds are now firmly placed in the production sector. Similar success has been achieved with trout, largemouth bass, yellow croaker, and red seabream [129].

Author Contributions: Conceptualization, E.M.; draft manuscript preparation, K.B.C. and E.M.; Resources, F.T.B., K.B.C. and E.M.; Writing—review and editing, F.T.B. and K.B.C.; Project administration, K.B.C. All authors have read and agreed to the published version of the manuscript.

Funding: This research received no external funding.

Data Availability Statement: All author-owned experimental data are available at f3fin.org (accessed on 11 November 2022).

Acknowledgments: The authors express gratitude to all those who participated in the mentioned f3fin projects and especially the Anthropocene Institute for its unwavering support throughout.

Conflicts of Interest: The authors declare no conflict of interest.

References

1. Davis, H.S. Some results of feeding experiments with trout fingerlings. *Trans. Am. Fish. Soc.* **1927**, *57*, 281–287. [CrossRef]
2. Davis, H.S. Cheaper trout diets. *Progress. Fish-Cult.* **1935**, *2*, 7–10. [CrossRef]
3. Vetter, H. Meat Substitutes for Feeding Trout by Heinz Vetter, Allgemeine Fischerei-Zeitung, Vol. 63, No. 2, Jan. 15, 1938. *Progress. Fish-Cult.* **1938**, *5*, 26–27. [CrossRef]
4. Tunison, A.V.; Brockway, D.R.; Shaffer, H.B.; Maxwell, J.M.; McKay, C.M.; Palm, C.E.; Webster, D.A. The nutrition of trout. Cortland Hatchery Report No. 12. *Fish Res. Bull.* **1943**, *5*, 26.
5. Leach, G.C. *Artificial Propagation of Brook Trout and Rainbow Trout, with Notes on Three Other Species*; No. 955; US Government Printing Office: Washington, DC, USA, 1923; p. 74.
6. Hayford, C.O.; Davis, N.J.; Davis, H.S. The use of dry foods in the diet of rainbow trout and results of overfeeding. *Progress. Fish-Cult.* **1936**, *3*, 7–10. [CrossRef]
7. Gutsell, J.S. Fingerling trout feeding experiments, Leetown, 1938. *Progress. Fish-Cult.* **1939**, *6*, 32–41. [CrossRef]
8. Gutsell, J.S. Frozen fish in hatchery diets may be dangerous. *Progress. Fish-Cult.* **1940**, *7*, 28–32. [CrossRef]
9. Frick, E.J. Raising of rainbow trout. *N. Am. Vet.* **1932**, *13*, 10–14.
10. Almy, L.H.; Robinson, R.K. Toxic action of ingested linseed meal on trout. *J. Biol. Chem.* **1920**, *43*, 97–112. [CrossRef]
11. Phillips, A.M. Meatless diets and anemia: The development of anemia in trout fed a synthetic diet and its cure by the feeding of fresh beef liver. *Progress. Fish-Cult.* **1940**, *7*, 11–13. [CrossRef]
12. Embody, C.G. Results of some trout feeding experiments carried on in the experimental hatching station of Cornell University. *Trans. Am. Fish. Soc.* **1918**, *48*, 26–33. [CrossRef]
13. Embody, C.G.; Gordon, M. A comparative study of natural and artificial foods of brook trout. *Trans. Am. Fish. Soc.* **1924**, *54*, 185–200. [CrossRef]
14. Davis, H.S. *Care and Diseases of Trout*; U.S. Department of the Interior: Washington, DC, USA, 1946; Volume 12, p. 98.
15. Wolf, L.E. Comparison of yeast and penicillin mat as supplements to dry-meal diets for brown trout. *Progress. Fish-Cult.* **1951**, *13*, 117–120. [CrossRef]
16. Phillips, A.M.; Blazer, G.C., Jr. The nutrition of trout V. Ingredients for trout diets. *Progress. Fish-Cult.* **1957**, *19*, 158–167. [CrossRef]
17. Hublou, W.F.; Wallis, J.; McKee, T.B.; Law, D.K.; Sinnhuber, R.O.; Yu, T.C. Development of the Oregon pellet diet. Fifth progress report on salmon diet experiments. *Fish Comm. Or. Res. Briefs* **1959**, *7*, 28–55.
18. Brockway, D.R. Fish food pellets show promise. *Progress. Fish-Cult.* **1953**, *15*, 92–93. [CrossRef]
19. Jeffries, E.R.; McKee, T.B.; Sinnhuber, R.O.; Law, D.K.; Yu, T.C. Third progress report on spring chinook diet experiments. *Fish Comm. Or. Res. Briefs* **1954**, *5*, 32–38.
20. Schumacher, R.E. Experimental feeding of a pelleted trout food to large fingerling brook, brown and rainbow trout, 1955–1956. *Progress. Fish-Cult.* **1958**, *20*, 53–57.
21. Nielsen, W.E.; Mazuranich, J.J. Dry diets for chinook salmon. *Progress. Fish-Cult.* **1959**, *21*, 86–88. [CrossRef]

22. Grassl, E.F. Pelleted dry rations for trout propagation in Michigan hatcheries. *Trans. Am. Fish. Soc.* **1956**, *86*, 307–322. [CrossRef]
23. Grassl, E.F. Possible value of continuous feeding of medicated dry diets to prevent and control pathogens in hatchery-reared trout. *Progress. Fish-Cult.* **1957**, *19*, 85–88. [CrossRef]
24. Grassl, E.F. Relation of a dry pelleted ration to nutritional anemia in brook and rainbow trout. *Progress. Fish-Cult.* **1958**, *20*, 62–65. [CrossRef]
25. Phillips, A.M., Jr.; Podoliak, H.A.; Poston, H.A.; Livingstone, D.L.; Brooks, H.E.; Pyle, E.E.; Hammer, G.L. *Dry Concentrates as Complete Fish Foods*; New York State Conservation Department: Albany, NY, USA, 1964; Volume 27, pp. 47–54.
26. Waite, D.; Buss, K. An automatic feeder for trout. *Progress. Fish-Cult.* **1963**, *25*, 52. [CrossRef]
27. Hardy, R.W. Diet preparation. In *Fish Nutrition*, 2nd ed.; Halver, J.E., Ed.; Academic Press: San Diego, CA, USA, 1989; pp. 475–548.
28. Hansen, L. The weak sustainability of the salmon feed transition in Norway—A bioeconomic case study. *Front. Mar. Sci.* **2019**, *6*, 764. [CrossRef]
29. Food and Agricultural Organization. *State of World Fisheries and Aquaculture 2022: Towards Blue Transformation*; FAO: Rome, Italy, 2022. [CrossRef]
30. Froehlich, H.E.; Jacobsen, N.S.; Essington, T.E.; Clavelle, T.; Halpern, B.S. Avoiding the ecological limits of forage fish for fed aquaculture. *Nat. Sustain.* **2018**, *1*, 298–303. [CrossRef]
31. McLean, E. Feed ingredients for sustainable aquaculture. In *Sustainable Food Science: A Comprehensive Approach*; Elsevier Inc.: Amsterdam, The Netherlands, 2023; *in press*.
32. Callet, T.; Médale, F.; Larroquet, L.; Surget, A.; Aguirre, P.; Kerneis, T.; Labbé, L.; Quillet, E.; Geurden, I.; Skiba-Cassy, S.; et al. Successful selection of rainbow trout (*Oncorhynchus mykiss*) on their ability to grow with a diet completely devoid of fishmeal and fish oil, and correlated changes in nutritional traits. *PLoS ONE* **2017**, *12*, e0186705. [CrossRef]
33. Haarstad, A.H.; Lavrutich, M.; Strypet, K.; Strøm, E. Multi commodity price risk hedging in the Atlantic salmon farming industry. Multi-commodity price risk hedging in the Atlantic salmon farming industry. *J. Commod. Mark.* **2022**, *25*, 100182. [CrossRef]
34. Pelletier, N.; Tyedmers, P. Feeding farmed salmon: Is organic better? *Aquaculture* **2007**, *272*, 399–416. [CrossRef]
35. Samuel-Fitwi, B.; Meyer, S.; Reckmann, K.; Schroeder, J.P.; Schulz, C. Aspiring for environmentally conscious aquafeed: Comparative LCA of aquafeed manufacturing using different protein sources. *J. Clean. Prod.* **2013**, *52*, 225–233. [CrossRef]
36. Basto-Silva, C.; Valente, L.M.P.; Matos, E.; Brandão, M.; Neto, B. Life cycle assessment of aquafeed ingredients. *Int. J. Life Cycle Assess.* **2018**, *23*, 995–1017. [CrossRef]
37. Glencross, B.D.; Bailey, J.; Berntssen, M.H.G.; Hardy, R.; MacKenzie, S.; Tocher, D.R. Risk assessment of the use of alternative animal and plant raw material resources in aquaculture feeds. *Rev. Aquac.* **2020**, *12*, 703–758. [CrossRef]
38. Lewis, S.G.; Alifano, A.; Boyle, M.; Mangel, M. Human rights and the sustainability of fisheries. In *Conservation for the Anthropocene Ocean*; Levin, P.S., Poe, M.R., Eds.; Academic Press: Cambridge, MA, USA, 2017; pp. 379–396.
39. Gilman, E.; Perez Roda, A.; Huntington, T.; Kennelly, S.J.; Suuronen, P.; Chaloupka, M.; Medley, P.A.H. Benchmarking global fisheries discards. *Sci. Rep.* **2020**, *10*, 14017. [CrossRef]
40. Al-Masroori, H.; Al-Oufi, H.S.; McLean, E.; McIlwain, J.L. Catches of lost fish traps (ghost fishing) from fishing grounds near Muscat, Sultanate of Oman. *Fish. Res.* **2004**, *68*, 407–414. [CrossRef]
41. World Trade Organization. *Implementing the WTO Agreement on Fisheries Subsidies. Challenges and Opportunities for Developing and Least-Developed Country Members*; WTO: Geneva, Switzerland, 2022; p. 23.
42. Parker, R.W.R.; Blanchard, J.L.; Gardner, C.; Green, B.S.; Hartmann, K.; Tyedmers, P.H.; Watson, R.A. Fuel use and greenhouse gas emissions of world fisheries. *Nat. Clim. Chang.* **2018**, *8*, 333–337. [CrossRef]
43. Boyd, C.E. Overview of aquaculture feeds: Global impacts of ingredient use. In *Feed and Feeding Practices in Aquaculture*; Davis, D.A., Ed.; Series in Food Science, Technology and Nutrition; Woodhead Publishing: Sawston, UK, 2015; pp. 3–25.
44. Eriksen, H.S. *Information Resources on Fish Welfare 1970-20903*; AWIC Resource Series No. 20; United States Department of Agriculture: Washington, DC, USA, 2003; p. 435.
45. Turchini, G.M.; Trushenski, J.T.; Glencross, B.D. Thoughts on the future of aquaculture nutrition: Realigning perspectives to reflect contemporary issues related to judicious use of marine resources in aquafeeds. *N. Am. J. Aquac.* **2019**, *81*, 13–39. [CrossRef]
46. McLean, E.; Alfrey, K.; Gatlin, D.M., III; Barrows, F.T. Responses of largemouth bass to fishmeal and fish oil-free diets. *Aquac. Res.* **2022**, *53*, 3036–3047. [CrossRef]
47. Alfrey, K.B.; Gatlin, D.M., III; Barrows, F.T.; McLean, E. Assessment of open-source, fish-free diets for pompano, *Trachinotus carolinus* (Perciformes, Carangidae), under hyposaline conditions. *Aquac. Res.* 2022, *in press*.
48. Mai, K.; Zhang, W. Feed developments in mariculture. In *Aquaculture in China: Success Stories and Modern Trends*, 3rd ed.; Gui, J.F., Tang, Q., Li, Z., Liu, J., De Silva, S.S., Eds.; John Wiley & Sons Ltd.: Hoboken, NJ, USA, 2018; pp. 451–462. 677p.
49. Suehs, B.; Alfrey, K.; Barrows, F.; Gatlin, D.M. III Evaluation of growth performance, condition indices and body composition of juvenile red drum (*Sciaenops ocellatus*) fed fishmeal- and fish-oil-free diets. *Aquaculture* **2022**, *551*, 737961. [CrossRef]
50. USDA. Livestock and Poultry: World Markets and Trade. 2022. Available online: https://apps.fas.usda.gov/psdonline/circulars/livestock_poultry.pdf (accessed on 24 October 2022).

51. Hertrampf, J.W.; Piedad-Pascual, F. *Handbook on Ingredients for Aquaculture Feeds*; Kluwer Academic Publishers: Amsterdam, The Netherlands, 2000; p. 624.
52. Viyakarn, V.; Watanabe, T.; Aoki, H.; Tsuda, H.; Sakamoto, H.; Okamoto, N.; Iso, N.; Satoh, S.; Takeuchi, T. Use of soybean meal as a substitute for fish meal in a newly developed soft-dry pellet for yellowtail. *Nippon. Suisan Gakkaishi* **1992**, *58*, 1991–2000. [CrossRef]
53. Rhodes, M.A.; Zhou, Y.; Davis, D.A. Use of Dried Fermented Biomass as a Fish Meal Replacement in Practical Diets of Florida Pompano, *Trachinotus carolinus*. *J. Appl. Aquac.* **2015**, *27*, 29–39. [CrossRef]
54. Rossi, W., Jr.; Moxely, D.; Buentello, A.; Pohlenz, C.; Gatlin, D.M. III Replacement of fishmeal with novel plant feedstuffs in the diet of red drum *Sciaenops ocellatus*: An assessment of nutritional value. *Aquac. Nutr.* **2013**, *19* (Suppl. S1), 72–81. [CrossRef]
55. Li, X.; Wei, X.; Guo, X.; Mi, S.; Hua, X.; Li, N.; Yao, J. Enhanced growth performance, muscle quality and liver health of largemouth bass (*Micropterus salmoides*) were related to dietary small peptides supplementation. *Aquac. Nutr.* **2020**, *26*, 2169–2177. [CrossRef]
56. Merrifield, D.L.; Olsen, R.E.; Myklebust, R.; Ringø, E. Dietary effect of soybean (*Glycine max*) products on gut histology and microbiota of fish. In *Soybean and Nutrition*; El-Shemy, H., Ed.; Tech Europe: Rijeka, Croatia, 2011; pp. 231–250.
57. Dersjant-Li, Y. The use of soy protein in aquafeeds. In Proceedings of the Avances en Nutrición Acuícola VI. Memorias del VI Simposium Internacional de Nutrición Acuícola, Cancun, Mexico, 3–6 September 2002.
58. Gyan, W.R.; Ayiku, S.; Yang, Q. Effects of replacing fishmeal with soybean products in fish and crustaceans performance. *J. Aquac. Res. Dev.* **2019**, *10*, 573.
59. Jannathulla, R.; Sravanthi, O.; Moomeen, S.; Gopikrishna, G.; Dayal, J.S. Microbial products in terms of isolates, whole-cell biomass, and live organisms as aquafeed ingredients: Production, nutritional values, and market potential—A review. *Aquac. Int.* **2021**, *29*, 623–650. [CrossRef]
60. Mugwanya, M.; Dawood, M.A.O.; Kimera, F.; Sewilam, H. Replacement of fish meal with fermented plant proteins in the aquafeed industry: A systematic review and meta-analysis. *Rev. Aquac.* **2022**, 1–27. [CrossRef]
61. Alagawany, M.; Taha, A.E.; Noreldin, A.; El-Tarabily, K.A.; El-Hack, M.E.A. Nutritional applications of species of *Spirulina* and *Chlorella* in farmed fish: A review. *Aquaculture* **2021**, *542*, 736841. [CrossRef]
62. Kissinger, K.R.; García-Ortega, A.; Trushenski, J.T. Partial fish meal replacement by soy protein concentrate, squid and algal meals in low fish-oil diets containing *Schizochytrium limacinum* for longfin yellowtail *Seriola rivoliana*. *Aquaculture* **2016**, *452*, 37–44. [CrossRef]
63. Guo, H.; Chen, C.; Yan, X.; Li, Y.; Wen, X.; You, C.; Monroig, Ó.; Tocher, D.R.; Wang, S. Effects of different dietary oil sources on growth performance, antioxidant capacity and lipid deposition of juvenile golden pompano *Trachinotus ovatus*. *Aquaculture* **2021**, *530*, 735923. [CrossRef]
64. Liang, C.; Zhao, X.; Jiao, L.; Shen, Y.; Luo, J.; Zhu, T.; Zhao, W.; Gen, Z.; Zhou, Q.; Jin, M. Effects of different lipid sources on growth performance, fatty acids composition in tissue and expression of genes related to lipid metabolism in largemouth bass (*Micropterus salmoides*). *Aquac. Rep.* **2022**, *23*, 101013. [CrossRef]
65. Meigs, H.; Barrows, F.T.; Sims, N.A.; Alfrey, K. Testing Diets without Fishmeal and Fish Oil for Kampachi. Responsible Seafood Advocate 24 August 2020. 2020. Available online: https://www.globalseafood.org/advocate/testing-diets-without-fishmeal-and-fish-oil-for-kampachi/ (accessed on 14 October 2022).
66. Stuart, K.R.; Barrows, F.T.; Silbernagel, C.; Alfrey, K.; Rotstein, D.; Drawbridge, M.A. Complete replacement of fish oil and fish meal in the diet of juvenile California yellowtail *Seriola dorsalis*. *Aquac. Res.* **2020**, *52*, 655–665. [CrossRef]
67. McLean, E.; Fredriksen, L.; Alfrey, K.; Craig, S.R.; Barrows, F.T. Performance of largemouth bass *Micropterus salmoides* (Lacépède, 1802), fed fishmeal- and fish oil-free diets. *J. Fish. Aquat. Stud.* **2020**, *8*, 6–10. [CrossRef]
68. McLean, E.; Fredriksen, L.; Alfrey, K.; Tlusty, M.F.; Barrows, F.T. Growth, integrity, and consumer acceptance of largemouth bass, *Micropterus salmoides* (Lacépède, 1802), fed marine resource-free diets. *Int. J. Fish. Aquat. Stud.* **2020**, *8*, 365–369. [CrossRef]
69. Riche, M.; Barrows, F.T.; Nilles, Z.; Alfrey, K.B.; Wills, P.S. *Replacement of Fish Oil with a High DHA Algal Oil in a Fishmeal-Free Diet Fed to Florida pompano Trachinotus carolinus*; Harbor Branch Oceanographic Institute: Fort Pierce, FL, USA, 2022; *to be submitted*.
70. AOAC. *Official Methods of Analysis*; Association of Official Analytical Chemists: Arlington, VA, USA, 2005.
71. Folch, J.; Lees, M.; Sloane Stanley, G.H. A simple method for the isolation and purification of total lipides from animal tissues. *J. Biol. Chem.* **1957**, *226*, 497–509. [CrossRef]
72. Morrison, W.R.; Smith, L.M. Preparation of fatty acid methyl esters and dimethylacetals from lipids with boron fluoride–methanol. *J. Lipid Res.* **1964**, *5*, 600–608. [CrossRef]
73. Archibeque, S.L.; Lunt, D.K.; Gilbert, C.D.; Tume, R.K.; Smith, S.B. Fatty acid indices of stearoyl-CoA desaturase do not reflect actual stearoyl-CoA desaturase enzyme activities in adipose tissues of beef steers finished with corn-, flaxseed-, or sorghum-based diets. *J. Anim. Sci.* **2005**, *83*, 1153–1166. [CrossRef]
74. Sheehan, D.C.; Hrapchak, B.B. *Theory and Practice of Histotechnology*, 2nd ed.; Battelle Press: Columbus, OH, USA, 1987; p. 481.
75. Gomes, E.F.; Rema, P.; Kaushik, S.J. Replacement of fish meal by plant proteins in the diet of rainbow trout (*Oncorhynchus mykiss*): Digestibility and growth performance. *Aquaculture* **1995**, *130*, 177–186. [CrossRef]
76. Kissil, G.W.; Lupatsch, I.; Higgs, D.A.; Hardy, R.W. Dietary substitution of soy and rapeseed protein concentrates for fish meal, and their effects on growth and nutrient utilization in gilthead seabream *Sparus aurata* L. *Aquac. Res.* **2000**, *31*, 595–601. [CrossRef]

77. Barrows, F.T.; Gaylord, T.G. Changing technologies, ingredients and formulations to replace fish meal in salmonid diets. In *Nutritional Biotechnology in the Food and Feed Industry*; Lyons, T.P.T., Jacques, K., Eds.; Nottingham University Press: Nottingham, UK, 2006; pp. 307–324.
78. Lunger, A.N.; McLean, E.; Gaylord, T.G.; Kuhn, D.; Craig, S.R. Taurine supplementation to alternative dietary proteins used in fish meal replacement enhances growth of juvenile cobia (*Rachycentron canadum*). *Aquaculture* **2007**, *271*, 401–410. [CrossRef]
79. Silva, J.; Espe, M.; Conceição, L.; Dias, J.; Valente, L. Senegalese sole juveniles (*Solea senegalensis* Kaup, 1858) grow equally well on diets devoid of fish meal provided the dietary amino acids are balanced. *Aquaculture* **2009**, *296*, 309–317. [CrossRef]
80. Davidson, J.; Barrows, F.T.; Kenney, P.B.; Good, C.; Schroyer, K.; Summerfelt, S.T. Effects of feeding a fishmeal-free versus a fishmeal-based diet on post-smolt Atlantic salmon *Salmo salar* performance, water quality, and waste production in recirculation aquaculture systems. *Aquac. Eng.* **2016**, *74*, 38–51. [CrossRef]
81. Lorez, E.K.; Sabioni, R.E.; Volkoff, H.; Cyrino, J.E.P. Growth performance, health, and gene expression of appetite-regulating hormones in Dourado *Salminus brasiliensis*, fed vegetable-based diets supplemented with swine liver hydrolysate. *Aquaculture* **2022**, *548*, 737640. [CrossRef]
82. Yadav, A.K.; Rossi, W., Jr.; Habte-Tsion, H.-M.; Kumar, V. Impacts of dietary eicosapentaenoic acid (EPA) and docosahexaenoic acid (DHA) level and ratio on the growth, fatty acids composition and hepatic-antioxidant status of largemouth bass (*Micropterus salmoides*). *Aquaculture* **2020**, *529*, 735683. [CrossRef]
83. Fracalossi, D.M.; Craig-Schmidt, M.C.; Lovell, R.T. Effect of dietary lipid sources on production of leukotriene B by head kidney of channel catfish held at different water temperatures. *J. Aquat. Anim. Health* **1995**, *6*, 242–250. [CrossRef]
84. Luzzana, U.; Scolari, M.; Campo Dall'Orto, B.; Caprino, F.; Turchini, G.; Orban, E.; Sinesio, F.; Valfre, F. Growth and product quality of European eel (*Anguilla tilizat*) as affected by dietary protein and lipid sources. *J. Appl. Ichthyol.* **2003**, *19*, 74–78. [CrossRef]
85. Karapanagiotidis, I.T.; Metsoviti, M.N.; Gkalogianni, E.Z.; Psofakis, P.; Asimaki, A.; Katsoulas, N.; Papapolymerou, G.; Zarkadas, I. The effects of replacing fishmeal by *Chlorella vulgaris* and fish oil by *Schizochytrium* sp. and *Microchloropsis gaditana* blend on growth performance, feed efficiency, muscle fatty acid composition and liver histology of gilthead seabream (*Sparus aurata*). *Aquaaculture* **2022**, *561*, 738709. [CrossRef]
86. Zhang, W.; Tan, B.; Liu, K.; Dong, X.; Yang, Q.; Chi, S.; Liu, H.; Zhang, S.; Wang, H. Effects of different dietary lipids on growth, body composition and lipid metabolism-related enzymes and genes in juvenile largemouth bass, *Micropterus salmoides*. *Aquac. Nutr.* **2019**, *25*, 1318–1326. [CrossRef]
87. Habte-Tsion, H.M.; Kolimadu, G.D.; Rossi, W.; Filer, K.; Kumar, V. Effects of *Schizochytrium* and micro-minerals on immune, antioxidant, inflammatory and lipid-metabolism status of *Micropterus salmoides* fed high- and low-fishmeal diets. *Sci. Rep.* **2020**, *10*, 7457. [CrossRef]
88. National Research Council. *Nutrient Requirements of Fish and Shrimp*; National Academies Press: Washington, DC, USA, 2011; p. 375.
89. Sprague, M.; Betancor, M.B.; Tocher, D.R. Microbial and genetically engineered oils as replacements for fish oil in aquaculture feeds. *Biotechnol. Lett.* **2017**, *39*, 1599–1609. [CrossRef]
90. Xu, H.; Turchini, G.M.; Francis, D.M.; Liang, M.; Mock, T.S.; Rombenso, A.; Ai, Q. Are fish what they eat? A fatty acid's perspective. *Prog. Lipid Res.* **2020**, *50*, 101064. [CrossRef]
91. Turchini, G.M.; Francis, D.S.; Du, Z.-Y.; Olsen, R.E.; Ringø, E.; Tocher, D.R. The lipids. In *Fish Nutrition*, 4th ed.; Hardy, R.W., Kaushik, S.J., Eds.; Academic Press: London, UK, 2022.
92. Tacon, A.G.J.; Lemos, D.; Metian, M. Fish for health: Improved nutritional quality of cultured fish for human consumption. *Rev. Fish. Sci. Aquac.* **2020**, *28*, 449–458. [CrossRef]
93. Fiorella, K.J.; Okronipa, H.; Baker, K.; Heilpern, S. Contemporary aquaculture: Implications for human nutrition. *Curr. Opin. Biotechnol.* **2021**, *70*, 83–90. [CrossRef] [PubMed]
94. Rasmussen, R.S.; Ostenfeld, T.H.; Rønsholdt, B.; McLean, E. Manipulation of end-product quality of rainbow trout with finishing diets. *Aquac. Nutr.* **2000**, *6*, 17–23. [CrossRef]
95. Parés-Sierra, G.; Durazo, E.; Ponce, M.A.; Badillo, D.; Correa-Reyes, G.; Viana, M.T. Partial to total replacement of fishmeal by poultry by-product meal in diets for juvenile rainbow trout (*Oncorhynchus mykiss*) and their effect on fatty acids from muscle tissue and the time required to retrieve the effect. *Aquac. Res.* **2014**, *45*, 1459–1469. [CrossRef]
96. Rønsholdt, B.; McLean, E. Quality characteristics of fresh rainbow trout as perceived by the Danish processing industry. *Aquac. Int.* **1999**, *7*, 117–127. [CrossRef]
97. Gatlin, D.M.; Barrows, F.T.; Brown, P.; Dabrowski, K.; Gaylord, T.G.; Hardy, R.W.; Herman, E.; Hu, G.; Krogdahl, A.; Nelson, R.; et al. Expanding the utilization of sustainable plant products in aquafeeds: A review. *Aquac. Res.* **2007**, *38*, 551–579. [CrossRef]
98. Tacon, A.G.J.; Hasan, M.R.; Subasinghe, R.P. Use of fishery resources as feed inputs to aquaculture development: Trends and policy implications. In *FAO Fisheries Circular 1018*; FAO: Rome, Italy, 2006.
99. Volpato, J.P.; Ribeiro, L.B.; Torezan, G.B.; da Silva, I.C.; de Oliveira Martins, I.; Genova, L.; de Oliveira, N.T.E.; Carvalho, S.T.; de Oliveira Carvalho, P.L.; Vasconcellos, R.S. Characterization of the variations in the industrial processing and nutritional variables of poultry by-product meal. *Poult. Sci.* **2022**, *101*, 101926. [CrossRef]
100. Galkanda-Arachchige, H.S.C.; Wilson, A.E.; Davis, D.A. Success of fishmeal replacement through poultry by-product meal in aquaculture feed formulations: A meta-analysis. *Rev. Aquac.* **2020**, *12*, 1624–1636. [CrossRef]

101. Steffens, W. Replacing fish meal with poultry by-product meal in diets for rainbow trout, Oncorhynchus mykiss. *Aquaculture* **1994**, *124*, 27–34. [CrossRef]
102. Siddik, M.A.B.; Chungu, P.; Fotedar, R.; Howieson, J. Bioprocessed poultry by-product meals on growth, gut health and fatty acid synthesis of juvenile barramundi, *Lates calcarifer* (Bloch). *PLoS ONE* **2019**, *14*, e0215025. [CrossRef]
103. Pham, H.D.; Siddik, M.A.B.; Phan, U.V.; Le, H.M.; Rahman, M.A. Enzymatic tuna hydrolysate supplementation modulates growth, nutrient utilisation and physiological response of pompano (*Trachinotus blochii*) fed high poultry-by product meal diets. *Aquac. Rep.* **2021**, *21*, 100875. [CrossRef]
104. Woodgate, S.L.; Wan, A.H.L.; Hartnett, F.; Wilkinson, R.G.; Davis, S.L. The tilization of European processed animal proteins as safe, sustainable and circular ingredients for global aquafeeds. *Rev. Aquac.* **2022**, *14*, 1572–1596. [CrossRef]
105. Gasco, L.; Gai, F.; Maricchiolo, G.; Genovese, L.; Ragonese, S.; Bottari, T.; Caruso, G. Fishmeal alternative protein sources for aquaculture feeds. In *Feeds for the Aquaculture Sector*; Springer Briefs in Molecular Science; Springer: Cham, Switzerland, 2018. [CrossRef]
106. Boyd, C.E.; McNevin, A.A.; Davis, R.P. The contribution of fisheries and aquaculture to the global protein supply. *Food Secur.* **2022**, *14*, 805–827. [CrossRef] [PubMed]
107. Hansen, A.; Hemre, G.-I. Effects of replacing fish meal and oil with plant resources in on-growing diets for Atlantic cod *Gadus morhua* L. *Aquac. Nutr.* **2013**, *19*, 641–650. [CrossRef]
108. Seong, T.; Kitagima, R.; Haga, Y.; Satoh, S. Non-fish meal, non-fish oil diet development for red sea bream, *Pagrus major*, with plant protein and graded levels of *Schizochytrium* sp.: Effect on growth and fatty acid composition. *Aquac. Nutr.* **2020**, *26*, 1173–1185. [CrossRef]
109. Li, X.; Zheng, S.; Ma, X.; Cheng, K.; Wu, G. Use of alternative protein sources for fishmeal replacement in the diet of largemouth bass (*Micropterus salmoides*). Part I: Effects of poultry by-product meal and soybean meal on growth, feed utilization, and health. *Amino Acids* **2021**, *53*, 33–47. [CrossRef] [PubMed]
110. Hong, Y.C.; Chu, J.H.; Kirby, R.; Sheen, S.S.; Chien, A. The effects of replacing fish meal protein with a mixture of poultry by-product meal and fermented soybean meal on the growth performance and tissue nutritional composition of Asian seabass (*Lates calcarifer*). *Aquac. Res.* **2021**, *52*, 4105–4115. [CrossRef]
111. Saxena, P.; Khanna, N. Animal feed formulation: Mathematical programming techniques. *CAB Rev.* **2014**, *9*, 035.
112. Uyeh, D.D.; Pamulapati, T.; Mallipeddi, R.; Park, T.; Asem-Hiablie, S.; Woo, S.; Kim, J.; Kim, Y.; Ha, Y. Precision animal feed formulation: An evolutionary multi-objective approach. *Anim. Feed Sci. Technol.* **2019**, *256*, 114211. [CrossRef]
113. Bailey, C.A. Chapter 21—Precision poultry nutrition and feed formulation. In *Animal Agriculture*; Bazer, F.W., Lamb, G.C., Wu, G., Eds.; Academic Press: London, UK, 2020; pp. 367–378.
114. Steinel, N.C.; Bolnick, D.I. Melanomacrophage centers as a histological indicator of immune function in fish and other poikilotherms. *Front. Immunol.* **2017**, *8*, 827. [CrossRef]
115. Bansemer, M.S.; Forder, R.E.; Howarth, G.S.; Suitor, G.M.; Bowyer, J.; Stone, D.A. The effect of dietary soybean meal and soy protein concentrate on the intestinal mucus layer and development of subacute enteritis in yellowtail kingfish (*Seriola lalandi*). *Aquac. Nutr.* **2014**, *21*, 300–310. [CrossRef]
116. Hermann, R.; Boissinger, K.; Krandick, L. Price premia for sustainability characteristics in foods: Measurement matters! In Proceedings of the System Dynamics and Innovation in Food Networks 2018, Innsbruck, Austria, 5–9 February 2018; pp. 28–37. [CrossRef]
117. Moreno-Rojas, J.M.; Tulli, F.; Messina, M.; Tibaldi, E.; Guillou, C. Stable isotope ratio analysis as a tool to discriminate between rainbow trout (*O. mykiss*) fed diets based on plant or fish-meal proteins. *Rapid Commun. Mass Spectrom.* **2008**, *22*, 3706–3710. [CrossRef] [PubMed]
118. Schwägele, F. Traceability from a European perspective. *Meat Sci.* **2005**, *71*, 164–173. [CrossRef] [PubMed]
119. Van der Zanden, M.J.; Hulshof, M.; Ridgway, M.S.; Rasmussen, J.B. Application of stable isotope techniques to trophic studies of age-0 smallmouth bass. *Trans. Am. Fish. Soc.* **1998**, *127*, 729–739. [CrossRef]
120. Pinnegar, J.K.; Polunin, N.V.C. Contributions of stable-isotope data to elucidating food webs of Mediterranean rocky littoral fishes. *Oecologia* **2000**, *122*, 399–409. [CrossRef] [PubMed]
121. Schlechtriem, C.; Focken, U.; Becker, K. Stable isotopes as a tool for nutrient assimilation studies in larval fish feeding on live food. *Aquat. Ecol.* **2004**, *38*, 93–100. [CrossRef]
122. Preston, N.P.; Smith, D.M.; Kellaway, D.M.; Bunn, S.E. The use of enriched ^{15}N as an indicator of the assimilation of individual protein sources from compound diets for juvenile *Penaeus monodon*. *Aquaculture* **1996**, *147*, 249–259. [CrossRef]
123. Gamboa-Delgado, J.; Le Vay, L. Natural stable isotopes as indicators of the relative contribution of soy protein and fish meal to tissue growth in Pacific white shrimp (*Litopenaeus vannamei*) fed compound diets. *Aquaculture* **2009**, *291*, 115–221. [CrossRef]
124. Wertz, A.; Mazumder, D.; Carter, C.G.; Codabaccus, M.B.; Fitzgibbon, Q.P.; Smith, G.S. Application of stable isotope analysis to evaluate the assimilation of protein sources in juvenile slipper lobsters (*Thenus australiensis*). *Aquacultuire* **2022**, *560*, 738570. [CrossRef]
125. Gamboa-Delgado, J.; Cañavate, J.P.; Zerolo, R.; Le Vay, L. Natural carbon stable isotope ratios as indicators of the relative contribution of live and inert diets to growth in larval Senegalese sole (*Solea senegalensis*). *Aquaculture* **2008**, *280*, 190–197. [CrossRef]

126. Zapata, D.B.; Lazo, J.P.; Herzka, S.Z.; Viana, M.T. The effect of substituting fishmeal with poultry by-product meal in diets for *Totoaba macdonaldi* juveniles. *Aquac. Res.* **2016**, *47*, 1778–1789. [CrossRef]
127. Tran, L.H.; Nhut, T.C.; Alfrey, K.B.; Barrows, F.T.; Kuhn, D.; McLean, E. Performance of Pacific whiteleg shrimp fed a fishmeal and fish oil-free diet under commercial conditions. *Int. J. Fish. Aquat. Stud.* **2022**, *10*, 33–41. [CrossRef]
128. McLean, E.; Tran, L.H.; Craig, S.R.; Alfrey, K.; Barrows, F.T. Complete replacement of fishmeal by soybean and poultry meals in whiteleg shrimp feeds: Growth and tolerance to EMS/AHPND and WSSV challenge. *Aquaculture* **2020**, *527*, 735383. [CrossRef]
129. Anonymous. Winners of Global Seafood Industry's Contest F3 Challenge Announced. 2022. Available online: https://www.aquafeed.com/newsroom/news/winners-of-global-seafood-industrys-contest-f3-challenge-announced/?utm_source=Aquafeed&utm_campaign=0a2c65f74e-EMAIL_CAMPAIGN_2019_06_26_19_COPY_01&utm_medium=email&utm_term=0_0e7f7c0399-0a2c65f74e-100222 (accessed on 6 October 2022).

fishes

Review

Nutritional Value of Dry Fish in Bangladesh and Its Potential Contribution to Addressing Malnutrition: A Narrative Review

Md. Hasan Al Banna [1], Abdullah Al Zaber [2], Nahidur Rahman [3], Md Abrar Mozahid Siddique [4], Muhammad Abu Bakr Siddique [5], John Elvis Hagan, Jr. [6,7,*], M. A. Rifat [8], Christiana Naa Atsreh Nsiah-Asamoah [9], Abdul-Aziz Seidu [10,11], Bright Opoku Ahinkorah [12] and Md Shafiqul Islam Khan [1]

1 Department of Food Microbiology, Faculty of Nutrition and Food Science, Patuakhali Science and Technology University, Patuakhali 8602, Bangladesh
2 Faculty of Public Health, University of Debrecen, 4028 Debrecen, Hungary
3 Department of Food Processing and Engineering, Chattogram Veterinary and Animal Sciences University, Chattogram 4225, Bangladesh
4 Department of Food Technology and Nutrition Science, Noakhali Science & Technology University, Noakhali 3814, Bangladesh
5 United Nations Children's Fund (UNICEF), Dhaka 1207, Bangladesh
6 Department of Health, Physical Education & Recreation, College of Education Studies, University of Cape Coast, Cape Coast PMB TF0494, Ghana
7 Neurocognition and Action-Biomechanics-Research Group, Faculty of Psychology and Sports Science, Bielefeld University, Postfach 10 01 31, 33501 Bielefeld, Germany
8 Department of Global Public Health, Karolinska Institutet, 17176 Stockholm, Sweden
9 Department of Clinical Nutrition and Dietetics, University of Cape Coast, Cape Coast PMB TF0494, Ghana
10 Department of Population and Health, University of Cape Coast, Cape Coast PMB TF0494, Ghana
11 College of Public Health, Medical and Veterinary Sciences, James Cook University, Townsville, QLD 4811, Australia
12 School of Public Health, Faculty of Health, University of Technology Sydney, Ultimo, NSW 2007, Australia
* Correspondence: elvis.hagan@ucc.edu.gh

Citation: Banna, M.H.A.; Al Zaber, A.; Rahman, N.; Siddique, M.A.M.; Siddique, M.A.B.; Hagan, J.E., Jr.; Rifat, M.A.; Nsiah-Asamoah, C.N.A.; Seidu, A.-A.; Ahinkorah, B.O.; et al. Nutritional Value of Dry Fish in Bangladesh and Its Potential Contribution to Addressing Malnutrition: A Narrative Review. *Fishes* **2022**, 7, 240. https://doi.org/10.3390/fishes7050240

Academic Editors: Marina Paolucci and Shunsuke Koshio

Received: 26 July 2022
Accepted: 5 September 2022
Published: 8 September 2022

Abstract: Understanding the linkage between the nutrient composition of foods and individuals' recommended nutrient intake is important to address malnutrition. Despite it being a traditional and popular food item in Bangladesh, the nutrient composition of dry fish has not been reviewed yet. This study used a narrative review to assess the nutrient composition of dry fish and estimated its potential contribution to addressing some common nutritional deficiencies among children and pregnant and lactating women in Bangladesh. Records were collected from different databases, including the Web of Science, Google Scholar, PubMed, ScienceDirect, Banglajol, and ResearchGate. Data were extracted from 48 articles containing 1128 entries regarding nutrient composition. Most of the nutrient analyses estimated the proximate composition, whereas vitamin, mineral, amino acid, and fatty acid compositions were scarce in the literature. We found that dry fish has high protein and mineral content and could contribute highly to meeting the recommended nutrient intake of protein, iron, zinc, and calcium for children and pregnant and lactating women. The summarized nutrient composition data could be useful for further research to observe how dry fish could be best utilized to address malnutrition in Bangladesh. This narrative review recommends that further nutrient analysis, with emphasis on vitamin, mineral, and fatty acid compositions.

Keywords: Bangladesh; children; dry fish; malnutrition; minerals; nutrients; protein; vitamins; women

1. Introduction

Bangladesh is a riverine country, blessed with vast fishery resources, inland as well as marine [1]. From the viewpoint of the natural gift of aquatic resources, the aquatic food system plays a vital role in the food culture, eating habits, and lifestyle of the people. Fresh fish has versatile nourishing properties, which include highly bioavailable protein, essential fatty acids, macro- and micro-minerals, and vitamins; therefore, it undergoes rapid microbial spoilage [2]. Thus, various preservation techniques such as drying, salting, chilling, freezing, and smoking are used to prevent microbial spoilage and keep up the nutrient quality with a view to storing throughout the year [3,4]. Among these techniques, drying is the most traditionally used method of fish processing and preservation in developing countries, including Bangladesh [5,6]. From nutritive aspects, dry fish consist of high-quality proteins, healthy fatty acids, including long-chain omega-3 fatty acids such as eicosapentaenoic acid (EPA) and docosahexaenoic acid (DHA), and are a unique source of essential nutrients such as iodine, zinc, copper, selenium, and calcium [7]. Generally, dry fish is a delicious and gastronomically nourishing food that provides high protein and low calories compared to other animal protein such as beef. For example, 100 g of dry fish contains approximately 80 g of protein and 300 calories of energy, whereas animal meat contains almost two times more calories but low protein compared to dry fish [7]. Moreover, dry fish is considered a healthy food item for individuals and an entirely natural product that contains almost equivalent omega-3 fatty acid content and antioxidant properties to fresh fish [7]. Dry fish contributes a big share of micronutrients to the diet of the low socioeconomic population groups in South and Southeast Asia [8].

Dry fish (locally known as Shutki) is a popular and traditional food item among Bangladeshi people because of its high nutritional value, good taste, aroma, and distinctive flavor [9]. Moreover, people prefer some fish species, for instance, Bombay duck and ribbon fish, as dried rather than fresh for consumption. Evidence shows that people often find it reasonable to include dry fish in their diet to avoid heart diseases, diabetes, and obesity [7]. Dry fish is of low cost, affordable to low socioeconomic groups, and usually consumed with vegetables, oil, and spices in mixed dishes along with the major staple rice; thus, it helps improve individuals' dietary diversity and nutrition security [1,7].

Nowadays, there are different areas of Bangladesh, including Charfashion in Bhola, Dublar Chor in Khulna, Kutubdia, Khuruskul, Moheskhali, Sonadia, and St. Martin Island in Cox's Bazar, and Alipur, Mohipur, Rangabali, and Kuakata in Patuakhali, where dry fish is commercially produced [1,9]. Here, mainly three categories of fish, including large fish, elongated fish, and small fish, are used for dry fish production [10]. About 20% of the artisanal catch is processed for dry fish production by the sun-drying method [11]. In addition to fresh fish and seafood products, dry fish has created potential market demand in Bangladesh and abroad [12]. In 2018–2019, Bangladesh exported a total of 2339.36 metric tons of dry fish to different countries, including India, Singapore, Hong Kong, Malaysia, the UK, the USA, and the United Arab Emirates, and earned approximately USD 4 million [13].

In Bangladesh, dry fish has market demand all around the year although their availability is somewhat more seasonal, mostly during the winter. Market demand for dry fish is high during spring and early monsoon when the supply of fresh fish from capture fisheries and aquaculture is the lowest [9]. There are regional and cultural differences in dry fish consumption; for instance, the highest consumption occurs in the Chattogram and Sylhet divisions, followed by moderate in the Dhaka division, low in the Barishal and Rajshahi divisions, and rare in the Khulna division [9]. The choice of type of dry fish consumption also varies geographically. For example, people from the Chattogram region consume mainly dried fish of marine origin, while people from Sylhet and Dhaka consume dry fish of both freshwater and marine origin.

In recent decades, Bangladesh has stepped forward to achieve self-sufficiency in food production; however, food and nutrition insecurity and malnutrition remain major public health issues [14]. Moreover, a recent study reported a moderate level of nutrition literacy among Bangladeshi adults [15]. According to Bangladesh Demographic and Health Survey 2014, the prevalence of childhood malnutrition is high (for instance, stunting: 36.1%, wasting: 14.3%, and underweight: 32.6%) [16,17]. Micronutrient deficiencies (such as iron deficiency anemia and zinc and calcium deficiencies) are still highly prevalent among children and women of reproductive age [17]. Therefore, the choice of a diet with high nutritional value is important to prevent the malnutrition burden. In this aspect, the inclusion of dry fish in the diet could be a consideration for improving dietary diversity and nutrient supply to the body.

Although drying causes some changes in fish flesh, it is still an excellent source of essential nutrients [7]. However, the use of spoiled raw fish, insecticides, and pesticides, poor hygiene and sanitation during preparation, the traditional way of processing, and long-term traditional storage conditions may lead to the nutrient loss and quality deterioration of dry fish [18–20]. A previous investigation conducted in the Chattogram region of Bangladesh identified that the nutritional value of dry fish undergoes deterioration with high storage time [18]. Another study in Bangladesh reported that deteriorative changes in dried fishes may result in browning reactions and develop rancidity when the moisture content is comparatively high [21]. Therefore, sufficient precautionary measures such as using fresh raw fish, proper drying, maintaining personal hygiene and sanitation, and proper storage and packaging must be taken into consideration during dry fish production.

There is a lack of scientific documentation and quantitative information on the nutrient composition of dry fish prepared from fish species captured on the coast of Bangladesh. The "Food Composition Table for Bangladesh" prepared by the Institute of Nutrition and Food Science (INFS), University of Dhaka, Bangladesh, contains 381 food items but only four dry fish items [22]. Exploration of food composition data is essential for providing basic information on various aspects of nutrition, understanding dietary choices, and developing basic tools to improve food and nutrition security [23]. Therefore, the purpose of our study was to perform a narrative review on the nutrient composition of dry fish in Bangladesh, summarize quantitative data on the nutrient composition of dry fish, and observe their potential contribution to meeting the recommended nutrient intake (RNI) of some nutrients with high public health importance for children and pregnant and lactating women in Bangladesh.

2. Materials and Methods

2.1. Data Sources and Search Strategy

To obtain the nutrient composition of dry fish prepared from fish species captured in Bangladesh, we conducted a narrative review with a literature search in the following databases: Web of Science, PubMed, Google Scholar, ScienceDirect, Banglajol (Bangladesh-based database), and ResearchGate. The following search strategy was used to collect records from the Web of Science and PubMed:

("Dried fish" OR "dry fish" OR fish) AND (Nutrient OR Composition OR "Nutrient composition" OR Vitamin OR Mineral OR Quality OR "Proximate composition" OR "Nutrient analysis") AND (Bangladesh)

Keywords, including "dry fish", "dried fish", nutrient composition, and Bangladesh, were used to conduct a manual record search in the databases of Google Scholar, ScienceDirect, Banglajol, and ResearchGate. No filter was applied while searching the Web of Science and PubMed. Database searching was conducted from June 2021 to December 2021. All citations were imported into Mendeley software and checked for duplication. Then, screening was conducted to identify eligible records.

2.2. Inclusion and Exclusion Criteria

Initially, titles and abstracts were screened using a checklist. The checklist consisted of three questions: (i) Is the article original research? (ii) Does the article contain the nutrient composition of dry fish? (iii) Does the article consider dry fish produced from fishes available in Bangladesh? Articles that qualified for initial screening were screened for full text. The full-text screening was conducted considering several criteria, including the number of fish species analyzed, name of the species, place of sample collection, number and types of nutrients considered for analysis, methods for nutrient analysis, and statistical representation of the data. In addition, the following inclusion criteria were followed: (1) publication date: no time restriction was applied, (2) language: no language restriction was applied, (3) laboratory methods: no article was excluded due to the study design or the laboratory methods used for nutrient analysis, and (4) sample type: only articles with dry fish nutrient composition data were considered. We excluded records if irrelevant to the research question, and not peer reviewed. Finally, articles that met the inclusion criteria were considered for data extraction and analysis.

2.3. Selection Process

Four researchers independently conducted the literature search and screened titles and abstracts. They also conducted full-text screening according to the screening strategy. Another reviewer assessed the variance of the number of records at each stage of the selection process. To minimize the bias, any discord in the selection process, such as database searching and inclusion and exclusion of the articles, was settled through a discussion among all the researchers. At first, the literature search was performed in different databases and sites. Then, all the records were imported into Mendeley software for duplication checking. After excluding the duplicates, an initial screening was conducted using the checklist. Articles that qualified for initial screening were then screened for full text. Finally, articles that qualified for the screening process and inclusion criteria were considered for data extraction.

2.4. Data Extraction

Four reviewers independently conducted data extraction from the selected articles. Data were extracted in a Microsoft Excel spreadsheet and included the name of the species (local, common, and scientific names) and nutrient composition per 100 g of edible dry fish. Some articles presented nutrient composition per 1 kg of weight or in other units, such as ppm or percentage. In these cases, we converted the unit into per 100 g of edible dry fish. In the original articles, nutrient compositions were often represented in values with standard deviation, as several samples were analyzed. However, we did not consider standard deviation during data extraction and synthesis. Some common and commercially important dry fishes were analyzed by different researchers, but we considered all the findings and included them in the data extraction spreadsheet. This means that several entries were considered for a single dry fish if reported in different articles. Fatty acid and lipid profiles showed a wide range of variation in results: individual fatty acid content, total saturated fatty acid (SFA), monounsaturated fatty acid (MUFA), polyunsaturated fatty acid (PUFA), EPA, DHA, total cholesterol, etc. We considered total SFA, MUFA, PUFA, EPA, DHA, and cholesterol. In the data extraction spreadsheet, we kept the cells blank for the following: missing value, no result, trace, not detected, and unreported information (such as English name and local name of the species). The data extraction spreadsheet is provided as Supplementary Material.

2.5. Assessment of Potential Contribution of Dry Fish to Addressing Malnutrition

We focused on four nutrients based on the availability of nutrient composition data and the nutritional importance of nutrients in the Bangladesh context, including protein, zinc, iron, and calcium. Protein was considered because of low protein consumption among children in Bangladesh. Micronutrients such as zinc, iron, and calcium were considered based on data showing both national and global deficiencies [24,25]. For each of the four nutrients, we presented a calculation where we compared the nutrient content of a given uncooked dry fish item to the daily RNI for women and children at different life stages. These calculations highlighted the relative variation in the nutrient composition and density (nutrient content per unit of dry fish) among the dry fish items. For every calculation, we considered five dry fish species or items that have the highest nutrient content according to the data extraction and synthesis. We also used two reference fish for comparison: Tilapia (*Oreochromis niloticus*) and Thai pangas (*Pangasianodon hypophthalmus*). Tilapia and Thai pangas were selected because they are the most commonly consumed fish, with the highest market accessibility in Bangladesh. Nutritional values fluctuate during processing and cooking, and other dietary factors influence the absorption of particular nutrients. As a result, these calculations are not meant to provide any individual dietary advice. Rather, they help provide an estimate of how certain dry fish contributes nutrients to the diet. The percentage of what a serving of fish covers for the RNI was calculated for pregnant women, lactating women, infants 6–12 months old, and children 1–2 years old [26].

While calculating how one serving of dry fish could meet the RNI of certain nutrients for pregnant and lactating women and children, we considered a daily serving of 50 g for pregnant and lactating women and 25 g for children based on a previously used method [27,28]. We assumed 10% bioavailability for iron [26]. The RNI for iron for pregnant women was estimated based on the FAO/WHO (2004) [26] value for women aged 19–50 years old, as no specific value for pregnant women is given. The daily value of 29.4 mg closely aligns with the Institute of Medicine's recommendation of 27 mg and the Indian Council of Medical Research's (ICMR) recommendation of 35 mg for pregnant women [29,30]. The RNI for protein for children 12–23 months old and pregnant and lactating mothers was directly received from the ICMR (2011) [29]; however, the ICMR does not directly mention the RNI for protein for infants 6–12 months old. Because of this, we considered the median body weight of boys and girls at 9 months of age, which is the average and median value between 6 and 12 months, and then calculated the average standard body weight [31]. The standard body weight was then multiplied by 1.69 to obtain the recommended daily protein intake [29]. For zinc, moderate bioavailability was assumed [26]. We calculated the daily zinc requirement by averaging the requirement across the three trimesters of pregnancy and the first 12 months of lactation, using a value of 7.5 mg for pregnancy and 8.5 mg for lactation. For the calcium requirements of the target populations, the FAO/WHO (2004) recommendation was followed [26]. The calculation showing how one serving of dry fish could meet the RNI of the target group is provided as Supplementary Material.

3. Results

Our search yielded 2139 articles. After screening titles and abstracts, 1939 records were excluded. After a full-text screening of the 198 records, we finally included 48 articles that had a nutrient composition of 1128 entries on dry fish from Bangladesh (Figure 1). Characteristics of the included studies [18,21,22,32–76] such as the number of samples analyzed, sample collection location, etc., are summarized in Table S1 (Supplementary Material).

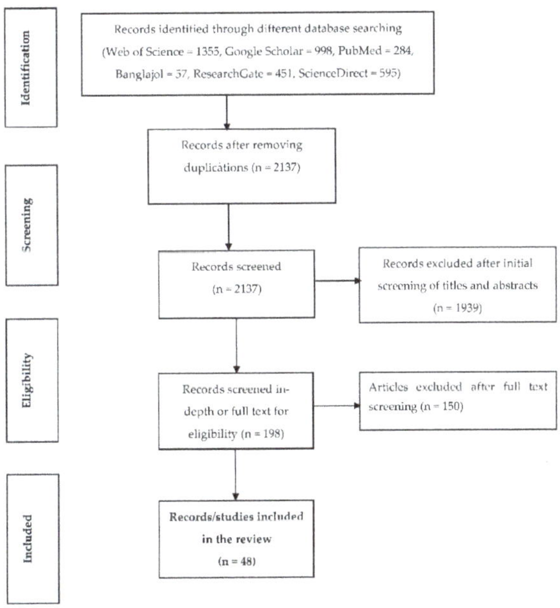

Figure 1. Selection process of the records.

3.1. Proximate Composition of Dry Fish Prepared from Fish Species Captured in Bangladesh

Of the total 1128 entries, 702 (62.23%) had the proximate composition of dry fish. Total protein content was determined in 14.54% (n = 164) of entries, and it was found that the average protein content was 56.63 g per 100 g of dry fish with the highest (77.68 g) in Churi (*Trichiurus savala*) and the lowest (26.73 g) in Ilish (*Tenualosa ilisha*). Total fat and ash content were analyzed in 15.70% (n = 177) and 15.25% (n = 172) of entries, respectively, and showed that Rupchanda (*Pampus chinensis*) and Tengra (*Mystus tengra*) contained the highest amount of fat and ash, respectively (Table 1).

Protein deficiency malnutrition is still prevalent among nutritionally vulnerable population groups such as children and pregnant and lactating women in Bangladesh. Being a good source of high-quality animal protein, dry fish can contribute to meeting individuals' daily protein requirements. For example, a serving of Ribbon fish (Churi) could fulfill 49.79% and 54.70% of daily recommended protein intake for pregnant and lactating mothers, respectively, and provides 100% of the recommended protein intake for children up to two years old. The potential contribution of dry fish to individuals' recommended protein intake is higher than that of fresh Thai pangas and Tilapia in Bangladesh (Figure 2).

Table 1. Analyzed nutrients, average nutrient content, and name of dry fish species with the highest and the lowest nutrient content.

Nutrients (Unit)	Number of Entries or Results *	Nutrient Content per 100 g of Dry Fish **			Species with the Highest Value: Local Name (Scientific Name)	Species with the Lowest Value: Local Name (Scientific Name)	Reference(s) (Highest Value, Lowest Value)
		Average Value	Highest Value	Lowest Value			
Proximate composition							
Energy (KJ)	9	362.67	412	318	Chela (*Salmostoma acinaces*)	Vetki (*Lates calcarifer*)	[22]
Moisture (g)	180	21.60	43.03	4.70	Rupchanda (*Stromateus chinensis*)	Chela (*Salmostoma acinaces*)	[22,37]
Protein (g)	164	56.63	77.68	26.73	Churi (*Trichiurus savala*)	Ilish (*Tenualosa ilisha*)	[39,61]
Fat (g)	177	9.73	27.74	0.47	Rupchanda (*Pampus chinensis*)	Datina koral (*Lates calcarifer*)	[57,61]
Ash (g)	172	14.29	36.20	0.16	Tengra (*Mystus tengra*)	Chingri (*Macrobrachium dayanum*)	[41,58]
Minerals							
Iron (mg)	50	13.15	45.10	2.80	Olua (*Coilia neglecta*)	Shol (*Channa striata*)	[40,73]
Zinc (mg)	38	4.31	19.30	0.23	Tengra (*Mystus tengra*)	Vetki (*Lates calcarifer*)	[22,65]
Calcium (mg)	44	954.61	3590	33.70	Chela (*Salmostoma acinaces*)	Rita (*Rit rita*)	[22,73]

Table 1. *Cont.*

Nutrients (Unit)	Number of Entries or Results *	Nutrient Content per 100 g of Dry Fish **			Species with the Highest Value: Local Name (Scientific Name)	Species with the Lowest Value: Local Name (Scientific Name)	Reference(s) (Highest Value, Lowest Value)
		Average Value	Highest Value	Lowest Value			
Phosphorus (mg)	23	810.08	2930	73	Khailsa (*Colisa fasciata*)	Taki (*Channa punctata*)	[32,62]
Magnesium (mg)	21	202.14	367.10	43.69	Churi (*Trichiurus lepturus*)	Shol (*Channa striata*)	[40,73]
Sodium (mg)	11	1133.91	3488	84	Pangas (*Pangasius pangasius*)	Vetki (*Lates calcarifer*)	[22,73]
Potassium (mg)	21	1060.56	1720.60	271	Vetki (*Lates calcarifer*)	Vetki (*Lates calcarifer*)	[22,74]
Manganese (mg)	21	0.58	1.11	0.15	Bata (*Cirrhina reba*)	Maitya (*Scomberomorus guttatum*)	[73,74]
Copper (mg)	39	0.26	1.11	0.02	Gura Chingri (*Leander styliferus*)	Rupchanda (*Pampus chinensis*)	[61,74]
Vitamins							
Folate (µg)	2	22	30	14	Fesha (*Engraulis tellara*)	Vetki (*Lates calcarifer*)	[22]
Thiamine (mg)	2	0.16	0.19	0.13	Fesha (*Engraulis tellara*)	Vetki (*Lates calcarifer*)	[22]
Riboflavin (mg)	2	0.68	0.86	0.50	Fesha (*Engraulis tellara*)	Vetki (*Lates calcarifer*)	[22]
Vitamin B_6 (mg)	2	0.43	0.48	0.38	Fesha (*Engraulis tellara*)	Vetki (*Lates calcarifer*)	[22]
Fatty acids							
Lauric acid (g)	13	6.21	11.80	2.50	Tuna (*Tunnus albacores*)	Sundori (*Bodianus neilli*)	[57]
Myristic acid (g)	14	10.15	18.80	0.17	Rangachoi (*Lutjanus indicus*)	Kauwa fish (*Megalaspis cordyla*)	[57,72]
Palmitic acid (g)	14	47.15	63.56	0.96	Koladia (*Otolithoides pama*)	Kauwa fish (*Megalaspis cordyla*)	[57,72]
Palmitoleic acid (g)	12	11.53	26.80	1.80	Bol koral (*Lates calcarifer*)	Koladia (*Otolithoides pama*)	[57]
Stearic acid (g)	14	7.07	16	0.64	Rup chanda (*Pampus chinensis*)	Kauwa fish (*Megalaspis cordyla*)	[57,72]
Oleic acid (g)	13	11.35	22.80	0.51	Bol koral (*Lates calcarifer*)	Kauwa fish (*Megalaspis cordyla*)	[57,72]

* Duplication (same species analyzed by several researchers) was considered. ** Values such as range, zero, not detected, missing values, and trace were not considered while calculating the average.

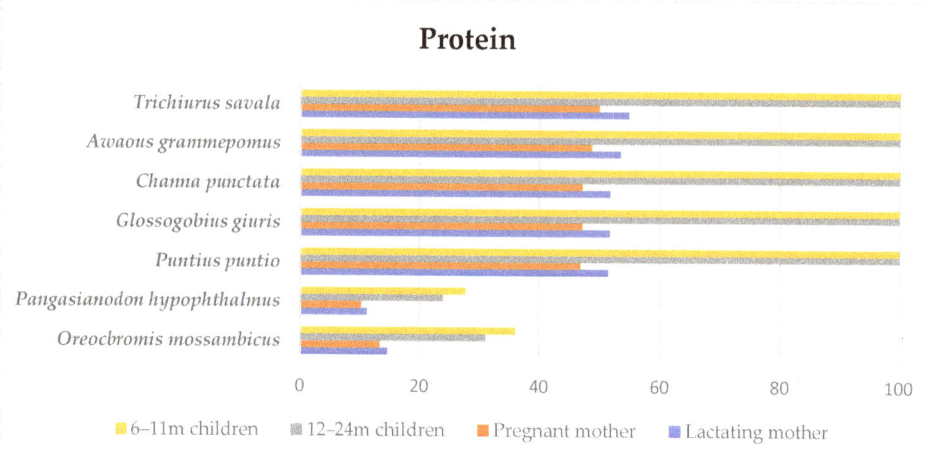

Figure 2. Potential contribution (%) of dry fish to the recommended nutrient intake (RNI) of protein for children and pregnant and lactating women in Bangladesh. Local names: Churi (*Trichiurus savala*), Bele (*Awaous grammepomus*), Taki (*Channa punctata*), Bele (*Glossogobius giuris*), Punti (*Puntius puntio*), Thai Pangas (*Pangasianodon hypophthalmus*), and Tilapia (*Oreocbromis mossambicus*).

3.2. Minerals

The content of minerals such as iron, zinc, calcium, phosphorus, magnesium, sodium, potassium, manganese, and copper was assessed in 23.76% (n = 268) of entries. The average mineral content with the highest and the lowest values is presented in Table 1.

Iron (Fe) was estimated in 4.43% (n = 50) of entries. The average Fe content was 13.15 mg per 100 g of dry fish which ranged from 45.10 mg (*Coilia neglecta*) to 2.80 mg (*Channa striata*) (Table 1). The bioavailability of iron is the extent to which dietary iron is absorbed by the body; therefore, highly bioavailable iron is good for health. Iron from dry fish (i.e, haem iron) has more bioavailability than non-haem iron and can meet iron demands at critical stages of the life cycle. Our analysis shows that a daily serving of Olua, Bata, or Loitta could fulfill 100% of the recommended intake of iron for children up to two years and lactating mothers (Figure 3). For a pregnant woman, a daily serving of Olua, Bata, and Churi meets 76.70%, 68.03%, and 52.72% of her daily iron needs, respectively (Figure 3). According to our data, the potential contribution of dry fish to individuals' recommended iron intake is higher than that of fresh Thai pangas and Tilapia in Bangladesh (Figure 3).

Zinc (Zn) was analyzed in 3.37% (n = 38) of entries, and we found that the average zinc content of dry fish was 4.31 mg (per 100 g of dry fish). The zinc content ranged from 0.23 mg to 19.30 mg per 100 g, and the highest amount was identified in dried Tengra (*Mystus tengra*). In many low- and middle-income countries including Bangladesh, zinc is deficient in diets. Thus, zinc-rich dry fish can contribute to reducing the gap. As demonstrated in Figure 4, a daily serving of dried Tengra provides 100% of the recommended intake of zinc for children aged up to two years and pregnant and lactating mothers. Like other nutrients, dry fish contributes much more to individuals' recommended zinc intake compared to reference fresh fish (Figure 4).

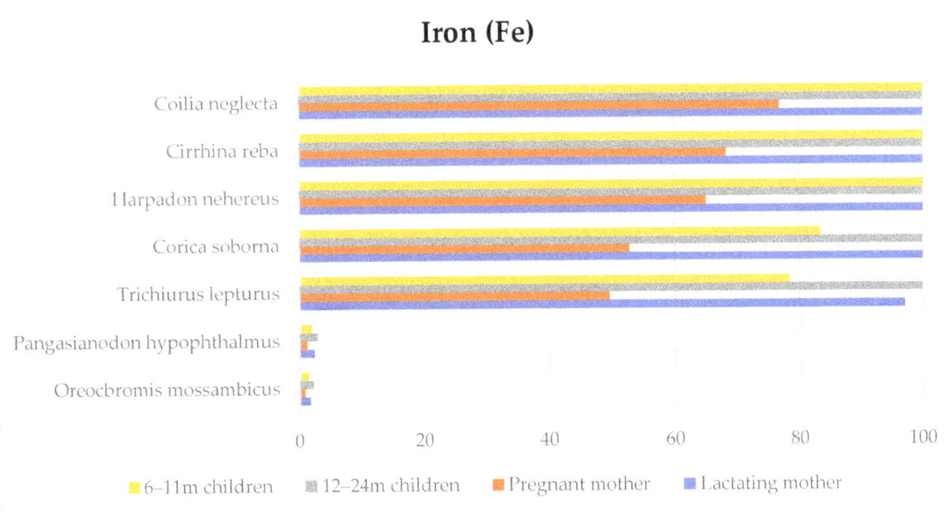

Figure 3. Potential contribution (%) of dry fish to the RNI of iron for children and pregnant and lactating women in Bangladesh. Local names: Olua (*Coilia neglecta*), Bata (*Cirrhina reba*), Loitta (*Harpadon nehereus*), Kachki (*Corica soborna*), Churi (*Trichiurus lepturus*), Thai Pangas (*Pangasianodon hypophthalmus*), and Tilapia (*Oreocbromis mossambicus*).

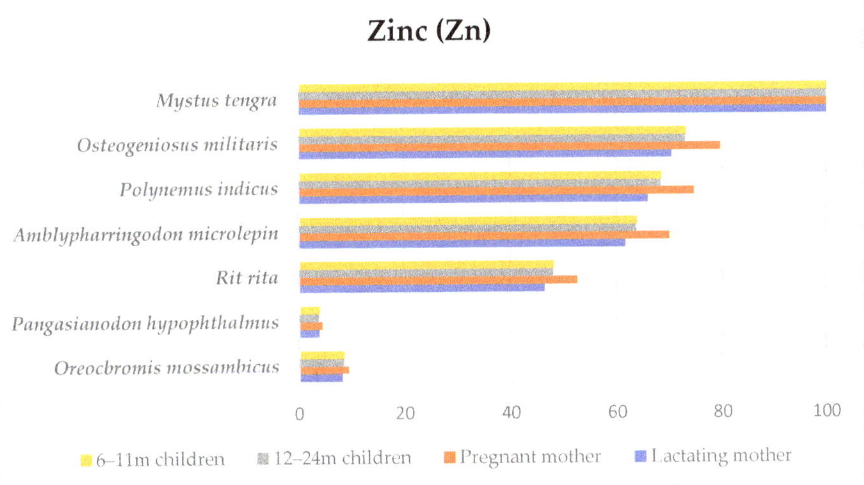

Figure 4. Potential contribution (%) of dry fish to the RNI of zinc for children and pregnant and lactating women in Bangladesh. Local names: Tengra (*Mystus tengra*), Kata mach (*Osteogeniosus militaris*), Lakhua (*Polynemus indicus*), Kachki (*Amblypharringodon microlepin*), Rita (*Rit rita*), Thai Pangas (*Pangasianodon hypophthalmus*), and Tilapia (*Oreocbromis mossambicus*).

In 3.90% ($n = 44$) of the entries, calcium (Ca) content was reported. According to the data analysis, the calcium content of dry fish varied from 3590 mg (dry Chela) to 33.70 mg (dry Rita) per 100 g of dry fish with an average of 954.61 mg. (Table 1). Figure 5 represents how one serving of dry fish could cover the recommended calcium intake of the target population. The calculation shows that one serving of dry fish could significantly meet the calcium requirement for women and children. We found one serving of the respective dry fishes (i.e., Chela, Olua, Bata, Punti, and Khailsa) could meet 100% of the daily dietary calcium requirement of the women and children. Similarly, dry fish are ahead in the case of potential contribution to the recommended intake of calcium compared to the referenced raw fish (Figure 5).

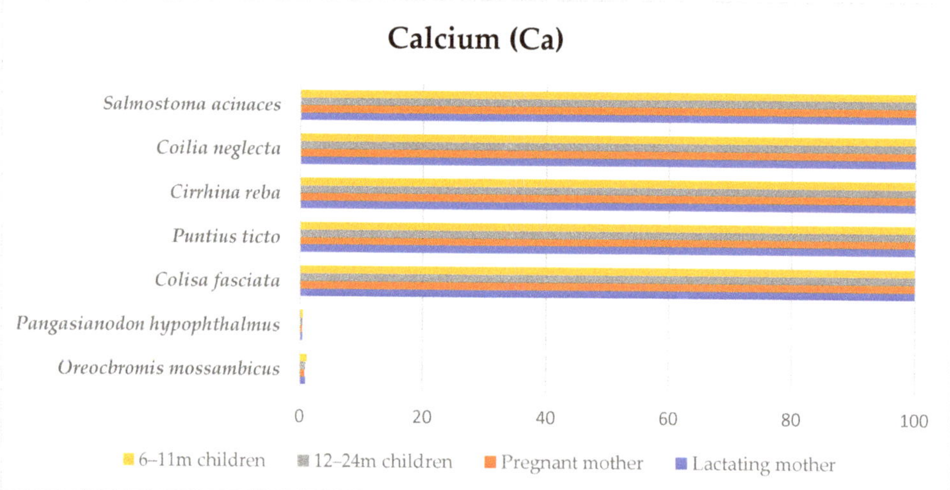

Figure 5. Potential contribution (%) of dry fish to the RNI of calcium for lactating and pregnant women and children in Bangladesh. Local names: Chela (*Salmostoma acinaces*), Olua (*Coilia neglecta*), Bata (*Cirrhina reba*), Punti (*Puntius ticto*), Khailsa (*Colisa fasciata*), Thai Pangas (*Pangasianodon hypophthalmus*), and Tilapia (*Oreocbromis mossambicus*).

3.3. Vitamins

Only 0.71% ($n = 8$) of entries were found to report the vitamin content of the dry fish. Data show that the average folate, thiamine, riboflavin, and vitamin B_6 content of the dry fish was 22 µg, 0.16 mg, 0.68 mg, and 0.43 mg per 100 g of dry fish, respectively. Based on the data, vitamin content was found to be higher in dried Fesha (*Engraulis tellara*) compared to other dried fishes. Moreover, a study reported that the vitamin D content of dried Giant sea perch (local name: Vetki) was 4.7 µg [22].

3.4. Fatty Acids and Amino Acids

Major fatty acid and amino acid contents of the dry fish were found in 13.30% ($n = 150$) of entries. The average fatty acid and amino acid content per 100 g of dry fish with the highest and lowest values is presented in Table 1. According to the data, the three most common fatty acids were palmitic acid, palmitoleic acid, and leic acid with an amount of 47.15 g, 11.53 g, and 11.35 g per 100 g of dry fish, respectively. The highest amount of palmitic acid and palmitoleic acid was found in the Koladia (*Otolithoides pama*) and Bol koral (*Lates calcarifer*) species, respectively (Table 1). Moreover, the PUFA content of dried Kauwa fish (*Megalaspis cordyla)* varied from 26.74% (traditional drying) to 30.45% (solar drying) in which docosahexaenoic acid and eicosapentaenoic acid were prominent [72]. The amino acid content of the dry fish shows that they contain all the essential amino acids, including sulfur-containing amino acids (e.g., methionine) which are lacking in plant protein and lysine which is absent in terrestrial meat proteins. According to the data, the highest amount of essential amino acids was found among three species, namely, Churi (*Lepturacanthus savala*), Coral (*Lates Calcarifer*), and Rupchanda (*Pampus chinensis*) (Table 1).

4. Discussion

The analysis of fish consumption data presented by the International Food Policy Research Institute (2006/7) shows that dried fish is one of the most frequently consumed categories of fish in Bangladesh [9]. Although dry fish plays a significant role in the diet and nutrition of the people in Bangladesh, there is a dearth of information about evidence-based documentation of the nutritive aspects of dry fish and its potential contribution to the recommended nutrient intake of vulnerable groups such as children and pregnant and lactating women. This study reviewed and accumulated the available nutrient composition data on dry fish produced in Bangladesh and represents its potential contribution to meeting the nutrient requirement of children and pregnant and lactating women. We found that most of the studies analyzed dry fish for proximate composition (i.e., moisture, protein, fat, and ash) rather than minerals, vitamins, and fatty acids. Our review suggests that future research should focus on the analysis of nutrients including vitamins, minerals, fatty acids, and amino acids. By documenting the available nutrient composition data on dry fish, our review provides a baseline resource for fisheries, nutrition researchers, and policymakers to better understand the need to include dry fish in food-based interventions to reduce malnutrition in Bangladesh.

Our analysis showed that different kinds of dry fish were able to meet the daily recommended intake of protein, iron, zinc, and calcium for children up to 2 years old and pregnant and lactating mothers. Additionally, dry fish contain more nutrients than the considered reference fresh fish (Thai pangas and Tilapia). A recent review, focusing on the global context, on fish nutrient composition conducted by Byrd et al. (2021) [28] also considered the RNI in the study. Generally, the RNI is a more conservative estimate than the estimated average requirement (EAR), which provides a concrete scientific basis for meeting the requirements of nearly all individuals in a group and the adequacy of diets [26].

The nutrient content of a particular fish species could vary from one habitat to another and season to season due to the variation in the amount and quality of food consumed by the fish and also their movement [77–79]. In our review, we found that the nutrient content of dried Loitta was analyzed in twelve studies; however, the crude protein content varied from 32.21 g to 67.21 g (per 100 g of dry fish). This inconsistency in the nutrient content of the same fish species might be due to the drying methods of raw fish (whether sun, solar, or mechanically dried), fish capture season, source of raw fish collection, and quality of the raw fish. Therefore, future comparative studies evaluating the effect of different drying techniques on nutrient composition are recommended to better understand the quantitative differences in the nutrient content of dry fish samples from the same species.

Evidence shows that the nutrient content of the dry fishes mainly varies due to the nutrient composition of their respective raw fishes and geographical differences [80]. According to the extracted data, dry fishes prepared from marine sources were more nutritious considering their proximate composition and fatty acid and amino acid contents. For instance, Churi (*Trichiurus savala*) had comparatively higher protein content, Rupchanda (*Pampus chinensis*) had higher fat content, Tuna (*Tunnus albacores*) had comparatively higher lauric acid content, and Churi (*Lepturacanthus savala*) had comparatively higher essential amino acid content than the other dry fishes. Marine fish species are rich in nutrients such as polyunsaturated fatty acids [81]; therefore, even after drying, they could retain a substantial amount of nutritive properties. A previous study reported no significant effect of drying on the fatty acid profile and composition of the dried cod heads [82]. Bangladeshi dry fish from freshwater sources contains higher protein than the respective freshwater dry fish in Northeast India as investigated by Ullah et al. (2016) [83]. Based on the extracted data, we found that mineral content was higher among small dry fish species, including Olua (*Coilia neglecta*), Tengra (*Mystus tengra*), Chela (*Salmostoma acinaces*), and Khailsa (*Colisa fasciata*), than the large dry fish. Several studies also reported that small fish and their dried products are good sources of minerals [28,84,85].

Sun drying is the most used fish drying technique in Bangladesh. It is a traditional preservation method of fish that is carried out in the open air using sunlight to evaporate the water and the airflow to carry away the vapor [86]. Though sun drying has several advantages, including a simple operation technique and economical convenience, it has some demerits too. The major constraints of traditional fish drying include dependency on weather, a long drying period (2–3 days), and the required hygienic handling of raw materials [86].

4.1. Hazards Associated with Dry Fish Production in Bangladesh and Some Recommendations

Dry fish is proven to have a higher nutritional profile that is important for public health. However, the quality of dried fish can be degraded due to various hazards in the production chain. Rasul and his colleagues (2020) [20] summarized the chemical and microbiological hazards of dried fish in Bangladesh. They reported that several dried fishes were contaminated with a high content of heavy metals (for example, Pb, Cd, and Cr) and pesticide residue (dichlorodiphenyltrichloroethane, heptachlor, endrin, aldrin, and dieldrin), and highly pathogenic *E. coli*, *Salmonella* sp., and *Vibrio* sp. were found in a few dried fish samples that may cause serious health hazards after consumption [20]. They also reported that lipid oxidation occurred in some dried fishes from Bangladesh which are responsible for the unpleasant flavors and odors. Sun-based drying affects polyunsaturated lipids and can promote lipid oxidation, which can reduce the nutritive value and functional quality and raise consumer health risks [7]. Two recent Bangladeshi studies conducted by Hoque et al. (2021) [87] and Rakib et al. (2021) [88] reported that heavy metals pose moderate-to-high health risks to the dry fish consumer. There are several factors associated with these health hazards which include traditional drying techniques, the use of harmful pesticides, anthropogenic contamination, atmospheric deposition, the lack of maintaining proper hygiene and sanitation, improper packaging and storage, and water pollution [20,88]. To minimize and prevent these hazards of dry fish production in Bangladesh, some recommendations include: (i) developing improved and cost-effective methods of fish drying [89], (ii) designing effective packaging and storage facilities, (iii) ensuring heavy metal decontamination strategies, (iv) organizing public health awareness programs for dry fish producers regarding basic hygiene and sanitation practices, the adverse effect of chemical contaminants in dry fish, and the importance of the quality of the raw materials used, and (v) providing training on safe dry fish production and waste management to dry fish producers.

The use of harmful chemicals in dry fish production is a special concern. In Bangladesh, dry fish producers usually use harmful chemicals (such as a mixture of organochlorine) to protect dry fish from insect infestation and to increase shelf life [19] without considering their deleterious impacts on human health. A study found the presence of harmful chemicals such as DDT and heptachlor in some dry fish samples in Bangladesh [90]. These pesticides are used to protect fish from insect infestation; however, they are associated with serious health problems, including cancer and non-allergic reactions, and environmental hazards [71]. Considering the situation, the following recommendations can be highlighted to prevent the use of harmful chemicals or insecticides during dry fish production: (i) Ensuring strict implementation of laws and policies related to harmful pesticide use. In many countries, including Sweden, Japan, and the USA, organochlorine chemicals have been banned due to their potential harm to human health. Therefore, the government of Bangladesh may strictly implement the updated pesticide legislation and policies for fishery products. (ii) Developing and implementing safe insect control strategies instead of using harmful insecticides. For instance, red pepper and turmeric (separately or mixed) have insect- and bacteria-repealing characteristics and thus can be used in fish drying. Evidence shows that the pretreatment of fish with 10–12% salt for 10–12 h can reduce infestation [91]. (iii) Sensitizing dry fish producers to the harmful impacts of pesticides on human health and the food system. (iv) Taking initiatives to prevent environmental pollution so that heavy metals and harmful chemicals cannot enter the aquatic ecosystem.

4.2. Implications for Practice

The findings of our study (i.e., up-to-date data on the nutrient composition of dried fish) can be useful to policymakers, public health practitioners, and nutrition experts for developing nutrition-based programs and interventions to improve the country's food and nutrition security. Such programs and interventions could encompass formulating dietary guidelines, updating the food composition table, promoting dry fish production and consumption, and formulating nutrition-education-related materials.

Again, an area that requires further exploration is assessing the shelf lives of dried fish during storage over a period of time to explore possible nutritional changes that may occur when dried fish is under storage. Assessing the impact of consuming dried fishes on growth and nutritional status among vulnerable populations is another key research area that has to be further explored.

Inclusion of nutritious food items in the diet is important for health and survival. Globally, the recent coronavirus (COVID-19) pandemic has imposed a new set of challenges on humans to maintain a diversified and healthy diet [92–94]. Energy, protein, and micronutrient (especially vitamin A, B complex, C, and D, zinc, iron, and selenium) deficiencies are associated with impaired immune function and an increased risk of infection and mortality among vulnerable populations as well as COVID-19 patients [94,95]. Evidence shows that zinc, magnesium, and vitamin C have a potential role in reducing the severity of the infection and inflammatory response associated with severe acute respiratory syndrome coronavirus 2 (SARS-CoV-2), while folate and vitamin D may have a role in antagonizing the entry of virus in host cells [96,97]. Zinc has the potential to reduce viral replication and increase immune responses as well as act as a prophylactic which might provide an additional shield against the initiation and progression of COVID-19 [98]. Since dry fish contain high nutritional properties, including protein and minerals (Fe, Zn, and Ca), the inclusion of dry fish in the diet could be a consideration especially for at-risk groups during this COVID-19 pandemic [99]. For instance, in Myanmar, the use of dry small fish powder provides an opportunity for accessible and acceptable forms of micronutrients required to improve the health status of young children during this pandemic [100]. The example from the Myanmar study [100] suggests the potential fortification of staple cereal and grain-based complementary foods with dried fish powder that can improve the nutrient quality of foods that are used to feed young children in most developing countries. A similar concept could also be replicated for Bangladesh. Dry fish of high nutritional quality could be considered for powder preparation and be used to improve the nutritional quality of complementary food for young children. Bangladesh is a disaster-prone country. As dry fish could be stored for a longer period of time, it could be included in the ration after any disaster, including floods and cyclones.

5. Conclusions

According to available information, dry fish possess high amounts of nutritional properties, especially protein, zinc, iron, and calcium. In particular, dry fish from marine sources and small fish species are high in protein, fatty acids, amino acids, and minerals. Dry fish significantly contributed to the recommended intake of protein, iron, zinc, and calcium for children up to two years and pregnant and lactating mothers. It is imperative that policymakers along with food and nutrition experts focus on promoting the nutritional value of dried fish and encouraging particularly vulnerable populations (children and pregnant and lactating women) to include dry fish in their diet. Further research in this area may emphasize analyzing vitamin and mineral composition rather than proximate analysis. Further research may highlight and analyze the essential fatty acid composition (omega-3, -6, and -9) of different species of fishes and its impact on improving the cognitive development and functioning/performance of children and the elderly, respectively.

Supplementary Materials: The following supporting information can be downloaded at: https://www.mdpi.com/article/10.3390/fishes7050240/s1, Table S1: Characteristics of included studies (N = 48), data extraction file, and calculation of RNI.

Author Contributions: M.H.A.B.: conceptualization, study design, literature search and screening of articles, writing—original draft (Abstract, Introduction, Results interpretation, Discussion, and Conclusion) and editing. A.A.Z.: literature search and screening of articles, data management and analysis. N.R.: literature search and screening of articles, data curation. M.A.M.S.: literature search and screening of articles. M.A.B.S., J.E.H.J., C.N.A.N.-A., A.-A.S., B.O.A. and M.S.I.K.: visualization, validation, writing—reviewing and editing. M.A.R.: conceptualization and supervision, study design, visualization, validation, writing—original draft (Materials and Methods), critical review for intellectual content, editing. All authors have read and agreed to the published version of the manuscript.

Funding: This research received no external funding. The article processing charge (APC) was funded by Bielefeld University, Germany.

Institutional Review Board Statement: Not applicable.

Data Availability Statement: Data are contained within the article or Supplementary Material.

Acknowledgments: The authors would like to express their humble gratitude to Bielefeld University, Germany, for providing financial support through the Institutional Open Access Publication Fund for the APC.

Conflicts of Interest: The authors declare no conflict of interest.

References

1. Hossain, M.A.R.; Belton, B.; Thilsted, S.H. *Dried Fish Value Chain in Bangladesh*; World Fish, Bangladesh and south Asia Office: Dhaka, Bangladesh, 2015; 122p.
2. Dehghani, S.; Hosseini, S.V.; Regenstein, J.M. Edible Films and Coatings in Seafood Preservation: A Review. *Food Chem.* **2018**, *240*, 505–513. [CrossRef] [PubMed]
3. Nwaigwe, U. Fish Preservation and Processing. *J. Food* **2017**, *2*, 1–31.
4. Ghaly, A.E.; Dave, D.; Budge, S.; Brooks, M.S. Fish Spoilage Mechanisms and Preservation Techniques. *Am. J. Appl. Sci.* **2010**, *7*, 859. [CrossRef]
5. Doe, P.; Olley, J. Drying and Dried Fish Products. In *Seafood: Resources, Nutritional Composition, and Preservation*; CRC Press: Boca Raton, FL, USA, 2020; pp. 125–145. ISBN 1003068413.
6. Marine, S.S.; Sayeed, M.A.; Barman, P.P.; Begum, R.; Hossain, M.M.; Alam, M.T. Traditional Methods of Fish Drying: An Explorative Study in Sylhet, Bangladesh. *Int. J. Fish. Sci. Aquac.* **2015**, *2*, 28–35.
7. Siddhnath; Ranjan, A.; Mohanty, B.P.; Saklani, P.; Dora, K.C.; Chowdhury, S. Dry Fish and Its Contribution towards Food and Nutritional Security. *Food Rev. Int.* **2020**, *38*, 508–536. [CrossRef]
8. Johnson, D. Research on the Social Economy of Dried Fish in South and Southeast Asia. 2020. Available online: https://agris.fao.org/agris-search/search.do?recordID=IN2022014370 (accessed on 25 August 2022).
9. Hossain, M.A.R.; Belton, B.; Thilsted, S.H. Preliminary Rapid Appraisal of Dried Fish Value Chains in Bangladesh. *World Fish Bangladesh Dhaka*. 2013. Available online: https://www.worldfishcenter.org/publications (accessed on 25 August 2022).
10. Paul, P.C.; Reza, M.S.; Islam, M.N.; Kamal, M. Quality Assessment of Traditionally Dried Marine Fish of Bangladesh. *Asian Food Sci. J.* **2018**, *5*, 44406. [CrossRef]
11. Begum, M.; Akter, T.; Minar, M.H. Effect of Salt and Garlic on the Quality and Microbial Content of Smoked Catfish (*Pangasianodon hypophthalmus*). *Int. J. Bio-Resour. Stress Manag.* **2012**, *3*, 464–467.
12. Siddique, M.A.M. Explaining the Role of Perceived Risk, Knowledge, Price, and Cost in Dry Fish Consumption within the Theory of Planned Behavior. *J. Glob. Mark.* **2012**, *25*, 181–201. [CrossRef]
13. Department of Fisheries (DoF). *Yearbook of Fisheries Statistics of Bangladesh, 2018–2019*; 2019. Available online: https://fisheries.portal.gov.bd/sites/default/files/files/fisheries.portal.gov.bd/page/4cfbb3cc_c0c4_4f25_be21_b91f84bdc45c/2020-10-20-11-57-8df0b0e26d7d0134ea2c92ac6129702b.pdf (accessed on 25 August 2022).
14. Ministry of Health and Family Welfare, Government of the People's Republic of Bangladesh. Advocacy Plan for Nutrition, Bangladesh 2019–2025. Available online: https://bnnc.portal.gov.bd/sites/default/files/files/bnnc.portal.gov.bd/publications/342ec9eb_7d02_41e3_a256_51b8ccbc0218/2021-02-08-12-45-3cab0c7dfbbcffa8a38be4b84957695a.pdf (accessed on 6 August 2022).
15. Al Banna, M.H.; Hamiduzzaman, M.; Kundu, S.; Ara, T.; Abid, M.T.; Brazendale, K.; Seidu, A.-A.; Disu, T.R.; Mozumder, N.H.M.R.; Frimpong, J.B. The Association Between Bangladeshi Adults' Demographics, Personal Beliefs, and Nutrition Literacy: Evidence from a Cross-Sectional Survey. *Front. Nutr.* **2022**, *9*, 867926. [CrossRef]
16. Das, S.; Gulshan, J. Different Forms of Malnutrition among under Five Children in Bangladesh: A Cross Sectional Study on Prevalence and Determinants. *BMC Nutr.* **2017**, *3*, 1. [CrossRef]

17. Ministry of Health and Family Welfare. *Second National Plan of Action for Nutrition 2016–2025*; Government of the People's Republic of Bangladesh: Dhaka, Bangladesh, 2017.
18. Siddique, M.A.M.; Aktar, M. Changes of Nutritional Value of Three Marine Dry Fishes (*Johnius dussumieri, Harpodon nehereus* and *Lepturacanthus savala*) during Storage. *Food Nutr. Sci.* **2011**, *2*, 1082–1087.
19. Bhuiyan, N.H.; Bhuiyan, H.R.; Nath, K.K.; Ahmed, K.; Hassan, T.; Bhuiyan, N.I. Organochlorine Insecticides (DDT and Heptachlor) in Dry Fish Available in Bangladesh: Seasonal Trends and Species Variability. *J. Chil. Chem. Soc.* **2009**, *54*, 278–281. [CrossRef]
20. Rasul, M.G.; Yuan, C.; Shah, A.K.M.A. Chemical and Microbiological Hazards of Dried Fishes in Bangladesh: A Food Safety Concern. *Food Nutr. Sci.* **2020**, *11*, 523–539. [CrossRef]
21. Mohammad, A.M.; Shafiqur, R.; Mohammad, N.A.K.; Md, S.R.; Shoji, U. Study on the Quality and Safety Aspect of Three Sun-Dried Fish. *Afr. J. Agric. Res.* **2013**, *8*, 5149–5155.
22. Shaheen, N.; Rahim, A.T.M.; Mohiduzzam, M.; Banu, C.P.; Bari, M.L.; Tukun, A.B.; Mannan, M.; Bhattacharjee, L.; Stadlmayr, B. *Food Composition Table for Bangladesh*; Institute of Nutrition and Food Science, Centre for Advanced Research in Sciences, University of Dhaka: Dhaka, Bangladesh, 2013.
23. Reksten, A.M.; Somasundaram, T.; Kjellevold, M.; Nordhagen, A.; Bøkevoll, A.; Pincus, L.M.; Rizwan, A.A.M.; Mamun, A.; Thilsted, S.H.; Htut, T. Nutrient Composition of 19 Fish Species from Sri Lanka and Potential Contribution to Food and Nutrition Security. *J. Food Compos. Anal.* **2020**, *91*, 103508. [CrossRef]
24. Beal, T.; Massiot, E.; Arsenault, J.E.; Smith, M.R.; Hijmans, R.J. Global Trends in Dietary Micronutrient Supplies and Estimated Prevalence of Inadequate Intakes. *PLoS ONE* **2017**, *12*, e0175554. [CrossRef]
25. Ferguson, E.; Chege, P.; Kimiywe, J.; Wiesmann, D.; Hotz, C. Zinc, Iron and Calcium Are Major Limiting Nutrients in the Complementary Diets of Rural Kenyan Children. *Matern. Child Nutr.* **2015**, *11*, 6–20. [CrossRef]
26. FAO/WHO. *Vitamin and Mineral Requirements in Human Nutrition*; Diamond Pocket Books (P) Ltd.: Delhi, India, 2004; ISBN 9241546123.
27. Bogard, J.R.; Thilsted, S.H.; Marks, G.C.; Wahab, M.A.; Hossain, M.A.R.; Jakobsen, J.; Stangoulis, J. Nutrient Composition of Important Fish Species in Bangladesh and Potential Contribution to Recommended Nutrient Intakes. *J. Food Compos. Anal.* **2015**, *42*, 120–133. [CrossRef]
28. Byrd, K.A.; Thilsted, S.H.; Fiorella, K.J. Fish Nutrient Composition: A Review of Global Data from Poorly Assessed Inland and Marine Species. *Public Health Nutr.* **2021**, *24*, 476–486. [CrossRef]
29. ICMR. *Dietary Guidelines for Indians*; National Institute of Nutrition: Hyderabad, India, 2011.
30. Stark, K.D.; Van Elswyk, M.E.; Higgins, M.R.; Weatherford, C.A.; Salem, N., Jr. Global Survey of the Omega-3 Fatty Acids, Docosahexaenoic Acid and Eicosapentaenoic Acid in the Blood Stream of Healthy Adults. *Prog. Lipid Res.* **2016**, *63*, 132–152. [CrossRef]
31. De Onis, M.; Onyango, A.W.; Borghi, E.; Garza, C.; Yang, H.; WHO Multicentre Growth Reference Study Group. Comparison of the World Health Organization (WHO) Child Growth Standards and the National Center for Health Statistics/WHO International Growth Reference: Implications for Child Health Programmes. *Public Health Nutr.* **2006**, *9*, 942–947. [CrossRef]
32. Akther, S.; Islam, M.B.; Rahman, H.; Bhuiyan, A.S. Proximate Composition and Nutritional Value of Fish Powder Composition (FPC) from Different Dried SIS Fishes in Bangladesh. *Res. J. Food Sci. Nutr.* **2018**, *3*, 9–14. [CrossRef]
33. Al-Reza, S.; Karmaker, S.; Hasan, M.; Roy, S.; Hoque, R.; Rahman, N. Effect of Traditional Fish Processing Methods on the Proximate and Microbiological Characteristics of Laubuka Dadiburjori during Storage at Room Temperature. *J. Fish. Aquat. Sci.* **2015**, *10*, 232.
34. Azam, K.; Basher, M.Z.; Ali, M.Y.; Asaduzzaman, M.; Hossain, M.M. Comparative Study of Organoleptic, Microbiological and Biochemical Qualities of Four Selected Dried Fish in Summer and Winter. *Pak. J. Biol. Sci.* **2003**, *6*, 2030–2033. [CrossRef]
35. Begum, M.; Akter, T.; Ahmed, A.T.A.; Khaleque, A. Evaluation of Quality and Storage Stability of Commercially Produced Dried Fish of Tengra (*Mystus vittatus*) with Laboratory Produced Dried Fish Products. *Int. J. Bio-Resour. Stress Manag.* **2011**, *2*, 246–249.
36. Begum, M.; Akter, T.; Uddin Ahmed, S.; Begum, M. The Effect of Lemon Leaves Extract Treatment on the Nutrient Content and Microbial Properties of Dried Chanda (*Chanda nama*). *J. Food Nutr. Res.* **2014**, *2*, 21–24.
37. Bhuiyan, M.N.H.; Dawlatana, M.; Bhuiyan, H.R.; Lucky, F.A.; Saha, B.K.; Bhuiyan, M.N.I. Quality Assessment of Dry Fishes of Bangladesh with the Special Emphasis on Harmful Additives. *Bangladesh J. Sci. Ind. Res.* **2009**, *44*, 311–318. [CrossRef]
38. Chakroborty, T.; Chakraborty, C.S. Comparative Analysis of Nutritional Composition and Microbial Quality of Salt-Smoke-Dried Mirror Carp (*Cyprinus carpio* Var. Specularis) during Storage at 22-28 C and 4 C. *Int. J. Food Sci. Nutr.* **2017**, *1*, 86–89.
39. Debnath, S.; Latifa, G.A.; Begum, M.; Obaida, M.A. Evaluation of Nutritional Values of Smoke Cured Riverine and Marine Hilsa (*Tenualosa ilisha*; Hamilton, 1882). *Bangladesh J. Zool.* **2018**, *46*, 177–184. [CrossRef]
40. Farid, F.B.; Latifa, G.A.; Chakraborty, S.C.; Nahid, M.N.; Begum, M. Combine Effect of *Curcuma longa* (Turmeric) Powder and Dry-Salt with Sun-Drying in Quality Changes of Three Freshwater Fish Species of Meghna River. *Eur. J. Biotechnol. Biosci.* **2016**, *4*, 42–47.
41. Farid, F.B.; Latifa, G.A.; Chakraborty, S.C.; Nahid, M.N.; Begum, M. Performance and Quality Assessment of Sun-Dried Salted *Channa punctatus* (Bloch, 1793) and *Mystus tengra* (Hamilton-Buchanan, 1822) during Refrigeration (4 C) Storage. *Int. J. Food Sci. Nutr.* **2017**, *2*, 157–162.
42. Flowra, F.A.; Tumpa, A.S.; Islam, M.T. Biochemical Analysis of Five Dried Fish Species of Bangladesh. *Univ. J. Zool. Rajshahi Univ.* **2012**, *31*, 9–11. [CrossRef]

43. Flowra, A.F.; Tumpa, S.A. Chemical Composition of Five Selected Dry Fish Species in Chalan Beel, Bangladesh. *DAV Int. J. Sci.* **2012**, *1*, 157–160.
44. Haque, E.; Kamruzzaman, M.; Islam, M.S.; Sarwar, T.; Rahman, S.S.; Karim, M.R. Assessment and Comparison of Quality of Solar Tunnel Dried Bombay Duck and Silver Pomfret with Traditional Sun Dried Samples. *Int. J. Nutr. Food Sci.* **2013**, *2*, 187–195. [CrossRef]
45. Hasan, M.M.; Shikha, F.H.; Hossain, M.I.; Kamal, M.; Islam, M.N.; Wahab, M.A. Quality Assessment of Traditional, Rotary and Solar Tunnel Dried Small Indigenous Fish Species Products. *Bangladesh J. Fish. Res.* **2006**, *10*, 73–84.
46. Hasan, M.M.; Rasul, M.G.; Ferdausi, H.J.; Trina, B.D.; Sayeed, A.; Shah, A.; Bapary, M.A.J. Comparison of Organoleptic and Chemical Characteristics of Some Traditional and Improved Dried Fish Products. *Res. J. Anim. Vet. Fish. Sci.* **2016**, *4*, 1–6.
47. Hasan, M.M.; Sayeed, M.A.; Khalil, I.; Kumar, K.; Tikadar, N.K.C.; Ahmed, Z. Nutritional Quality Assessment of Dried Baim Fishes in Sylhet Sadar Upazila. *Int. J. Fish. Aquat. Stud.* **2018**, *6*, 363–367.
48. Hossain, M.M.; Hossain, M.D.; Noor, M.A.; Haque, A.; Kabir, M.A. Quality Aspects of Some Dried Fish Products Collected from Different Super Shops of Dhaka City in Bangladesh. *J. Sylhet Agric. Univ.* **2015**, *2*, 283–287.
49. Hossain, M.N.; Jamil, M.G.M.; Mia, M.M.; Uddin, M.N.; Mansur, M.A. Studies on the Proximate Composition, Quality and Heavy Metal Concentration of Two Sun-Dried Marine Fish (Sun-Dried Silver Pomfret and Sun-Dried Perch) of Cox's Bazar District of Bangladesh. *J. Environ. Sci. Nat. Resour.* **2017**, *10*, 25–32. [CrossRef]
50. Huque, R.; Islam, M.; Khatun, A. Microbiological Quality Improvement of Dried Fish by Gamma Irradiation and Assessment of Food Value upon Irradiation with Respect to Biochemical Aspect. *Int. Res. J. Pharm. Appl. Sci.* **2013**, *3*, 1–5.
51. Imtiaz, M.M.; Haque, S.A.; Sarker, M.R.; Mufty, M.M.; Ali, M.M.; Parvez, M.S.; Reza, M.S.; Bhuyan, M.S.; Rahman, M.A. Quality Assessment of Improved Sun Dried Ribbon Fish (*Trichiurus haumela*) by Salt and Turmeric Powder Treatment. *Int. Inst. Chem. Biol. Environ. Eng.* **2017**, *4*, 162–167.
52. Islam, M.T.; Kamal, M.; Islam, M.N.; Neazuddin, M.; Mehbub, M.F. Organoleptic, Biochemical and Bacteriological Aspects of the Low Cost Tunnel Dried Fish Products. *Bangladesh J. Fish. Res.* **2006**, *10*, 175–183.
53. Islam, D.; Mustafa, M.G.; Wahed, M.A.; Khaleque, M.A.; Naser, M.N.; Islam, M.M. Comparative Assessment of Traditional Sun Dried and Solar Tunnel Dried Pomfret (*Pampus argenteus*) under Different Storage Conditions. *Bangladesh J. Zool.* **2007**, *35*, 331–340.
54. Islam, M.J.; Hossain, M.S.; Mian, S. Nutritive Value of Dried and Heat Processed Mola Fish (*Amblypharyngodon mola*) Products. *Int. J. Nat. Sci.* **2012**, *2*, 43–48. [CrossRef]
55. Islam, M.T.; Ahmed, S.; Sultana, M.A.; Tumpa, A.S.; Flowra, F.A. Nutritional and Food Quality Assessment of Dried Fishes in Singra Upazila under Natore District of Bangladesh. *Trends Fish. Res.* **2013**, *2*, 2319–4758.
56. Islam, M.M.; Abedin, M.T.; Haque, M.A.; Hayder, S.; Sujon, M.N.; Alam, M.A.; Sayed, M.A. Evaluation of Nutritional Status of Small Indigenous Fish Species under Climate Change Found in Northern Region of Bangladesh. *Int. J. Fish. Aquat. Stud.* **2020**, *8*, 51–58.
57. Islam, T.; Shoeb, M.; Islam, M.N.; Islam, R.; Rhaman, M.M.; Nahar, N. Proximate Analysis and Nutrient Content Estimation of Marine Fish Species of Saint Martin's Island in Bangladesh. *J. Bangladesh Acad. Sci.* **2020**, *44*, 191–195. [CrossRef]
58. Jahan, S.N.; Bayezid, M.A.; Islam, B.; Siddique, M.A.B.; Karmokar, P.K.; Flowra, F.A. Biochemical Quality Assessment of Fish Powder. *Am. J. Food Nutr.* **2017**, *5*, 110–114. [CrossRef]
59. Jahan, S.N.; Akhtar, S. Quality Assessment of Oven Dried and Traditional Sun Dried Channa Punctatus. *Int. J. Food Sci. Biotechnol.* **2019**, *4*, 94. [CrossRef]
60. Jamil, M.G.M.; Hossain, M.N.; Mia, M.M.; Mansur, M.A.; Uga, S. Studies on the Proximate Composition, Quality and Heavy Metal Concentration of Sun-Dried Bombay Duck and Sun-Dried Ribbon Fish of Cox's Bazar District of Bangladesh. *J. Environ. Sci. Nat. Resour.* **2017**, *10*, 55–60. [CrossRef]
61. Kar, M.; Hoq, M.E.; Islam, M.S.; Islam, M.M.; Meghla, N.T.; Suravi, M.; Kabir, H. Monitoring of Proximate Composition, Heavy Metal Concentrations and Pesticide Residues in Marine Dried Fish Available in the Coastal Region of Bangladesh. *Grassroot J. Nat. Resour.* **2020**, *3*, 30–41. [CrossRef]
62. Mahmud, N.; Al-Fuad, S.; Satya, S.I.; Al Mamun, A.; Ahmed, S.; Karim, A.; Islam, M.; Ferdaus, J.; Islam, S.; Sakib, N. Development and Biochemical Composition Assessment of Fish Powders from Bangladeshi Indigenous Fish Species and Shelf-Life Characteristics Evaluation during 90 Days of Room Temperature (27–30 °C) Storage. *Food Nutr. Sci.* **2019**, *10*, 963. [CrossRef]
63. Majumdar, B.C.; Afrin, F.; Rasul, M.; Khan, M.; Shah, A.K.M. Comparative Study of Physical, Chemical, Microbiological and Sensory Aspects of Some Sun Dried Fishes in Bangladesh. *Braz. J. Biol. Sci.* **2017**, *4*, 323–331. [CrossRef]
64. Majumdar, B.C.; Afrin, F.; Rasul, M.G.; Shaha, D.C.; Shah, A.K.M.A. Changes in Physicochemical, Microbiological, and Sensory Properties of Sun-Dried Mystus Vittatus During Storage at Ambient Temperature. *Fishes* **2018**, *3*, 32. [CrossRef]
65. Mansur, M.A.; Uddin, M.N.; Jamil, M.G.M.; Mia, M.M.; Karmakar, M. Quality and Safety Aspect of Some Traditionally Processed Freshwater Fish and Fishery Products of Mymensingh District in Bangladesh. *Int. J. Curr. Res.* **2017**, *9*, 61867–61872.
66. Nurullah, M.; Kamal, M.; Wahab, M.A.; Islam, M.N.; Reza, M.S.; Thilsted, S.H.; Mazid, M.A. Quality Assessment of Traditional and Solar Tunnel Dried SIS (Small Indigenous Fish Species) Products. *Bangladesh J. Fish. Res.* **2006**, *10*, 63–72.

67. Prodhan, S.; Kamrujjaman, M.; Hossain, M.A.; Islam, M.S. A comparative study of effects of radiation (1kGy) and heat treatment (50 C) on the bio-chemical and mineral contents of dried spotted Spanish Mackerel, *Scomberomorus guttatus* (Bloch & Schneider 1801). *Bangladesh J. Zool.* **2010**, *38*, 205–212.
68. Rahman, M.; Siddique, M.F.; Barman, A.C.; Riar, M.G.S. Preparation and Quality Aspect of Traditional Fish Dried Product (Shidhil) of Bangladesh. *Int. J. Zool. Stud.* **2016**, *1*, 28–34.
69. Rana, M.M.; Chakraborty, S.C.; Saeid, A. Comparative Studies of Nutritional, Microbial and Organoleptic Properties of Different Indigenous Dried Fish from Local Market in Bangladesh. *Adv. J. Chem. Sect. A* **2020**, *3*, 318–327.
70. Rasul, M.G.; Majumdar, B.C.; Afrin, F.; Bapary, M.A.J.; Shah, A.K.M.A. Biochemical, Microbiological, and Sensory Properties of Dried Silver Carp (*Hypophthalmichthys molitrix*) Influenced by Various Drying Methods. *Fishes* **2018**, *3*, 25. [CrossRef]
71. Rasul, M.G.; Majumdar, B.C.; Afrin, F.; Jahan, M.; Yuan, C.; Shah, A.K.M.A. Physico-Chemical, Microbiological and Sensory Changes in Sun-Dried Trichogaster Fasciata During Storage. *Turk. J. Agric. Sci. Technol.* **2019**, *7*, 1568–1574. [CrossRef]
72. Rasul, M.G.; Kabir, I.E.; Yuan, C.; Shah, A.K.M.A. Research Article effects of drying methods on physico-chemical, microbiological and sensory properties of torpedo scad (*Megalaspis cordyla*). *J. Microbiol. Biotechnol. Food Sci.* **2021**, *10*, e2796. [CrossRef]
73. Sharif, A.K.M.; Mustafa, A.I.; Mirza, A.H.; Safiullah, S. Trace Metals in Tropical Marine Fish from the Bay of Bengal. *Sci. Total Environ.* **1991**, *107*, 135–142. [CrossRef]
74. Sharif, A.K.M.; Mustafa, A.I.; Amin, M.N.; Safiullah, S. Trace Element Concentrations in Tropical Marine Fish from the Bay of Bengal. *Sci. Total Environ.* **1993**, *138*, 223–234. [CrossRef]
75. Siddiky, M.N.I.; Bosu, A.; Roy, B.C.; Sarker, S.K.; Moniruzzaman, M. Proximate Composition Analysis of Five Important Dried Sea Fish and Evaluate Their Nutritive Value. *Int. J. Nat. Soc. Sci.* **2017**, *4*, 103–110.
76. Sultana, S.; Hossain, M.I.; Shikha, F.H.; Islam, M.M.; Kamal, M. Quality Assessment of Rotating and Solar Tunnel Dried Marine Fish Product. *Bangladesh J. Fish. Res.* **2008**, *12*, 121–128.
77. Abdullahi, S.A. Investigation of Nutritional Status of Chrysichthys Nigrodigitatus Bayrus Filamentous and Auchenoglanis Occidentals' Family Barigdae. *J. Arid Zone Fish.* **2001**, *1*, 39–50.
78. Effiong, B.N.; Fakunle, J.O. Proximate and Mineral Composition of Some Commercially Important Fishes in Lake Kainji, Nigeria. *J. Basic Appl. Sci. Res.* **2011**, *1*, 2497–2500.
79. Rasul, M.G.; Jahan, I.; Yuan, C.; Sarkar, M.S.I.; Bapary, M.A.J.; Baten, M.A.; Shah, A.K.M.A. Seasonal Variation of Nutritional Constituents in Fish of South Asian Countries: A Review. *Fundam. Appl. Agric.* **2021**, *6*, 193–209.
80. Rasul, M.G. Chemical Composition and Nutritional Value of Dried Fish in Bangladesh. *Egypt. J. Aquat. Biol. Fish.* **2021**, *25*, 379–399. [CrossRef]
81. Islam, R.; Islam, M.D.M.; Shoeb, M.; Nahar, N. Fatty Acid Compositions in Marine Fish Samples of Bangladesh. *J. Nutr. Food Sci.* **2018**, *8*, 2. [CrossRef]
82. Salaudeen, M.M. Quality Analysis of Dried Cod (*Gadus morhua*) Heads along the Value Chain from Iceland to Nigeria. United Nations University, Fisheries Training Programme, Iceland [Final Project]. 2014. Available online: http://www.unuftp.is/static/fellows/document/mutiat13prf.pdf (accessed on 25 August 2022).
83. Ullah, N.; Hazarika, P.; Handique, P.J. Biochemical Quality Assessment of Ten Selected Dried Fish Species of North East India. *Int. Adv. Res. J. Sci. Eng. Technol.* **2016**, *3*, 30–33. [CrossRef]
84. Kawarazuka, N.; Béné, C. The Potential Role of Small Fish Species in Improving Micronutrient Deficiencies in Developing Countries: Building Evidence. *Public Health Nutr.* **2011**, *14*, 1927–1938. [CrossRef] [PubMed]
85. Roos, N.; Wahab, M.A.; Chamnan, C.; Thilsted, S.H. The Role of Fish in Food-Based Strategies to Combat Vitamin A and Mineral Deficiencies in Developing Countries. *J. Nutr.* **2007**, *137*, 1106–1109. [CrossRef]
86. Alam, A.K.M.N. *Low-Cost Processing of Fish in Coastal Bangladesh*; BGD/97/017 Field Doc: 05/2005; Food and Agriculture Organization of the United Nations: Dhaka, Bangladesh, 2005.
87. Hoque, M.; Tamanna, F.; Hasan, M.; Al Banna, M.; Mondal, P.; Prodhan, M.D.H.; Rahman, M.; van Brakel, M.L. Probabilistic Public Health Risks Associated with Pesticides and Heavy Metal Exposure through Consumption of Common Dried Fish in Coastal Regions of Bangladesh. *Environ. Sci. Pollut. Res.* **2022**, *29*, 20112–20127. [CrossRef]
88. Rakib, M.R.J.; Jolly, Y.N.; Enyoh, C.E.; Khandaker, M.U.; Hossain, M.B.; Akther, S.; Alsubaie, A.; Almalki, A.S.A.; Bradley, D.A. Levels and Health Risk Assessment of Heavy Metals in Dried Fish Consumed in Bangladesh. *Sci. Rep.* **2021**, *11*, 14642. [CrossRef]
89. Mithun, B.D.; Hoque, M.S.; Van Brakel, M.L.; Hasan, M.M.; Akter, S.; Islam, M.R. Comparative Quality Assessment of Traditional vs. Improved Dried Bombay Duck (*Harpodon nehereus*) under Different Storage Conditions: Solar Chimney Dryer a Low-cost Improved Approach for Nutritional Dried Fish. *Food Sci. Nutr.* **2021**, *9*, 6794–6805. [CrossRef]
90. Hasan, M.N.; Islam, H.M.R.; Akter, R.; Mahmud, Y.; Ahmed, K.K.U.; Siddiquee, S. Determination of Dichlorodiphenyltrichloroethane Residues Levels in Commercial Marine Dry Fish from Different Regions of Bangladesh. *Annu. Res. Rev. Biol.* **2014**, *4*, 2722–2729. [CrossRef]
91. Nowsad, A. *Participatory Training of Trainers: A New Approach Applied in Fish Processing;* Bangladesh Fisheries Research Forum: Dhaka, Bangladesh, 2007; 328p.
92. Kundu, S.; Al Banna, M.H.; Sayeed, A.; Sultana, M.S.; Brazendale, K.; Harris, J.; Mandal, M.; Jahan, I.; Abid, M.T.; Khan, M.S.I. Determinants of Household Food Security and Dietary Diversity during the COVID-19 Pandemic in Bangladesh. *Public Health Nutr.* **2021**, *24*, 1079–1087. [CrossRef]

93. Banna, M.; Al, H.; Sayeed, A.; Kundu, S.; Kagstrom, A.; Sultana, M.; Begum, M.R.; Khan, M.S.I. Factors Associated with Household Food Insecurity and Dietary Diversity among Day Laborers amid the COVID-19 Pandemic in Bangladesh. *BMC Nutr.* **2022**, *8*, 25. [CrossRef]
94. Naja, F.; Hamadeh, R. Nutrition amid the COVID-19 Pandemic: A Multi-Level Framework for Action. *Eur. J. Clin. Nutr.* **2020**, *74*, 1117–1121. [CrossRef] [PubMed]
95. Akhtar, S.; Das, J.K.; Ismail, T.; Wahid, M.; Saeed, W.; Bhutta, Z.A. Nutritional Perspectives for the Prevention and Mitigation of COVID-19. *Nutr. Rev.* **2021**, *79*, 289–300. [CrossRef] [PubMed]
96. Pinnawala, N.U.; Thrastardottir, T.O.; Constantinou, C. Keeping a Balance during the Pandemic: A Narrative Review on the Important Role of Micronutrients in Preventing Infection and Reducing Complications of COVID-19. *Curr. Nutr. Rep.* **2021**, *10*, 200–210. [CrossRef] [PubMed]
97. Eskander, M.; Razzaque, M.S. Can Maintaining Optimal Magnesium Balance Reduce the Disease Severity of COVID-19 Patients? *Front. Endocrinol. (Lausanne)* **2022**, *13*, 501. [CrossRef] [PubMed]
98. Razzaque, M.S. COVID-19 Pandemic: Can Zinc Supplementation Provide an Additional Shield against the Infection? *Comput. Struct. Biotechnol. J.* **2021**, *19*, 1371–1378. [CrossRef]
99. Johnson, D.; Thilsted, S.H.; Belton, B. Dried Fish in a COVID-19 World. Available online: https://www.worldfishcenter.org/blog/dried-fish-covid-19-world (accessed on 14 May 2022).
100. Rizaldo, Q.; Morris, H. Dried Small Fish Powder Provides Opportunity for Child Health in Myanmar. Available online: https://www.worldfishcenter.org/blog/dried-small-fish-powder-provides-opportunity-child-health-myanmar (accessed on 14 May 2022).

MDPI
St. Alban-Anlage 66
4052 Basel
Switzerland
Tel. +41 61 683 77 34
Fax +41 61 302 89 18
www.mdpi.com

Fishes Editorial Office
E-mail: fishes@mdpi.com
www.mdpi.com/journal/fishes

www.ingramcontent.com/pod-product-compliance
Lightning Source LLC
Chambersburg PA
CBHW061841210926
51850CB00006B/944
9783036572789